MOTOR CORTEX IN VOLUNTARY MOVEMENTS

A DISTRIBUTED SYSTEM FOR DISTRIBUTED FUNCTIONS

FRONTIERS IN NEUROSCIENCE

Series Editors
Sidney A. Simon, Ph.D.
Miguel A.L. Nicolelis, M.D., Ph.D.

Published Titles

Apoptosis in Neurobiology
Yusuf A. Hannun, M.D., Professor of Biomedical Research and Chairman/Department of Biochemistry and Molecular Biology, Medical University of South Carolina
Rose-Mary Boustany, M.D., tenured Associate Professor of Pediatrics and Neurobiology, Duke University Medical Center

Methods for Neural Ensemble Recordings
Miguel A.L. Nicolelis, M.D., Ph.D., Professor of Neurobiology and Biomedical Engineering, Duke University Medical Center

Methods of Behavioral Analysis in Neuroscience
Jerry J. Buccafusco, Ph.D., Alzheimer's Research Center, Professor of Pharmacology and Toxicology, Professor of Psychiatry and Health Behavior, Medical College of Georgia

Neural Prostheses for Restoration of Sensory and Motor Function
John K. Chapin, Ph.D., Professor of Physiology and Pharmacology, State University of New York Health Science Center
Karen A. Moxon, Ph.D., Assistant Professor/School of Biomedical Engineering, Science, and Health Systems, Drexel University

Computational Neuroscience: Realistic Modeling for Experimentalists
Eric DeSchutter, M.D., Ph.D., Professor/Department of Medicine, University of Antwerp

Methods in Pain Research
Lawrence Kruger, Ph.D., Professor of Neurobiology (Emeritus), UCLA School of Medicine and Brain Research Institute

Motor Neurobiology of the Spinal Cord
Timothy C. Cope, Ph.D., Professor of Physiology, Emory University School of Medicine

Nicotinic Receptors in the Nervous System
Edward D. Levin, Ph.D., Associate Professor/Department of Psychiatry and Pharmacology and Molecular Cancer Biology and Department of Psychiatry and Behavioral Sciences, Duke University School of Medicine

Methods in Genomic Neuroscience
Helmin R. Chin, Ph.D., Genetics Research Branch, NIMH, NIH
Steven O. Moldin, Ph.D, Genetics Research Branch, NIMH, NIH

Methods in Chemosensory Research
Sidney A. Simon, Ph.D., Professor of Neurobiology, Biomedical Engineering, and Anesthesiology, Duke University
Miguel A.L. Nicolelis, M.D., Ph.D., Professor of Neurobiology and Biomedical Engineering, Duke University

The Somatosensory System: Deciphering the Brain's Own Body Image
Randall J. Nelson, Ph.D., Professor of Anatomy and Neurobiology,
University of Tennessee Health Sciences Center

The Superior Colliculus: New Approaches for Studying Sensorimotor Integration
William C. Hall, Ph.D., Department of Neuroscience, Duke University
Adonis Moschovakis, Ph.D., Institute of Applied and Computational Mathematics, Crete

New Concepts in Cerebral Ischemia
Rick C. S. Lin, Ph.D., Professor of Anatomy, University of Mississippi Medical Center

DNA Arrays: Technologies and Experimental Strategies
Elena Grigorenko, Ph.D., Technology Development Group, Millennium Pharmaceuticals

Methods for Alcohol-Related Neuroscience Research
Yuan Liu, Ph.D., National Institute of Neurological Disorders and Stroke, National Institutes of Health
David M. Lovinger, Ph.D., Laboratory of Integrative Neuroscience, NIAAA

***In Vivo* Optical Imaging of Brain Function**
Ron Frostig, Ph.D., Associate Professor/Department of Psychobiology,
University of California, Irvine

Primate Audition: Behavior and Neurobiology
Asif A. Ghazanfar, Ph.D., Primate Cognitive Neuroscience Lab, Harvard University

Methods in Drug Abuse Research: Cellular and Circuit Level Analyses
Dr. Barry D. Waterhouse, Ph.D., MCP-Hahnemann University

Functional and Neural Mechanisms of Interval Timing
Warren H. Meck, Ph.D., Professor of Psychology, Duke University

Biomedical Imaging in Experimental Neuroscience
Nick Van Bruggen, Ph.D., Department of Neuroscience Genentech, Inc.,
South San Francisco
Timothy P.L. Roberts, Ph.D., Associate Professor, University of Toronto

The Primate Visual System
John H. Kaas, Department of Psychology, Vanderbilt University
Christine Collins, Department of Psychology, Vanderbilt University

Neurosteroid Effects in the Central Nervous System
Sheryl S. Smith, Ph.D., Department of Physiology, SUNY Health Science Center

Modern Neurosurgery: Clinical Translation of Neuroscience Advances
Dennis A. Turner, Department of Surgery, Division of Neurosurgery, Duke University Medical Center

Sleep: Circuits and Functions
Pierre-Hervé Luoou, Université Claude Bernard Lyon I, Lyon, France

Methods in Insect Sensory Neuroscience
Thomas A. Christensen, Arizona Research Laboratories, Division of Neurobiology, University of Arizona, Tucson, AZ

MOTOR CORTEX IN VOLUNTARY MOVEMENTS

A DISTRIBUTED SYSTEM FOR DISTRIBUTED FUNCTIONS

EDITED BY

Alexa Riehle and Eilon Vaadia

CRC Press is an imprint of the
Taylor & Francis Group, an **informa** business

CRC Press
Taylor & Francis Group
6000 Broken Sound Parkway NW, Suite 300
Boca Raton, FL 33487-2742

First issued in paperback 2019

ISBN-13: 978-0-8493-1287-8 (hbk)
ISBN-13: 978-0-367-39339-7 (pbk)

Library of Congress Card Number 2004057046

Library of Congress Cataloging-in-Publication Data

Motor cortex in voluntary movements : a distributed system for distributed functions / edited by Alexa Riehle and Eilon Vaadia.
p. cm.
Includes bibliographical references and index.
ISBN 0-8493-1287-6 (alk. paper)
1. Motor cortex. 2. Human locomotion. I. Riehle, Alexa. II. Vaadia, Eilon. III. Series.

QP383.15.M68 2005
612.8′252—dc22 2004057046

Visit the Taylor & Francis Web site at
http://www.taylorandfrancis.com

and the CRC Press Web site at
http://www.crcpress.com

Methods & New Frontiers in Neuroscience

Our goal in creating the **Methods & New Frontiers in Neuroscience** series is to present the insights of experts on emerging experimental techniques and theoretical concepts that are or will be at the vanguard of the study of neuroscience. Books in the series cover topics ranging from methods to investigate apoptosis to modern techniques for neural ensemble recordings in behaving animals. The series also covers new and exciting multidisciplinary areas of brain research, such as computational neuroscience and neuroengineering, and describes breakthroughs in classical fields such as behavioral neuroscience. We want these to be the books every neuroscientist will use in order to graduate students and postdoctoral fellows when they are looking for guidance to start a new line of research.

Each book is edited by an expert and consists of chapters written by the leaders in a particular field. Books are richly illustrated and contain comprehensive bibliographies. Chapters provide substantial background material relevant to the particular subject; hence, they are not only "methods" books. They contain detailed tricks of the trade and information as to where these methods can be safely applied. In addition, they include information about where to buy equipment and about Web sites that are helpful in solving both practical and theoretical problems.

We hope that as the volumes become available, the effort put in by us, by the publisher, by the book editors, and by the individual authors will contribute to the further development of brain research. The extent to which we achieve this goal will be determiend by the utility of these books.

Sidney A. Simon, Ph.D.
Miguel A.L. Nicolelis, M.D., Ph.D.
Series Editors

Preface

Voluntary movement is undoubtedly the overt basis of human behavior. Without movement we cannot walk, nourish ourselves, communicate, or interact with the environment. This is one of the reasons why the motor cortex was one of the first cortical areas to be explored experimentally. Historically, the generation of motor commands was thought to proceed in a rigidly serial and hierarchical fashion. The traditional metaphor of the piano presents the premotor cortex "playing" the upper motoneuron keys of the primary motor cortex (M1), which in turn activate with strict point-to-point connectivity the lower motoneurons of the spinal cord. Years of research have taught us that we may need to reexamine almost all aspects of this model. Both the premotor and the primary motor cortex project directly to the spinal cord in highly complex overlapping patterns, contradicting the simple hierarchical view of motor control. The task of generating and controlling movements appears to be subdivided into a number of subtasks that are accomplished through parallel distributed processing in multiple motor areas. Multiple motor areas may increase the behavioral flexibility by responding in a context-related way to any constraint within the environment. Furthermore, although more and more knowledge is accumulating, there is still an ongoing debate about what is represented in the motor cortex: dynamic parameters (such as specific muscle activation), kinematic parameters of the movement (for example, its direction and speed), or even more abstract parameters such as the context of the movement. Given the great scope of the subject considered here, this book focuses on some new perspectives developed from contemporary monkey and human studies. Moreover, many topics receive very limited treatment.

Section I, which includes the first two chapters, uses functional neuroanatomy and imaging studies to describe motor cortical function. The objective of Chapter 1 is to describe the major components of the structural framework employed by the cerebral cortex to generate and control skeletomotor function. **Dum and Strick** focus on motor areas in the frontal lobe that are the source of corticospinal projections to the ventral horn of the spinal cord in primates. These cortical areas include the primary motor cortex (M1) and the six premotor areas that project directly to it. The results presented lead to an emerging view that motor commands can arise from multiple motor areas and that each of these motor areas makes a specialized contribution to the planning, execution, or control of voluntary movement. The purpose of Chapter 2 is to provide an overview of the contribution of functional magnetic resonance imaging (fMRI) to some of the prevailing topics in the study of motor control and the function of the primary motor cortex. **Kleinschmidt and Toni** claim that in several points the findings of functional neuroimaging seem to be in apparent disagreement with those obtained with other methods, which cannot always be attributed to insufficient sensitivity of this noninvasive technique. In part, it may reflect the indirect and spatio-temporally imprecise nature of the fMRI signal, but these studies remain informative by virtue of the fact that usually the whole brain is covered. Not only does fMRI reveal plausible brain regions for the control of localized effects, but the distribution of response foci and the correlation of effects observed at many different sites can assist in the guidance of detailed studies at the mesoscopic or microscopic spatio-temporal level. A prudently modest view might conclude that fMRI is at present primarily a tool of exploratory rather than explanatory value.

Section II provides a large overview of studies about neural representations in the motor cortex. Chapter 3 focuses on the neuromuscular evolution of individuated finger movements. **Schieber, Reilly, and Lang** demonstrate that rather than acting as a somatotopic array of upper motor neurons, each controlling a single muscle that moves a single finger, neurons in the primary motor cortex (M1) act as a spatially distributed network of very diverse elements, many of which have outputs that diverge to facilitate multiple muscles acting on different fingers. This biological control of a complex peripheral apparatus initially may appear unnecessarily complicated compared to the independent control of digits in a robotic hand, but can be understood as the result of concurrent evolution of the peripheral neuromuscular apparatus and its descending control from the motor cortex. Chapter 4 deals with simultaneous movements of the two arms, as a simple example of complex movements, and may serve to test whether and how the brain generates unique representations of complex movements from their constituent elements. **Vaadia and Cardoso de Oliveira** present evidence that bimanual representations indeed exist, both at the level of single neurons and at the level of neuronal populations (in local field potentials). They further show that population firing rates and dynamic interactions between the hemispheres contain information about the bimanual movement to be executed. In Chapter 5, **Ashe** discusses studies with respect to the debate as to whether the motor cortex codes the spatial aspects (kinematics) of motor output, such as direction, velocity, and position, or primarily controls, muscles, and forces (dynamics). Although the weight of evidence is in favor of M1 controlling spatial output, the effect of limb biomechanics and forces on motor cortex activity is beyond dispute. The author proposes that the motor cortex indeed codes for the most behaviorally relevant spatial variables and that both spatial variables and limb biomechanics are reflected in motor cortex activity. Chapter 6 starts with the important issue of how theoretical concepts guide experimental design and data analysis. **Scott** describes two conceptual frameworks for interpreting neural activity during reaching: sensorimotor transformations and internal models. He claims that sensorimotor transformation have been used extensively over the past 20 years to guide neurophysiological experiments on reaching, whereas internal models have only recently had an impact on experimental design. Furthermore, the chapter demonstrates how the notion of internal models can be used to explore the neural basis of movement by describing a new experimental tool that can sense and perturb multiple-joint planar movements. Chapter 7 deals with the function of oscillatory potentials in the motor cortex. **MacKay** notes that from their earliest recognition, oscillatory EEG signals in the sensorimotor cortex have been associated with stasis: a lack of movement, static postures, and possibly physiological tremor. It is now established that

10-, 20-, and 40-Hz motor cortical oscillations are associated with constant, sustained muscle contractions, again a static condition. Sigma band oscillations of about 14 Hz may be indicative of maintained active suppression of a motor response. The dynamic phase at the onset of an intended movement is preceded by a marked decrease in oscillatory power, but not all frequencies are suppressed. Fast gamma oscillations coincide with movement onset. Moreover, there is increasing evidence that oscillatory potentials of even low frequencies (4–12 Hz) may be linked to dynamic episodes of movement. Most surprisingly, the 8-Hz cortical oscillation — the neurogenic component of physiological tremor — is emerging as a major factor in shaping the pulsatile dynamic microstructure of movement, and possibly in coordinating diverse actions performed together. In Chapter 8, **Riehle** discusses the main aspects of preparatory processes in the motor cortex. Preparation for action is thought to be based on central processes, which are responsible for maximizing the efficiency of motor performance. A strong argument in favor of such an efficiency hypothesis of preparatory processes is the fact that providing prior information about movement parameters or removing time uncertainty about when to move significantly shortens reaction time. The types of changes in the neuronal activity of the motor cortex, and their selectivity during preparation, are portrayed and compared with other cortical areas that are involved in motor behavior. Furthermore, linking motor cortical activity directly to behavioral performance showed that the trial-by-trial correlation between single neuron firing rates and reaction time revealed strong task-related cortical dynamics. Finally, the cooperative interplay among neurons, expressed by precise synchronization of their action potentials, is illustrated and compared with changes in the firing rate of the same neurons. New concepts including the notion of coordinated ensemble activity and their functional implication during movement preparation are discussed. In the last chapter of Section II, Chapter 9, **Jeannerod** poses the question of the role of the motor cortex in motor cognition. The classical view of the primary motor cortex holds that it is an area devoted to transferring motor execution messages that have been elaborated upstream in the cerebral cortex. More recently, however, experimental data have pointed to the fact that the relation of motor cortex activity to the production of movements is not as simple as was thought on the basis of early stimulation experiments. This revision of motor cortical function originated from two main lines of research, dealing first with the plasticity of the somatotopic organization of the primary motor cortex, and second with its involvement in cognitive functions such as motor imagery.

Section III is mainly concerned with motor learning. Chapter 10 explores various conditions of mapping between sensory input and motor output. **Brasted and Wise** claim that studies on the role of the motor cortex in voluntary movement usually focus on standard sensorimotor mapping, in which movements are directed toward sensory cues. Sensorimotor behavior can, however, show much greater flexibility. Some variants rely on an algorithmic transform between the location of the cue and that of the target. The well-known "antisaccade" task and its analogues in reaching serve as special cases of such transformational mapping, one form of nonstandard mapping. Other forms of nonstandard mapping differ strongly: they are arbitrary. In arbitrary sensorimotor mapping, the cue's location has no systematic spatial relationship with the response. The authors explore several types of arbitrary mapping, with emphasis on the neural basis of learning. In Chapter 11, **Shadmehr, Donchin, Hwang, Hemminger, and Rao** deal with internal models that transform the desired movement into a motor command. When one moves the hand from one point to another, the brain guides the arm by relying on neural structures that estimate the physical dynamics of the task. Internal models are learned with practice and are a fundamental part of voluntary motor control. What do internal models compute, and which neural structures perform that computation? The authors approach these questions by considering a task where the physical dynamics of reaching movements are altered by force fields that act on the hand. Many studies suggest that internal models are sensorimotor transformations that map a desired sensory state of the arm into an estimate of forces; i.e., a model of the inverse dynamics of the task. If this computation is represented as a population code via a flexible combination of basis functions, then one can infer activity fields of the bases from the patterns of generalization. Shadmehr and colleagues provide a mathematical technique that facilitates this inference by analyzing trial-by-trial changes in performance. Results suggest that internal models are computed with bases that are directionally tuned to limb motion in intrinsic coordinates of joints and muscles, and this tuning is modulated multiplicatively as a function of static position of the limb. That is, limb position acts as a gain field on directional tuning. Some of these properties are consistent with activity fields of neurons in the motor cortex and the cerebellum. The authors suggest that activity fields of these cells are reflected in human behavior in the way that we learn and generalize patterns of dynamics in reaching movements. In the last chapter of Section III, Chapter 12, **Padoa-Schioppa, Bizzi, and Mussa-Ivaldi** address the question of the cortical control of motor learning. In robotic systems, engineers coordinate the action of multiple motors by writing computer codes that specify how the motors must be activated for achieving the desired robot motion and for compensating unexpected disturbance. Humans and animals follow another path. Something akin to programming is achieved in nature by the biological mechanisms of synaptic plasticity — that is, by the variation in efficacy of neural transmission brought about by past history of pre- and post-synaptic signals. However, robots and animals differ in another important way. Robots have a fixed mechanical structure and dimensions. In contrast, the mechanics of muscles, bones, and ligaments change in time. Because of these changes, the central nervous system must continuously adapt motor commands to the mechanics of the body. Adaptation is a form of motor learning. Here, a view of motor learning is presented that starts from the analysis of the computational problems associated with the execution of the simplest gestures. The authors discuss the theoretical idea of internal models and present some evidence and theoretical considerations suggesting that internal models of limb dynamics may be obtained by the combination of simple modules or "motor primitives." Their findings suggest that the motor cortical areas include neurons that process well-acquired movements as well as neurons that change their behavior during and after being exposed to a new task.

The last section, Section IV, is devoted to the reconstruction of movements using brain activity. For decades, science fiction authors anticipated the view that computers can be made to communicate directly with the brain. Now, a rapidly expanding science community is making this a reality. In Chapter 13, **Carmena and Nicolelis** present and discuss the recent research in the field of brain–machine interfaces (BMI) conducted mainly on nonhuman primates. In fact, this research field has supported the contention that we are at the brink of a technological revolution, where artificial devices may be "integrated" in the multiple sensory, motor, and cognitive representations that exist in the primate brain. These studies have demonstrated that animals can learn to utilize their brain activity to control the displacements of computer cursors, the movements of simple and elaborate robot arms, and, more recently, the reaching and grasping movements of a robot arm. In addition to the current research performed in rodents and primates, there are also preliminary studies using human subjects. The ultimate goal of this emerging field of BMI is to allow human subjects to interact effortlessly with a variety of actuators and sensory devices through the expression of their voluntary brain activity, either for augmenting or restoring sensory, motor, and cognitive function. In the last chapter, Chapter 14, **Pfurtscheller, Neuper, and Birbaumer** deal with BMIs, which transform signals originating from the human brain into commands that can control devices or applications. BCIs provide a new nonmuscular communication channel, which can be used to assist patients who have highly compromised motor functions, as is the case with patients suffering from neurological diseases such as amyotrophic lateral sclerosis (ALS) or brainstem stroke. The immediate goal of current research in this field is to provide these users with an opportunity to communicate with their environment. Present-day BCI systems use different electrophysiological signals such as slow cortical potentials, evoked potentials, and oscillatory activity recorded from scalp or subdural electrodes, and cortical neuronal activity recorded from implanted electrodes. Due to advances in methods of signal processing, it is possible that specific features automatically extracted from the electroencephalogram (EEG) and electrocorticogram (ECoG) can be used to operate computer-controlled devices. The interaction between the BCI system and the user, in terms of adaptation and learning, is a challenging aspect of any BCI development and application.

It is the increased understanding of neuronal mechanisms of motor functions, as reflected in this book, that led to the success of BCI. Yet, the success in tapping and interpreting neuronal activity and interfacing it with a machine that eventually executes the subject's intention is amazing, considering the limited understanding we have of the system as a whole.

Perhaps ironically, the proof of our understanding of motor cortical activity will stem from how effectively we, as external observers of the brain, can tap into it and make use of it.

Alexa Riehle
Eilon Vaadia

Dedication

---

to Hanns-Günther Riehle

Editors

Alexa Riehle received a B.Sc. degree in biology (main topic: deciphering microcircuitries in the frog retina) from the Free University, Berlin, Germany, in 1976, and a Ph.D. degree in neurophysiology (main topic: neuronal mechanisms of temporal aspects of color vision in the honey bee) from the Biology Department of the Free University in 1980.

From 1980 to 1984, she was a postdoctoral fellow at the National Center for Scientific Research (CNRS) in Marseille, France (main topic: neuronal mechanisms of elementary motion detectors in the fly visual system). In 1984, she moved to the Cognitive Neuroscience Department at the CNRS and has been mainly interested since then in the study of cortical information processing and neural coding in cortical ensembles during movement preparation and execution in nonhuman primates.

Eilon Vaadia graduated from the Hebrew University of Jerusalem (HUJI) in 1980 and joined the Department of Physiology at Hadassah Medical School after postdoctoral studies in the Department of Biomedical Engineering at Johns Hopkins University Medical School in Baltimore, Maryland.

Vaadia studies cortical mechanisms of sensorimotor functions by combining experimental work (recordings of multiple unit activity in the cortex of behaving animals) with a computational approach. He is currently the director of the Department of Physiology and the head of the Ph.D. program at the Interdisciplinary Center for Neural Computation (ICNC) at HUJI, and a director of a European advanced course in computational neuroscience.

Contributors

James Ashe
Veterans Affairs Medical Center
Brain Sciences Center
University of Minnesota
Minneapolis, Minnesota

Emilio Bizzi
Department of Brain and Cognitive Sciences
Massachusetts Institute of Technology
Cambridge, Massachusetts

Niels Birbaumer
Institute of Medical Psychology and Behavioral Neurobiology
Eberhard-Karls-University of Tübingen
Tübingen, Germany

Peter J. Brasted
Laboratory of Systems Neuroscience
National Institute of Mental Health
National Institutes of Health
Bethesda, Maryland

Simone Cardoso de Oliveira
German Primate Center
Cognitive Neuroscience Laboratory
Göttingen, Germany

Jose M. Carmena
Center for Neuroengineering
Department of Neurobiology
Duke University Medical Center
Durham, North Carolina

Opher Donchin
Laboratory for Computational Motor Control
Department of Biomedical Engineering
Johns Hopkins School of Medicine
Baltimore, Maryland

Richard P. Dum
Department of Neurobiology
University of Pittsburgh School of Medicine
Pittsburgh, Pennsylvania

Sarah E. Hemminger
Laboratory for Computational Motor Control
Department of Biomedical Engineering
Johns Hopkins School of Medicine
Baltimore, Maryland

Eun-Jung Hwang
Laboratory for Computational Motor Control
Department of Biomedical Engineering
Johns Hopkins School of Medicine
Baltimore, Maryland

Marc Jeannerod
Institute of Cognitive Sciences
National Center for Scientific Research (ISC-CNRS)
Bron, France

Andreas Kleinschmidt
Cognitive Neurology Unit
Department of Neurology
Johann Wolfgang Goethe University
Frankfurt am Main, Germany

Catherine E. Lang
University of Rochester
Department of Neurology
Rochester, New York

William A. MacKay
Department of Physiology
University of Toronto
Toronto, Ontario, Canada

Ferdinando A. Mussa-Ivaldi
Departments of Physiology,
Physical Medicine and Rehabilitation,
and Biomedical Engineering
Northwestern University
Chicago, Illinois

Christa Neuper
Ludwig Boltzmann Institute of Medical
Informatics and Neuroinformatics
Graz University of Technology
Graz, Austria

Miguel A.L. Nicolelis
Department of Neurobiology
Duke University Medical Center
Durham, North Carolina

Camillo Padoa-Schioppa
Department of Neurobiology
Harvard Medical School
Boston, Massachusetts

Gert Pfurtscheller
Laboratory of Brain–Computer
Interfaces
Graz University of Technology
Graz, Austria

Ashwini K. Rao
Columbia University Medical Center
Program in Physical Therapy
Neurological Institute
New York, New York

Karen T. Reilly
University of Rochester
Department of Neurology
Rochester, New York

Alexa Riehle
Mediterranean Institute for Cognitive
Neuroscience
Natinoal Center for Scientific Research
(INCM-CNRS)
Marseille, France

Marc H. Schieber
University of Rochester
Department of Neurology
Rochester, New York

Stephen H. Scott
Centre for Neuroscience Studies
Department of Anatomy and Cell
Biology
Canadian Institutes of Health Research
Group in Sensory-Motor Systems
Queen's University
Kingston, Ontario

Reza Shadmehr
Laboratory for Computational Motor
Control
Department of Biomedical Engineering
Johns Hopkins School of Medicine
Baltimore, Maryland

Peter L. Strick
Veterans Affairs Medical Center for the
Neural Basis of Cognition
Department of Neurobiology
University of Pittsburgh
Pittsburgh, Pennsylvania

Ivan Toni
F.C. Donders Center for Cognitive
Neuroimaging
Nijmegen, The Netherlands

Eilon Vaadia
Department of Physiology
Hadassah Medical School
The Hebrew University
Jerusalem, Israel

Steven P. Wise
Laboratory of Systems Neuroscience
National Institute of Mental Health
National Institutes of Health
Bethesda, Maryland

Table of Contents

SECTION I Functional Neuroanatomy and Imaging

SECTION II Neuronal Representations in the Motor Cortex

SECTION III Motor Learning and Performance

SECTION IV Reconstruction of Movements Using Brain Activity

Section I

Functional Neuroanatomy and Imaging

1 Motor Areas in the Frontal Lobe: The Anatomical Substrate for the Central Control of Movement

Richard P. Dum and Peter L. Strick

CONTENTS

0-8493-1287-6/05/$0.00+$1.50

1.1 INTRODUCTION

The objective of this chapter is to describe the major components of the structural framework employed by the cerebral cortex to generate and control skeletomotor function. We will focus on motor areas in the frontal lobe that are the source of corticospinal projections to the ventral horn of the spinal cord in primates. These cortical areas include the primary motor cortex (M1) and the six premotor areas that project directly to M1. We will begin by examining anatomical and physiological evidence that demonstrates how each of these cortical areas directly accesses spinal cord mechanisms involved in the generation and control of movement. This evidence suggests that all these cortical areas have some direct involvement in movement execution. Then we will examine how the pattern of cortical and subcortical inputs could shape the functional role of each cortical area in motor control. We will show that each of these cortical areas receives a unique pattern of cortical and subcortical input. Taken together, these results have led to an emerging view that motor commands can arise from multiple motor areas and that each of these motor areas makes a specialized contribution to the planning, execution, or control of voluntary movement. In this chapter, we will describe some of the relevant anatomical and physiological evidence that has led to this viewpoint.

Given the breadth of the subject considered here, our review will focus on new perspectives developed from contemporary primate studies. Even with this focus, many topics will receive limited treatment. For instance, the physiological and behavioral studies that provide evidence of differential involvement of each motor area in the generation and control of movement are beyond the scope of this chapter. For further insight into the historical development of this field and a broader coverage of related issues, numerous reviews on this and related topics are available.[1–11] In addition, the corticospinal system has been the subject of a recent book.[12]

1.2 FUNCTIONAL ANATOMY

1.2.1 Primary Motor Cortex

The primary motor cortex (M1) owes its name to the fact that thresholds for evoking movement with electrical stimulation are lower here than in any other cortical region.[13–15] (For historical review, see Reference 12.) Anatomically, M1 corresponds

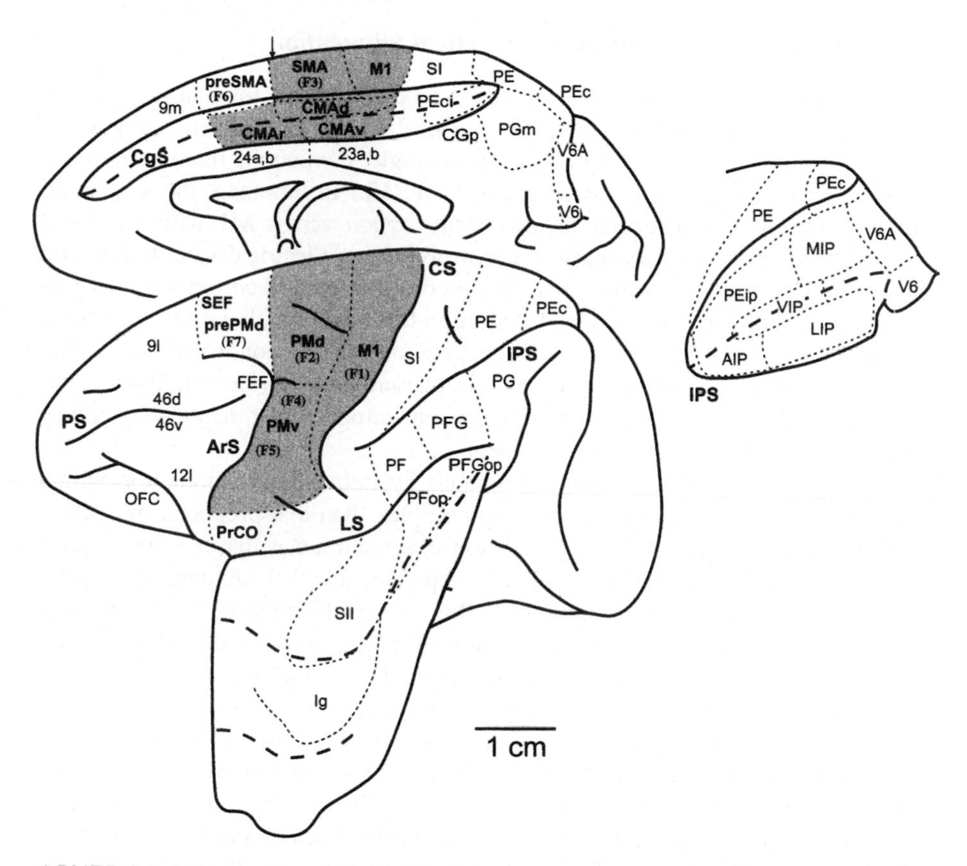

FIGURE 1.1 Identification of cortical areas in the macaque monkey. The cingulate sulcus (CgS), lateral sulcus (LS), and intraparietal sulcus (IPS) are unfolded and each fundus is indicated by a *dashed line.* The borders between cytoarchitectonic areas are delineated with *dotted lines.* M1 and the premotor areas are *shaded.* Abbreviations: AIP, LIP, MIP, VIP: anterior, lateral, medial, and ventral intraparietal areas; ArS: arcuate sulcus; CGp: posterior cingulate gyrus; CMAd, CMAv, CMAr: dorsal, ventral, and rostral cingulate motor areas; CS: central sulcus; F1 to F7: cytoarchitectonic areas in the frontal lobe according to Matelli et al.[77,248]; FEF: frontal eye fields; Ig: granular insular cortex; M1: primary motor cortex; OFC: orbital frontal cortex; PMd: dorsal premotor area; PMv: ventral premotor area; PrCO: precentral opercular cortex; prePMd: pre-premotor area, dorsal; preSMA: presupplementary motor area; PS: principal sulcus; SEF: supplementary eye field; SI: primary somatosensory cortex; SII: secondary somatosensory cortex; SMA: supplementary motor area; PE, PEc, PEci, PF, PFG, PFop, PG, PGm, Pgop: parietal areas after Pandya and Selzer[249]; V6A, V6: posterior parietal areas after Galletti et al.[177]; 9m, 91, 46d, 46v, 121: prefrontal areas after Walker[181] and Barbas and Pandya.[186]

to cytoarchitectonic area 4, which is identified by the presence of giant pyramidal cells in cortical layer V.[16–18] Based on these definitions, M1 is located in the anterior bank of the central sulcus and on the adjacent caudal portion of the precentral gyrus (Figure 1.1). (For more complete reviews, see References 4,5,9,12.)

1.2.1.1 Organization Based on Intracortical Stimulation

Our view of the organization of M1 as based on electrical stimulation has evolved with advances in stimulation techniques. Classically, surface stimulation suggested that M1 contained a "motor map" that was a single, contiguous representation of the body.[14,15] (For reviews, see References 4 and 12.) In this map, the leg, trunk, arm, and face formed a medial to lateral procession across M1 with the distal musculature of each limb located in the central sulcus. Electrical stimulation with microelectrodes inserted into the cortex lowered the amount of current necessary to evoke movement by a factor of 100.[19] Although this advance allowed a much more detailed exploration of the cortex, intracortical stimulation confirmed the overall somatotopy of leg, arm, and face representation described by surface stimulation.[19–32] Thus, electrical stimulation of M1 generated a somatotopic motor map with relatively sharp boundaries between major body parts.

The organization of movements generated by intracortical stimulation within each major body part, however, was more complex than that produced by surface stimulation (Color Figure 1.2).* A consistent observation was that the same movement could be evoked at multiple, spatially separate sites.[22–32] Although this observation precluded an orderly somatotopy, the general features of this map were reproducible. Within the arm representation of macaque monkeys, distal limb movements (fingers and wrist) tended to form a central core that was surrounded by a horseshoe of proximal limb movements (elbow and shoulder) (Color Figure 1.2A).[22,33] Some intermingling of distal and proximal limb movements occurred at the borders. This organizational structure has been confirmed with single-pulse, stimulus-triggered averaging (Color Figure 1.2B).[34] The presence of multiple representations of an individual movement/muscle in M1 has been proposed as an arrangement that allows a muscle to engage in multiple synergies with other muscles acting at the same or different joints. (See Reference 35.)

Other studies utilizing intracortical stimulation[20,26,28,32] reported even more complex patterns of muscle activation. For example, stimulation at some sites in M1 evoked reciprocal activation of wrist antagonists, whereas at other sites it caused their co-contraction.[26] Some stimulus locations evoked movements of several joints at barely differing thresholds. Thus, multiple-joint movements could also be evoked by relatively localized stimulation. These more complex relationships may allow "automatic" coordination of postural stabilization of the proximal limb during object manipulation by the distal limb musculature.

More recently, long trains (0.5 to 1.0 sec) of supra-threshold intracortical stimulation have been reported to evoke coordinated forelimb movements in the awake primate (Color Figure 1.2C).[36] Each stimulation site produced a stereotyped posture in which the arm moved to the same final position regardless of its posture at the initiation of stimulation. In the most complex example, the monkey formed a frozen pose with the hand in a grasping position in front of the open mouth. The map of final hand location in the workspace in front of the monkey included both M1 and the premotor cortex (Color Figure 1.2C). In many respects, these results were a more

---

* Please see color insert following page 170.

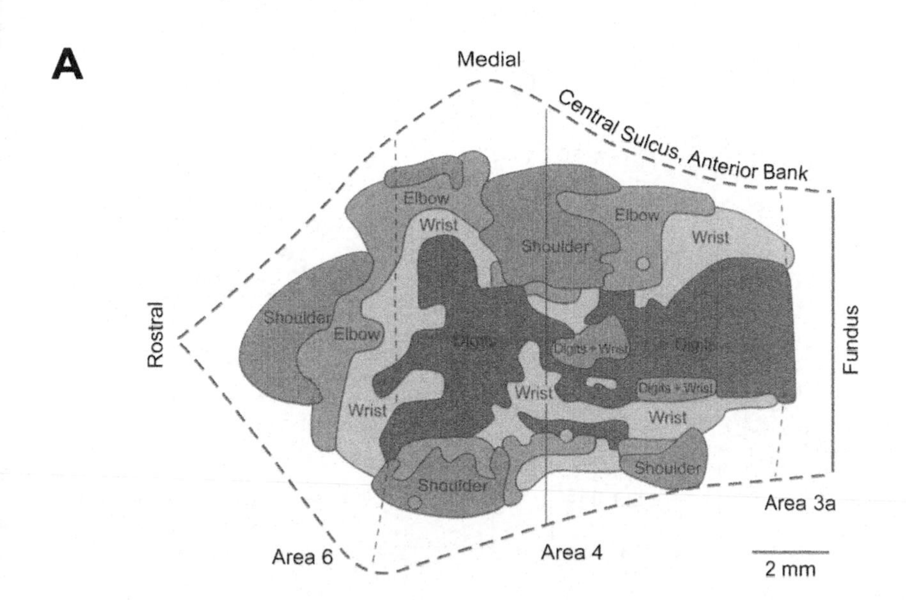

FIGURE 1.2 (see color figure) Intracortical stimulation maps of M1 in macaque monkeys. Note that in each map, hand movements form a central core (*red*). (A) Summary map of the movements evoked by intracortical stimulation (2–30 μA) in an awake macaque monkey. (Adapted with permission from Reference 22.) (B) Summary map of muscle representation in M1 derived from stimulus-triggered averages of rectified EMG activity (15 μA at 15 Hz) in an awake monkey. Sites that influenced only proximal muscles are indicated by *light shading*, those that influenced only distal muscles by *dark shading*, and those sites that influenced both proximal and distal muscles by *intermediate shading*. Sites of significant stimulus-triggered averages of rectified EMG activity for the shorthead of biceps (BIS, *blue*) and extensor digitorum communis (EDC, *red*) are indicated with size-coded dots (3, 4, 5, 6 S.D. levels above pre-trigger level baseline activity). (Adapted with permission from Reference 34.) (C) Summary of hand and arm postures produced by long train (0.5 sec), high intensity (25–150 μA) intracortical stimulation in M1, the PMd, and the PMv of an awake monkey. *Arm* sites evoked postures involving the arm but without changes in the configuration of the hand. *Hand + arm* indicates sites where stimulation evoked postures involving both the hand and arm. *Hand to mouth* indicates sites that evoked grasp-like movements of the hand which was brought to the mouth. *Bimodal/defensive* indicates sites where neurons received visual input and stimulation moved the arm into a defensive posture. See text for further explanation. (Adapted with permission from Reference 36.)

detailed equivalent of observations made initially by Ferrier[37] who reported that in M1 "long-continued stimulation brings the hand to the mouth, and at the same time the angle of the mouth is retracted and elevated." The interpretation of these complex movements is limited by the fact that intracortical stimulation primarily activates neurons trans-synaptically, and thereby enlarges its sphere of activation.[38,39] (See also References 40,41.) At the extreme, long stimulus trains and high stimulus intensities open the route for interactions at multiple levels, including local, cortical, subcortical, and spinal. Thus, intracortical stimulation is unable to determine the

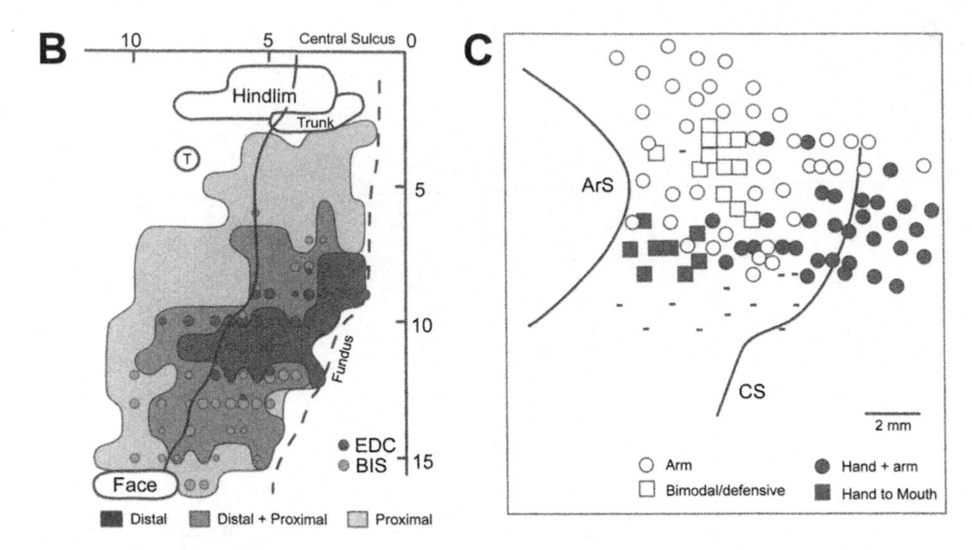

FIGURE 1.2 (continued)

output structure of M1 unambiguously or to ascertain the functional organization of a cortical motor area.

1.2.1.2 Output of Single Corticomotoneuronal Cells

A more focused approach to examining the output structure of M1 has been to determine the axonal branching patterns of single corticospinal neurons. Both physiological and anatomical studies provide evidence that single corticospinal neurons may have a rather widespread influence in the spinal cord. A substantial proportion of corticospinal neurons (43%) innervates several segments of the spinal cord.[42] Reconstruction of individual corticospinal axons filled with an intracellular tracer reveals terminal arbors located in as many as four separate motor nuclei.[43] Thus, a single corticospinal axon can directly influence several muscles.

These anatomical observations are consistent with the results of studies employing the spike-triggered averaging technique to examine the divergence of single corticomotoneuronal (CM) cells.[44–49] (For review see Reference 6.) In this technique, electromyographic (EMG) activity of a sampled muscle was averaged following each action potential of a single CM cell. Averaged muscle activity exhibiting facilitation or suppression at a short latency after the spike was considered to indicate a connection between the CM cell and the muscle's motoneurons. Most CM cells (71%) produced post-spike effects in two or more muscles (mean = 3.1, maximum 10 of 24.[49] Many of the post-spike effects were confined to distal muscles (45%) and some were found in proximal muscles (10%). Remarkably, the remaining 45% of CM neurons produced post-spike effects in both distal and proximal muscles. This result strongly suggests that single CM neurons can influence muscles at both proximal and distal joints.

The size of the branching patterns of individual CM cells appears to be related to the muscles they innervate. CM cells that influence both proximal and distal muscles have wider branching patterns than those that project to either proximal or distal muscles.[49] In addition, half of the CM cells that facilitate intrinsic hand muscles targeted just one of the muscles sampled.[48] These observations suggest that CM cells have more restricted branching to distal muscles than they do to proximal muscles. Lemon and colleagues[50–52] have emphasized, on the basis of electrophysiological data from macaque and squirrel monkeys, that direct CM projections are important for the control of grasp. Although Schieber[35] has argued that restricted branching is not a requirement for producing individuated finger movements, the restricted branching of some CM cells suggests that they may be specialized to control individual finger muscles.

The limited branching patterns of some CM neurons as well as the observation that small clusters of CM neurons tend to innervate the same motoneuron pool[42,46] may explain why intracortical stimulation can evoke contractions of a single muscle at threshold.[19] This raises the possibility that a framework for muscle representation exists at the level of small clusters of neurons. On the other hand, the highly divergent projections of many CM neurons are consistent with some of the more complex, multiple-joint movements observed with other variations of the intracortical stimulation technique.[26,36] Thus, adjustment of the parameters of intracortical stimulation may promote access to different structural features of the output organization of M1 as well as other portions of the motor system.

1.2.1.3 Peripheral Input to M1

Another type of map within M1 concerns the responses of its neurons to peripheral somatosensory stimulation. In both New and Old World primates, neurons in the caudal part of the forelimb representation of M1 were activated by peripheral input predominantly from cutaneous afferents.[25,53–55] In contrast, neurons in the rostral part of the M1 forelimb representation were driven by peripheral afferents originating largely from muscles or joints. A similar segregation of peripheral input has been observed in the hindlimb representation of M1 in the macaque.[24] Strick and Preston[54] have proposed that the segregation of peripheral inputs within M1 may represent a functional specialization designed to solve tasks demanding high levels of sensory–motor integration. For example, the portion of the hand representation in M1 that receives largely cutaneous input may be specialized to control finger coordination during object manipulation. Thus, the internal organization of M1 is quite complicated and may include multiple, overlapping maps of sensory input and motor output.

1.2.2 Premotor Areas

The identification and characterization of the premotor cortex has been the subject of some controversy and considerable revision over the last century.[2,9,15,56–61] The term "premotor cortex" was originally applied to the portion of agranular cortex (area 6) located anterior to M1 (Figure 1.1).[56,62] However, this cytoarchitectonically designated premotor cortex turned out to be functionally heterogeneous. For example, electrical stimulation of area 6 on the medial wall revealed a complete motor map of the body in a region that has been subsequently subdivided into the supplementary motor area (SMA) and presupplementary motor area (preSMA) (Figure 1.1).[15,63] (See below.) On the lateral surface, attempts to define the boundaries of the premotor cortex using electrical stimulation or cytoarchitectonic criteria failed to produce a consensus.[9,61]

1.2.2.1 Identification by Direct Projections to M1

A more recent approach for determining the location of premotor cortex has been based on its neuroanatomical connections. The premotor cortex in non-human primates has been operationally defined as consisting of those regions in the frontal lobe that have direct projections to M1 (For review see References 9,59,60,64–66.) According to this definition, the frontal lobe contains at least six spatially separate premotor areas (Figures 1.1 and 1.3A). For example, the arm representation of M1 receives projections from two rostrally adjacent regions on the lateral surface: the ventral premotor area (PMv) and the dorsal premotor area (PMd) (Figure 1.3A). The PMv is located in the portion of area 6 that is lateral to the arcuate spur and extends rostrally into the posterior bank of the inferior limb of the arcuate sulcus. The PMd occupies the portion of area 6 that is medial to the fundus of the arcuate spur and caudal to the genu of the arcuate sulcus. Its caudal extent typically includes the cortex within the superior precentral sulcus (Figures 1.1, 1.3A, and 1.4).

Four premotor areas are located on the medial wall of the hemisphere (Figures 1.1, 1.3A, and 1.4). These premotor areas include the SMA and three motor areas located within the cingulate sulcus: the rostral, dorsal, and ventral cingulate motor areas (CMAr, CMAd, and CMAv). The SMA is confined to the portion of area 6 on the mesial surface of the superior frontal gyrus that lies between the arcuate genu rostrally and the hindlimb representation in M1 caudally. The CMAr is located within area 24c on the dorsal and ventral banks of the cingulate sulcus at levels largely anterior to the genu of the arcuate sulcus. The CMAd occupies area 6c on the dorsal bank of the cingulate sulcus at levels caudal to the genu of the arcuate sulcus. The CMAv lies on the ventral bank of the cingulate sulcus in area 23c, mostly at the same levels as the CMAd. Thus, the premotor cortex, as defined by its anatomical connections to M1, is more complicated than previously recognized (for review see References 2,3,8,15,57,62) and is composed of multiple, spatially separate premotor areas (Figures 1.1, 1.3, and 1.4).[59,60,67–69] (See also References 70–76.)

The portion of area 6 (area 6aB)[17] that lies dorsal and anterior to the genu of the arcuate sulcus can no longer be considered as part of the premotor cortex because it lacks direct connections with M1. In fact, the connections of these rostral portions of area 6 suggest that they are more properly considered regions of the prefrontal cortex (see below). On the medial wall, this rostral portion of area 6 (area F6[77,78]) has been recognized as a separate functional region and termed the preSMA (Figures 1.1 and 1.4).[65,79,80] Similarly, on the lateral surface, the rostral portion of area 6 (area F7[77,78]) has been termed the prePMd (Figures 1.1 and 1.4). (For review see Reference

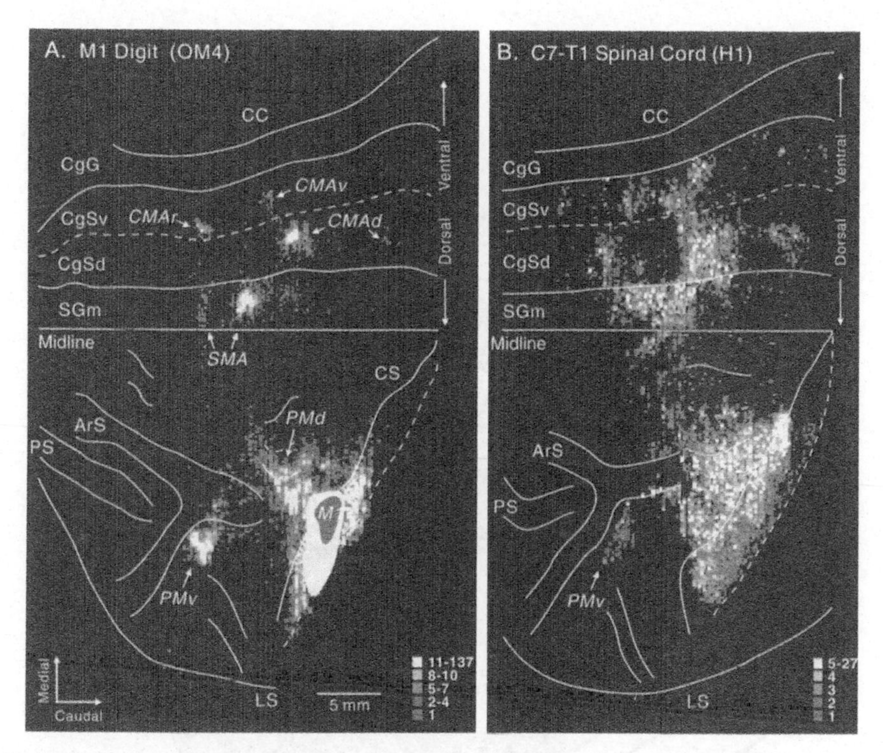

FIGURE 1.3 Identification of premotor areas in the frontal lobe. (A) Premotor areas project to M1. An unfolded map of the frontal lobe depicts the density of labeled neurons after WGA–HRP injections into the physiologically identified digit representation of M1 in the macaque monkey. (For details of the unfolding and the determination of cell density, see Dum and Strick.[60]) The medial wall is unfolded and reflected upward from the midline so that it appears upside down. The lip of each sulcus (*solid line*) and its fundus (*dashed line*) are indicated. The labeled neurons in the PMv (*arrow*) are located in the posterior bank of the arcuate sulcus and have been projected to the surface. This projection to the surface artificially increases the displayed density. (B) Premotor areas project to the spinal cord. An unfolded map of the frontal lobe shows the density of labeled corticospinal neurons after injections of a fluorescent tracer into the C7–T1 segments of the spinal cord. Abbreviations: CC: corpus callosum; CgSd: dorsal bank of the cingulate sulcus; CgSv: ventral bank of the cingulate sulcus; SGm: medial superior frontal gyrus. (Reproduced with permission from Reference 64.)

66.) Thus, the current definition of premotor cortex includes multiple premotor areas located in the caudal half of area 6 as well as in additional regions within the cingulate sulcus that were historically considered part of the limbic cortex.[9]

1.2.2.2 Somatotopic Organization Based on Connections with M1

The somatotopic organization of the premotor areas has been evaluated based of their projections to the arm, leg, and face representations of M1.[59,60,64,67–69,71–76,81,82] A number of general conclusions have come from these studies. Some premotor

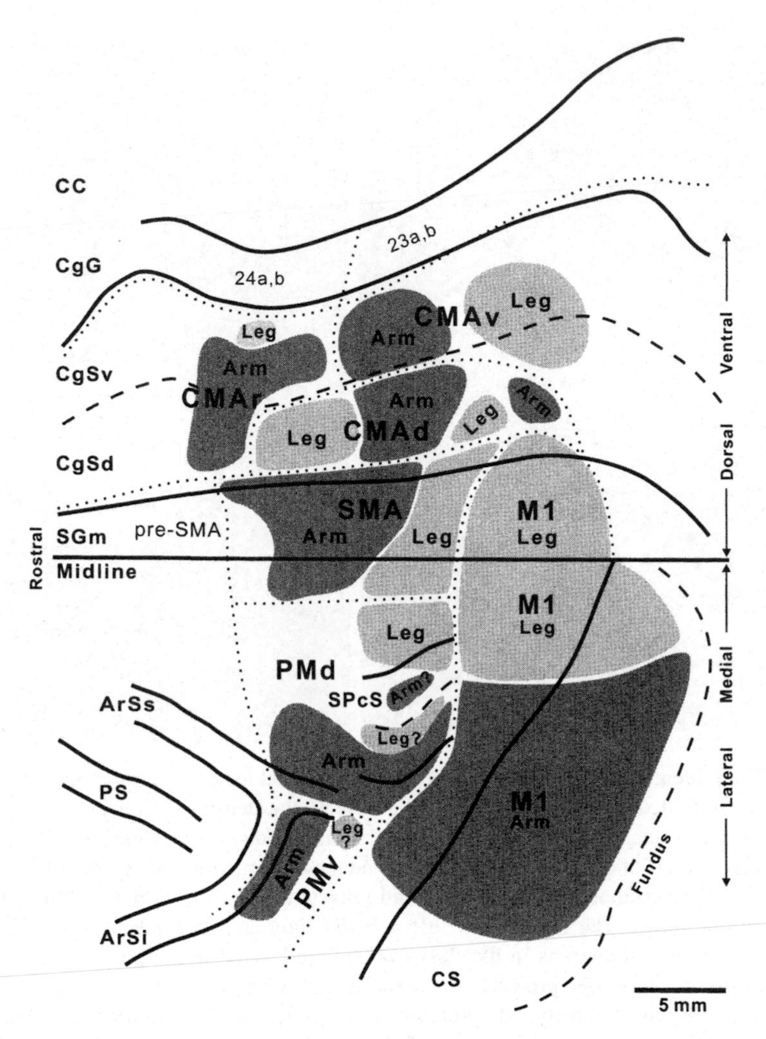

FIGURE 1.4 Somatotopy of corticospinal projections. In this map, the location of the arm representations in M1 and the premotor areas are based on the origin of neurons that project to upper and lower cervical segments. The location of the leg representations in each cortical area is based on the origin of neurons that project to lower lumbosacral segments. For conventions and abbreviations see Figures 1.1 and 1.3. ArSi: arcuate sulcus, inferior limb; ArSs: arcuate sulcus, superior limb. (Adapted with permission from Reference 84. Also adapted with permission from Reference 85.)

areas lack a complete representation of the body (e.g., the PMd lacks a face area). Indeed, complete maps of the body can only be defined for the SMA, CMAv, and CMAr. On the other hand, the arm has the most widespread and robust representation within each of the premotor areas. Overall, the major representations within each premotor area originate from distinct, non-overlapping regions.

1.2.2.3 Corticospinal Output

Russell and DeMyer[83] first demonstrated that area 6 contributes about the same number of axons to the pyramids as does area 4. However, the importance of corticospinal projections from the premotor areas has only been appreciated recently. With the advent of retrograde and anterograde neuronal tracing techniques, numerous authors were able to demonstrate that each premotor area has direct access to the spinal cord (Figures 1.3B and 1.4).[59,60,84,85] (See also References 86–93.) The distribution of corticospinal neurons in the premotor areas that projected to cervical segments of the spinal cord corresponded remarkably well to the distribution of neurons in the premotor areas that projected directly to the arm representation in M1 (Figures 1.3A and 1.3B). These results suggest that each premotor area has the potential to influence the generation and control of movement directly at the level of the spinal cord, as well as at the level of the primary motor cortex.

Numerically, the overall contribution of the premotor areas to the corticospinal tract is equivalent to or greater than that of M1. This is most apparent for corticospinal projections to the cervical segments of the spinal cord. After tracer injections confined to the cervical segments (arm representation), the percentage of the total number of corticospinal neurons in the frontal lobe that originated in the premotor areas was always equal to or greater than the percentage of corticospinal neurons in M1 (premotor mean = 56%, range 50–70%, n = 6).[60,84,85] For tracer injections confined to the lumbosacral segments (leg representation), the percentage of corticospinal neurons in the frontal lobe that originated in the premotor areas was less than the percentage of corticospinal neurons in M1 (premotor mean = 43%, range 39–46%, n = 2).[85] These observations reinforce the view that the arm representation within the premotor areas is more robustly developed than is the leg representation.

In other measures of the relative strength of corticospinal projections, M1 clearly dominates but the premotor areas still make significant contributions. For example, each premotor area had some localized regions in which the density of corticospinal neurons was equivalent to that found in M1. In fact, the relative density of corticospinal neurons in the SMA, CMAd, CMAv and PMd was similar to that found in M1.[60] (See also References 84,85.) With respect to the distribution of large and small corticospinal neurons, most large corticospinal neurons (79%) were concentrated in M1.[60] The remaining large corticospinal neurons were located in the PMv, PMd, SMA and CMAd.[60] Large corticospinal neurons, which comprise less than 20 percent of the total,[60,88,94,95] are thought to be especially important for mediating corticomotoneuronal synapses. (See Reference 11.) Taken together, the observations on the number, density, and size of corticospinal neurons indicate that the premotor areas make a substantial contribution to the corticospinal system.

1.2.2.4 Somatotopic Organization Based on Corticospinal Output: Forelimb and Hindlimb Representation

Because cervical segments of the spinal cord are known to control arm movements and lumbosacral segments are known to control leg movements, the "arm" and the "leg" representations of a cortical area also can be identified on the basis of the origin of their projections to the cervical or lumbosacral segments of the spinal cord, respectively. This is possible because only 0.2% of corticospinal neurons branch and innervate both the cervical and lumbosacral levels of the spinal cord.[85] Corticospinal projections from all of the premotor areas displayed a high degree of topographic organization. The origin of corticospinal neurons in the premotor areas that projected to cervical or to lumbar segments of the spinal cord corresponded remarkably well to the origin of neurons in the premotor areas that projected directly to the M1 arm or to the M1 leg representations, respectively (Figures 1.3A, 1.3B, and 1.4).[59,60,76,84,85] Thus, the origins of corticospinal and cortico-cortical projections to M1 are in the somatotopic register.

Five premotor areas projected to the cervical and to the lumbosacral segments of the spinal cord (Figure 1.4). In the PMd, SMA, CMAd, and CMAv, the origin of projections to cervical segments did not overlap with the origin of projections to the lumbosacral segments. In the CMAr, the arm and leg representations were not as clearly separated, whereas in the PMv, most of the corticospinal neurons projected only to the upper cervical segments.[84] Thus, at least four premotor areas contained arm and leg representations that appear to be as distinct as those found in M1.

1.2.2.5 Somatotopic Organization Based on Corticospinal Output: Proximal and Distal Arm Representation

The topography of the "proximal" and "distal" arm representations has been examined by injecting different fluorescent tracers into upper cervical and lower cervical segments of the spinal cord.[84,85] In general, lower cervical segments are primarily involved in the control of the hand and wrist muscles, whereas upper cervical segments are largely involved in the control of the neck, elbow, and shoulder muscles. (He et al.[84] have discussed the topographic organization of the spinal cord motor nuclei.) All of the premotor areas projected to upper and lower cervical segments, but only 5% of corticospinal neurons innervated both the upper and lower cervical segments.[85] In each premotor area, the densest concentrations of corticospinal neurons that projected to upper cervical segments were separate from the densest concentrations of neurons that projected to lower cervical segments.[84,85] This same pattern was also evident in M1. These results suggest that some of the premotor areas have proximal and distal representations of the arm that are as distinct as those in M1.

One measure of the importance of each premotor area in the control of distal versus proximal arm movements is the relative amount of cortex projecting to the lower versus upper cervical segments.[84,85] Within M1, the region that projects to lower cervical segments is equal in size to the region that projects to upper cervical segments. This result suggests that the hand representation in M1 is expanded relative to the actual physical proportion of the arm that is occupied by the hand. The expansion of the hand representation has been viewed as a reflection of the special role that M1 retains in the generation and control of highly skilled hand movements.[12,15,96,97]

1.2.2.6 Organization Based on Intracortical Stimulation

The anatomical framework outlined above firmly establishes that the premotor areas are important components in the central mechanisms of skeletomotor control. Intracortical stimulation with microelectrodes has been used to assess the potential of each premotor area to generate movements and to construct a map of the body parts represented in each area. Significantly, intracortical stimulation has evoked movement in each of the premotor areas. Typically, the average threshold for evoking movement with intracortical stimulation in a premotor area is somewhat higher than that in M1, and the probability of evoking movement at any given site is lower in the premotor areas than in M1.[64,76,98–101] In most respects, the body maps produced by intracortical stimulation within the premotor areas are congruent with the topographic organization revealed by anatomical methods.

Electrical stimulation of the SMA generated a complete map of the body with a rostral to caudal orientation of its face, arm, and leg representations (Figure 1.5).[63,80,89,98,99,102,103] Overall, this somatotopy was consistent with the body map based on the SMA's projections to M1 and on corticospinal projections to different segmental

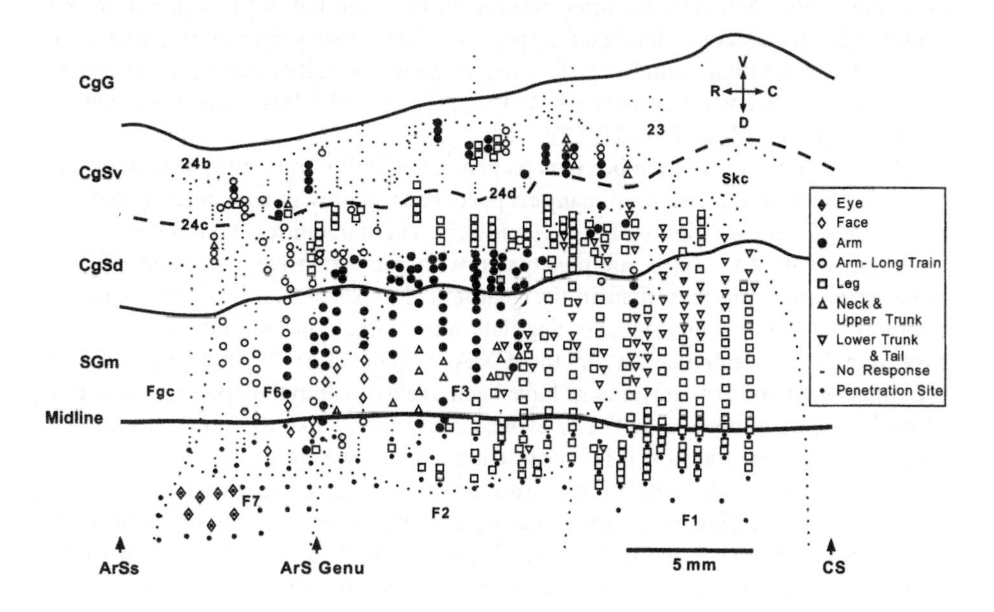

FIGURE 1.5 Intracortical stimulation map of the medial wall of the hemisphere. The medial wall is unfolded and reflected upward to display the medial wall in an "upside down" orientation. The boundaries between cytoarchitectonic areas (*dotted lines*), the fundus of the cingulate sulcus (*dashed line*), and the lips of the cingulate sulcus (*solid lines*) are indicated. Movements were evoked by short- or long-train intracortical stimulation in a macaque monkey. All movements were contralateral to the stimulated hemisphere. For conventions and abbreviations see Figures 1.1 and 1.3. ArSs: rostral limit of the superior limb of the arcuate sulcus; Fgc: frontal granular cortex; Skc: somatic koniocortex. (Adapted from Reference 99.)

levels (Figures 1.4 and 1.5) (see above).[67,75,76,84,85] The reported organization of body parts within the face, arm, and leg representations has been less consistent, perhaps due to the fact that complex movements involving multiple joints or noncontiguous joints were evoked at some sites in the SMA. Nevertheless, sites within the arm representation of the SMA that evoked movements of distal joints tended to be located ventral to sites where movements of proximal joints were evoked.[80,89,98,99] Correspondingly, the origin of corticospinal neurons projecting to lower cervical segments tended to be located ventral to the origin of corticospinal neurons projecting to upper cervical segments.[85]

Intracortical stimulation reinforced the distinction between the SMA and the preSMA. Intracortical stimulation with parameters that were effective in the SMA did not evoke movement in the preSMA which lies just rostral to the SMA on the medial wall of the hemisphere (Figures 1.1 and 1.5).[79,80,99,104–107] Movements of the arm and rarely the face were evoked at some sites within the preSMA when higher currents and longer pulse trains were applied.[80,99,102,106] The movements were also different in character from those evoked in the SMA. Movements elicited in the preSMA were typically slow, involved multiple joints, and resembled natural postural movements. PreSMA neurons often responded to visual but not to somatosensory stimuli, whereas SMA neurons had the opposite characteristics, responding to somatosensory but not visual stimuli.[79,80] The requirement for higher currents and longer stimulus trains is consistent with the fact that the preSMA lacks direct projections to the spinal cord[60,85] and to M1.[59,60,74,78,79]

The major features of the body maps generated with intracortical stimulation in the cingulate motor areas are in many respects consistent with anatomically defined somatotopy (compare Figures 1.3, 1.4, and 1.5). The intracortical stimulation maps, however, are more fractured, are punctuated with nonresponsive areas, and reflect a lower sampling frequency than in the anatomical experiments. Movements of the arm and leg were elicited in each cingulate motor area, but face movements were evoked only, and infrequently, in the CMAr (Figure 1.5).[98,99,101–103] In addition, proximal and distal arm movements have been evoked within the arm representation of each cingulate motor area.[99] The evoked movements, like those elicited in M1, were usually limited to fast, brief contractions at a single joint.

Longer pulse trains and higher currents were required to evoke movements within the CMAr. This observation is congruent with the relatively low density of corticospinal neurons found in this area (Figure 1.3B). Thorough exploration of the CMAr was limited to one animal, where arm and leg movements were found to be somewhat intermingled.[99] Similarly, the origins of corticospinal projections to the cervical and lumbar segments are somewhat overlapping in the CMAr.[85] In the region corresponding to the CMAd on the dorsal bank of the cingulate sulcus, leg and trunk movements were found rostrally, just ventral to the arm representation in the SMA.[99,103] Thus, the orientation of the body map in the CMAd is reversed compared to the one in the SMA, just as was predicted by the origin of corticospinal projections to the cervical and lumbar segments (Figure 1.4).[85,108] Arm movements were consistently evoked in the rostral portion of the region corresponding to the CMAv (Figure 1.5)[99] (see also Reference 109), but few penetrations have been made in the caudal portion of the CMAv where a leg representation was reported to be located.[72,76,82,85,108] Thus, there is a reasonable correspondence between the maps generated by intracortical stimulation and those generated using anatomical methods.

Systematic mapping of the PMd with intracortical stimulation has been limited to one study,[101] although numerous studies have reported the results of partial explorations of this region.[22,36,76,110] In general, leg movements were evoked in the region of the PMd that was medial to the superior precentral sulcus (dimple) and arm movements were evoked in the region that was lateral to this sulcus.[101] Distal and proximal arm movements were evoked within this region.[22,101,110] Within the PMd, the threshold for evoking movements is highest rostrally and decreases caudally.[101,111,112] These results parallel the increase in the density of corticospinal neurons in the caudal portion of the PMd.[60,84] Eye movements were evoked by stimulation in the prePMd which lies just rostral to the PMd.[112] Thus, here again, the border between the prePMd and the PMd defined by intracortical stimulation corresponds to the border defined using connections to M1 and the spinal cord.[60,64,76,84]

Intracortical stimulation has defined arm and face representations within the PMv.[101] Distal arm movements dominated the portion of the arm representation that was buried medially within the inferior limb of the posterior bank of the arcuate sulcus (Figure 1.6).[64,100,101] This portion of the PMv projected almost exclusively to upper cervical segments of the spinal cord.[60,84,113] The distal movements evoked by stimulation at this site must therefore be mediated either by propriospinal connections from upper to lower cervical segments (for discussions see References 84,114–116) or through connections with the hand area of M1.[117,118] Proximal arm movements tended to be evoked on the surface near the arcuate spur.[100,101] The PMv face representation was located lateral to the arm representation both on the surface and within the posterior bank of the arcuate sulcus (not shown in Figure 1.6).[101] Information regarding the internal organization of the face area is limited, although laryngeal muscles appear to be represented laterally along the inferior limb of the arcuate sulcus.[119]

In summary, intracortical stimulation evokes body movements from each premotor area. These stimulation effects could be mediated directly via corticospinal efferents from each premotor area or indirectly by projections from each premotor area to M1 and henceforth via the corticospinal efferents from M1. Examination of this issue is limited to a short report.[120] In this study, the arm and vibrissae representations in the SMA and M1 of the owl monkey were mapped with intracortical stimulation. Following removal of M1, intracortical stimulation of the SMA could still evoke movements with stimulus currents that were in the range of prelesion values. This observation suggests that electrical activation of corticospinal efferents in the SMA is sufficient to generate muscle contraction. This conclusion reinforces the view that independent and parallel pathways for motor control originate in the SMA and M1.

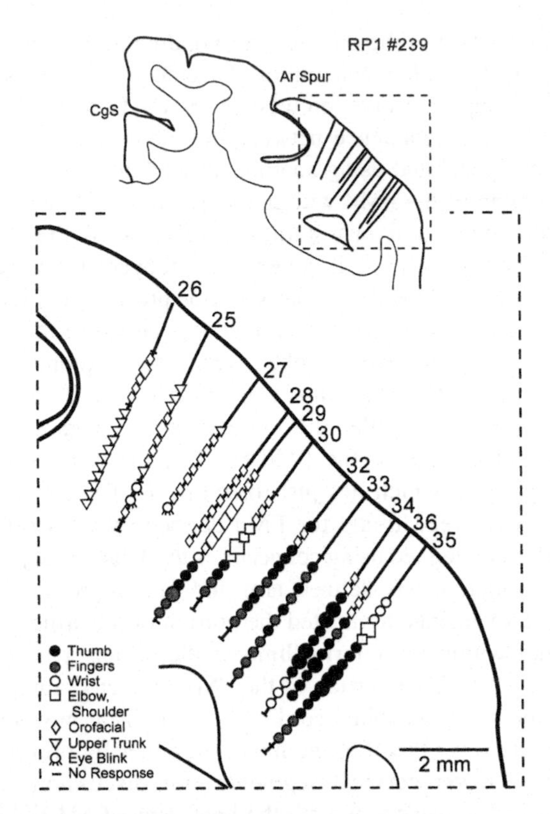

FIGURE 1.6 Intracortical stimulation map of the arm representation of the PMv. A cross-section from a macaque brain (*Macaca nemestrina*) illustrates the location of electrode penetrations and the movements evoked at each stimulation site within the posterior bank of the inferior limb of the arcuate sulcus. Thresholds for evoking movement are indicated by symbol size (large symbols = #15 μA, small symbols = 16–50 μA). (Reproduced with permission from Reference 64.)

1.2.3 Corticospinal Terminations

1.2.3.1 Primary Motor Cortex

The pattern of corticospinal terminations is one indicator of a cortical area's potential influence on different spinal mechanisms. (A complete discussion of this issue is provided by Kuypers.[1]) Corticospinal efferents from M1 project to the intermediate zone (laminae V–VIII) of the spinal cord where interneurons that innervate motoneurons are located as well as directly to the portions of the ventral horn where motoneurons are located (Figure 1.7).[1,97,121–129] On the other hand, corticospinal projections from somatosensory and posterior parietal cortex terminate primarily in the dorsal horn.[1,121,122,124–126] Evidence from physiological studies suggests that these corticospinal efferents modulate neural processing in ascending somatosensory pathways. (For review, see References 12,130.) Thus, both anatomical and physiological

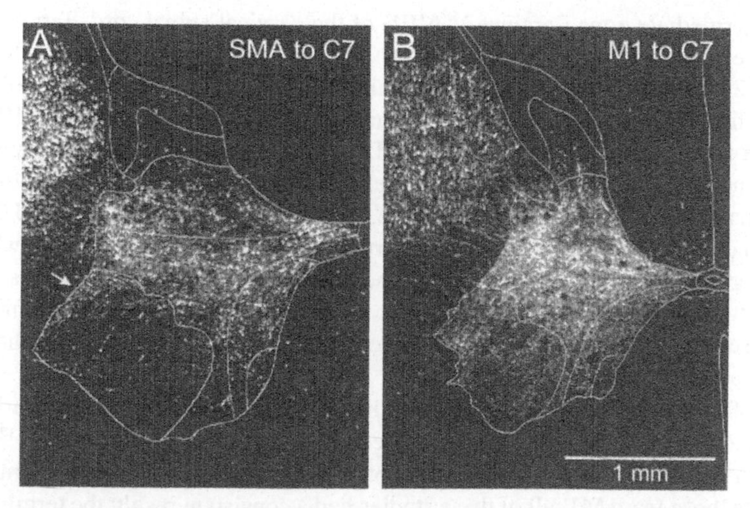

FIGURE 1.7 Corticospinal terminations in C7 of a macaque monkey. Digital photomicrographs of spinal cord sections viewed under dark-field illumination with polarized light. The gray matter and spinal laminae are outlined. (A) SMA efferents terminate densely in intermediate zone of the gray matter of the cervical spinal cord. *Arrow* points to terminations in the dorsolateral part of lamina IX that contains motoneurons. (B) M1 efferents terminate in the same regions as do SMA efferents. Compared to SMA terminations, M1 terminations are denser and more extensive in lamina IX, and extend further into the base of the dorsal horn. (Adapted with permission from Reference 9.)

evidence suggest that the pattern of corticospinal terminations reflects the differential involvement of these cortical areas in motor output or in somatosensory processing.

The extent of M1 terminations within the motor nuclei of the spinal cord changes during development[129,131,132] and varies between different species.[1,133] This variation appears to correlate with an animal's manual dexterity.[127,132–136] For example, cebus monkeys, which grasp small objects and manipulate tools with a modified "precision grip,"[137–139] have abundant direct projections from M1 to spinal motor nuclei.[127] In contrast, squirrel monkeys, which pick up small items with all fingers grasping in concert,[139,140] have sparse monosynaptic corticospinal terminations that are located remotely on motoneuron dendrites.[50,127] Thus, the extent of monosynaptic projections from M1 to spinal motoneurons appears to be part of the neural substrate necessary to make highly skilled and relatively independent movements of the fingers.[50,127,129] (For review, see References 1,12.)

1.2.3.2 Premotor Areas

The relationship between the terminations of corticospinal efferents and the motor, sensory, and interneuronal systems of the spinal cord has been studied only for premotor areas on the medial wall of the hemisphere.[97,128,141,142] In general, the pattern of corticospinal terminations from the SMA was quite similar to that of M1 (Figure 1.7).[97,128] The densest terminations of efferents from the SMA and M1 were located in the intermediate zone (laminae V–VIII) of the cervical spinal cord. Terminations here were concentrated at three locations: (1) the dorsolateral portion of laminae V–VII; (2) the dorsomedial portion of lamina VI at the base of the dorsal columns; and (3) the ventromedial portion of lamina VII and adjacent lamina VIII. The cingulate motor areas (CMAr, CMAd, CMAv) also terminated most densely within the intermediate zone.[128,142] However, the density of their terminations was noticeably lower than those from the SMA. In addition, terminations from the CMAr and CMAd were concentrated in the dorsolateral portions of the intermediate zone whereas CMAv terminations were most dense in the dorsomedial portions.[128,142] This differential pattern of terminations suggests that the CMAr, CMAd, and CMAv innervate specific sets of spinal interneurons and thereby influence different spinal mechanisms for controlling forelimb movements.

All of the medial wall premotor areas, like M1, had terminations that overlapped motor nuclei in the ventral horn of the cervical segments (Figure 1.7).[97,128,141,142] Although the terminations from the premotor areas were less dense over lamina IX than were those from M1, all of these studies had a consistent result: the terminations of premotor areas that do overlap lamina IX were concentrated over the motor nuclei innervating muscles of the fingers and wrist. Furthermore, the presence of monosynaptic projections onto motoneurons innervating muscles of the distal forelimb has been confirmed electrophysiologically for the SMA.[97] This result implies that the presence of anterogradely labeled terminations over spinal motoneurons is an indication of direct corticomotoneuronal connections. Thus, not only the SMA but also the CMAd, CMAv, and CMAr appear to project directly to motoneurons controlling the distal forelimb. In summary, these results suggest that the premotor areas have the anatomical substrate required to influence the generation and control of limb movement, particularly of the hand. This influence is mediated by pathways that are parallel to and independent of those originating in M1.

1.3 CORTICAL INPUTS TO THE MOTOR AREAS

The recognition that the premotor areas as well as M1 project to the spinal cord suggests that motor commands may arise from multiple cortical areas. We have proposed that each cortical area in the frontal lobe that projects to the spinal cord could operate as a separate efferent system for the control of specific aspects of motor behavior.[59] Obviously, identification of the cortical and subcortical inputs to these motor areas could provide some insight into their functional contributions to motor control.

Analysis of inputs to M1 and the premotor areas is complicated by several factors. First, the characterization of the premotor areas is still evolving and thus, their precise borders remain controversial.[60,77,82] (For review, see Reference 66.) For instance, some initial examinations of the inputs to the SMA actually studied the rostrally adjacent region that is now termed the preSMA. Second, the representations of the face, arm, and leg within a cortical area may receive different sets of cortical inputs.[73,75,76,78,82,143–146] For example, the arm representations of M1 and the SMA have robust connections with the PMv whereas their leg representations do not.[75,76,78] Thus, discrepancies between studies may result from differences in the body representation actually injected. Third, the boundaries and identification of the cortical areas projecting to the motor areas are still evolving. For instance, area PE in the parietal lobe projects to several premotor areas, but these projections tend to originate from separate portions of this parietal area.[81,146,148,149] These results suggest that PE may not be a single homogeneous area. Thus, precise localization and identification of the cortical areas injected with tracers and the cortical areas containing labeled neurons are essential for valid comparisons between different experiments.

Another aspect of comparing the inputs to each motor area is judging the relative importance of various inputs. A small cortical region may receive input from 40 to 70 cytoarchitectonically recognized cortical areas in the ipsilateral hemisphere alone.[82,144,150,151] However, quantitative analysis of all of the inputs to a single cortical site has rarely been attempted.

To minimize the problems of cortical identification and strength of input, we focused our analysis on studies that examined the inputs to the arm representation of motor areas in macaque monkeys. We then transformed the results of these studies onto a standarized map of the frontal and parietal lobes (Figure 1.1). Next, we pooled the results from recent publications and assigned a "strength" to specific connections based on the relative number of labeled neurons and the consistency with which a projection was observed in all studies (Tables 1.1 and 1.2). Even with these constraints, we found considerable variation in the results among studies. Consequently, our synthesis of these results reflects our consensus derived from multiple studies and may not always fit with the data reported in an individual study.

1.3.1 Primary Motor Cortex

1.3.1.1 Frontal Cortex

Cortical input to M1 is entirely confined to cortical regions in the frontal and parietal lobes that, like M1, are the origin of projections to the spinal cord (Figure 1.8, Table 1.1; see Figure 1.1 for area identification). These corticospinal tract (CST) projecting areas include all the premotor areas in the frontal lobe (defined above) and portions of the superior parietal lobe (SPL). M1 has no substantial connections with the prefrontal, pre-premotor or limbic cortex.

1.3.1.2 Parietal Cortex

The densest and most extensive of projections from the parietal lobe to M1 originate in the posterior portions of the SPL (Figure 1.8, Table 1.2; see Figure 1.1 for area identification). This input arises in area PE on the lateral surface of the postcentral gyrus and area PEip in the lateral portion of the dorsal bank of the intraparietal sulcus.[59,68,71,73,76,147,152–155] M1 also receives strong inputs from the primary (SI) and secondary (SII) somatosensory cortices. The origin of SI projections to M1 is surprisingly widespread although their density is more modest than those from area PE (Table 1.2). The strength of projection from the subdivisions of SI is greater for those regions (e.g., areas 1 and 2) that are at a "later" stage in processing of the cutaneous and proprioceptive afferent information than area 3b, which is at an "earlier" stage of processing. Nevertheless, area 3a does have substantial input to

TABLE 1.1
Ipsilateral Cortical Input from the Frontal Lobe to M1, the Premotor Areas and the Pre-Premotor Areas

Cortical Area	M1	PMdc	PMv	SMA	CMAd	CMAv	CMAr	Pre-PMd	Pre-SMA
Motor									
M1	Inj. Site	xxx	xxx	xxx	xx	xx	x		
PMd	xxx	Inj. Site	xx	xxx	xx	x	x	xxx	xx
PMv	xxx	x	Inj. Site	xxx	x	x	xx	x	xx
SMA	xxx	xxx	xxx	Inj. Site	xxx	xxx	xx	x	x
CMAd	xx	xxx	x	xxx	Inj. Site	xxx	xxx		
CMAv	xx	xxx	xx	xxx	xxx	Inj. Site	xxx	?	
CMAr	xx	xx	xx	xxx	xxx	xxx	Inj. Site	xxx	xxx
Pre-Premotor									
PreSMA		xx	x	xx	?	x	xx	xxx	Inj. Site
PrePMd		xx		x	?	xx	xx	Inj. Site	xxx
PrCO			xxx			?	xx		?
Prefrontal									
Area 46d		?				xx	x	xxx	x
Area 46v			xx			x	x		xx
Area 9m							?	xx	xxx
Area 91						?	?	x	
FEF									x
SEF								x	
Limbic									
Area 24a,b	?	?	?	x	x	xx	xx	?	xxx
Area 23a,b	?					x	x	?	
25, 29, 30							?		
Orbital Frontal Cortex									
TF, 28, 35						?	x		
Prostriata							?		

Note: xxx = major input, xx = moderate input, x = weak input, ? = weak input that was less than 1% of total input and/or not observed in every case.

M1.[71,73,154–156] SII has heavy projections to M1 in most studies,[68,71,73,154,155] but these projections were less substantial when injections were confined to the surface of the precentral gyrus.[59,76]

As noted earlier, neurons in M1 exhibit short latency responses to activation of cutaneous and proprioceptive receptors. However, the route by which this somatosensory input reaches M1 remains controversial. Lesion of the dorsal columns, a major ascending pathway to the parietal cortex, extinguishes the responsiveness of

TABLE 1.2
Ipsilateral Cortical Input from the Parietal Lobe to M1, the Premotor Areas and the Pre-Premotor Areas

Cortical Area	M1	PMdc	PMv	SMA	CMAd	CMAv	CMAr	Pre-PMd	Pre-SMA
				Superior Parietal Lobule					
Area 3a	xx		x		xx	x		?	
Area 3b	x								
Area 1	xx		?		x	?			
Area 2	xx		?	xx	x	x			
5 (PE)	xx		xx	xxx	x	xx			
PEip	xxx	xxx	xx	xx	xx	xx	?		
MIP	x	xxx	?	?		xxx	?	x	
PEc		xxx		?	x		?	xx	
PEci		xx		xx	x	x	x	xx	
V6A								xxx	
PGm		x	?				?	xxx	
CGp		x				?	?	xxx	
				Inferior Parietal Lobule					
AIP	x		xxx						
VIP	x		?					?	
LIP	?		?						
7b (PF)	?	?	xxx				x	?	
PFG	?	?	?			xxx	x	?	x
7a (PG)		?	?			x		xx	x
				Lateral Sulcus					
PFop	x		?		xx	xxx	xx		
PGop			?			x			
SII	xx		xxx		xx	x	xxx		
Ig			xx		xx	x	xxx		x
Idg						x	xx		
RI			?						
				Temporal					
STS			?						x

Note: xxx = major input, xx = moderate input, x = weak input, ? = weak input that was less than 1% of total input and/or not observed in every case.

M1 neurons to peripheral input.[157] Removal of the primary and secondary somatosensory areas, as well as removal of the cerebellum, does not.[5,158] Some authors have proposed that M1 receives input directly from a thalamic region innervated by a portion of the dorsal column pathway (for references and discussion, see Reference 5), but firm anatomical evidence for such a neural circuit remains elusive.[159] Although the major route by which short latency somatosensory information reaches M1 has

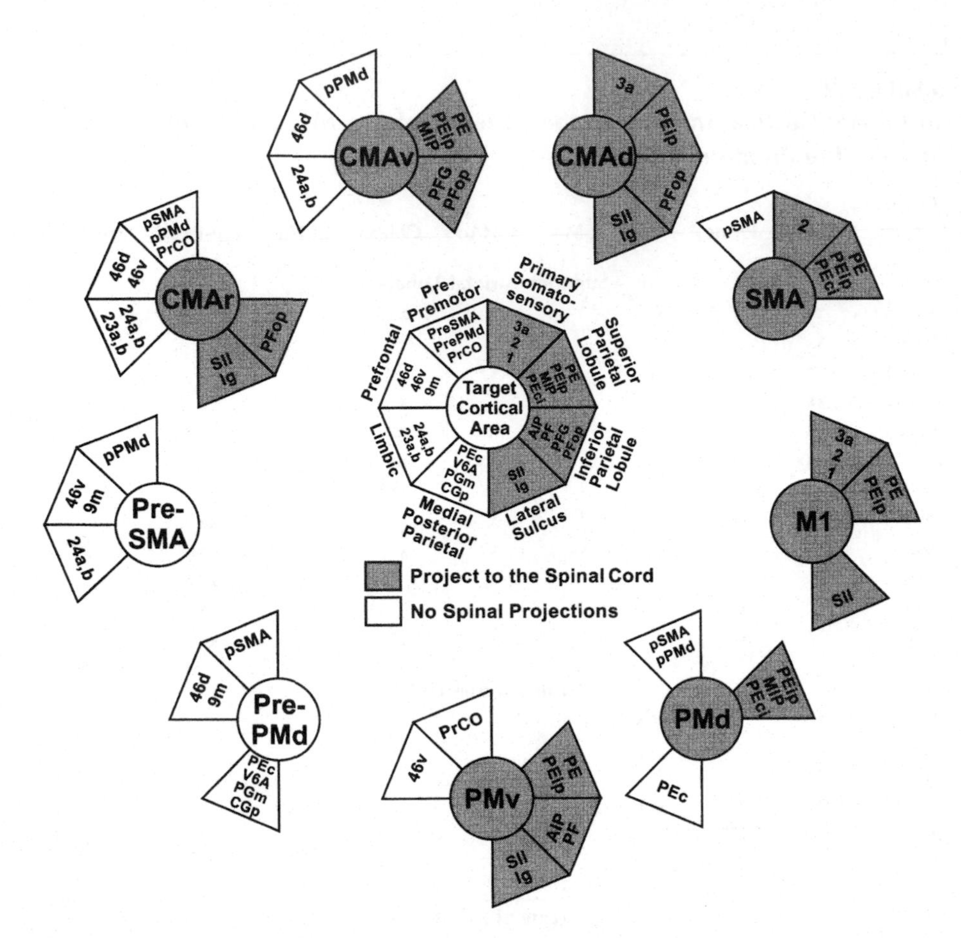

FIGURE 1.8 Major ipsilateral cortical inputs to M1, the premotor areas and the pre-premotor areas in macaque monkeys. The premotor areas (*gray shading*) have reciprocal connections with M1 and project to the spinal cord, whereas pre-premotor areas (*no shading*) do not. All 25 of the major cortical inputs are grouped into 8 categories, reflecting morphological location and proposed functional similarity (see key, Tables 1.1 and 1.2). Sources of cortical input are divided into cortical regions that project to the spinal cord (*gray shading*) and those that do not (*no shading*). Note that each of the profiled cortical areas receives a unique signature of extrinsic cortical inputs.

not been determined, the interconnections between SI and M1 may be necessary for an animal to learn a new motor skill under the guidance of somatosensory cues.[5,160]

1.3.2 Premotor Areas

1.3.2.1 Interconnections among the Motor Areas

In general, the premotor areas are richly interconnected (Figure 1.8, Table 1.1). Several features of the connections among the premotor areas stand out. First, the SMA, of all the premotor areas, has the densest and most balanced reciprocal connections with every other premotor area as well as with M1. Second, the interconnections among the cingulate motor areas (CMAd, CMAv, CMAr) on the medial wall are dense and equal. Third, the PMd and the PMv on the lateral surface have strong, reciprocal connections with M1 and the SMA. On the other hand, the connections between the PMd and PMv are more limited. The PMv is connected to the caudal portion of the PMd,[58,59,68,72,81] but only sparsely to its rostral portions.[58,161,162] The restricted connections between the PMd and the PMv may reflect differences in the body representation in each area. The PMd was reported to project mainly to the shoulder representation of M1, whereas the PMv projected largely to the digit representation in M1.[73] Fourth, the projections from the lateral motor areas (PMd, PMv, M1) to the cingulate motor areas, particularly the CMAv and the CMAr, tend to be relatively weak.[82,163,164] These patterns of connectivity suggest that the PMd and PMv are fundamentally distinct from each other and from the motor areas on the medial wall. The SMA with its broad, balanced connectivity is ideally situated to coordinate and integrate information flowing among the motor areas in the frontal lobe.

1.3.2.2 Parietal Cortex

Parietal lobe input to the premotor areas appears to follow a general trend. Most of the parietal lobe input to the premotor areas originates from posterior portions of the parietal lobe including the SPL (area 5, as described by Brodmann[16]), the inferior parietal lobule (area 7, as described by Brodmann[16]) and the secondary somatosensory cortex (SII) (Figure 1.8, Table 1.2). These areas are thought to be concerned with the highest levels of somatosensory processing and to participate in multimodal sensory integration, spatial attention, or visuomotor control.[165] Every premotor area in the frontal lobe is richly interconnected with parts of at least one of these posterior parietal areas. On the other hand, only the SMA receives dense input from any of the subdivisions of the primary somatosensory cortex. This input to the SMA originates in area 2, which is thought to be at an intermediate stage of somatosensory processing (Figure 1.8, Table 1.2).[59,68,70,72,78,81,146,148,155,161–163]

Portions of the superior parietal lobule project to all the premotor areas except for the CMAr. On the postcentral gyrus, the lateral portion of area PE targets the PMv, whereas its more medial portions project heavily to the SMA and CMAv. In the most caudal portion of the postcentral gyrus, area PEc supplies dense input to the PMd. Laterally within the intraparietal sulcus, area PEip has the most widespread projections. Area PEip targets five premotor areas, including the PMd, PMv, SMA, CMAd, and CMAv (Figure 1.8, Table 1.2). Despite this apparently broad divergence, the origin of projections from PEip to the PMv and PMd as well as M1 tend to arise from separate location.[59,147,149,162] Medially in the intraparietal sulcus, area MIP projects densely to the PMd and the CMAv. In the caudal portion of the cingulate sulcus, area PEci provides strong input to the SMA and the PMd.[59,78,146,148,149,166] Thus, although the premotor areas are broadly targeted by SPL projections, each premotor area can be distinguished by its unique mixture of inputs from the SPL (Figure 1.8, Table 1.2).

Projections from subdivisions of area 7 (AIP, PF, PFG, PFop) are primarily restricted to the PMv and the three cingulate motor areas (Figure 1.8, Table 1.2).

Subdivisions of area 7 are characterized by more complex forms of somatosensory processing and the presence of multimodal information that integrates visual and somatosensory inputs.[167–169] Area PFop is the only IPL subdivision with strong links to more than one premotor area. It has heavy projections to the CMAd, CMAv, and CMAr. AIP in the anterior portion of the ventral bank of the intraparietal sulcus and area PF on the adjacent IPL provide dense input to the PMv.[59,68,81,149,161,162] Area PFG, located just caudal to area PF, projects to the CMAv.[82,163,170] Within the lateral sulcus, SII projects densely to the PMv, CMAd, and CMAr, as well as M1.[58,59,70,149,162,163,171–174] Taken together, these observations indicate that each premotor area receives a unique pattern of input from the various subdivisions of the parietal lobe and provides another basis for differentiating the individual premotor areas.

The outputs from the medial posterior parietal cortex to the frontal motor areas arise from regions that are one to two synapses removed from the primary visual cortex. These regions include the PEc on the most caudal portion of the superior parietal gyrus, area V6A buried on the anterior bank of the parietal occipital sulcus, area PGm on the medial wall just rostral to the parietal occipital sulcus and the posterior cingulate gyrus (CGp) (Figure 1.1). The projections from these regions are restricted to the prePMd and PMd. The differential distribution of the projections from this posterior parietal region to regions of the PMd and prePMd suggests that these frontal lobe areas may require further parcellation. This is likely to be especially evident as more physiological studies are designed to explore the visual–spatial capabilities of these regions.[146,148,175]

In general, the density of projections from these "visual areas" in posterior parietal cortex increases as one proceeds from the caudal border of the PMd to the prePMd.[146–149] Some have argued that these projections provide a neural substrate for the visual guidance of reaching movements to objects in extrapersonal space.[146–149] It is unclear, however, whether the visual information provided by these regions is sufficient for the accurate localization of targets. For instance, area V6A has the most direct visual input to the motor areas in the frontal lobe. V6A neurons have large receptive fields located in the periphery of the visual field. Such fields seem better suited to alert the motor system and shift attention to a particular quadrant of space[176–178] than to drive the visuomotor transformation required to reach out and grasp an object.

1.3.2.3 Pre-Premotor Cortex

Three cortical areas — the preSMA, the prePMd, and the precentral opercular cortex (PrCO)[179] — reside at the junction between the prefrontal cortex and the premotor areas in the frontal lobe (Figure 1.1). All three areas are located in subdivisions of area 6. The prePMd and preSMA are part of area 6a.[17] The PrCO is part of area 6b.[17] At one time or another, each of them has been considered to be a motor area and part of the broad term — premotor cortex. (For prePMd, see Reference 66 for review; for preSMA, see Reference 8; for PrCO, also known as motor proisocortex [ProM], see Reference 58). None of these areas projects directly to M1, and therefore we do not consider any of them to be a premotor area.[9,59,64,163] Instead, the prePMd, preSMA, and PrCO are one step removed from M1 and have connections with at least one premotor area (Figure 1.8, Table 1.1). For example, all of these frontal lobe areas are interconnected with the CMAr.[58,78,82,107,144,164,180] Each area is also connected with the adjacent premotor area — PrCO with the PMv, prePMd with the PMd, and preSMA with the SMA.[78,107,148,161,162,180] However, the significance of the preSMA–SMA connection and the prePMd–PMd connection is unclear. Only immediately adjacent regions in these cortical areas appear to be interconnected and at times these interconnections do not appear especially dense. The organization of these connections clearly requires further investigation.

1.3.2.4 Prefrontal Cortex

Only three premotor areas — the PMv, the CMAr, and the CMAv — receive substantial input from the dorsolateral prefrontal cortex (Walker's area 46[181]). The vast majority of these projections originate in the dorsal (area 46d) and ventral (area 46v) banks of the principal sulcus (Figures 1.1 and 1.8; Table 1.1).[58,59,74,81,82,143,162] The PMv is the target of dense prefrontal projections that originate from area 46v in a topographic manner.[59,74,81,162,182,183] The CMAv and the CMAr are the targets of projections from both the dorsal and ventral banks of principal sulcus (areas 46d and 46v). Additional projections to the CMAr arise from the medial and lateral portions of area 9.[82,166] These observations indicate that specific portions of the prefrontal cortex selectively target just three of the seven motor areas in the frontal lobe. These projections of area 46 to the premotor areas link the prefrontal cortex to cortical regions with direct access to the primary motor cortex and spinal mechanisms of motor control. Connections of the prefrontal cortex with the motor system appear to be tightly focused and designed to provide specialized information about particular aspects of cognitive and executive functions.

1.3.2.5 Limbic Cortex

Limbic input provides a route for emotional and affective influences over motor behavior. Such influences may include the direction of attention toward sensory stimuli and the expression of the motivational–affective response to noxious stimuli, as well as the potential integration of autonomic responses.[184,185] Strong projections from various portions of the limbic cortex are limited to the cingulate motor areas and the PMv (Figures 1.1 and 1.8; Tables 1.1 and 1.2). The insular cortex provides the most widespread projections to the premotor areas. The granular insular cortex targets the PMv, CMAd, CMAv, and CMAr[58,59,68,81,144,149,162,163,166,186] and the dysgranular insular cortex targets the CMAv and CMAr.[144] From the cingulate gyrus, areas 24a,b and 23a,b send robust projections to the CMAv and the CMAr, whereas area 24b has additional weak projections to the SMA and CMAd.[78,82,107,163,164] The CMAr and to a lesser degree the CMAv are also the target of weak, scattered projections from a wide variety of cortical areas including cingulate (e.g., areas 25, 29, 30), orbitofrontal (POdg, OFdg, OFg), and temporal (e.g., TPdg, TF, areas 28 and 35) cortex.[143,144] Thus, widespread regions of the limbic cortex target the CMAr and the CMAv. These connections provide the cingulate motor areas and to a lesser degree the PMv with access to a wide spectrum of information about the state of the entire organism as well as an integrated view of the body in space.

1.3.3 Summary of Cortical Connections

We have summarized the pattern of major cortical inputs to nine cortical areas in the frontal lobe: M1, the premotor areas and two pre-premotor areas (Figure 1.8). Two major conclusions are evident from this analysis. First, the premotor areas are distinguished from the two pre-premotor areas (preSMA, prePMd) on the basis of three major characteristics. All the premotor areas (1) have reciprocal connections with M1, (2) project directly to the spinal cord, and (3) receive input from regions of the parietal lobe that also project directly to the spinal cord. In contrast, the pre-SMA and pre-PMd (1) are not connected to M1, (2) do not project to the spinal cord, and (3) do not receive dense projections from parietal areas with corticospinal projections. These distinctions provide an anatomical basis for assigning the pre-premotor areas to a hierarchical class that is separate from the premotor areas. The results of imaging studies provide considerable support for this conclusion.[66]

The second major conclusion from our analysis is that each of the nine cortical areas in the frontal lobe has a unique constellation of inputs from other areas of the cerebral cortex. Because M1 and all of the premotor areas project directly to the spinal cord, we have proposed that each premotor area may be a "nodal point for parallel pathways to the spinal cord." As a consequence, each of these cortical areas may operate as a functionally distinct system that differentially generates and/or controls specific aspects of motor behavior.[59]

1.4 SUBCORTICAL INPUTS

The basal ganglia and cerebellum are the major subcortical systems that target the cortical motor areas in the frontal lobe. Their outputs reach the frontal lobe via subdivisions of the ventrolateral thalamus. Efferents from the basal ganglia and cerebellum terminate in separate sets of these thalamic nuclei.[187,188] Physiological evidence confirms that convergence of pallidal and cerebellar input on single thalamic neurons is limited (~5%).[189,190] Efferents from the globus pallidus terminate in the rostral portions of the ventrolateral thalamus including ventralis anterior pars parvocellularis (VApc), ventralis lateralis pars oralis (VLo) and the rostral portion of ventralis lateralis pars caudalis (VLcr) (Figure 9) (terminology according to Olszewski[191]).[187,188,192–195] In contrast, efferents from the deep cerebellar nuclei terminate in the caudal portions of the ventrolateral thalamus including ventralis posterior lateralis pars oralis (VPLo), the caudal part of ventralis lateralis pars caudalis (VLcc), ventralis lateralis pars postrema (VLps), and area X (Figure 1.9).[187,188,196–204] These thalamic nuclei project to multiple regions in the frontal lobe including M1 and all the premotor areas (Tables 1.3 and 1.4). Thus, in principle, one can infer the subcortical inputs of each cortical motor area from the origin of its thalamocortical projections.

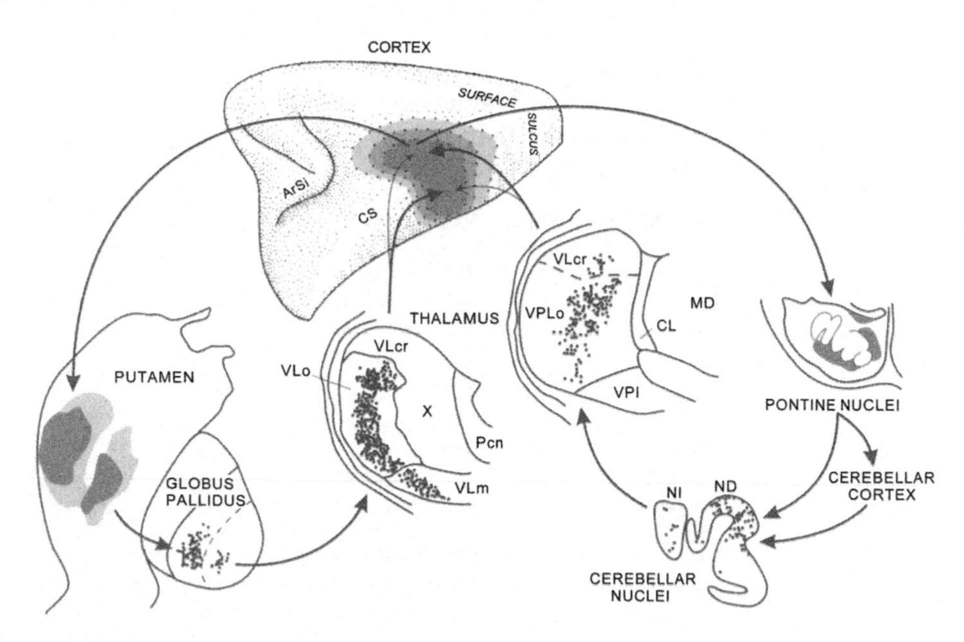

FIGURE 1.9 Cortical–subcortical "loops" of the basal ganglia and the cerebellum. Neurons in the globus pallidus project via the ventrolateral thalamus mainly to the sulcal portion of M1. Neurons in the dentate nucleus of the cerebellum project via the ventrolateral thalamus mainly to the surface portion of M1. M1 projects to the pontine nuclei and the putamen to complete the loops. Dark to light shading in M1 indicates hand, elbow, and shoulder representations. The location of labeled neurons in the globus pallidus and the deep cerebellar nuclei was determined by retrograde transneuronal transport of virus from M1. The *shading* in the diagram of the pontine nuclei indicates the location of terminations from M1. The M1 terminations in the putamen are shaded according to the intensity of anterograde labeling. (Adapted with permission from Reference 9.)

In general, multiple nuclei in the ventrolateral thalamus project to each cortical motor area (Tables 1.3 and 1.4). As a consequence, all of the cortical motor areas except the CMAr appear to be the target of both basal ganglia and cerebellar output. The CMAr appears to be the target largely of pallidal input (VApc and VLo).[163,164,166] In addition, the balance of pallidal and cerebellar thalamocortical projections may vary between different portions of a single cortical area.[205–207] For example, within the hand representation of M1, the rostral portion on the crest of the precentral gyrus receives its densest input from VPLo, a target of cerebellar efferents (Figure 1.9). In contrast, the caudal portion of the hand representation in the depth of the central sulcus receives its densest input from VLo, a target of pallidal efferents. These results suggest that pallidal and cerebellar inputs may differentially influence specific motor areas as well as distinct regions within a single motor area. (For discussion, see Reference 206.)

Even though each thalamic nucleus projects to more than one cortical motor area, the neurons projecting to different cortical areas originate from largely separate regions within the nucleus.[207–210] (See, however, References 211,212.) In fact, tha-

TABLE 1.3
Thalamocortical Connections: Thalamic Nuclei That Receive Pallidal and/or Nigral Afferents

Cortical Area	VApc	VLo	VLcr	VLm
M1	?	xxx	x	xx
PMd	x	xx	x	x
PMv	xx	x	?	x
SMA	x	xxx	x	xx
CMAd	x	xxx		x
CMAv	?	xx		?
CMAr	xxx	xxx		x
PreSMA	xxx	xx	xx	xx
PrePMd	xxx	x	x	x

Note: xxx = major input, xx = moderate input, x = weak input, ? = weak input observed in a few cases.

TABLE 1.4
Thalamocortical Connections: Thalamic Nuclei That Receive Cerebellar Afferents

Cortical Area	X	VLcc	VPLo	LP*	Pul. o*
M1	?	xx	xxx	x	x
PMd		xx	xx	?	?
PMv	xxx	xx	xx		
SMA	?	xx	x	?	?
CMAd		?	xx		
CMAv		xx	xx		
CMAr		?	?		
PreSMA	xxx	xx		?	?
PrePMd	xxx	xx		?	?

Note: xxx = major input, xx = moderate input, x = weak input, ? = weak input observed in a few cases.

* Cerebellar afferents confined to a few scattered patches in the most rostral portions of the nuclei.

lamic neurons rarely branch to innervate more than one cortical area.[207,209,210,212,213] Thus, thalamic nuclei appear to contain specific subregions, each of which projects to a separate cortical motor area.

The presence of distinct basal ganglia and cerebellar territories in the ventrolateral thalamus raises the question of how these subcortical inputs become integrated within a cortical area. Although this issue has not been systematically investigated,[214–216] both pallidal and cerebellar receiving thalamic nuclei terminate heavily in the deeper cortical layers (layers III, V) with layer III projections being denser. These thalamocortical projections may provide a topographically organized input that influences corticocortical, corticothalamic, and corticostriatal circuits. Current evidence suggests that only pallidal receiving thalamic nuclei terminate superficially in cortical layer I.[214] These layer I terminations tend to be more broadly distributed than are terminations in the deeper layers.[214,216] Layer I terminations may provide a more global influence and assist in synchronizing the activity of cortical neurons with dendrites extending into layer I.[217,218] Thus, pallidal and cerebellar circuits may differentially access and influence some cortical laminae, but the question of whether pallidal and cerebellar circuits converge onto the same cortical columns remains to be addressed.

Because M1 and the premotor areas provide input to the basal ganglia via the striatum[219,220] and to the cerebellum via the pontine nuclei,[221–224] these cortical areas have the potential to form "closed" loops with the basal ganglia and cerebellum (Figure 1.9).[219,220,225] Important aspects of the organization of these circuits remain to be examined. For example, the extent to which related cortical areas provide convergent input into basal ganglia and cerebellar loops with the cerebral cortex remains to be defined. The number of distinct closed-loop circuits in each system also remains to be determined.

To examine these and other related issues, we have employed retrograde transneuronal transport of neurotropic viruses[226–231] to define the topographic organization of the basal ganglia and cerebellar outputs to the cortical motor areas. Injections of virus were made into different cortical motor areas. The survival time was set to allow retrograde transport of herpes simplex virus type I (HSV1) from the injection site to first-order neurons in the thalamus, and then retrograde transneuronal transport from these first-order neurons to "second-order" neurons in the basal ganglia and deep cerebellar nuclei.

Neurons labeled by retrograde transneuronal transport of virus after injections into the arm representation of M1 were found dorsally in the dentate at mid-rostrocaudal levels (Color Figure 1.10B).[229,232] Injections into the leg area also labeled neurons in a dorsal portion of the dentate, but in this case at more rostral levels of the nucleus (Color Figure 1.10A). Likewise, virus injections into the face area labeled neurons dorsally in the dentate, but at more caudal levels of the nucleus (Color Figure 1.10C). This rostral to caudal arrangement of the origin of projections to the leg, arm, and face representations in M1 corresponds well with the somatotopy previously proposed for the dentate.[201,203,233–235] It is also clear that large portions of the dentate nucleus were not labeled following virus injections into M1. As a consequence, the body map generated by the projections to M1 occupied only a portion of the dorsal third of the nucleus.[229,232]

Some of the dentate regions that were not labeled after virus injections into M1 did contain labeled neurons after injections into the PMv and other premotor

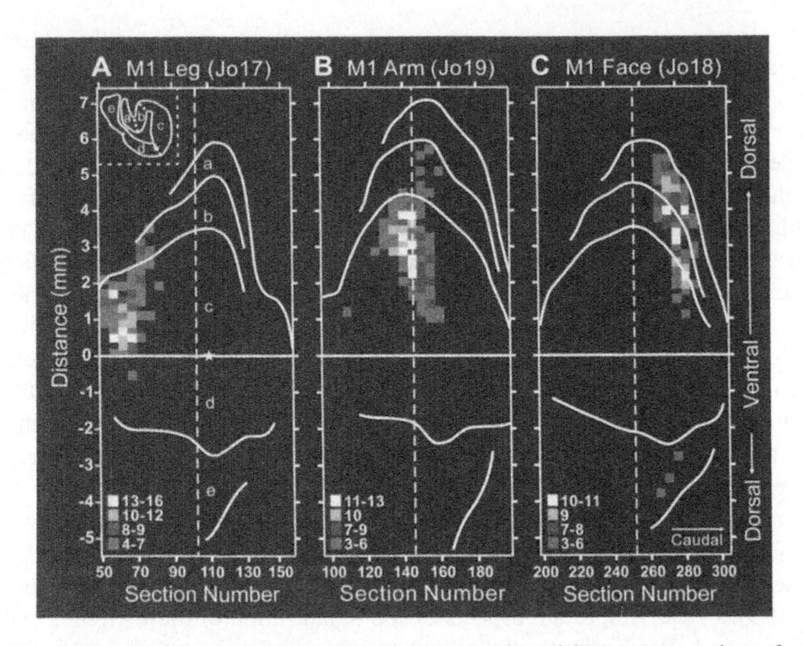

FIGURE 1.10 (see color figure) Somatotopic organization of dentate output channels to M1. Unfolded maps of the dentate illustrate the neurons labeled after HSV1 injections into the (A) leg, (B) arm, and (C) face representations of M1. These maps of the dentate were created by unfolding serial coronal sections through the nucleus. Inset in part A illustrates a coronal section of the dentate where each segment in the unfolded map is identified. The *dashed vertical line* indicates the rostrocaudal center of the nucleus. (Adapted with permission from Reference 232.)

areas.[236–238] Perhaps more importantly, other regions of the dentate nucleus contained labeled neurons after transneuronal transport of virus from prefrontal and posterior parietal areas of the cortex.[239,240] Clearly, a substantial portion of the output from the dentate targets nonmotor areas of cortex.[239–241] Furthermore, the dentate contains distinct topographically organized maps of outputs to motor and nonmotor areas of cortex (Figure 1.11).[232]

An analogous organization is found in the globus pallidus (Figure 1.12).[228,229] Injection of virus into the arm representation of M1 labeled two dense clusters of neurons in the globus pallidus — one in the inner segment and another in the outer segment (Figure 1.12A).[229] Virus injections into the leg representation result in similar clusters of labeled neurons that are dorsal and rostral to those projecting to M1 arm. Virus injections into the face representation of M1 labeled neurons that are located ventral and caudal to those projecting to M1 arm. This pattern of labeled neurons suggests that the globus pallidus contains a rostral to caudal, dorsal to ventral body map with respect to the leg, arm, and face representation. This somatotopic organization is consistent with the body maps described in physiological studies.[242] As in the dentate nucleus, large portions of the globus pallidus were not labeled following virus

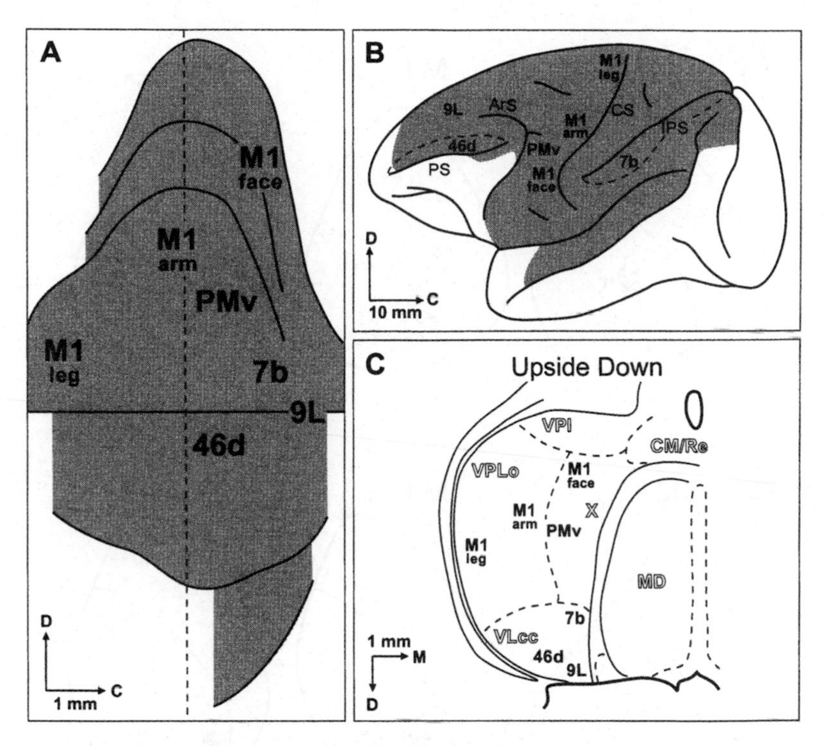

FIGURE 1.11 Topography within the cerebello-thalamocortical circuit. (A) Dentate output channels. The origins of the peak density of dentate projections to selected cortical areas are labeled. (B) Selected cortical targets of cerebello-thalamocortical circuits. The *shading* identifies cortical regions (lateral hemisphere only) that project to the cerebellum via the pons.[221–224] (C) Origin of selected cortical projections from the ventrolateral thalamus. The cortical regions indicated receive input from regions of the ventrolateral thalamus that lie within the termination zone of cerebellar efferents. The thalamus has been turned upside down to indicate the match between its topography and that of the dentate. Abbreviations: see Figures 1.1, 1.3, and 1.4; CM/Re: nucleus centrum medianum/nucleus reuniens; IPS: intraparietal sulcus; MD: nucleus medialis dorsalis; VLcc: caudal portion of the nucleus ventralis lateralis, pars caudalis; VPI: nucleus ventralis posterior inferior; VPLo: nucleus ventralis posterior lateralis, pars oralis; X: Area X. (Adapted with permission from Reference 232.)

injections into M1. Some of these unlabeled areas contained labeled neurons after virus injections into premotor areas, whereas others contained labeled neurons only after virus injections into prefrontal areas of the cortex.[243,244] Thus, the globus pallidus contains a topographically organized map of outputs to motor and nonmotor areas of the cortex.

We have termed the clusters of neurons in subcortical nuclei that project to a specific cortical area an "output channel."[245] Each of these output channels, whether originating in the basal ganglia or cerebellum, appears to follow a simple rule: namely, if a cortical area has dense projections to the input stage of basal ganglia

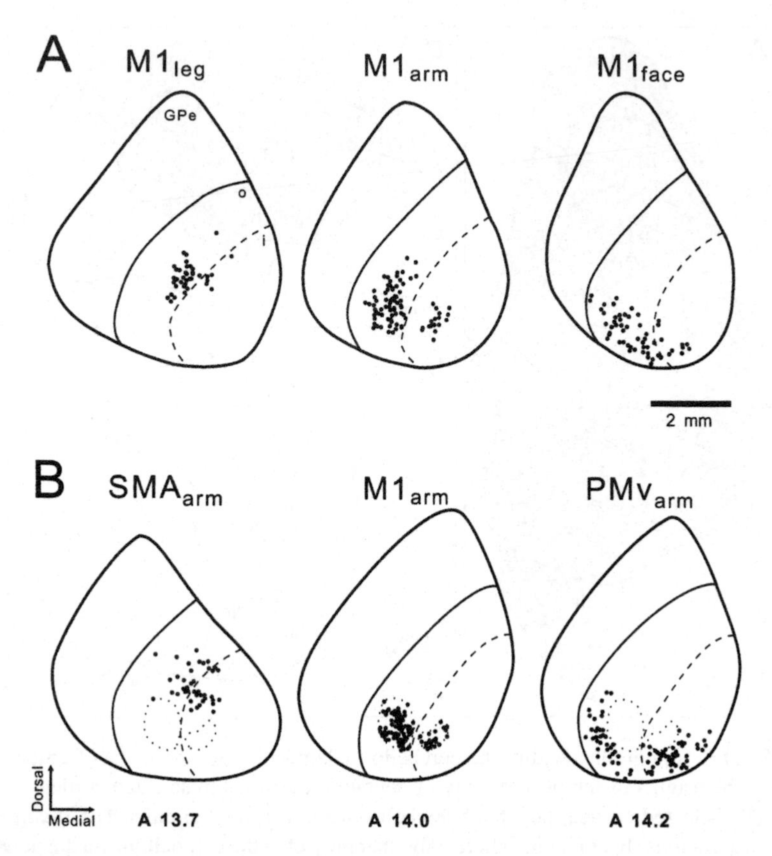

FIGURE 1.12 Somatotopic organization of pallidal output channels to M1. (A) Neurons in the internal segment of the globus pallidus (GPi) were labeled by retrograde transneuronal transport of HSV1 following injections into the leg, arm, or face representations of M1 in monkeys. Labeled neurons (*dots*) are indicated for two coronal sections separated by 0.5 mm. (Adapted with permission from Reference 229.) (B) Neurons in the GPi were labeled following injections of HSV1 into the arm representations of the SMA, M1, or PMv of monkeys. Labeled neurons (*dots*) are indicated for two or three coronal sections near the same stereotaxic level (A 14.0). For comparison, the *dotted lines* indicate the region of the GPi containing neurons labeled from M1. The *thin solid line* indicates the border between the internal and external segments of GPi. The *dashed line* indicates the border between the inner (i) and outer (o) portions of the GPi. (Adapted with permission from Reference 228.)

or cerebellar processing, then the cortical area is generally a target of efferents from a distinct basal ganglia or cerebellar output channel. This rule implies that areas of the cerebral cortex participate in multiple closed-loop circuits with the basal ganglia (Figure 1.9). Not every interaction between cortex and the basal ganglia follows this general plan. For example, regions of primary somatic sensory cortex are known to provide input to the basal ganglia, but they do not appear to be the target of basal ganglia output.[246,247] However, our evidence suggests that closed-loop circuits characterize many of the interconnections between cerebral cortex and the basal ganglia and cerebellum. Therefore, closed loops may represent a fundamental unit of basal ganglia and cerebellar interconnections with the cerebral cortex.[231,232,243]

1.5 SUMMARY AND CONCLUSIONS

A new perspective has emerged about the organization and function of the cortical areas concerned with the control of movement. Classically, the generation of motor commands was thought to proceed in a serial, hierarchical fashion. The output of the premotor cortex was viewed as being funneled to M1 which served as the final common pathway for the central control of movement.[1,4] It is now clear that the frontal lobe contains at least six spatially separate premotor areas. Each of these premotor areas, like M1, projects directly to the spinal cord. In fact, at least as many corticospinal neurons originate from the premotor areas as originate from M1. Thus, each premotor area appears to have the potential to influence the control of movement not only at the level of the primary motor cortex, but also more directly at the level of the spinal cord.[59,60]

Why does the frontal lobe contain all of these premotor areas? Although there is as yet no definitive answer to this question, current data from anatomical, physiological, behavioral, and imaging studies suggest that each premotor area is concerned with a specific aspect of movement planning, preparation, and execution. Thus, the task of generating and controlling movement appears to be broken up into a number of subtasks that are accomplished through parallel distributed processing in multiple motor areas. Multiple motor areas may thereby decrease response time and increase response flexibility. In any event, the cortical control of movement is achieved by multiple motor areas, all of which send signals to the spinal cord.

ACKNOWLEDGMENTS

This work was supported by the Veterans Affairs Medical Research Service, and by U.S. Public Health Service grant #24328 (PLS).

REFERENCES

1. Kuypers, H.G.J.M., Anatomy of the descending pathways, in *Handbook of Physiology, Section I: The Nervous System, Vol. II: Motor Control, Part I*, Brooks, V.B., Ed., American Physiological Society, Bethesda, MD, 1981, 567.
2. Wise, S.P., The primate premotor cortex fifty years after Fulton, *Behav. Brain Res.*, 18, 79, 1985.
3. Wiesendanger, M., Recent developments in studies of the supplementary motor area of primates, *Rev. Physiol. Biochem. Pharmacol.*, 103, 1, 1986.

4. Hepp-Reymond, M.C., Functional organization of motor cortex and its participation in voluntary movements, in *Comparative Primate Biology, Vol. 4: Neurosciences*, Alan R. Liss, 1988, 501.
5. Asanuma, H., *The Motor Cortex*, New York, Raven Press, 1989.
6. Cheney, P.D., Fetz, E.E., and Mewes, K., Neural mechanisms underlying corticospinal and rubrospinal control of limb movements, *Prog. Brain Res.*, 87, 213, 1991.
7. Georgopoulos, A., Higher order motor control, *Annu. Rev. Neurosci.*, 14, 361, 1991.
8. Tanji, J., The supplementary motor area in the cerebral cortex, *Neurosci. Res.*, 19, 251, 1994.
9. Dum, R.P. and Strick, P.L., The corticospinal system, a structural framework for the central control of movement, in *Handbook of Physiology, Section 12: Exercise, Regulation and Integration of Multiple Systems*, Rowell, L.B. and Shepard, J.T., Eds., American Physiological Society, New York, 1996, 217.
10. Geyer, S. et al., Functional neuroanatomy of the primate isocortical motor system, *Anat. Embryol.*, 202, 443, 2000.
11. Rizzolatti, G. and Luppino, G., The cortical motor system, *Neuron*, 31, 889, 2001.
12. Porter, R. and Lemon, R.N., *Corticospinal Function and Voluntary Movement*, Oxford University Press, Oxford, 1993.
13. Leyton, S.S.F. and Sherrington, C.S., Observations on the excitable cortex of the chimpanzee, orangutan and gorilla, *Q. J. Exp. Physiol.*, 11, 135, 1917.
14. Penfield, W. and Boldrey, E., Somatic motor and sensory representation in the cerebral cortex of man as studied by electrical stimulation, *Brain*, 60, 389, 1937.
15. Woolsey, C.N. et al., Patterns of localization in precentral and "supplementary" motor area and their relation to the concept of a premotor area, *Assoc. Res. Nerv. Ment. Dis.*, 30, 238, 1952.
16. Brodmann, K. *Vergleichende Lokalisationslehre der Grosshirnrinde in ihren Prinzipien dargestellt auf grund des Zellenbaues*, Barth, Leipzig, 1909.
17. Vogt, C. and Vogt, O., Allgemeinere Ergebnisse unserer Hirnforschung, *J. Psychol. Neurol. (Leipzig)*, 25, 277, 1919.
18. Von Bonin, G. and Bailey, P., *The Neocortex of Macaca Mulatta*, University of Illinois Press, Urbana, 1947.
19. Asanuma, H. and Rosen, I., Topographical organization of cortical efferent zones projecting to distal forelimb muscles in the monkey, *Exp. Brain Res.*, 14, 243, 1972.
20. Andersen, P., Hagan, P.J., Phillips, C.G., and Powell, T.P., Mapping by microstimulation of overlapping projections from area 4 to motor units of the baboon's hand, *Proc. R. Soc. Lond. (Biol.),* 188, 31, 1975.
21. Jankowska, E., Padel, Y., and Tanaka, R., Projections of pyramidal tract cells to alphamotoneurones innervating hind-limb muscles in the monkey, *J. Physiol. (Lond.),* 249, 637, 1975.
22. Kwan, H.C., MacKay, W.A., Murphy, J.T., and Wong, Y.C., Spatial organization of precentral cortex in awake primates. II. Motor outputs, *J. Neurophysiol.*, 41, 1120, 1978.
23. McGuinness, E., Sivertsen, D., and Allman, J.M., Organization of the face representation in macaque motor cortex, *J. Comp. Neurol.*, 193, 591, 1980.
24. Wise, S.P. and Tanji, J., Supplementary and precentral motor cortex, contrast in responsiveness to peripheral input in the hindlimb area of the unanesthetized monkey, *J. Comp. Neurol.,* 195, 433, 1981.
25. Strick, P.L. and Preston, J.B., Two representations of the hand in area 4 of a primate. I. Motor output organization, *J. Neurophysiol.*, 48, 139, 1982.

26. Humphrey, D.R. and Reed, D.J., Separate cortical systems for control of joint movement and joint stiffness, reciprocal activation and coactivation of antagonist muscles, *Adv. Neurol.,* 39, 347, 1983.
27. Gould, H.J., III et al., The relationship of corpus callosum connections to electrical stimulation maps of motor, supplementary motor, and the frontal eye fields in owl monkeys, *J. Comp. Neurol.*, 247, 297, 1986.
28. Humphrey, D.R., Representation of movements and muscles within the primate precentral motor cortex, historical and current perspectives, *Fed. Proc.,* 45, 2687, 1986.
29. Huang, C.S., Sirisko, M.A., Hiraba, H., Murray, G.M., and Sessle, B.J., Organization of the primate face motor cortex as revealed by intracortical microstimulation and electrophysiological identification of afferent inputs and corticobulbar projections, *J. Neurophysiol.*, 59, 796, 1988.
30. Sato, K.C. and Tanji, J., Digit-muscle responses evoked from multiple intracortical foci in monkey precentral motor cortex, *J. Neurophysiol.*, 62, 959, 1989.
31. Nudo, R.J. et al., Neurophysiological correlates of hand preference in primary motor cortex of adult squirrel monkeys, *J. Neurosci.,* 121, 2918, 1992.
32. Donoghue, J.P., Leibovic, S., and Sanes, J.N., Organization of the forelimb area in squirrel monkey motor cortex, representation of digit, wrist, and elbow muscles, *Exp. Brain Res.*, 89, 1, 1992.
33. Murphy, J.T. et al., Spatial organization of precentral cortex in awake primates. III. Input-output coupling, *J. Neurophysiol.*, 41, 1132, 1978.
34. Park, M.C., Belhaj-Saif, A., Gordon, M., and Cheney, P.D., Consistent features in the forelimb representation of primary motor cortex in rhesus macaques, *J. Neurosci.*, 21, 2784, 2001.
35. Schieber, M.H., Constraints on somatotopic organization in the primary motor cortex, *J. Neurophysiol.,* 86, 2125, 2001.
36. Graziano, M.S., Taylor, C.S., and Moore, T., Complex movements evoked by microstimulation of precentral cortex, *Neuron,* 34, 841, 2002.
37. Ferrier, D., Experiments on the brain of monkeys, *Proc. R. Soc. Lond.,* 23, 409, 1875.
38. Jankowska, E., Padel, Y., and Tanaka, R., The mode of activation of pyramidal tract cells by intracortical stimuli, *J. Physiol. (Lond.),* 249, 617, 1975.
39. Asanuma, H., Arnold, A., and Zarzecki, P., Further study on the excitation of pyramidal tract cells by intracortical microstimulation, *Exp. Brain Res.,* 26, 443, 1976.
40. Murphy, J.T. et al., Activity of primate precentral neurons during voluntary movements triggered by visual signals, *Brain Res.*, 236, 429, 1982.
41. Lemon, R.N., Muir, R.B., and Mantel, G.W., The effects upon the activity of hand and forearm muscles of intracortical stimulation in the vicinity of corticomotor neurons in the conscious monkey, *Exp. Brain Res.*, 66, 621, 1987.
42. Shinoda, Y., Zarzecki, P., and Asanuma, H., Spinal branching of pyramidal tract neurons in the monkey, *Exp. Brain Res.*, 34, 59, 1979.
43. Shinoda, Y., Yokota, J., and Futami, T., Divergent projection of individual corticospinal axons to motoneurons of multiple muscles in the monkey, *Neurosci. Lett.*, 23, 7, 1981.
44. Fetz, E.E. and Cheney, P.D., Postspike facilitation of forelimb muscle activity by primate corticomotoneuronal cells, *J. Neurophysiol.,* 44, 751, 1980.
45. Cheney, P.D. and Fetz, E.E., Functional classes of primate corticomotoneuronal cells and their relation to active force, *J. Neurophysiol.*, 44, 773-791, 1980.
46. Cheney, P.D., Fetz, E.E., and Palmer, S.S., Patterns of facilitation and suppression of antagonist forelimb muscles from motor cortex sites in the awake monkey, *J. Neurophysiol.*, 53, 805, 1985.

47. Kasser, R.J. and Cheney, P.D., Characteristics of corticomotoneuronal postspike facilitation and reciprocal suppression of EMG activity in the monkey, *J. Neurophysiol.*, 53, 959, 1985.
48. Buys, E.J. et al., Selective facilitation of different hand muscles by single corticospinal neurones in the conscious monkey, *J. Physiol. (Lond.)*, 381, 529, 1986.
49. McKiernan, B.J. et al., Corticomotoneuronal postspike effects in shoulder, elbow, wrist, digit, and intrinsic hand muscles during a reach and prehension task, *J. Neurophysiol.*, 80, 1961, 1998.
50. Maier, M.A. et al., Direct and indirect corticospinal control of arm and hand motoneurons in the squirrel monkey (*Saimiri sciureus*), *J. Neurophysiol.*, 78, 721, 1997.
51. Lemon, R.N. et al., The importance of the cortico-motoneuronal system for control of grasp, *Novartis Found. Symp.*, 218, 202, 1998.
52. Nakajima, K. et al., Striking differences in transmission of corticospinal excitation to upper limb motoneurons in two primate species, *J. Neurophysiol.*, 84, 698, 2000.
53. Lamour, Y., Jennings, V.A., and Solis, H., Functional characteristics and segregation of cutaneous and non-cutaneous neurons in monkey precentral motor cortex (MI), *Soc. Neurosci. Abstr.*, 6, 158, 1980.
54. Strick, P.L. and Preston, J.B., Two representations of the hand in area 4 of a primate. II. Somatosensory input organization, *J. Neurophysiol.*, 48, 150, 1982.
55. Picard, N. and Smith, A.M., Primary motor cortical activity related to the weight and texture of grasped objects in the monkey, *J. Neurophysiol.*, 68, 1867, 1992.
56. Fulton, J.F., Definition of the "motor" and "premotor" areas, *Brain*, 58, 311, 1935.
57. Humphrey, D.R., On the cortical control of visually directed reaching, contribution by nonprecentral motor areas, in *Posture and Movement*, Talbot, R.E. and Humphrey, D.R., Eds., Raven Press, New York, 1979, 51.
58. Barbas, H. and Pandya, D.N., Architecture and frontal cortical connections of the premotor cortex (area 6) in the rhesus monkey, *J. Comp. Neurol.*, 256, 211, 1987.
59. Dum, R.P. and Strick, P.L., Premotor areas, nodal points for parallel efferent systems involved in the central control of movement, in *Motor Control, Concepts and Issues*, Humphrey, D.R. and Freund, H.-J., Eds., London, Wiley, 1991, 383.
60. Dum, R.P. and Strick, P.L., The origin of corticospinal projections from the premotor areas in the frontal lobe, *J. Neurosci.*, 11, 667, 1991.
61. Wise, S.P. et al., What are the specific functions of the different cortical motor areas? in *Motor Control, Concepts and Issues*, Humphrey, D.R. and Freund, H-J., Eds., London, Wiley, 1991, 463.
62. Fulton, J.F., *Physiology of the Nervous System*, New York, Oxford University Press, 1949.
63. Penfield, W. and Welch, K., Supplementary motor area of the cerebral cortex, *Arch. Neurol. Psychiat.*, 66, 289, 1951.
64. Dum, R.P. and Strick, P.L., Motor areas in the frontal lobe of the primate, *Physiol. Behav.*, 77, 677, 2002.
65. Picard, N. and Strick, P.L., Motor areas of the medial wall, a review of their location and functional activation, *Cereb. Cortex*, 6, 342, 1996.
66. Picard, N. and Strick, P.L., Imaging the premotor areas, *Cur. Opin. Neurobiol.*, 11, 663, 2001.
67. Muakkassa, K.F. and Strick, P.L., Frontal lobe inputs to primate motor cortex, evidence for four somatotopically organized 'premotor' areas, *Brain Res.*, 177, 176, 1979.
68. Godschalk, M. et al., Cortical afferents and efferents of monkey postarcuate area, an anatomical and electrophysiological study, *Exp. Brain Res.*, 56, 410, 1984.

69. Strick, P.L., How do the basal ganglia and cerebellum gain access to the cortical motor areas? *Behav. Brain Res.*, 18, 107, 1985.
70. Künzle, H., Cortico-cortical efferents of primary motor and somatosensory regions of the cerebral cortex, *Neuroscience,* 3, 25, 1978.
71. Ghosh, S., Brinkman, C., and Porter, R.A., Quantitative study of the distribution of neurons projecting to the precentral motor cortex in the monkey (*M. fascicularis*), *J. Comp. Neurol.*, 259, 424, 1987.
72. Morecraft, R.J. and Van Hoesen, G.W., Cingulate input to the primary and supplementary motor cortices in the rhesus monkey, evidence for somatotopy in areas 24c and 23c, *J. Comp. Neurol.*, 322, 471, 1992.
73. Tokuno, H. and Tanji, J., Input organization of distal and proximal forelimb areas in the monkey primary motor cortex, a retrograde double labeling study, *J. Comp. Neurol.*, 333, 199, 1993.
74. Lu, M.-T., Preston, J.B., and Strick, P.L., Interconnections between the prefrontal cortex and the premotor areas in the frontal lobe, *J. Comp. Neurol.*, 341, 375, 1994.
75. Tokuno, H. and Inase, M., Direct projections from the ventral premotor cortex to the hindlimb region of the supplementary motor area in the macaque monkey, *Neurosci. Lett.*, 171, 159, 1994.
76. Hatanaka, N. et al., Somatotopic arrangement and corticocortical inputs of the hindlimb region of the primary motor cortex in the macaque monkey, *Neurosci. Res.*, 40, 9, 2001.
77. Matelli, M., Luppino, G., and Rizzolatti, G., Architecture of superior and mesial area 6 and the adjacent cingulate cortex in the macaque monkey, *J. Comp. Neurol.*, 311, 445, 1991.
78. Luppino, G. et al., Corticocortical connections of area F3 (SMA-Proper) and area F6 (Pre-SMA) in the macaque monkey, *J. Comp. Neurol.*, 338, 114, 1993.
79. Matsuzaka, Y., Aizawa, H., and Tanji, J., A motor area rostral to the supplementary motor area (presupplementary motor area) in the monkey, neuronal activity during a learned motor task, *J. Neurophysiol.*, 68, 653, 1992.
80. Inase, M. et al., Corticostriatal and corticosubthalamic input zones from the presupplementary motor area in the macaque monkey, comparison with the input zones from the supplementary motor area, *Brain Res.*, 833, 191, 1999.
81. Matelli, M. et al., Afferent and efferent projections of the inferior area 6 in the macaque monkey, *J. Comp. Neurol.*, 251, 281, 1986.
82. Morecraft, R.J. and Van Hoesen, G.W., Frontal granular cortex input to the cingulate (M3), supplementary (M2) and primary (M1) motor cortices in the rhesus monkey, *J. Comp. Neurol.*, 337, 669, 1993.
83. Russell, J.R. and DeMyer, W., The quantitative cortical origin of pyramidal axons of *Macaca mulatta, Neurology (Minneapolis)* 11, 96, 1961.
84. He, S.Q., Dum, R. P., and Strick, P.L., Topographic organization of corticospinal projections from the frontal lobe, motor areas on the lateral surface of the hemisphere, *J. Neurosci.*, 13, 952, 1993.
85. He, S.Q., Dum, R.P., and Strick, P.L., Topographic organization of corticospinal projections from the frontal lobe, motor areas on the medial wall of the hemisphere, *J. Neurosci.*, 15, 3284, 1995.
86. Catsman-Berrevoets, C.E. and Kuypers, H.G.J.M., Cells of origin of cortical projections to dorsal column nuclei, spinal cord and bulbar medial reticular formation in the rhesus monkey, *Neurosci. Lett.*, 3, 245, 1976.

87. Biber, M.P., Kneisley, L.W., and LaVail, J.H., Cortical neurons projecting to the cervical and lumbar enlargements of the spinal cord in young and adult rhesus monkeys, *Exp. Neurol.*, 59, 492, 1978.
88. Murray, E.A. and Coulter, J.D., Organization of corticospinal neurons in the monkey, *J. Comp. Neurol.*, 195, 339, 1981.
89. Macpherson, J.M. et al., Microstimulation of the supplementary motor area (SMA) in the awake monkey, *Exp. Brain Res.*, 45, 410, 1982.
90. Toyoshima, K. and Sakai, H., Exact cortical extent of the origin of the corticospinal tract (CST) and the quantitative contribution to the CST in different cytoarchitectonic areas. A study with horseradish peroxidase in the monkey, *J. Hirnforsch.*, 23, 257, 1982.
91. Keizer, K. and Kuypers, H.G., Distribution of corticospinal neurons with collaterals to the lower brain stem reticular formation in monkey (*Macaca fascicularis*), *Exp. Brain Res.*, 74, 311, 1989.
92. Nudo, R.J. and Masterton, R.B., Descending pathways to the spinal cord. III. Sites of origin of the corticospinal tract, *J. Comp. Neurol.*, 296, 559, 1990.
93. Galea, M.P. and Darian-Smith, I., Multiple corticospinal neuron populations in the macaque monkey are specified by their unique cortical origins, spinal terminations, and connections, *Cereb. Cortex*, 4, 166, 1994.
94. Jones, E.G., and Wise, S.P., Size, laminar and columnar distribution of efferent cells in the sensory-motor cortex of monkeys, *J. Comp. Neurol.*, 175, 391, 1977.
95. Humphrey, D.R. and Corrie, W.S., Properties of pyramidal tract neuron system within a functionally defined subregion of primate motor cortex, *J. Neurophysiol.*, 41, 216, 1978.
96. Lawrence, D.G. and Kuypers, H.G.J.M., The functional organization of the motor system in the monkey. I. The effects of bilateral pyramidal lesions, *Brain*, 91, 1, 1968.
97. Maier, M.A. et al., Differences in the corticospinal projection from primary motor cortex and supplementary motor area to macaque upper limb motoneurons, an anatomical and electrophysiological study, *Cereb. Cortex*, 12, 281, 2002.
98. Mitz, A.R. and Wise, S.P., The somatotopic organization of the supplementary motor area, intracortical microstimulation mapping, *J. Neurosci.*, 7, 1010, 1987.
99. Luppino, G. et al., Multiple representations of body movements in mesial area 6 and adjacent cingulate cortex, an intracortical microstimulation study in the macaque monkey, *J. Comp. Neurol.*, 311, 463, 1991.
100. Hepp-Reymond, M-C. et al., Force-related neuronal activity in two regions of the primate ventral premotor cortex, *Can. J. Physiol. Pharm.*, 72, 571, 1994.
101. Godschalk, M. et al., Somatotopy of monkey premotor cortex examined with microstimulation, *Neurosci. Res.*, 23, 269, 1995.
102. Akazawa, T. et al., A cortical motor region that represents the cutaneous back muscles in the macaque monkey, *Neurosci. Lett.*, 282, 125, 2000.
103. Takada, M. et al., Organization of inputs from cingulate motor areas to basal ganglia in macaque monkey, *Eur. J. Neurosci.*, 14, 1633, 2001.
104. Rizzolatti, G. et al., Neurons related to reaching-grasping arm movements in the area 6 (area 6a beta), *Exp. Brain Res.*, 82, 337, 1990.
105. Alexander, G.E. and Crutcher, M.D., Preparation for movement, neural representations of intended direction in three motor areas of the monkey, *J. Neurophysiol.*, 64, 133, 1990.
106. Inase, M. et al., Origin of thalamocortical projections to the presupplementary motor area (pre-SMA) in the macaque monkey, *Neurosci. Res.*, 25, 217, 1996.

107. Wang, Y. et al., Spatial distribution of cingulate cells projecting to the primary, supplementary, and pre-supplementary motor areas, a retrograde multiple labeling study in the macaque monkey, *Neurosci. Res.*, 39, 39, 2001.
108. Hutchins, K.D., Martino, A.M., and Strick, P.L., Corticospinal projections from the medial wall of the hemisphere, *Exp. Brain Res.*, 71, 667, 1988.
109. Cadoret, G. and Smith, A.M., Comparison of the neuronal activity in the SMA and the ventral cingulate cortex during prehension in the monkey, *J. Neurophysiol.*, 77, 153, 1997.
110. Mitz, A.R. and Humphrey, D.R., Intracortical stimulation in pyramidotomized monkeys, *Neurosci. Lett.*, 64, 59, 1986.
111. Fogassi, L. et al., Visual responses in the dorsal premotor area F2 of the macaque monkey, *Exp. Brain Res.*, 128, 194, 1999.
112. Fujii, N., Mushiake, H., and Tanji, J., Rostrocaudal distinction of the dorsal premotor area based on oculomotor involvement, *J. Neurophysiol.*, 83, 1764, 2000.
113. Martino, A.M. and Strick, P.L., Corticospinal projections originate from the arcuate premotor area, *Brain Res.*, 404, 307, 1987.
114. Maier, M.A. et al., Does a C3-C4 propriospinal system transmit corticospinal excitation in the primate? An investigation in the macaque monkey, *J. Physiol. (Lond.)*, 511, 191, 1998.
115. Alstermark, B. et al., Disynaptic pyramidal excitation in forelimb motoneurons mediated via C(3)-C(4) propriospinal neurons in the *Macaca fuscata*, *J. Neurophysiol.*, 82, 3580, 1999.
116. Kirkwood, P.A., Maier, M.A., and Lemon, R.N., Interspecies comparisons for the C3-C4 propriospinal system, unresolved issues, *Adv. Exp. Med. Biol.*, 508, 299, 2002.
117. Cerri, G., Shimazu, H., Maier, M.A., and Lemon, R.N., Facilitation from ventral premotor cortex of primary motor cortex outputs to macaque hand muscles, *J. Neurophysiol.*, 90, 832, 2003.
118. Shimazu, H. et al., Macaque ventral premotor cortex exerts powerful facilitation of motor cortex outputs to upper limb motoneurons, *J. Neurosci.*, 24, 1200, 2004.
119. Hast, M.H. et al., Cortical motor representation of the laryngeal muscles in *Macaca mulatta*, *Brain Res.*, 73, 229, 1974.
120. Hahm, J. et al., Parallel cortical pathways for the control of movement, *Soc. Neurosci. Abstr.*, 18, 216, 1992.
121. Kuypers, H.G.J.M., Central cortical projections to motor and somato-sensory cell groups, *Brain*, 83, 161, 1960.
122. Liu, C.N. and Chambers, W.W., An experimental study of the cortico-spinal system in the monkey (*Macaca mulatta*), *J. Comp. Neurol.*, 123, 257, 1964.
123. Kuypers, H.G.J.M. and Brinkman, J., Precentral projections of different parts of the spinal intermediate zone in the rhesus monkey, *Brain Res.*, 24, 29, 1970.
124. Coulter, J.D. and Jones, E.G., Differential distribution of corticospinal projections from individual cytoarchitectonic fields in the monkey, *Brain Res.*, 129, 335, 1977.
125. Cheema, S.S., Rustioni, A., and Whitsel, B.L., Light and electron microscopic evidence for a direct corticospinal projection to superficial laminae of the dorsal horn in cats and monkeys, *J. Comp. Neurol.*, 225, 276, 1984.
126. Ralston, D.D. and Ralston, H.J., III, The terminations of corticospinal tract axons in the macaque monkey, *J. Comp. Neurol.*, 242, 325, 1985.
127. Bortoff, G.A. and Strick, P.L., Corticospinal terminations in two new-world primates, further evidence that corticomotoneuronal connections provide part of the neural substrate for manual dexterity, *J. Neurosci.*, 13, 5105, 1993.

128. Dum, R.P. and Strick, P.L., Spinal cord terminations of the medial wall motor areas in macaque monkeys, *J. Neurosci.*, 16, 6513, 1996.
129. Armand, J., Olivier, E., Edgley, S.A., and Lemon, R.N., Postnatal development of corticospinal projections from motor cortex to the cervical enlargement in the macaque monkey, *J. Neurosci.*, 17, 251, 1997.
130. Phillips, C.G. and Porter, R., *Corticospinal Neurons: Their Role in Movement*, Monograph of the Physiological Society, No. 34, London, Academic Press, 1977.
131. Kuypers, H.G.J.M., Corticospinal connections, postnatal development in the rhesus monkey, *Science*, 138, 678, 1962.
132. Olivier, E., Edgley, S.A., Armand, J., and Lemon, R.N., An electrophysiological study of the postnatal development of the corticospinal system in the macaque monkey, *J. Neurosci.*, 17, 267, 1997.
133. Heffner, R. and Masterton, R.B., Variation in form of the pyramidal tract and its relationship to digital dexterity, *Brain Behav. Evol.*, 12, 161, 1975.
134. Phillips, C.G., Evolution of the corticospinal tract in primates with special reference to the hand, in *Proceedings of the 3rd International Congress on Primatology, Zürich*, Vol. 2, Karger, Basel, 1971, 2.
135. Heffner, R.S. and Masterton, R.B., The role of the corticospinal tract in the evolution of human digital dexterity, *Brain Behav. Evol.*, 23, 165, 1983.
136. Lawrence, D.G. and Hopkins, D.A., The development of motor control in the rhesus monkey, evidence concerning the role of corticomotoneuronal connections, *Brain*, 99, 235, 1976.
137. Antinucci, F. and Visalberghi, E., Tool use in *Cebus apella*, a case study, *Int. J. Primatol.*, 7, 349, 1986.
138. Westergaard, G.C. and Fragaszy, D.M., The manufacture and use of tools by capuchin monkeys (*Cebus apella*), *Zoo. Biol.*, 4, 317, 1987.
139. Costello, M.B. and Fragaszy, D.M., Prehension in *Cebus* and *Saimiri*. I. Grip type and hand preference, *Am. J. Primotol.*, 15, 235, 1988.
140. Fragaszy, D.M., Preliminary quantitative studies of prehension in squirrel monkeys (*Saimiri sciureus*), *Brain Behav. Evol.*, 23, 81, 1983.
141. Rouiller, E.M. et al., Evidence for direct connections between the hand region of the supplementary motor area and cervical motoneurons in the macaque monkey, *Eur. J. Neurosci.*, 8, 1055, 1996.
142. Morecraft, R.J. et al., Segregated parallel inputs to the brachial spinal cord from the cingulate motor cortex in the monkey, *Neuro Report*, 8, 3933, 1997
143. Bates, J.F. and Goldman-Rakic, P.S., Prefrontal connections of medial motor areas in the rhesus monkey, *J. Comp. Neurol.*, 336, 211, 1993.
144. Morecraft, R.J. and Van Hoesen, G.W., Convergence of limbic input to the cingulate motor cortex in the rhesus monkey, *Brain Res. Bull.*, 45, 209, 1998.
145. Tokuno, H. et al., Reevaluation of ipsilateral corticocortical inputs to the orofacial region of the primary motor cortex in the macaque monkey, *J. Comp. Neurol.*, 389, 34, 1997.
146. Matelli, M. et al., Superior area 6 afferents from the superior parietal lobule in the macaque monkey, *J. Comp. Neurol.*, 402, 327, 1998.
147. Johnson, P.B. et al., Cortical networks for visual reaching, physiological and anatomical organization of frontal and parietal lobe arm regions, *Cereb. Cortex*, 6, 102, 1996.
148. Marconi, B. et al., Eye–hand coordination during reaching. I. Anatomical relationships between parietal and frontal cortex, *Cereb. Cortex*, 11, 513, 2001.

149. Tanne-Gariepy, J., Rouiller, E.M., and Boussaoud, D., Parietal inputs to dorsal versus ventral premotor areas in the macaque monkey, evidence for largely segregated visuomotor pathways, *Exp. Brain Res.,* 145, 91, 2002.
150. Morecraft, R.J., Rockland, K.S., and Van Hoesen, G.W., Localization of area prostriata and its projection to the cingulate motor cortex in the rhesus monkey, *Cereb. Cortex,* 10, 192, 2000.
151. Simonyan, K. and Jürgens, U., Cortico-cortical projections of the motorcortical larynx area in the rhesus monkey, *Brain Res.,* 949, 23, 2002.
152. Strick, P.L. and Kim, C.C., Input to primate motor cortex from posterior parietal cortex (area 5). I. Demonstration by retrograde transport, *Brain Res*., 157, 325, 1978.
153. Vogt, B.A. and Pandya, D.N., Cortico-cortical connections of somatic sensory cortex (areas 3, 1 and 2) in the rhesus monkey, *J. Comp. Neurol.,* 177, 179, 1978.
154. Leichnetz, G.R., Afferent and efferent connections of the dorsolateral precentral gyrus (area 4, hand/arm region) in the macaque monkey, with comparisons to area 8, *J. Comp. Neurol.,* 254, 460, 1986.
155. Darian-Smith, C. et al., Ipsilateral cortical projections to areas 3a, 3b, and 4 in the macaque monkey, *J. Comp. Neurol*., 335, 200, 1993.
156. Huerta, M.F. and Pons, T.P., Primary motor cortex receives input from area 3a in macaques, *Brain Res*., 537, 367, 1990.
157. Brinkman, J., Bush, B.M., and Porter, R., Deficient influence of peripheral stimuli on precentral neurones in monkeys with dorsal column lesions, *J. Physiol. (Lond.),* 276, 27, 1978.
158. Asanuma, H. and Mackel, R., Direct and indirect sensory input pathways to the motor cortex; its structure and function in relation to learning of motor skills, *Jpn. J. Physiol*., 39, 1, 1989.
159. Tracey, D. J. et al., Thalamic relay to motor cortex, afferent pathways from brain stem, cerebellum, and spinal cord in monkeys, *J. Neurophysiol*., 44, 532, 1980.
160. Pavlides, C., Miyashita, E., and Asanuma, H., Projection from the sensory to the motor cortex is important in learning motor skills in the monkey, *J. Neurophysiol*., 70, 733, 1993.
161. Kurata, K., Corticocortical inputs to the dorsal and ventral aspects of the premotor cortex of macaque monkeys, *Neurosci. Res.,* 12, 263, 1991.
162. Ghosh, S. and Gattera, R., A comparison of the ipsilateral cortical projections to the dorsal and ventral subdivisions of the macaque premotor cortex, *Somatosens. Mot. Res*., 12, 359, 1995.
163. Dum, R.P. and Strick, P.L., Cingulate motor areas, in *Neurobiology of Cingulate Cortex and Limbic Thalamus*, Vogt, B.A. and Gabriel, M., Eds., Birkhauser, Boston, 1993, 415.
164. Hatanaka, N. et al., Thalamocortical and intracortical connections of monkey cingulate motor areas, *J. Comp. Neurol.,* 462, 121, 2003.
165. Lynch, J.C., The functional organization of posterior parietal association cortex, *Behav. Brain Sci.,* 3, 485, 1980.
166. Van Hoesen, G.W., Morecraft, R.J., and Vogt, B.A., Connections of the monkey cingulate cortex, in *Neurobiology of cingulate cortex and limbic thalamus*, Vogt, B.A. and Gabriel, M., Eds., Birkhauser, Boston, 1993, 249.
167. Hyvarinen, J., Posterior parietal lobe of the primate brain, *Physiol. Rev.*, 62, 1060, 1982.
168. Sakata, H. et al., Neural mechanisms of visual guidance of hand action in the parietal cortex of the monkey, *Cereb. Cortex*, 5, 429, 1995.

169. Murata, A. et al., Selectivity for the shape, size, and orientation of objects for grasping in neurons of monkey parietal area AIP, *J. Neurophysiol.*, 83, 2580, 2000.
170. Cavada, C. and Goldman-Rakic, P., Posterior parietal cortex in rhesus monkey, II. Evidence for segregated corticocortical networks linking sensory and limbic areas with the frontal lobe, *J. Comp. Neurol.*, 287, 422, 1989.
171. Baleydier, C. and Mauguiere, F., The duality of the cingulate gyrus in monkey, neuroanatomical study and functional hypothesis, *Brain*, 103, 525, 1980.
172. Baleydier, C. and Mauguiere, F., Network organization of the connectivity between parietal area 7, posterior cingulate cortex and medial pulvinar nucleus. A double fluorescent tracer study in monkey, *Exp. Brain Res.*, 66, 385, 1987.
173. Petrides, M. and Pandya, D.N., Projections to the frontal cortex from the posterior parietal region in the rhesus monkey, *J. Comp. Neurol.*, 228, 105, 1984.
174. Vogt, B.A. and Pandya, D.N., Cingulate cortex of the rhesus monkey, II. Cortical afferents, *J. Comp. Neurol.*, 262, 271, 1987.
175. Matelli, M. and Luppino, G., Parietofrontal circuits for action and space perception in the macaque monkey, *Neuroimage*, 14, S27, 2001.
176. Shipp, S., Blanton, M., and Zeki, S., A visuo-somatomotor pathway through superior parietal cortex in the macaque monkey, cortical connections of areas V6 and V6A, *Eur. J. Neurosci.*, 10, 3171, 1998.
177. Galletti, C. et al., Functional demarcation of a border between areas V6 and V6A in the superior parietal gyrus of the macaque monkey, *Eur. J. Neurosci.*, 8, 30, 1996.
178. Galletti, C. et al., Brain location and visual topography of cortical area V6A in the macaque monkey, *Eur. J. Neurosci.*, 11, 575, 1999.
179. Roberts, T.S. and Akert, K., Insular and opercular cortex and its thalamic projection in *Macaca mulatta, Schweiz. Archiv. Neurol. Neurochir. Psychiat.*, 92, 1, 1963.
180. Luppino, G. et al., Prefrontal and agranular cingulate projections to the dorsal premotor areas F2 and F7 in the macaque monkey, *Eur. J. Neurosci.*, 17, 559, 2003.
181. Walker, A., A cytoarchitectural study of the prefrontal area of the macaque monkey, *J. Comp. Neurol.*, 73, 59, 1940.
182. Barbas, H. and Mesulam, M.M., Cortical afferent input to the principalis region of the rhesus monkey, *Neuroscience*, 15, 619, 1985.
183. Wang, Y. et al., Spatial distribution and density of prefrontal cortical cells projecting to three sectors of the premotor cortex, *NeuroReport* 13, 1341, 2002.
184. Papez, J.W., A proposed mechanism of emotion, *Arch. Neurol. Psychiat.*, 38, 725, 1937.
185. Vogt, B.A. and Gabriel, M., Eds., *Neurobiology of Cingulate Cortex and Limbic Thalamus*, Birkhauser, Boston, 1993.
186. Barbas, H. and Pandya, D.N., Architecture and intrinsic connections of the prefrontal cortex in the rhesus monkey, *J. Comp. Neurol.*, 286, 353, 1989.
187. Ilinsky, I.A. and Kultas-Ilinsky, K., Sagittal cytoarchitectonic maps of the *Macaca mulatta* thalamus with a revised nomenclature of the motor-related nuclei validated by observations on their connectivity, *J. Comp. Neurol.*, 262, 331, 1987.
188. Percheron, G., Francois, C., Talbi, B., Yelnik, J., and Fenelon, G., The primate motor thalamus, *Brain Res. Rev.*, 22, 93, 1996.
189. Nambu, A., Yoshida, S., and Jinnai, K., Projection on the motor cortex of thalamic neurons with pallidal input in the monkey, *Exp. Brain Res.*, 71, 658, 1988.
190. Nambu, A., Yoshida, S., and Jinnai, K., Movement-related activity of thalamic neurons with input from the globus pallidus and projection to the motor cortex in the monkey, *Exp. Brain Res.*, 84, 279, 1991.

191. Olszewski, J., *The Thalamus of the Macaca Mulatta: An Atlas for Use with the Stereotaxic Instrument*, S. Karger AG, Basel, 1952.
192. Nauta, W.J.H. and Mehler, W.R., Projections of the lentiform nucleus in the monkey, *Brain Res.*, 1, 3, 1966.
193. Kuo, J.S. and Carpenter, M.B., Organization of pallidothalamic projections in rhesus monkey, *J. Comp. Neurol.*, 151, 201, 1973.
194. Kim, R. et al., Projections of the globus pallidus and adjacent structures, an autoradiographic study in the monkey, *J. Comp. Neurol.*, 169, 263, 1976.
195. DeVito, J.L. and Anderson, M.E., An autoradiographic study of efferent connections of the globus pallidus in Macaca mulatta, *Exp. Brain Res.*, 46, 107, 1982.
196. Kusama, T., Mabuchi, M., and Sumino, T., Cerebellar projections to the thalamic nuclei in monkeys, *Proc. Jpn. Acad.*, 47, 505, 1971.
197. Kievit, J. and Kuypers, H.G.J.M., Organization of the thalamo-cortical connections to the frontal lobe in the rhesus monkey, *Exp. Brain Res.*, 29, 299, 1977.
198. Batton, R.R. et al., Fastigial efferent projections in the monkey. An autoradiographic study, *J. Comp. Neurol.*, 174, 281, 1977.
199. Chan-Palay, V., *Cerebellar Dentate Nucleus: Organization, Cytology and Transmitters*, Springer-Verlag, Berlin, 1977.
200. Percheron, G., The thalamic territory of cerebellar afferents and the lateral region of the thalamus of the macaque in stereotaxic ventricular coordinates, *J. Hirnforsch.*, 18, 375, 1977.
201. Stanton, G.B., Topographical organization of ascending cerebellar projections from the dentate and interposed nuclei in *Macaca mulatta*. An anterograde degeneration study, *J. Comp. Neurol.*, 190, 699, 1980.
202. Kalil, K., Projections of the cerebellar and dorsal column nuclei upon the thalamus of the rhesus monkey, *J. Comp. Neurol.*, 195, 25, 1981.
203. Asanuma, C., Thach, W.T., and Jones, E.G., Distribution of cerebellar terminations and their relation to other afferent terminations in the ventral lateral thalamic region of the monkey, *Brain Res.*, 286, 237, 1983.
204. Asanuma, C., Thach, W.R., and Jones, E.G., Anatomical evidence for segregated focal groupings of efferent cells and their terminal ramifications in the cerebellothalamic pathway of the monkey, *Brain Res.*, 286, 267, 1983.
205. Matelli, M. et al., Thalamic input to inferior area 6 and area 4 in the macaque monkey, *J. Comp. Neurol.*, 280, 468, 1989.
206. Holsapple, J.W., Preston, J.B., and Strick, P.L., The origin of thalamic inputs to the "hand" representation in the primary motor cortex, *J. Neurosci.*, 11, 2644, 1991.
207. Matelli, M. and Luppino, G., Thalamic input to mesial and superior area 6 in the macaque monkey, *J. Comp. Neurol.*, 372, 59, 1996.
208. Schell, G.R. and Strick, P.L., The origin of thalamic inputs to the arcuate premotor and supplementary motor areas, *J. Neurosci.*, 4, 539, 1984.
209. Darian-Smith, C., Darian-Smith, I., and Cheema, S.S., Thalamic projections to sensorimotor cortex in the macaque monkey, use of multiple retrograde fluorescent tracers, *J. Comp. Neurol.*, 299, 17, 1990.
210. Shindo, K., Shima, K., and Tanji, J., Spatial distribution of thalamic projections to the supplementary motor area and the primary motor cortex, a retrograde multiple labeling study in the macaque monkey, *J. Comp. Neurol.*, 357, 98, 1995.
211. Rouiller, E.M. et al., Cerebellothalamocortical and pallidothalamocortical projections to the primary and supplementary motor cortical areas, a multiple tracing study in macaque monkeys, *J. Comp. Neurol.*, 345, 185, 1994.

212. Rouiller, E.M. et al., Origin of thalamic inputs to the primary, premotor, and supplementary motor cortical areas and to area 46 in macaque monkeys, a multiple retrograde tracing study, *J. Comp. Neurol.,* 409, 131, 1999.
213. Inase, M. and Tanji, J., Thalamic distribution of projection neurons to the primary motor cortex relative to afferent terminal fields from the globus pallidus in the macaque monkey, *J. Comp. Neurol.*, 353, 415, 1995.
214. Nakano, K. et al., An autoradiographic study of cortical projections from motor thalamic nuclei in the macaque monkey, *Neurosci. Res.*, 13, 119, 1992.
215. Nakano, K. et al., Cortical connections of the motor thalamic nuclei in the Japanese monkey, *Macaca fuscata, Stereotact. Funct. Neurosurg*., 60, 42, 1993.
216. McFarland, N.R. and Haber, S.N., Convergent inputs from thalamic motor nuclei and frontal cortical areas to the dorsal striatum in the primate, *J. Neurosci*., 20, 3798, 2000.
217. Herkenhan, M., New perspectives on the organization and evolution of nonspecific thalamocortical projections, in *Cerebral Cortex, Sensory-motor Areas and Aspects of Cortical Connectivity*, Jones, E.G., and Peters. A., Eds., Plenum, New York, 1986, 403.
218. Jones, E.G., The thalamic matrix and thalamocortical synchrony, *Trends Neurosci*., 24, 595, 2001.
219. Alexander, G.E., DeLong, M.R., and Strick, P.L., Parallel organization of functionally segregated circuits linking basal ganglia and cortex, *Ann. Rev. Neurosci.,* 9, 357, 1986.
220. Parent, A. and Hazrati, L.N., Functional anatomy of the basal ganglia. I. The cortico-basal ganglia-thalamo-cortical loop, *Brain Res. Rev*., 20, 91, 1995.
221. Brodal, P., The corticopontine projection in the rhesus monkey. Origin and principles of organization, *Brain,* 101, 251, 1978.
222. Glickstein, M., May, J.G., III, and Mercier, B.E., Corticopontine projection in the macaque, the distribution of labeled cortical cells after large injections of horseradish peroxidase in the pontine nuclei, *J. Comp. Neurol.,* 235, 343, 1985.
223. Wiesendanger, R., Wiesendanger, M., and Ruegg, D.G., An anatomical investigation of the corticopontaine projection in the primate (*Macaca fascicularis* and *Saimiri sciureus*). II. The projection from frontal and parental association areas, *Neuroscience*, 4, 747, 1979.
224. Schmahmann, J.D. and Pandya, D.N., Anatomic organization of the basilar pontine projections from prefrontal cortices in rhesus monkey, *J. Neurosci.,* 17, 438, 1997.
225. Strick, P.L., Dum, R.P., and Picard, N., Macro-organization of circuits connecting the basal ganglia with the cortical motor areas, in *Models of Information Processing in the Basal Ganglia*, Houk, J., Ed., MIT Press, Boston, 1995, 117.
226. Zemanick, M.C., Strick, P.L., and Dix, R.D., Direction of transneuronal transport of herpes simplex virus 1 in the primate motor system is strain-dependent, *Proc. Nat. Acad. Sci. U.S.A.*, 88, 8048, 1991.
227. Strick, P.L. and Card, J.P., Transneuronal mapping of neural circuits with alpha herpesviruses, in *Experimental Neuroanatomy, A Practical Approach*, Bolam, J.P., Ed., Oxford University Press, Oxford, 1992, 81.
228. Hoover, J.E. and Strick, P.L., Multiple output channels in the basal ganglia, *Science*, 259, 819, 1993.
229. Hoover, J.E. and Strick, P.L., The organization of cerebellar and basal ganglia outputs to primary motor cortex as revealed by retrograde transneuronal transport of herpes simplex virus type 1, *J. Neurosci.,* 19, 1446, 1999.
230. Kelly, R.M. and Strick, P.L., Rabies as a transneuronal tracer of circuits in the central nervous system, *J. Neurosci. Meth.,* 103, 63, 2000.
231. Kelly, R.M. and, Strick, P.L., Cerebellar loops with motor cortex and prefrontal cortex of a nonhuman primate, *J. Neurosci*., 23, 8432, 2003.

232. Dum, R.P. and Strick, P.L., An unfolded map of the cerebellar dentate nucleus and its projections to the cerebral cortex, *J. Neurophysiol.*, 89, 634, 2003.
233. Allen, G.I., Gilbert, P.F., and Yin, T.C., Convergence of cerebral inputs onto dentate neurons in monkey, *Exp. Brain Res.*, 32, 151, 1978.
234. Rispal-Padel, L., Cicirata, F., and Pons, C., Cerebellar nuclear topography of simple and synergistic movements in the alert baboon (*Papio papio*), *Exp. Brain Res.*, 47, 365, 1982.
235. Thach, W.T. et al., Cerebellar nuclei, rapid alternating movement, motor somatotopy, and a mechanism for the control of muscle synergy, *Rev. Neurol. (Paris)*, 149, 607, 1993.
236. Akkal, D., Dum, R.P., and Strick, P.L., Cerebellar and pallidal inputs to the supplementary motor area (SMA), *Soc. Neurosci. Abstr.*, 27, 825, 2001.
237. Dum, R.P. and Strick, P.L., Pallidal and cerebellar inputs to the digit representations of the dorsal and ventral premotor areas (PMd and PMv), *Soc. Neurosci. Abstr.*, 25, 1925, 1999.
238. Middleton F.A. and Strick P.L., New concepts regarding the organization of basal ganglia and cerebellar output, in *Integrative and Molecular Approach to Brain Function*, Ito, M., and Miyashita, Y., Eds., Elsevier Science, Amsterdam, 1996, 253.
239. Middleton, F.A. and Strick, P.L., Cerebellar "projections" to the prefrontal cortex of the primate, *J. Neurosci.*, 21, 700, 2001.
240. Clower, D.M., West, R.A., Lynch, J.C., and Strick, P.L., The inferior parietal lobule is the target of output from the superior colliculus, hippocampus and cerebellum, *J. Neurosci.*, 21, 6283, 2001.
241. Middleton, F.A. and Strick, P.L., Anatomical evidence for cerebellar and basal ganglia involvement in higher cognitive function, *Science*, 266, 458, 1994.
242. DeLong, M.R., Crutcher, M.D., and Georgopoulos, A.P., Primate globus pallidus and subthalamic nucleus, functional organization, *J. Neurophysiol.*, 53, 530, 1985.
243. Middleton, F.A. and Strick, P.L., Cerebellar output, motor and cognitive channels. *Trends Cognit. Sci.*, 2, 348, 1998.
244. Middleton, F.A. and Strick, P.L., Basal ganglia "projections" to the prefrontal cortex of the primate, *Cereb. Cortex*, 12, 926, 2002.
245. Strick, P.L., Hoover, J.E., and Mushiake, H., Evidence for "output channels" in the basal ganglia and cerebellum, in *Role of the Cerebellum and Basal Ganglia in Voluntary Movement*, Mano, N., Hamada, I., and DeLong, M.R., Eds., Elsevier Science, Amsterdam, 1993, 171.
246. Künzle, H., Projections from the primary somatosensory cortex to the basal ganglia and thalamus in the monkey, *Exp. Brain Res.*, 30, 481, 1977.
247. Flaherty, A. and Graybiel, A.M., Two input systems for body representations in the primate striatal matrix, experimental evidence in the squirrel monkey, *J. Neurosci.*, 13, 1120, 1993.
248. Matelli, M., Luppino, G., and Rizzolatti, G., Patterns of cytochrome oxidase activity in the frontal agranular cortex of the macaque monkey, *Behav. Brain Res.*, 18, 125, 1985.
249. Pandya, D.N. and Selzer, B., Intrinsic connections and architectonics of posterior parietal cortex in the rhesus monkey, *J. Comp. Neurol.*, 204, 196, 1982.

2 Functional Magnetic Resonance Imaging of the Human Motor Cortex

Andreas Kleinschmidt and Ivan Toni

CONTENTS

2.1 INTRODUCTION

Using the title of this chapter as a search command in Medline gives more than 1000 hits, and the number of false negatives probably largely exceeds that of false positives. This points to the vast number of functional neuroimaging studies that have reported motor cortex activation, but it does not help to decide whether these studies have advanced our knowledge of the functional organization and response properties of motor cortex. In relation to findings from other techniques in the neurosciences, the authors of this chapter are tempted to acknowledge that the contribution to understanding the motor cortex that has come from neuroimaging is small yet significant, in particular with respect to the human motor cortex. In collating some of the findings that may be seen to provide such a contribution, it has nonetheless been necessary to constrain somewhat arbitrarily the number and type of studies that are considered in detail.

0-8493-1287-6/05/$0.00+$1.50

A first major constraint that we decided to apply was to focus on studies that used magnetic resonance as the functional imaging modality. This is not motivated by the overwhelmingly greater number of studies using functional magnetic resonance imaging (fMRI) techniques rather than others. In particular, and as a second constraint, we will not cover the many studies in the wake of the fMRI avalanche that have dealt with feasibility and methodological optimization. This first constraint we introduced is motivated instead by the superior spatiotemporal resolution and sensitivity of fMRI compared to other imaging modalities that can be applied noninvasively in human subjects. In fact, the sensitivity of fMRI is even good enough to permit analyses that invert the usual direction of inference, i.e., from the neurophysiological signal to behavior.[1] Naturally, many of the issues previously addressed by other imaging modalities have been revisited using fMRI, and in part this has been merely confirmatory, but in part this has also resulted in more detailed findings.

Nonetheless, there is an important downside to this because fMRI is not only exquisitely sensitive to the hemodynamic signals associated with the neural activity related to movements, but unfortunately also to direct effects of motion. First, the necessity of retaining the organ of interest within the rigid imaging grid precludes studies with, for instance, free natural movements such as walking. Techniques that can apply head-mounted or even telecommunicating devices (such as electrical or optical recordings[2]) or that allow for an interval between the activating paradigm and data collection (such as positron emission tomography [PET] with slow tracers like fluoro-deoxy-glucose[3]) offer distinct advantages in this respect although they also fight artifacts and limitations of sensitivity and resolution. Second, within the bore of the magnetic resonance scanner overt limb motion or even mere changes in muscle tone readily translate into shifts of the brain relative to the machine's imaging coordinates. Even slight shifts result in devastating effects on image quality that are far more complex than the mere displacement accounts for and thus not readily compensated for by simple realignment algorithms. Accordingly, there have been relatively few successful studies on movements of facial, proximal, or axial muscles. There are some noteworthy studies on respiration[4,5] and facial functions such as swallowing or speaking,[6–8] but most of the work with fMRI has dealt with movements of the distal upper extremity that are associated with so little artifact that the existing correction tools can handle it without compromising data quality. In other words, and as a third constraint, our review will mostly cover hand function.

As a fourth constraint we will not consider in this review those many valuable studies that have integrated the imaging of motor cortex activation into a clinical context, be that the issue of presurgical mapping or that of postlesional plasticity, or the influence of other disease conditions or pharmacological manipulations on task-related motor cortex activity.

Finally, as a fifth constraint, and despite the multitude of "motor" areas in the brain[9] we will focus on studies dealing with or involving effects on activity in the primary motor cortex. The functional behavior of other motor areas will nonetheless often be mentioned along these lines in the context of paradigms that are associated with, but not only with, primary motor cortex activity.

Even when implementing all these constraints, we are certain to have missed relevant studies in the abundant literature and we apologize for these omissions. The purpose and hopefully the result of this chapter is to provide the reader with an overview of the contribution of fMRI to some of the prevailing topics in the study of motor control and of primary motor cortex function. In several points, the findings with functional neuroimaging will seem to be in apparent disagreement with those from other modalities. This cannot always be related to insufficient sensitivity of this noninvasive modality. In part, it may reflect the indirect and spatio-temporally imprecise nature of the fMRI signal, but these studies remain informative by virtue of the fact that usually the whole brain is covered. This does not only provide a plausibility control for localized effects, but the distribution of response foci and the relation of effects observed at these different sites can assist the guidance of detailed studies at the mesoscopic or microscopic spatio-temporal level. Even when denying any single current neuroscience method a gold standard status, an adequately modest view should probably conclude that fMRI currently is mostly a tool of exploratory rather than explanatory value.

2.2 SIGNALS IN fMRI

Along with visual stimulation, voluntary or paced movements have belonged to the first experimental conditions used to evoke and observe fMRI responses in the human brain.[10] The mainstream of functional activation studies by MRI relies on the blood-oxygenation-level-dependent (BOLD) contrast although other techniques that measure task-associated changes in blood flow or — via contrast agents — in blood volume can also be used for functional imaging and can even offer distinct advantages in some settings. Simply put, the basis of the BOLD contrast is that a neural activity increase results in a blood flow increase that exceeds the concomitant increase in oxygen consumption. This means that more blood flows through the capillaries without that proportionately as much more oxygen is being extracted from it. As a consequence, and somewhat counterintuitively, the blood in the postcapillary vasculature will become hyper-oxygenated during activation and thus will contain less deoxyhemoglobin than before. As opposed to diamagnetic oxyhemoglobin, deoxyhemoglobin is paramagnetic and causes more and more signal loss the longer it takes to record the echo. Accordingly, wherever in the brain this decrease in deoxyhemoglobin concentration occurs during an "activated" as opposed to a "resting" state, there will be an image signal increase in the corresponding voxel, the so-called BOLD response. Of course, the change in deoxyhemoglobin concentration is not the only physiological effect occurring during activation, and BOLD contrast fMRI sequences can also be sensitive to other effects, such as changes in blood volume or flow velocity. However, a number of studies have used simultaneous transcranial optical absorption measurements, so-called near-infrared spectroscopy (NIRS), during fMRI to validate task-related deoxyhemoglobin concentration changes as the physiological basis of the BOLD fMRI response.[11–13]

Apart from the problem of confidently relating the signal changes observed in fMRI to changes in a single physiological parameter, there remains the problem that deoxyhemoglobin is only an indirect index of neural activity. The mechanisms that link this parameter to neural activity are still not fully understood, although some progress has been made in recent years. A more superficial but, for some purposes, more relevant concern is to simply understand the coupling functions in terms of the spatial and temporal dispersion that the BOLD signal change exhibits in relation to neural activity changes.[14] Most of the work characterizing the coupling functions between neural activity and blood flow and metabolism in man has used visual stimulation as the functional challenge. This means that the related findings are heavily dominated by the behavior of calcarine cortex, a cortical area with a peculiar cyto- and myelo-architectonic organization that is not representative of other neocortical or even primary sensory areas.[15–17] If one considers the huge neurochemical and neurovascular heterogeneity in the central nervous system,[15] the usual assumption of generic hemodynamic and metabolic response properties across different brain regions needs to be taken with caution or even to be addressed analytically.[18]

With these constraints in mind, it can be said that the BOLD response generally occurs with 2 to 3 sec latency even after very brief neural events, peaks after approximately 6 sec, and usually takes more than the rise time to decay back to baseline level. This signal increase is often followed by an "undershoot" that may take 10 or easily even more seconds before the signal asymptotically recovers baseline level. Several laboratories have observed an "early dip," i.e., an actual signal decrease during the initial latency period, and this has also been shown for the motor cortex.[19] Because this signal is also found in optical imaging studies and because in animal fMRI studies it has been used to resolve functional architecture at the columnar level, there has been considerable hope that it might be useful in human fMRI studies to obtain mapping results at a higher spatial specificity.[20,21] Yet, many laboratories have found it difficult to reproduce this dip at all, and even those that have, consistently observed lower amplitudes in relation to the positive BOLD response than seems to be the case in fMRI studies with laboratory animals. Conversely, it has been established in humans that the spatial specificity of the early components of the positive BOLD response is sufficient to map, for instance, ocular dominance columns in primary visual cortex.[22] Another hope related to the early dip has been that since it occurs earlier than the positive response, it might also preserve temporal information on a finer scale. Again, this is compromised by the overall weakness of this signal, whereas there have been encouraging results from studying in more detail the temporal information contained in the envelope of the positive BOLD response.[23] It is therefore not surprising that the findings reviewed in the following sections are almost entirely based on the positive BOLD response.

Over and above the uncertainties regarding the coupling between neural activity and blood flow and metabolism, and between blood oxygenation changes and fMRI signal, there remain open questions as to the precise nature or component out of the orchestrated spectrum of neural activity that drives these effects. Both classical studies and, in a more direct way, recent work with simultaneous fMRI and electrophysiological recordings point to synaptic activity instead of action potentials as the source underlying hemodynamic responses.[24] Synaptic activity arises mainly from intracortical connectivity, with some contribution by afferents from distant neurons. Synapses can be excitatory or inhibitory, and the metabolic demands from their activity may be comparable in magnitude but the effect they have on their targets is sign-inverted and probably often differs in efficiency. Although there has been some attempt to elucidate their relative contribution to blood flow regulation, this issue remains far from resolved.[25]

Despite all these uncertainties regarding the nature of the fMRI signal, it has been widely used in the neurosciences in the past decade. There are conceptually different ways of using the BOLD response to study brain function and they will all be touched on in the following sections. At a first level, the response can be used for the simple purpose of mapping, i.e., showing a responsiveness of neural tissue in association with a task as opposed to rest. This approach has been used in the context of studies on response lateralization and somatotopical representation (see the next two sections). At a second level, the response can be used to determine response properties, as related by analogy to stimulus–response functions. The relation between fMRI signal and movement parameters is covered in a section on motor response properties. The sections thereafter deal with further aspects of the topic, such as acute (attentional) or long-term modulation of responses (learning), other sources of primary motor cortex activation than overt movement (sensation, imagery), and findings related to cognitive states such as motor intention and preparation.

2.3 LATERALIZATION AND HANDEDNESS

One of the basic observations in functional neuroimaging during simple unilateral hand movements is that the strongest associated activation is observed in the contralateral primary motor cortex (M1). This corresponds to the decussation of the pyramidal tract as the main output of M1 and mirrors the clinical deficit observed after lesions of this tract or its cortical origin. Yet, in addition to some PET studies that have been conducted, early fMRI studies also observed activation in M1 ipsilateral to the moving hand.[26] In a very detailed study, Dassonville et al.[27] quantified this degree of lateralization. They studied predictably and unpredictably visually cued finger movement sequences and computed for a given voxel significance threshold the contra- and ipsilaterally activated volumes as well as their ratios, i.e., lateralization indices. With this analysis, they observed larger contralateral activation volumes for dominant than for nondominant hand movements. This effect was only significant in a region of interest covering M1 but not in other distant motor areas. It was not accounted for by behavioral differences in that response times and error rates were matched between hands. Interestingly, the degree of lateralization of primary motor cortex activation during dominant hand usage was related to the degree of handedness. This effect was driven by weaker ipsilateral activations in those subjects with strong behavioral lateralization. A similar observation regarding ipsilateral activation as a function of dominant vs. nondominant hand movements was made by Singh et al.,[28] although they pointed out that this effect was stronger in regions presumably covering premotor rather than M1.

The study by Dassonville found no significant effect of handedness on contralateral activation volume and no interaction of dominance with handedness. However, an earlier study from the same laboratory had shown a handedness effect. Kim et al.[29] reported that while the right motor cortex was activated mostly during contralateral finger movements in both right-handed and left-handed subjects, the left motor cortex was substantially more active during ipsilateral movements, and that this effect was more pronounced in right-handed than in left-handed subjects. This pointed to a hemispheric asymmetry, with the left motor cortex contributing more to finger movements of either side, and suggested that this asymmetry was stronger in right-handed people. A similar result was obtained in a later study by Li et al.,[30] whereas a somewhat different pattern was reported by Singh et al.[31] In their study, an ipsilateral precentral (premotor) region showed a hand dominance effect only in right-handed but not in left-handed subjects.

Several observations have been added to these initial findings on ipsilateral activation during finger movements and have contributed to assessing its functional significance. When comparing the locations of response peaks in M1 activation for a given hemisphere in greater spatial detail, it was found that the peak during ipsilateral index tapping did not colocalize with that during contralateral index tapping. Instead, it was shifted ventrally, laterally, and anteriorly by about 1 cm in each direction.[32] Whenever it could be elicited, the peak during ipsilateral index tapping was shifted. However, this distinct focus, which may pertain to the premotor cortex, was also activated in half of the cases during contralateral finger movement, even if it was not the dominating peak.

Another study tested not only for activation but also deactivation effects, i.e., reduced fMRI signal during a motor task.[33] Using a sequential finger-to-thumb opposition task, ipsilateral activation was observed in about one third of the participants. However, the authors also found ipsilateral deactivation, and this was equally inconsistent across all subjects. Interestingly, the authors showed by a conjunction analysis that these deactivated regions strongly overlapped with those activated during contralateral task execution and were localized to the primary motor hand representation, whereas the ipsilateral activations were localized in adjacent regions. The authors hypothesized this effect to result from transcallosal inhibition, but another study that confirmed their observation in normal subjects could also reproduce it in patients with congenital callosal agenesis.[34] This does not rule out the role of transcallosal inhibition in the normally organized brain but nonetheless fails to provide positive evidence for it, and the issue is therefore still unresolved.

An interesting finding in this context was provided by Hsieh et al.,[35] who also reproduced ipsilateral deactivation in healthy controls but found this effect greatly dampened or even in part reverted to an activation in patients with severe brachial plexus injury contralateral to the hand studied. This does not clarify the source of ipsilateral deactivation but it suggests that the functional meaning may be to reduce activity in the hand contralateral to the one executing the task. The need for such a "silencing" would be reduced in patients with peripheral nerve damage and accordingly compromised motor abilities.

Regarding ipsilateral activation, many laboratories have found it difficult to consistently observe it, especially when applying simpler or more distal hand motor tasks than in the studies described above.[36,37] This agrees with findings from studies explicitly addressing response lateralization as a function of motor proficiency or movement type. Ipsilateral activation was found to be stronger both for regular finger movement sequences executed with the nondominant hand and random finger movements executed with the dominant hand as opposed to regular finger movement sequences executed with the dominant hand.[38] This effect occurred not only in M1 but in several cortical and subcortical areas and was related to complexity (probed by random as opposed to regular sequence) and/or familiarity of the movements (probed by nondominant as opposed to dominant hand). A greater relevance of ipsilateral and in particular left-sided M1 for more complex movements is also suggested by experiments studying the disruptive effects of transcranial magnetic stimulation on movement execution.[39]

In a similar vein, Solodkin at al.[40] found that left- and right-handed subjects had similar activation patterns with strong lateralization during single-finger movements, but that these patterns comprised more areas with greater volumes and expressed less lateralization for sequential finger movements, and particularly so in left-handed subjects. Of note, they did observe ipsilateral M1 activation in left-handed but not in right-handed subjects for the simple movement, but they also found greater contralateral M1 activation in left-handed than in right-handed subjects for this movement type.

Despite differences in the detailed findings, it seems fair to summarize that ipsilateral motor cortex activation appears to be more prevalent when studying the nondominant hand or left-handed subjects. The question remains what functional significance M1 activation has for ipsilateral finger movements. The influence of complexity or effort might suggest that ipsilateral activation corresponds to involuntary mirror movements occurring in the hand contralateral to the one driven by the task instruction. Although mirror movements can occur even in healthy adult subjects under certain conditions, one would then also expect the ipsilateral activation foci to mirror the contralateral ones. There seems to be sufficient evidence reviewed above suggesting that this is not the case. Moreover, one would then expect that unilateral brain lesions should not affect ipsilateral motor behavior, but should only result in a loss of contralateral mirror movements. However, it has been shown that on detailed kinematic analysis even very distal movements are affected by ipsilateral brain damage.[41] Interestingly, this effect is stronger in the case of left hemisphere damage and for proximal movements. While this does point to an ipsilateral contribution to upper-limb movements, this type of study cannot relate the ipsilateral contribution to any one of the several motor areas contained in each hemisphere.

One potential explanation that would relate the ipsilateral contribution to finger movements to M1 comes from the small fraction of pyramidal tract fibers that do not cross. This fraction could result in a lateralized but bihemispheric control of distal finger movements. Again, one would then (although with less confidence) expect mirroring ipsilateral foci, which is not the case. Moreover, it has been established that the more distal the muscle is, the less bilateral the pyramidal tract innervation becomes. Accordingly, a more recent study found only nonprimary motor areas activated during distal ipsilateral movements, while the primary motor cortex was spared or even deactivated.[42] Conversely, proximal movements were associated with ipsilateral activation in both primary and nonprimary motor areas. In particular, the authors noted a joint contra- and ipsilateral response focus in the precentral gyrus that they assigned to premotor cortex.

Reconsidering the data presented above, it seems that the precentral foci during ipsilateral movement are indeed different from those related to the identical contralateral movement. It is less clear whether they belong to different parts of M1 or to the more anterior premotor cortex. If they belong to the premotor cortex, one could interpret the above findings to reflect a greater bilaterality of premotor corticospinal innervation and a left hemispheric predominance for movement that increases with complexity. However, if they belong to M1, one could account for the activation foci by proximal coinnervation. In this latter case, the ipsilateral effect would conceivably also be enhanced by movement complexity, and it would express the greater bilateral control of more proximal muscles. This view would be compatible with the observation that while the location of the dominant contralateral M1 focus is not mirrored in the ipsilateral cortex, the ipsilateral activations do in part mirror the minor contralateral foci.

The interpretation of the various findings discussed above is stuck at the level of anatomical analysis, which is still not detailed enough to allow for the confident discrimination between effects in M1 and those in the premotor areas. For that reason, the issue of ipsilateral activation has in recent years been advanced by experiments combining functional neuroimaging with transcranial magnetic stimulation, which are beyond the scope of this chapter.[43]

2.4 SOMATOTOPY

The "Jacksonian march" during the propagation of a focal seizure or the relation between lesion topography and concomitant distribution of paresis are long-standing clinical observations which have suggested that movement of different body parts is related to the activity of spatially distinct brain regions. In a more explicit and experimental way, the pioneering work on electrical stimulation during open brain surgery established the notion of somatotopy in the human motor cortex, i.e., the systematic and orderly representation of the body along the medio-lateral extent of the precentral gyrus. In textbooks, this is usually represented as the so-called homunculus of the primary sensory and motor cortices, with the knee bent approximately into the interhemispheric fissure and the more cranial body parts rolled out laterally along the convexity, with the exception of an inverted and thus upright face representation. Despite the high illustrative value of these cartoons, they have somewhat clouded a more precise understanding of what somatotopy in M1 could mean in functional terms.

Functional imaging of the activation during voluntary movements has produced findings that are congruent with those from stimulation studies, at least on a coarse spatial scale. Using fMRI, this was first addressed by Rao et al.[44] in a study that also confirmed previous PET findings by showing some degree of intralimb somatotopy for the upper extremity.

One of the key features of the historical cartoons that contributes to their poignancy is the distortion of the homunculus with respect to the proportions of the human body. This largely corresponds in anatomical terms to the sizes of motor units and in functional terms to the degree of differentiation and proficiency of movement for different parts of the body. Accordingly, the hand occupies a long stretch of cortical surface, and the cartoon features an orderly representation of individual fingers with the thumb at the lateral and the little finger at the medial end of this overall hand representation.

This view was challenged by Sanes et al.,[45] who showed that various types of wrist and finger movements were associated with distributed, but between themselves largely overlapping, activation patterns within this overall hand representation. Yet, this study left several questions open. One question is whether the absence of significant fMRI activation in a voxel can be taken as evidence for a lack of task-related neural activity therein. Obviously, this is not the case because the fMRI method is far too insensitive. A second question is whether at a given significance threshold the observation of qualitatively very similar activation patterns for different movements is good evidence against somatotopy. Again, the answer is no. The observation of overlap argues only against segregation, but the entire previous literature on finger somatotopy never suggested segregation in the first place. What this experiment did not address was whether there is a quantitative difference between activations along the hand motor representation as a function of which fingers are being moved. In other words, the conceptual mistake had been to address somatotopy by a mapping procedure instead of a study of cortical response properties.

The historical accounts of responses to stimulation indeed suggest only a quantitative difference. Foerster[46] summarized his experience in the following way: "Every so-called focus contains not only preferentially motor elements of the body part assigned to it, but also contains elements for neighboring body parts; however, these are fewer in number and less excitable [...]. Hence, the so-called thumb focus is not an absolute focus, but also contains motor elements for the other fingers and the hand, which are intertwined with the thumb elements. Yet the thumb elements outnumber the others and, of all the elements contained in the same area, show the lowest excitation threshold" (translated by the authors). Foerster also commented on the huge variability of responses elicited from a given stimulation site and related this to the tiring of certain elements that would make the other, initially less dominant elements step into the foreground — a notion we would refer to as adaptation today. Penfield and Rasmussen[47] reported similar observations and also emphasized that "in most cases movement appears at more than one joint simultaneously." They stated that their cartoon only referred to those rare cases when movement appeared at only one joint, although they also stated that grouped responses often involved neighboring fingers.

The two issues raised above were readdressed in an fMRI study performed by one of the authors[48] at considerably higher spatial resolution than was the case with Sanes and colleagues.[45] We found that any type of hand or finger movement tested was associated with almost complete and continuous activation along the entire stretch of the cortical hand representation. Second, when comparing activation strength by using a different finger task instead of rest as the control condition, the response maxima for the experimental movement tasks concorded with the classical somatotopic representation as it had been illustrated by the homunculus cartoon. Although there was some indication of somatotopy when contrasting finger movements against rest (lateral or medial extensions of the significant activation band for thumb and little finger movements, respectively, at given significance thresholds), the clear-cut demonstration of somatotopy was improved by directly contrasting different finger movements performed in alternating blocks. One of the reasons for this may be that spatially less specific effects — for instance, in the locally draining vasculature or from partial-volume effects in single voxels — arise when contrasting against rest. In that sense, this approach resembled those used in the visual system for retinotopic mapping where there is continuous visual stimulation that slowly changes its position in the visual field.[49] Alternatively, this may also reflect the fact that similar types of proximal coinnervation for stabilization were recruited for both the experimental and the control task, and that nonspecific activations were thus canceled out. Because we found no segregation but only relative predominance, we proposed to think of the contribution of somatatopy to functional organization of M1 as a "gradient." In other words, movements of different fingers are associated with extensive neural activations throughout the entire hand representation (and beyond), but the peaks for different fingers are in systematic accord with the homunculus cartoon. We believe that such a description presents a safeguard not only against overinterpreting the historical homunculus cartoons, but also against seeing more segregation in contemporary fractured or mosaic patterns obtained in nonhuman primates than the methods applied in those studies can positively affirm.

The finding of a somatotopic gradient in the M1 hand representation has since been revisited and reproduced by several research groups. In one case, there was a claim of somatotopy but the actual layout found did not correspond to the classical homunculus cartoon,[50] with index movements represented more laterally than thumb movements. In all the other cases, however, the somatotopy did match the classical findings.[51–53] Overall, one can summarize that this finding is most readily obtained for thumb movements, which is in accordance with a spatially low-pass filtered view on the observations made in nonhuman primates.[54,55] Furthermore, it is most readily observed by contrasting thumb against little finger movements, which expresses the distance between the sites where either of the two dominates the representation of the other. Alternatively, this can be seen to reflect the degree to which we manage to move a single finger in as much isolation as possible. Although each of these studies added some aspect of refinement or some degree of more detailed quantification and thus further corroborated the experimental proof and characterization of somatotopy in the motor hand representation, none of these studies advanced our understanding from a functional perspective of why this should be the case. The unanswered question is what good does it serve the brain to represent information in a topographical fashion.

One of the potential benefits from such a functional architecture is segregation, and this makes sense for unique solutions. In the visual system for instance, a dot that is present in one spot of the visual field is not present in another spot, and accordingly the processing of this information may be aided by spatially separating the neural populations that code for these different spots in the visual field. Because a dot may have a certain size and thus cover a certain extent of the visual field, it also makes sense to organize those representations that code for one spot in the immediate vicinity of those coding for the adjacent spots. This in itself presents a sufficient functional benefit to justify a retinotopic layout of the primary visual cortex, but does this predict any such benefit for the motor system? The example in the visual system deals with unambiguous information in the physical environment as collapsed into a two-dimensional visual field that is first (and most precisely) represented by virtue of a mere optical projection at the retinal level. In fact, the propagation of a veridical primary retinotopic representation into the central nervous system progressively degrades, distorts, and fractures the relative contribution of retinotopy to the individual neural response properties.

Seeking the analogy with the visual system may be good for the somatosensory system, and the studies on somatosensory somatotopy provide experimental evidence in favor of this notion.[53,56–58] Yet, the analogy seems to hold less well for the motor system. In fact, from a motor control perspective, a somatotopic layout does not make a lot of sense for the hand representation. Motor acts involve concerted activity changes across a wide range of different muscles. Many of these muscles affect movements of more than one joint, and movement in many joints is affected by more than one muscle. Accordingly, the neural pathways involved display high degrees of divergence and convergence.[59]

If there are so many facts arguing against a somatotopical organization of M1, why then should consistent experimental evidence show that a contribution from this feature is nonetheless detectable in the functional organization of the M1? In compiling the functional benefit from topical organization as above for the sensory systems, we have not yet mentioned one additional important factor. Man is in motion and so are objects in the world, and one of the cardinal functions of the sensory and motor systems is to optimize the related neural processes. Using the visual system again for illustration, if we think of a dot at one spot in the visual field and assume that it is moving, then it will be at another spot of the visual field at a later time point. To reconstruct the trajectory of the dot requires interactions of those neurons that code the spot where it is first, with those that code where it appears thereafter.

Obviously, mere connection of these neuronal units is a prerequisite, but is not in itself sufficient for the perceptual success of this functional interaction. In addition, the preservation of precise temporal information is required to determine whether the dot moved one way or the other or whether these are in fact the two ends of a bar that just appeared behind an occluding surface. The preservation of temporal information can be achieved by a high speed of information relay between neuronal units, and the nervous system has two ways of doing this. One involves the degree of myelination, and works well for instance for the corticospinal tract. However, this strategy is costly in terms of the volume required by such a heavily myelinated pathway. In projection pathways, this may not pose a constraint, but for intra- or inter-areal associative communication pathways this may be disadvantageous. Alternatively, short pathways offer a strategy of rapid communication that is less costly but that cannot be applied uniformly if it has to deal with interconnecting each position with every other position for a two-dimensional cortical sheet.

To achieve the best functional result despite this problem requires knowledge of which neural units must be closely connected and which neural units can interact by sparser or longer association fibers. If we think of the dot again that appears sequentially at different positions of the visual field, there is no absolute prediction of its subsequent position by means of its previous one. The subsequent position could be anywhere in the visual field, but there is an obvious probability distribution of where this will be. This probability distribution provides a meaningful *a priori* hypothesis for how to wire the horizontal interactions in a visual area, and it corresponds to a retinotopic map layout. This teleological consideration is also related to theoretical work that has addressed the connectivity patterns by which neural elements can serve the functional needs of segregation and of integration while maximizing their proficiency (or complexity) in information processing.[60] In other words, structural features likely reflect the interaction of general optimization principles (minimizing wiring length) with the implementation of a given function.[61]

Yet, what is the relevance of this consideration for the somatotopical layout of M1? The fundamental difference between the motor system and the visual (or somatosensory) system is that in the former there is almost no functional benefit from segregation. In other words, for a given neural unit the probability distribution of interactions with other neural units does not present distinct peaks, as in sensory systems. Translated into movements, this means that in real life there is virtually no such thing as a single muscle/single joint movement for which it would make sense to implement a segregated representation. This does not mean that single neurons cannot elicit motor actions, but only that it is virtually impossible to hardwire functional demands into the response properties of single neurons (see however Brecht et al.[62]). A classical debate in motor control research has been whether muscles or movements are "represented" in the primary motor cortex. In a way, this is equivalent to asking whether a piano functions by playing a sonata or striking chords. In other words, this debate is conceptually related to the debate about "grandmother neurons" in the visual system. In a more contemporary view, functionally meaningful motor acts arise from the concerted activity of many neural units, just as a sonata arises from the effect of playing a certain complex pattern of piano keys. However, in contrast with the concert pianist, we do not know which sonata is on the program today and under which biophysical circumstances we will need to play it. These prospective uncertainties impose on the motor control system the need to retain a high flexibility that permits adaptation to variable demands by virtue of flexibly associating the available neural units into a customized pattern that will yield the required action result. If we assume that movements and their direction are coded by population activity, the functional performance of such constantly regrouped populations would benefit from single keys striking several chords and single chords being accessible via several keys, i.e., from members of a neural ensemble with a high degree of divergence and convergence.[63]

However, the need for flexible adaptation to varying functional demands does not necessarily mean that the probability distribution for interactions between neural elements in the motor cortex is flat. What are the determinants that pattern this probability distribution? One factor comes from purely motor considerations. The probability for interactions is higher within limbs than across limbs, and facial movements will again be fairly independent of those in upper and lower extremities. This is not to say that for instance ballistic arm movements will not require coordinated output to trunk and leg muscles, but just that the predictability is lower than for within-limb coordination. The functional anatomical consequences of this pseudo-segregation have lured many researchers into the jargon of face, arm/hand,

and leg "areas," as if these were strictly segregated distinct representations. The data available only suggest that embedded into one continuous motor representation a greater degree of spatial separation and a lower degree of overlap can be found for the representation of different movements across than within these body parts.

The next question is whether within an upper limb representation the probability distribution of interactions is flat. This is clearly not the case, and may account for the relative ease with which it has been found that proximal movements elicit more medial activation than distal movements, and vice versa. It should be noted that in natural contexts these movements differ not only in terms of the musculature involved, but also in terms of movement type and purpose. This is no longer the case when considering different types of hand and finger movements. Lesions of the primary motor hand representation affect our capability for performing individuated finger movements.[64] At the same time, the representation of these movements is associated with activity across a wide stretch of M1, not only in terms of mediolateral extent, but also representation is enhanced by a characteristic cortical surface expansion that has been labeled the "hand knob." The experimental evidence reviewed in this section has undoubtedly clarified the fact that, qualitatively, even simple and relatively isolated finger movements are associated with activation effects that span the entire range of the hand representation. At the same time, there is evidence, recently confirmed by magnetoencephalography,[65] that in quantitative terms, or when analyzed as the center of gravity, thumb movements are represented more laterally than index movements, and index movements more laterally than little finger movements.

The degree of separation is not sufficient to make isolated or even predominant finger paresis after focal motor cortex lesions a frequent clinical observation. Quite to the contrary, such clinical findings have been extremely rare.[66] This is different from the "dropping hand" phenomenon that can occasionally be observed after M1 lesions and that to some extent mimics peripheral radial nerve palsy. The observation of such cases is in accordance with our general reasoning here because the radial nerve innervates muscles that implement extension across several joints of the arm (elbow, wrist, phalanges), and the synergy of these movements in natural contexts may result in a higher degree of interaction of the underlying neural units. Yet, this association is orthogonal to the general notion of somatotopy because it spans proximal to distal upper limb movements.

So what can within-hand somatotopy mean from a functional perspective? It must be emphasized that the observation of within-hand somatotopy does not depend on the body parts containing the muscles involved. Relatively isolated finger movements can be performed using either forearm or small hand muscles. If somatotopy were to code for whether the muscles are more proximal or more distal one would not expect this finding. Conversely, the detection of somatotopy is related to which part of the body will manifest the effects of coordinated muscular activity. The question then is whether this aspect influences the probability distribution of interactions. We believe that it does, not in the sense of pure motor interactions (for the reasons outlined above), but in the sense of sensorimotor interactions.

The functional significance of distal upper extremity movements lies in establishing our proficiency in finely graded manipulations. These manipulations rely heavily on feedback from cutaneous and proprioceptive sensory afferents, and accordingly there are dense fiber connections from the primary somatosensory to the primary motor cortex.[67] The somatotopical layout of the primary somatosensory cortex is clear-cut, and the reasons why this is associated with a functional benefit have been outlined above. Following the general reasoning of optimizing interaction by minimizing the associative fiber path lengths, one would expect a bias to cluster those neural units in the primary motor cortex that may engender, for instance, thumb movements (and concomitant sensory thumb stimulation) in spatial proximity to those neural units in M1 that will receive and process this sensory stimulation and thus be informative for optimizing the muscular activity pattern. Experimental support for this notion comes from findings in the forelimb representation of the nonhuman primate motor cortex. Using electrodes both for the recording of responses to somatosensory stimulation and for eliciting movements by stimulation, Rosén and Asanuma[68] observed that in roughly half of the units that could be driven by somatosensory stimuli, the receptive fields of these units colocalized to the parts of the hand displaying the motor response when this unit was stimulated.

In the view presented here, the feature of within-hand somatotopy in the primary motor cortex arises merely as a repercussion of functional principles that guide the somatotopical layout of the primary somatosensory cortex. In fact, one concern with those fMRI data that have shown an influence of somatotopy on the activation patterns in the primary motor hand representation has been that this might reflect concomitant tactile activation from the usual finger-tapping type task used in these studies. In that case, the results would be driven by the fraction of neurons in M1 that have somatosensory response properties. However, this could be made unlikely by a study in which finger opposition with and without actual touch were compared and no difference between the two activation patterns was observed in M1.[69] The possibility that remains is that the finding of somatotopy was driven by the proprioceptive input during different isolated finger movements. Yet, this mechanism would not explain the aforementioned findings from intracortical microstimulation experiments or magnetoencephalography,[65] and it thus appears more likely that there is also a somatotopical gradient in the movement-related functional architecture of M1.

We believe that the view presented in the previous paragraph, albeit speculative, accounts for the entirety of the currently available experimental observations. It should be noted that the contribution of interactions with primary somatosensory cortex to the probability distribution of interactions for neural units in the motor hand representation is only one of several factors. It can be seen to compete with other factors related to movement execution which in their own right do not drive the connectivity and thus the functional cortical architecture toward a somatotopical layout. Since somatotopical segregation is not compatible with fundamental features of motor control, the effect of this somatosensory factor can only be a somatotopical gradient superimposed onto a more complex layout.

It has been observed by many researchers that when viewed on a horizontal brain section the somatotopical foci in the somatosensory cortex appear at more lateral positions than the corresponding somatotopical centers of gravity in the motor cortex. However, when taking the angulation of the central sulcus into account, the corresponding foci are observed at exactly those positions that allow them to be connected by the shortest possible fiber paths. It is tempting to speculate how motor development and practice affect the functional organization of the hand representation. Its anatomical configuration into a "hand knob" can be seen as a simple cortical surface expansion if medio-lateral expansion is constrained during evolution, but the neuronal computing demands increase with dexterity. However, it can also be seen as a way to minimize the path lengths of those fibers that provide connections within the hand representation, and thus that allow for a greater degree of functional flexibility. From this perspective, one might expect a decrease in the functional neuroimaging correlate of somatotopy during maturation of motor proficiency, and less of such a gradient in the representation of the dominant hand.[70] This would establish a dissociation of behavior from functional brain architecture because individuated finger movements appear to be performed as well by the dominant as by the nondominant hand.[71] It is worth noting that all the fMRI studies discussed above have been carried out in adults using their dominant hand.

In conclusion, one could say that somatotopy in the motor cortex codes which body part will show the effect of movement, and that demonstrating a somatotopy defined in this way already points to the role of somatosensory somatotopy as its functional source. In this view, one can reestablish the analogy to the visual system for the sensorimotor system as a whole. Starting from an unambiguous situation at the sensory periphery (the body surface) information is relayed along spinal and subcortical relays into a first cortical representation level, from which it infiltrates motor structures and interacts with motor output streams. Due to this interaction with the other (dominant) principles governing the functional organization of the motor areas, the contribution of somatotopy becomes diluted and distorted along this path. This view obviously still represents a gross simplification, but it can also account for the progressively blurred or even functionally redefined observations of somatotopy in motor areas that are upstream from the primary motor cortex or that pertain to other motor circuits than the corticospinal tract and contribute to the planning, initiation, and execution of voluntary movements.[72–76]

2.5 MOTOR RESPONSE PROPERTIES

2.5.1 Rate and Complexity Effects

The findings discussed in the previous section were related to mapping, i.e., to delineating where along the cortex responses can be observed during a given type of movement. At the same time, the issue of somatotopy served to illustrate the limitations of this approach. The set of studies considered next deals with the functional response properties at activated sites. The most obvious approach to response properties lies in studies that determine the relation between fMRI signal and the properties of motor output.

One of the easiest and most potent ways of manipulating movement-related brain activity is by increasing the rate at which a given movement is performed. Two early studies reported mainly linear increases in fMRI responses in contralateral primary motor cortex with higher movement rates. One experiment involved repetitive movements of the index finger at 1, 2, and 3 Hz,[77] the other flexion–extension movements of digits 2 to 5 at rates of 1, 2, 3, 4, or 5 Hz.[78] The observation of a rate-dependent

BOLD response increase was confirmed by Wexler et al.[79] Jäncke and colleagues,[80] who studied dominant index tapping from 0.5 to 5 Hz, added the observation that at low frequencies linearity may be disrupted, presumably because of a different mode of movement execution at frequencies below 1 Hz.

This notion of a qualitative change of movement-related brain activity, despite identical task instruction for a motor production, was pursued in another study by the same group, comparing activation strength during regular (rhythmic) and irregular pacing of finger movements. Although the average frequency of movements executed over time was 1.5 Hz in both conditions, greater activation during irregular pacing was found not only in M1 but also in several other motor structures activated by this task (supplementary motor area [SMA], cerebellum).[81] At the behavioral level, a later movement execution relative to the cue signal was noted for irregular pacing.

This finding underscores that the study of rate effects is only meaningful in experimental settings where aspects of cueing and preparation, as well as motor execution do not change with rate. Although this appears intuitively clear, there is to date not that much positive experimental evidence for the role of such confounds. For instance, while the primary motor cortex displayed significant rate effects, it did not appear to be affected by the complexity of the temporal movement sequence induced by a fixed cueing sequence. Using an identical regular cueing sequence, Mayville et al.[82] asked their subjects to perform either synchronized (on the beat) or syncopated (off the beat) finger movements. No significant effect could be detected in M1, while in the premotor and subcortical areas syncopated as opposed to synchronized movement was associated with greater activity. Another study compared conditions where movements were to be filled into the inter-cue gaps, or where a pause was required after every alternating cue.[83] Again, M1 only showed a rate effect but, in contrast with many other motor areas, was not sensitive to the manipulation of complexity.

Although the cue-response relation, automaticity, and temporal complexity changed in the studies by Mayville et al.[82] and Nakai et al.,[83] the predictability of cueing did not because each of the sequences was perfectly regular. When comparing brain activation during unpredictable vs. predictable cueing of motor reactions, Dassonville et al.[84] found that activity was related to reaction time and thus was greater for the unpredictable condition throughout the premotor, cingulate, supplementary motor area, the presupplementary motor area, and the superior parietal lobule, but M1 was the only structure spared from this effect. Largely similar findings were obtained in other studies.[85,86] No effect in M1 was found when studying the influence of stimulus–response compatibility at the set or the element level or as an interaction. Virtually all of the brain structures participating in stimulus-driven behavior were affected one way or the other, and only activation in M1 was found to be robust against the experimental manipulations.[87]

In addition to the temporal sequence or the association with cues, complexity can also be addressed at the level of coordination. While maintaining formally matched motor output over time, Ehrsson et al.[88] compared a "nonsynergistic" coordination pattern with a natural synergistic pattern (opening and closing the fist) and found greater activity across several nonprimary frontoparietal areas and the cerebellum, but not in M1. Regarding bimanual coordination, there is one report of greater activity not only in the supplementary motor area, but also in the primary sensorimotor cortex during antiphasic as opposed to phasic fist clenching, but this is in contradiction to several studies that instead have mostly emphasized the involvement of medial wall structures without observing effects in M1.[89]

While the influence of many candidate variables on M1 activity remains difficult to extract from the existing neuroimaging data, the rate effect has been robust in the hands of many different laboratories and seems to manifest most clearly in primary as opposed to higher order motor areas. When applying more sophisticated analyses than the usual estimated linear models, the influence of rate shows complex properties, but as a first approximation it can still be regarded as roughly linear.[90] In terms of the anatomical structures, rate effects have also been studied (and in part observed) in mesial wall motor cortical areas;[91] and in terms of the movement type, rate effects in M1 have also been described using articulation instead of finger movements.[92] Overall, it appears that, of the motor areas, the primary cortex displays the most prominent effect of movement rate. It is less clear what this actually means.

For a hemodynamic signal, as in fMRI, one can expect two types of rate effect on the cortical response focus: (1) the amount of signal change in a given cluster of voxels could increase, and (2) the number of voxels that constitute this cluster at a given significance threshold of signal change could increase. Both observations have been made experimentally, and the latter poses some interpretation problems. As stated above, an important but not often explicitly stated assumption in manipulating the rate is that other movement properties remain unaffected. This is certainly not the case, and there is no study available yet that has dissociated the relative contributions of effects related to rate vs. the other affected parameters. While the observation of a greater areal extent of significant fMRI signal change at a higher rate could simply be a spillover from a qualitatively constant single cortical focus, it could equally well point to the recruitment of additional neural populations in adjacent tissue. At higher movement rates, these could reflect the increased necessity of stabilization or proximal coinnervation. To address this point would require simultaneous multiple-muscle electromyography (EMG) recordings, which have not been carried out so far.

As another movement parameter apart from rate, amplitude has been found to correlate positively with the BOLD fMRI response in M1 and, as a more indirect sign, with the significance levels for responses in the SMA proper and in the premotor and postcentral areas, the insula, and the cerebellum.[25] This was studied for two movement amplitudes of index finger extension at 1 Hz, and the authors used offline EMG recordings to rule out a qualitative change in the pattern of muscles recruited as a function of amplitude.

2.5.2 Force Effects

The experimental manipulation of movement amplitude inevitably translates into changes of velocity, acceleration, and force, each parameter that is correlated with neural activity in M1. Force has in fact been the parameter of interest in a number of studies, usually studying responses under different isometric force levels. For right-hand power grip, Cramer et al.[93] reported increases of both activation volume and amplitude (percent signal change) in the left M1, and also, to a less reliable extent, in the right M1 and the supplementary motor area. This expands the findings from previous studies using index flexion[79] and squeezing[94] at different force levels, where the effect of force on M1 activity was mainly reflected in the activation volume and not, or not so much, in the signal change within given voxels. In a similar vein, Dai et al.[95] had also found that increased force of hand squeezing translated into a greater number of voxels showing significant activation and a greater signal change per voxel, not only in the M1 but in several other areas, including the sensory cortex and the cerebellum. That study benefited additionally from electromyographic recordings, which demonstrated that during these force increases forearm muscles showed similar linear increases in surface EMG signal as those observed in fMRI, thus arguing for a close relationship between cerebral synaptic activity and peripheral muscular output. In addition to two agonist muscles, however, the authors also recorded from an antagonist muscle, and also found a linear relation of EMG signal to force level. As discussed above in reference to rate effects, this latter finding underscores the difficulty in interpreting fMRI responses in relation to agonist force level and suggests that they might in fact reflect coinnervation in antagonistic or even proximal muscles.[96] The latter possibility would conceivably affect both the topography of responses in the contralateral motor strip and the degree of response laterality observed at different force levels.

Power grip as studied in most of the previously mentioned studies on force-related effects is distinct from precision grip. Ehrsson et al.[97] illustrated the corresponding neural difference that manifests in the associated cerebral activation patterns. Precision as opposed to power grip involved less activity in the primary motor cortex, but stronger bilateral activations in the ipsilateral ventral premotor areas, the rostral cingulate motor area, and at several locations in the posterior parietal and prefrontal cortices. In subsequent studies, the same group showed that, in contrast to the behavior during the power grip, the activity in the contralateral primary sensorimotor cortex, as well as in the inferior parietal, ventral premotor, supplementary, and cingulate motor areas, increased when the force of the precision grip was lowered such that it became barely sufficient to hold a given object without letting it slip.[98,99] Another interesting recent observation from low force pinch grip is that while the contralateral side shows a weak activation response, the fMRI signal in the ipsilateral M1 decreases. This parallels the ipsilateral reduction of cortical excitability, shown by studies with transcranial magnetic stimulation.[100]

The most obvious disruption in the relation between M1 and muscle activity has been demonstrated by studying relaxation.[101] Monitoring muscle activity by EMG, an activation in contralateral primary and supplementary motor areas could be observed whenever the subjects initiated voluntary muscle relaxation in their arms. In the supplementary motor region, activation during relaxation was even stronger than during movement.

2.6 SENSATION AND ATTENTION

These latter findings illustrate that parameters of motor output are one determinant, but not the only determinant, of activity levels in M1. But what are the other determinants? For example, what can account for greater activity when the actual force executed in a precision grip is being reduced? In this situation, force adjustment parts from automatic regulation, and the sensory feedback from cutaneous and proprioceptive inputs becomes very important. These inputs can, for instance, detect an impending slipping of an object and thus guide an up-regulation in force. Somatosensory responses in the primary motor cortex have been observed for a long time, but have not yet received due attention. A different line of recent work has provided further compelling evidence of somatosensory processing in M1.[102] The illusion of hand movement one can generate by vibrating a wrist tendon on one arm can be transferred to the other hand by skin contact. Naito and colleagues[102] found that this transferred kinesthetic illusion was associated with activation in a network of areas that included M1 contralateral to the nonvibrated hand. From detailed anatomical analyses involving the use of probabilistic cytoarchitectonic maps it appeared that this effect occurred in the posterior portion and not in the primary somatosensory or anterior M1. This was corroborated by transcranial magnetic stimulation measurements that showed a time course of motor cortex excitability changes that paralleled the time course of the perceptual illusion.

Somatosensory processing can contribute to synaptic activity in M1 but does this account for the aforementioned findings of primary motor cortex activation during precision grip? Another possible interpretation builds on attentional modulation of cortical activity. In functional neuroimaging studies, the amount of attentional resources allocated to movements has usually been manipulated by comparing single-task motor settings with distracting, i.e., dual-task settings. The rationale here is that voluntary movement involves activation of those neural representations that are appropriate for an intended result. This is a process of selection that is guided or at least improved by attentional mechanisms that may originate in parietal or frontal areas, the same areas that are also challenged during demanding nonmotor tasks. Indeed, these areas do show profound activity modulations in response to attention to action, as do their connectivity patterns to more executive (premotor) areas downstream.[103]

Two studies have shown that this effect propagates into M1,[104,105] and interestingly, both studies emphasize that these attentional effects were localized to the posterior strip of Brodmann area 4 that is located more deeply in the central sulcus. This raises an interesting question with respect to the routing of attentional modulation. One conceivable hypothesis is that this could be mediated via the gating of somatosensory processes and the related information flux into M1; another (and not necessarily mutually exclusive) hypothesis is that the attentional modulation reflects input from premotor areas. In the latter case, it is unclear why the anterior portion that neighbors premotor Brodmann area 6 should be spared. Obviously, the streams along which attentional effects manifest will also critically depend on the types of movement and distraction chosen for the particular experiment. A final residual confound in these studies is that, on detailed kinematic analysis, the movements actually executed may change when attention is distracted to other goals. Yet, there is good reason to believe that these subtle changes cannot account for the aforementioned neuroimaging results.

Despite these open questions, a rather global view on attentional effects in the motor system thus strongly resembles that in the visual system, where attentional effects can be detected by functional neuroimaging as early as in the lateral geniculate nucleus, but where they become more pronounced the deeper one advances into the hierarchically organized processing levels.[106] In other words, the cortical substrates of sensation and action on the one hand are not segregated from those of cognitive processes on the other hand. Instead, they appear interpenetrated and colocalized into the same cortical hierarchy, but with inverted gradients of their relative contribution to local activity. The gradients reflect the synaptic distance of the areas from the receptor and effector sheets that interface with the external physical world. As opposed to the visual system, however, it remains much more obscure in the motor system whether the attentional modulations observed should be considered the neural correlates of attention-dependent changes in motor behavior.

2.7 IMAGERY

Another parallel to the visual system exists with respect to studies on the influence of imagery on cortical activity. We have so far detailed how properties of motor output can be traced in corresponding activity changes in M1, and we have reviewed evidence showing that factors over and above motor output contribute to and modulate M1 activity. But a question that has tickled many researchers' minds for a long time has been to what extent does the mental simulation or rehearsal of movements share neural activity patterns with actual movement execution. This is clearly the case for many of the upstream areas involved in motor planning and preparation, but the contribution of M1 has remained controversial.

Several of the earlier and in part anecdotal reports on motor imagery studied by fMRI were negative with respect to M1 activation. However, Roth et al.[107] reported M1 activation during mental execution of a finger-to-thumb opposition task in four of six subjects studied, in addition to effects in the premotor and supplementary motor areas. This was also found in a more detailed study the same year by Porro et al.[108] using the same motor task but with a control task of visual imagery. The same group later studied the issue of ipsilateral motor cortex activation during motor imagery.[109] The ipsilateral effect was just significant in a region-of-interest group analysis and, due to the analytical methods applied, it remains unknown in how many of the individuals it was significant. Yet, the analysis provided convincing data showing that the motor imagery effect occurred in those ipisilateral voxels that were also active during motor performance. Furthermore, the observation of greater effects in the left hemisphere (i.e., of ipsilateral activation in the case of left-hand movements) parallels the pattern observed for ipsilateral activation during overt motor performance. This notion receives further support from studies that have found a similar congruence for the activations during executed and imagined movements in relation to the somatotopic representation of hand, foot, and tongue movements.[110,111] Together, these studies suggest that motor imagery yields quantitatively smaller but qualitatively similar patterns of activation in the M1 as motor execution.

However, not all laboratories have obtained this type of result. Across different but probably comparably sensitive approaches there have also been recent studies that reported negative findings regarding M1 activation during motor imagery.[112–114] How can these apparent discrepancies be accounted for? A simple view would conclude that some experiments or laboratories simply did not have enough sensitivity or power, or that they applied too conservative statistical thresholds to detect effects in M1. This appears rather unlikely because the degree of signal change in those studies that did observe activation in M1, and its relation to the signal change during movement execution, should have been readily detectable by other laboratories as well.

The opposite approach would be to assume that the cases with positive findings are accounted for by movement execution during (and despite the instruction of) motor imagery. In contrast with the visual system, which can easily be deprived of input, the situation is far more complicated for motor output. There is continuous output, and imagery easily elicits electromyographic (EMG) activity above this resting level. Voluntary relaxation, which in itself may activate M1 as discussed above, and suppression of such imagery-induced activity is difficult to achieve and usually requires training. The problem of controlling for involuntary movement during imagery has been noted by several groups. One way of ensuring "pure" imagery would be to perform extensive EMG monitoring. EMG has indeed been used in the context of imagery studies. In the study by Porro et al.,[109] EMG recordings showed some activity increases during the imagery condition in roughly half of the subjects. However, these recordings only covered two sites and were obtained off-line. In that sense, it is doubtful whether the lack of correlation observed with the fMRI findings is conclusive. Similarly, Lotze et al.[115] used EMG to train subjects via biofeedback to minimize muscular activity during imagery, but obtained no EMG recordings during the fMRI sessions. It is therefore not clear to what extent the learned pattern may have progressively decayed during those sessions. In that sense, it is probably fair to say that to date no study that has reported M1 activation during motor imagery has provided sufficient support for the claimed absence of muscular activity during that condition.

So far, one of the few studies using on-line EMG recordings during fMRI of motor imagery was that of Hanakawa et al.,[116] and they did not find significant M1 activation. However, they used an interesting analytical approach. Instead of qualitatively mapping activation under different conditions with a somewhat arbitrary threshold, the authors addressed the quantitative relation of activation effects under imagery and execution of movement. They determined areas with movement-predominant activity, imagery-predominant activity, and activity common to both movement and imagery modes of performance (movement-and-imagery activity). The movement-predominant activity included the primary sensory and motor areas, the parietal operculum, and the anterior cerebellum, which had little imagery-related activity (–0.1~0.1%), and the caudal premotor areas and Brodmann area 5, which had mild to moderate imagery-related activity (0.2~0.7%). Many frontoparietal areas and the posterior cerebellum demonstrated movement-and-imagery activity. Imagery-predominant areas included the precentral sulcus at the level of the middle frontal gyrus and the posterior superior parietal cortex/precuneus.

One of us used a different approach for dissociating the effects of motor imagery and actual movements during fMRI measurements.[114] In this study, subjects were presented with drawings of hands and asked to quickly report whether they were seeing a left hand or a right hand, regardless of the angle of rotation of each stimulus from its upright position (rotation). Several psychophysical studies[117–119] have demonstrated that subjects solve this task by imagining their own hand moving from its current position into the stimulus orientation for comparison. This motor imagery task was paired with a task known to evoke visual imagery, in which subjects were presented with typographical characters and asked to quickly report whether they were seeing a canonical letter or its mirror image, regardless of its rotation.[120] Behavioral (reaction times) and neural correlates (BOLD) of motor and visual imagery were quantified on a trial-by-trial basis, while EMG recordings controlled for muscular activity during task performance. Using a fast event-related fMRI protocol, imagery load was parametrically manipulated from trial to trial, while the type of imagery (motor, visual) was blocked across several trials. This experimental design permitted to isolate modulations of neural activity driven by motor imagery, over and above generic imagery- and performance-related effects. In other words, the distribution of neural variance was assessed along multiple dimensions, namely the overall effects of task performance, the specific effects of motor imagery, and the residual trial-by-trial variability in reaction times unaccounted for by the previous factors. With this approach, it was found that portions of posterior parietal and precentral cortex increased their activity as a function of mental rotation *only* during the motor imagery task. Within these regions, parietal cortex was visually responsive, whereas dorsal precentral cortex was not. Crucially, fMRI responses around the knob of the central sulcus, i.e., around the hand representation of primary sensorimotor cortex,[56,121] correlated significantly with the actual motor responses, but neither showed any relationship with stimulus rotation, nor did they distinguish between motor and visual imagery. This result indicates that, at the mesoscopic level of analysis by fMRI, putative primary motor cortex deals with movement execution, rather than motor planning. However, it remains to be seen whether this finding is limited to a precise experimental context, namely implicit motor imagery, or whether it represents a general modus operandi of the human M1.

So should one conclude that those studies that did find activations in M1 during motor imagery were confounded, e.g., by associated motor output? Let us again turn to the visual system for an analogy. In visual cortices, sensory effects are readily detected in early areas and become progressively difficult to follow the deeper one ascends into the cortical hierarchy. Conversely, the participation of primary visual cortex in mental imagery has been far more difficult to demonstrate and does not reach the strength of effects that visual imagery evokes in higher-order areas.[122] So far, this nicely parallels the pattern described in the studies by Hanakawa et al.[116] and Gerardin et al.[112] that were discussed above. Nonetheless, there is now a consensus that the primary visual cortex can participate in imagery, and this may depend on specific aspects of the paradigm employed, such as the requirement of processing capacities that are best represented at this cortical level. If we attempt to transfer this analogy to the sensorimotor system we must analyze in greater depth the paradigms employed across the various motor imagery studies.

Motor imagery can be carried out in a predominantly visual mode (imagining seeing one's own hand moving) or in a kinesthetic mode (imagining the proprioceptive sensations one would experience if one moved the hand in the mentally simulated way). Indeed, it seems to be the case that only studies employing the latter strategy have reported robust effects in M1. This means that the fMRI responses would then be accounted for, not necessarily by the executive neural elements in M1, but by those dealing with proprioceptive input in the context of movement. Psychophysically, it was found that motor imagery affects the illusory perception of movement created by a purely proprioceptive stimulus.[102] However, in a related functional neuroimaging experiment, the authors found no M1 response during imagery and, accordingly, overlap of activations from these two conditions was confined to nonprimary motor areas. It should be noted that this experiment was carried out using PET, and it may therefore have suffered from sensitivity or spatial resolution limitations. At the same time, the authors reproduced their finding of M1 activation from the illusion, and it hence seems unlikely that this should be accounted for by the movement illusion rather than by proprioceptive processing. The issue therefore awaits further investigation.

2.8 PREPARATION, READINESS, AND OTHER TEMPORAL ASPECTS

Early investigations on the properties of motor cortices in awake, behaving macaques were particularly interested in testing the capability of this region to encode action plans.[123,124] Accordingly, several studies used delayed response tasks, under the assumption that bridging a temporal gap between cue and response requires sustained preparation of motor responses driven by internal representations.[125,126] Conceptually, the emphasis here is on the crucial distinction between sensory and motor activities — time-locked to stimuli and responding on a moment-by-moment basis — and preparatory activity, dependent on more persistent neural activity.[127] Empirically, these pioneering studies in nonhuman primates directly manipulated the temporal aspects of delayed response tasks, using delay-related neural responses as an index of the maintenance of motor representations, in the context of controlled stimulus-response associations.[128,129]

With the advent of modern neuroimaging techniques, there has been a surge of studies investigating the role of different motor cortices in preparing a given movement. However, the initial PET investigations characterized neural correlates of response preparation by comparing conditions either involving or not involving motor preparation,[130–133] rather than by following the electrophysiological approach of isolating specific delay-related neural activity.[134–136] Those early neuroimaging studies suggested that central regions (i.e., those impinging on the central sulcus and the precentral gyrus) are involved in preparing movements whose timing and type are fully specified by an external cue. However, these inferences are limited in at least two respects. First, it is possible to isolate preparatory activity by directly comparing trials with and without a preparatory component, other factors being equal. In other words, one needs to assume that movement preparation is a stand-alone cognitive module, indifferent to the selection and execution components of the sensorimotor process. But response selection appears to be significantly influenced by the possibility of preparing a response before a trigger cue.[137,138] We have already discussed how this issue might have confounded a series of studies on rate effects in the motor cortex. In the context of motor preparation, it is possible to overcome this limitation by isolating specific delay-related activity, while accounting for selection and execution components of the sensorimotor process.[139–141] A second point that deserves to be mentioned concerns the nature of the information processes implemented by frontal regions during the transformation of sensory stimuli into motor responses. Although it might be important to define which regions are implicated in movement preparation, neuroimaging studies have usually avoided addressing the crucial question of how a given cerebral region contributes to the preparatory process.

A few notable exceptions to this consideration come from fMRI studies trying to investigate the dynamics of the BOLD signal to gather temporal information from the pattern of hemodynamic responses evoked by a given motor task. The rationale behind this approach is to extract the sequence of neural events occurring during a given motor task in order to map different cerebral regions onto different stages of a given cognitive process. The study of Wildgruber et al.[142] was one of the first to address this issue, in the context of self-generated movements known to engage mesial motor cortical regions earlier than lateral central regions. Their results showed consistent temporal precedence of the onset of the BOLD response in a mesial ROI (putative SMA) as compared to a lateral ROI (putative M1). However, these data do not allow one to infer that the temporal offset is neural in nature. It might equally well be the case that mesial and lateral regions have different neurovascular coupling properties. This potential confound was considered in a follow-up study by Weilke et al.[143] by analyzing responses to two motor tasks, namely self-generated movements and externally triggered movements. The authors found a temporal shift of the BOLD response between the rostral portion of SMA and M1 of 2000 msec during the self-generated movements compared to only 700 msec during the externally triggered movements. In an elegant study by Menon et al.,[144] the authors used intersubject variability in reaction times to dissociate neural from vascular delays in the BOLD responses measured across visual and motor brain regions. By correlating the difference in fMRI response onset of pairs of regions (visual cortex–supplementary motor area; supplementary motor area–primary motor cortex) with the reaction times on a subject-by-subject basis, the authors showed that reaction time differences could be predicted by BOLD delays between SMA and M1, but not between V1 and SMA. In other words, the authors localized the source of visuomotor processing delays to the motor portion of the sensorimotor chain bringing visual information to the motor cortex. However, one could argue that the observations of these reports[143,144] crucially depend on how BOLD delays are measured. In both studies, the authors fitted a linear regression to the initial uprising portion of the BOLD response. The intercept of this regressed line with "zero intensity" was taken as the onset point of the BOLD response. Therefore, this measure of response onset depends crucially on defining a stable baseline. For primary sensory or motor regions it is conceivable to define baseline as the absence of sensory stimuli or motor responses, but this criterion would not be appropriate for higher order cerebral regions. There is a further difficulty with this approach, namely, how to disentangle changes in response magnitude from changes in response latency. The study by Weilke et al.,[143] as well as other studies using "time-resolved" fMRI,[145,146] deals with this issue simply by scaling BOLD responses of different areas or conditions to the unit range, thus avoiding the task of effectively accounting for changes in response latency induced by changes in response magnitude. An elegant alternative approach to this problem was suggested by Henson et al.[147]: explicitly estimating response latency via the ratio of two basis functions used to fit BOLD responses in the General Linear Model; namely, the inverse of the ratio between a "canonical" hemodynamic response function[148] and its partial derivative with respect to time (temporal derivative).

To summarize, the studies on the contribution of M1 to movement preparation reviewed here agree in suggesting that this cortical region is mainly involved in the executive stages of the sensorimotor chain. This role seems to fit into the more general perspective of the organization of the parieto-frontal system, with parietal areas involved in evaluating the potential motor significance of sensory stimuli,[141,149] frontal areas involved in preparing movements as a function of their probability,[150,151] and central regions focused on executing the actual movement.[114,139]

2.9 PLASTICITY AND MOTOR SKILL LEARNING

Although the available neuroimaging data on motor imagery and movement preparation might suggest that the contribution of M1 to sensorimotor tasks is limited to movement performance, one should not neglect that neural responses are dynamic in nature and vary over time. Accordingly, it could be argued that, during overlearned situations, the contribution of M1 to cognitive aspects of sensorimotor tasks is reduced to a minimum. However, this scenario might not be true in the context of learning. For instance, it has been shown that motor cortex contributions to the performance of a given task appear to change dramatically as a function of learning.[152–154] Following 5 weeks of daily practice in the performance of a thumb-finger opposition sequence, Karni et al.[152] reported an increase in the number of task-related voxels along the precentral gyrus and the anterior bank of the central sulcus. However, there were no differences in the actual signal intensity measured during performance of a trained and an untrained sequence. This result is quite puzzling, given that the cortical point spread function of vascular signals related to neural activity has been estimated at around 4 mm,[155] i.e., for small voxels (<4 mm) a change in signal intensity should result in a change in signal extent and vice versa. Furthermore, the findings of Karni et al.[152] appeared to be in conflict with subsequent reports. For instance, De Weerd et al.[156] report a *reduction* in the number of responsive M1 voxels following extensive practice in motor sequence learning, while Muller et al.[157] report changes in premotor but not in motor cortex during proficient performance of a sequence of finger movements. Irrespective of these conflicting findings, it is relevant to emphasize that, in order to ascribe a crucial role to motor cortex, it is essential to disentangle neural responses genuinely associated with learning from other time-dependent phenomena like habituation, fatigue, and motor adjustments during action repetitions.[158] For instance, both Karni et al.[152] and De Weerd et al.[156] used sequences of thumb–finger opposition movements, but it remains unclear whether the motor components of this motorically complex task can be adequately characterized by error rate and performance speed alone. Other studies of motor learning have relied on simpler finger flexions,[154,159–162] but this procedure, per se, is obviously not sufficient to guarantee an appropriate level of control.

For instance, in Toni et al.,[161] motor sequence learning was compared to a passive visual condition, thus preventing a distinction between time-dependent and motor learning-dependent changes in neural responses. This potential confound was explicitly addressed in Toni et al.[162] by comparing brain activity during performance of two visuomotor tasks, one learned before and the other during the scanning session. This approach allowed the authors to assess learning-related effects not confounded by behavioral effects, since the mean reaction times in the two conditions did not change differentially as a function of time, despite a strong time-dependent decrease common to both conditions. There were no specific learning-related changes in motor cortex. This finding was confirmed in other related studies,[159,163] although it should be emphasized that the focus of these papers was on learning visuomotor associations, rather than motor skills. In this latter respect, Ramnani et al.[164] have studied the learning of an extremely well-controlled motor response, namely the eye-blink reflex. The authors reported specific learning-related increases in the BOLD signal in the ventral sector of the precentral gyrus, in the region containing a motor representation of the face.[164] This result is in agreement with previous PET studies concerned with motor skill learning as assessed by the serial reaction time task.[154,165]

In summary, there appear to be contributions of M1 to motor learning, reflecting genuine changes in neuronal processing rather than spurious byproducts of changes in motor output. However, this does not imply that M1 plays a general role in motor learning, as documented by the studies on the acquisition of novel sensorimotor associations.

2.10 GENERAL PROBLEMS OF (AND PERSPECTIVES FOR) FMRI OF MOTOR FUNCTION

In this chapter, we have reviewed some of the findings on M1 function obtained by functional neuroimaging in humans. We have also reviewed some of the basic experimental approaches that functional neuroimaging can take: mapping, measuring stimulus-response functions or context-dependent modulations, and analyzing time-resolved response sequences. Naturally, this chapter is not exhaustive. Resting state fluctuations, pharmacological manipulations, and analyses of functional or effective connectivities were not covered or were barely touched upon, though they offer interesting prospects. Yet, in addition to providing some sort of overview, we hope this chapter can help achieve a better assessment of the strengths and limitations of magnetic resonance as a tool in the neurosciences in general and in the study of motor control in particular. Sensorimotor function is associated with a distributed neural substrate, and the fact that fMRI readily covers the entire brain is helpful in this respect. Highly focalized research techniques are hypothesis-driven, at least in terms of the location they target, and this impairs the potential for new discoveries. In that sense, even if fuzzy, a picture of activity in the entire brain may help to generate and then test novel hypotheses, apart from offering plausibility controls for ongoing studies. Furthermore, the fact that the contrast agent exploited in the BOLD contrast is endogenous and thus permanently present permits a true neurophysiological recording that can go beyond evoked responses.

However, the spatiotemporal response function, i.e., the dispersion of the fMRI response in time and space, poses the most relevant limitation for this type of recording. In other words, fMRI hits hard biological limits, not technical ones, even though technical difficulties are abundant and not yet always fully mastered. Studies in the visual system with fMRI have established that dedicated acquisition and analysis techniques can resolve much smaller functional cortical units than in the currently available motor studies discussed here.[166] This increase in spatial resolution is of interest because, in contrast to earlier neuroimaging techniques, fMRI experiments readily generate highly significant findings in single subjects.

Many of the topics discussed in the previous sections illustrated that one of the major shortcomings of functional neuroimaging studies still lies in the uncertainties of anatomical labeling. Each brain is different, but previous neuroimaging techniques required normalizing the data into a common standard stereotactic space so as to perform averaging of voxel-based signals from roughly homologous brain areas across subjects. These group analyses then had sufficient statistical power and the advantage of ensuring some degree of generality in terms of volume coverage and intersubject variability. Yet, the price paid for this procedure was at the level of anatomical analysis. Even if a spatial normalizing technique incorporates nonlinear algorithms that warp one gyrification pattern rather well into another, the correspondence of actual brain areas becomes blurred by these procedures, and accordingly probabilistic atlases are the closest one can get to reality in this setting. In the previous sections, it has become obvious that such maps can indeed be helpful in tentatively assigning fMRI responses to certain areas, but often enough, even probabilistic statements leave painful uncertainties as to which areas we are obtaining effects from.

But what defines an area as charted in an atlas? The set of neuroanatomical criteria range from cyto- and myeloarchitectonic features to densities and laminar distributions of receptors and other neurochemical markers.[167] In the case of M1, recent detailed analyses have demonstrated a considerable degree of variability both between different brains and within individual brains, i.e., between hemispheres.[168] Moreover, similar methods have rather recently unveiled the fact that, regarding Brodmann area 4, we are actually dealing with two architectonically distinct areas instead of one.[169] If form follows function, we must also assume different response properties of these two areas, and we have discussed some of the evidence from functional neuroimaging that this may indeed be the case. However, these conclusions were based on relating functional findings from one or several brains to a database formed from many other and thus different brains. The *desideratum* at this stage is to map areas in individual subjects, namely those subjects in whom we can also obtain physiological observations, thus avoiding the limitations introduced by intersubject variability. Whether this will be achieved by morphological or functional criteria is not yet clear, but in any case this will need to be done in a noninvasive fashion and will thus require imaging techniques.[170,171] The exquisite sensitivity of magnetic resonance to a whole range of biophysical parameters suggests that it will take center stage in this promising effort.

ACKNOWLEDGMENTS

Andreas Kleinschmidt is funded by the Volkswagen Foundation. We thank Ulf Ziemann for helpful comments on the manuscript.

REFERENCES

1. Dehaene, S. et al., Inferring behavior from functional brain images, *Nat. Neurosci.*, 1, 549, 1998.
2. Miyai, I. et al., Cortical mapping of gait in humans: a near-infrared spectroscopic topography study, *NeuroImage*, 14, 1186, 2001.
3. Mishina, M. et al., Cerebellar activation during ataxic gait in olivopontocerebellar atrophy: a PET study, *Acta Neurol. Scand.*, 100, 369, 1999.
4. Evans, K.C., Shea, S.A., and Saykin, A.J., Functional MRI localisation of central nervous system regions associated with volitional inspiration in humans, *J. Physiol.*, 520 Pt. 2, 383, 1999.
5. McKay, L.C. et al., Neural correlates of voluntary breathing in humans, *J. Appl. Physiol.*, 95, 1170, 2003.
6. Birn, R.M. et al., Event-related fMRI of tasks involving brief motion, *Hum. Brain Mapp.*, 7, 106, 1999.
7. Mosier, K.M. et al., Lateralization of cortical function in swallowing: a functional MR imaging study, *AJNR Am. J. Neuroradiol.*, 20, 1520, 1999.
8. Preibisch, C. et al., Event-related fMRI for the suppression of speech-associated artifacts in stuttering, *NeuroImage*, 19, 1076, 2003.
9. Roland, P.E. and Zilles, K., Functions and structures of the motor cortices in humans, *Curr. Opin. Neurobiol.*, 6, 773, 1996.
10. Kwong, K.K. et al., Dynamic magnetic resonance imaging of human brain activity during primary sensory stimulation, *Proc. Nat. Acad. Sci. U.S.A.*, 89, 5675, 1992.
11. Kleinschmidt, A. et al., Simultaneous recording of cerebral blood oxygenation changes during human brain activation by magnetic resonance imaging and near-infrared spectroscopy, *J. Cereb. Blood Flow Metab.*, 16, 817, 1996.
12. Toronov, V. et al., The roles of changes in deoxyhemoglobin concentration and regional cerebral blood volume in the fMRI BOLD signal, *NeuroImage*, 19, 1521, 2003.
13. Franceschini, M.A. et al., Hemodynamic evoked response of the sensorimotor cortex measured noninvasively with near-infrared optical imaging, *Psychophysiology*, 40, 548, 2003.

14. Menon, R.S. and Goodyear, B.G., Submillimeter functional localization in human striate cortex using BOLD contrast at 4 Tesla: implications for the vascular point-spread function, *Magn. Reson. Med.*, 41, 230, 1999.
15. Barone, P. and Kennedy, H., Non-uniformity of neocortex: areal heterogeneity of NADPH-diaphorase reactive neurons in adult macaque monkeys, *Cereb. Cortex*, 10, 160, 2000.
16. Elston, G.N., Tweedale, R., and Rosa, M.G., Cortical integration in the visual system of the macaque monkey: large-scale morphological differences in the pyramidal neurons in the occipital, parietal and temporal lobes, *Proc. R. Soc. Lond. B Biol. Sci.*, 266, 1367, 1999.
17. Zheng, D., LaMantia, A.S., and Purves, D., Specialized vascularization of the primate visual cortex, *J. Neurosci.*, 11, 2622, 1991.
18. Ciuciu, P. et al., Unsupervised robust nonparametric estimation of the hemodynamic response function for any fMRI experiment, *IEEE Trans. Med. Imaging*, 22, 1235, 2003.
19. Yacoub, E. and Hu, X., Detection of the early decrease in fMRI signal in the motor area, *Magn. Reson. Med.*, 45, 184, 2001.
20. Sheth, S.A. et al., Columnar specificity of microvascular oxygenation and volume responses: implications for functional brain mapping, *J. Neurosci.*, 24, 634, 2004.
21. Kim, S.G. and Duong, T.Q., Mapping cortical columnar structures using fMRI, *Physiol. Behav.*, 77, 641, 2002.
22. Menon, R.S. et al., Ocular dominance in human V1 demonstrated by functional magnetic resonance imaging, *J. Neurophysiol.*, 77, 2780, 1997.
23. Bellgowan, P.S., Saad, Z.S., and Bandettini, P.A., Understanding neural system dynamics through task modulation and measurement of functional MRI amplitude, latency, and width, *Proc. Nat. Acad. Sci. U.S.A.*, 100, 1415, 2003.
24. Logothetis, N.K., The neural basis of the blood–oxygen-level-dependent functional magnetic resonance imaging signal, *Phil. Trans. R. Soc. Lond B Biol. Sci.*, 357, 1003, 2002.
25. Waldvogel, D. et al., The relative metabolic demand of inhibition and excitation, *Nature*, 406, 995, 2000.
26. Kim, S.G. et al., Functional magnetic resonance imaging of motor cortex: hemispheric asymmetry and handedness, *Science*, 261, 615, 1993.
27. Dassonville, P. et al., Functional activation in motor cortex reflects the direction and the degree of handedness, *Proc. Nat. Acad. Sci. U.S.A.*, 94, 14015, 1997.
28. Singh, L.N. et al., Functional MR imaging of cortical activation of the cerebral hemispheres during motor tasks, *AJNR Am. J. Neuroradiol.*, 19, 275, 1998.
29. Kim, S.G. et al., Functional imaging of human motor cortex at high magnetic field, *J. Neurophysiol.*, 69, 297, 1993.
30. Li, A. et al., Ipsilateral hemisphere activation during motor and sensory tasks, *Am. J. Neuroradiol.*, 17, 651, 1996.
31. Singh, L.N. et al., Comparison of ipsilateral activation between right and left handers: a functional MR imaging study, *NeuroReport*, 9, 1861, 1998.
32. Cramer, S.C. et al., Activation of distinct motor cortex regions during ipsilateral and contralateral finger movements, *J. Neurophysiol.*, 81, 383, 1999.
33. Allison, J.D. et al., Functional MRI cerebral activation and deactivation during finger movement, *Neurology*, 54, 135, 2000.
34. Reddy, H. et al., An fMRI study of the lateralization of motor cortex activation in acallosal patients, *NeuroReport*, 11, 2409, 2000.

35. Hsieh, J.C. et al., Loss of interhemispheric inhibition on the ipsilateral primary sensorimotor cortex in patients with brachial plexus injury: fMRI study, *Ann. Neurol.*, 51, 381, 2002.
36. Rao, S.M. et al., Functional magnetic resonance imaging of complex human movements, *Neurology*, 43, 2311, 1993.
37. Boecker, H. et al., Functional cooperativity of human cortical motor areas during self-paced simple finger movements: a high-resolution MRI study, *Brain*, 117 Pt. 6, 1231, 1994.
38. Mattay, V.S. et al., Hemispheric control of motor function: a whole brain echo planar fMRI study, *Psychiatry Res.*, 83, 7, 1998.
39. Chen, R. et al., Involvement of the ipsilateral motor cortex in finger movements of different complexities, *Ann. Neurol.*, 41, 247, 1997.
40. Solodkin, A. et al., Lateralization of motor circuits and handedness during finger movements, *Eur. J. Neurol.*, 8, 425, 2001.
41. Hermsdorfer, J. and Goldenberg, G., Ipsilesional deficits during fast diadochokinetic hand movements following unilateral brain damage, *Neuropsychologia*, 40, 2100, 2002.
42. Nirkko, A.C. et al., Different ipsilateral representations for distal and proximal movements in the sensorimotor cortex: activation and deactivation patterns, *NeuroImage*, 13, 825, 2001.
43. Kobayashi, M. et al., Ipsilateral motor cortex activation on functional magnetic resonance imaging during unilateral hand movements is related to interhemispheric interactions, *NeuroImage*, 20, 2259, 2003.
44. Rao, S.M. et al., Somatotopic mapping of the human primary motor cortex with functional magnetic resonance imaging, *Neurology*, 45, 919, 1995.
45. Sanes, J.N. et al., Shared neural substrates controlling hand movements in human motor cortex, *Science*, 268, 1775, 1995.
46. Foerster, O., Motorische felder and bahnen, in *Handbuch der Neurologie*, 6, 56, 1936.
47. Penfield, W. and Rasmussen, T., *The Cerebral Cortex of Man: A Clinical Study of Localization of Function*, Macmillan, New York, 1950.
48. Kleinschmidt, A., Nitschke, M.F., and Frahm, J., Somatotopy in the human motor cortex hand area: a high-resolution functional MRI study, *Eur. J. Neurosci.*, 9, 2178, 1997.
49. Engel, S.A. et al., fMRI of human visual cortex, *Nature*, 369, 525, 1994.
50. Lotze, M. et al., fMRI evaluation of somatotopic representation in human primary motor cortex, *NeuroImage*, 11, 473, 2000.
51. Dechent, P. and Frahm, J., Functional somatotopy of finger representations in human primary motor cortex, *Hum. Brain Mapp.*, 18, 272, 2003.
52. Beisteiner, R. et al., Finger somatotopy in human motor cortex, *NeuroImage*, 13, 1016, 2001.
53. Hlustik, P. et al., Somatotopy in human primary motor and somatosensory hand representations revisited, *Cereb. Cortex*, 11, 312, 2001.
54. Nudo, R.J. et al., Neurophysiological correlates of hand preference in primary motor cortex of adult squirrel monkeys, *J. Neurosci.*, 12, 2918, 1992.
55. Schieber, M.H. and Hibbard, L.S., How somatotopic is the motor cortex hand area? *Science*, 261, 489, 1993.
56. Maldjian, J.A. et al., The sensory somatotopic map of the human hand demonstrated at 4 Tesla, *NeuroImage*, 10, 55, 1999.

57. Blankenburg, F. et al., Evidence for a rostral-to-caudal somatotopic organization in human primary somatosensory cortex with mirror-reversal in areas 3b and 1, *Cereb. Cortex*, 13, 987, 2003.
58. Kurth, R. et al., fMRI shows multiple somatotopic digit representations in human primary somatosensory cortex, *NeuroReport*, 11, 1487, 2000.
59. Schieber, M.H., Constraints on somatotopic organization in the primary motor cortex, *J. Neurophysiol.*, 86, 2125, 2001.
60. Sporns, O., Tononi, G., and Edelman, G.M., Theoretical neuroanatomy and the connectivity of the cerebral cortex, *Behav. Brain Res.*, 135, 69, 2002.
61. Klyachko, V.A. and Stevens, C.F., Connectivity optimization and the positioning of cortical areas, *Proc. Nat. Acad. Sci. U.S.A.*, 100, 7937, 2003.
62. Brecht, M. et al., Whisker movements evoked by stimulation of single pyramidal cells in rat motor cortex, *Nature*, 427, 704, 2004.
63. Schieber, M.H., Muscular production of individuated finger movements: the roles of extrinsic finger muscles, *J. Neurosci.*, 15, 284, 1995.
64. Lang, C.E. and Schieber, M.H., Differential impairment of individuated finger movements in humans after damage to the motor cortex or the corticospinal tract, *J. Neurophysiol.*, 90, 1160, 2003.
65. Beisteiner, R. et al., Magnetoencephalography indicates finger motor somatotopy, *Eur. J. Neurosci.*, 19, 465, 2004.
66. Schieber, M.H., Somatotopic gradients in the distributed organization of the human primary motor cortex hand area: evidence from small infarcts, *Exp. Brain Res.*, 128, 139, 1999.
67. Stepniewska, I., Preuss, T.M., and Kaas, J.H., Architectonics, somatotopic organization, and ipsilateral cortical connections of the primary motor area (M1) of owl monkeys, *J. Comp. Neurol.*, 330, 238, 1993.
68. Rosén, I. and Asanuma, H., Peripheral afferent inputs to the forelimb area of the monkey motor cortex: input–output relations, *Exp. Brain Res.*, 14, 257, 1972.
69. Jansma, J.M., Ramsey, N.F., and Kahn, R.S., Tactile stimulation during finger opposition does not contribute to 3D fMRI brain activity pattern, *NeuroReport*, 9, 501, 1998.
70. Hammond, G., Correlates of human handedness in primary motor cortex: a review and hypothesis, *Neurosci. Biobehav. Rev.*, 26, 285, 2002.
71. Hager-Ross, C. and Schieber, M.H., Quantifying the independence of human finger movements: comparisons of digits, hands, and movement frequencies, *J. Neurosci.*, 20, 8542, 2000.
72. Nitschke, M.F. et al., Somatotopic motor representation in the human anterior cerebellum: a high-resolution functional MRI study, *Brain*, 119 Pt. 3, 1023, 1996.
73. Fink, G.R. et al., Multiple nonprimary motor areas in the human cortex, *J. Neurophysiol.*, 77, 2164, 1997.
74. Rijntjes, M. et al., A blueprint for movement: functional and anatomical representations in the human motor system, *J. Neurosci.*, 19, 8043, 1999.
75. Scholz, V.H. et al., Laterality, somatotopy and reproducibility of the basal ganglia and motor cortex during motor tasks, *Brain Res.*, 879, 204, 2000.
76. Mayer, A.R. et al., Somatotopic organization of the medial wall of the cerebral hemispheres: a 3 Tesla fMRI study, *NeuroReport*, 12, 3811, 2001.
77. Schlaug, G. et al., Cerebral activation covaries with movement rate, *NeuroReport*, 7, 879, 1996.

78. Rao, S.M. et al., Relationship between finger movement rate and functional magnetic resonance signal change in human primary motor cortex, *J. Cereb. Blood Flow Metab.*, 16, 1250, 1996.
79. Wexler, B.E. et al., An fMRI study of the human cortical motor system response to increasing functional demands, *Magn. Reson. Imaging*, 15, 385, 1997.
80. Jäncke, L. et al., A parametric analysis of the "rate effect" in the sensorimotor cortex: a functional magnetic resonance imaging analysis in human subjects, *Neurosci. Lett.*, 252, 37, 1998.
81. Lutz, K. et al., Tapping movements according to regular and irregular visual timing signals investigated with fMRI, *NeuroReport*, 11, 1301, 2000.
82. Mayville, J.M. et al., Cortical and subcortical networks underlying syncopated and synchronized coordination revealed using fMRI, *Hum. Brain Mapp.*, 17, 214, 2002.
83. Nakai, T. et al., A functional magnetic resonance imaging study of internal modulation of an external visual cue for motor execution, *Brain Res.*, 968, 238, 2003.
84. Dassonville, P. et al., Effects of movement predictability on cortical motor activation, *Neurosci. Res.*, 32, 65, 1998.
85. Jancke, L. et al., Cortical activations during paced finger-tapping applying visual and auditory pacing stimuli, *Brain Res. Cogn. Brain Res.*, 10, 51, 2000.
86. Harrington, D.L. et al., Specialized neural systems underlying representations of sequential movements, *J. Cogn. Neurosci.*, 12, 56, 2000.
87. Dassonville, P. et al., The effect of stimulus-response compatibility on cortical motor activation, *NeuroImage*, 13, 1, 2001.
88. Ehrsson, H.H., Kuhtz-Buschbeck, J.P., and Forssberg, H., Brain regions controlling nonsynergistic versus synergistic movement of the digits: a functional magnetic resonance imaging study, *J. Neurosci.*, 22, 5074, 2002.
89. Toyokura, M. et al., Relation of bimanual coordination to activation in the sensorimotor cortex and supplementary motor area: analysis using functional magnetic resonance imaging, *Brain Res. Bull.*, 48, 211, 1999.
90. Berns, G.S., Song, A.W., and Mao, H., Continuous functional magnetic resonance imaging reveals dynamic nonlinearities of "dose–response" curves for finger opposition, *J. Neurosci.*, 19, RC17, 1999.
91. Deiber, M.P. et al., Mesial motor areas in self-initiated versus externally triggered movements examined with fMRI: effect of movement type and rate, *J. Neurophysiol.*, 81, 3065, 1999.
92. Wildgruber, D., Ackermann, H., and Grodd, W., Differential contributions of motor cortex, basal ganglia, and cerebellum to speech motor control: effects of syllable repetition rate evaluated by fMRI, *NeuroImage*, 13, 101, 2001.
93. Cramer, S.C. et al., Motor cortex activation is related to force of squeezing, *Hum. Brain Mapp.*, 16, 197, 2002.
94. Thickbroom, G.W. et al., Isometric force-related activity in sensorimotor cortex measured with functional MRI, *Exp. Brain Res.*, 121, 59, 1998.
95. Dai, T.H. et al., Relationship between muscle output and functional MRI-measured brain activation, *Exp. Brain Res.*, 140, 290, 2001.
96. Dettmers, C. et al., Relation between cerebral activity and force in the motor areas of the human brain, *J. Neurophysiol.*, 74, 802, 1995.
97. Ehrsson, H.H. et al., Cortical activity in precision- versus power-grip tasks: an fMRI study, *J. Neurophysiol.*, 83, 528, 2000.
98. Ehrsson, H.H., Fagergren, E., and Forssberg, H., Differential fronto-parietal activation depending on force used in a precision grip task: an fMRI study, *J. Neurophysiol.*, 85, 2613, 2001.

99. Kuhtz-Buschbeck, J.P., Ehrsson, H.H., and Forssberg, H., Human brain activity in the control of fine static precision grip forces: an fMRI study, *Eur. J. Neurosci*, 14, 382, 2001.
100. Hamzei, F. et al., Reduction of excitability ("inhibition") in the ipsilateral primary motor cortex is mirrored by fMRI signal decreases, *NeuroImage*, 17, 490, 2002.
101. Toma, K. et al., Activities of the primary and supplementary motor areas increase in preparation and execution of voluntary muscle relaxation: an event-related fMRI study, *J. Neurosci.*, 19, 3527, 1999.
102. Naito, E., Roland, P.E., and Ehrsson, H.H., I feel my hand moving: a new role of the primary motor cortex in somatic perception of limb movement, *Neuron*, 36, 979, 2002.
103. Rowe, J. et al., Attention to action in Parkinson's disease: impaired effective connectivity among frontal cortical regions, *Brain*, 125, 276, 2002.
104. Johansen-Berg, H. and Matthews, P.M., Attention to movement modulates activity in sensori-motor areas, including primary motor cortex, *Exp. Brain Res.*, 142, 13, 2002.
105. Binkofski, F. et al., Neural activity in human primary motor cortex areas 4a and 4p is modulated differentially by attention to action, *J. Neurophysiol.*, 88, 514, 2002.
106. Pessoa, L., Kastner, S., and Ungerleider, L.G., Neuroimaging studies of attention: from modulation of sensory processing to top-down control, *J. Neurosci.*, 23, 3990, 2003.
107. Roth, M. et al., Possible involvement of primary motor cortex in mentally simulated movement: a functional magnetic resonance imaging study, *NeuroReport*, 7, 1280, 1996.
108. Porro, C.A. et al., Primary motor and sensory cortex activation during motor performance and motor imagery: a functional magnetic resonance imaging study, *J. Neurosci.*, 16, 7688, 1996.
109. Porro, C.A. et al., Ipsilateral involvement of primary motor cortex during motor imagery, *Eur. J. Neurosci.*, 12, 3059, 2000.
110. Stippich, C., Ochmann, H., and Sartor, K., Somatotopic mapping of the human primary sensorimotor cortex during motor imagery and motor execution by functional magnetic resonance imaging, *Neurosci. Lett.*, 331, 50, 2002.
111. Ehrsson, H.H., Geyer, S., and Naito, E., Imagery of voluntary movement of fingers, toes, and tongue activates corresponding body-part-specific motor representations, *J. Neurophysiol.*, 90, 3304, 2003.
112. Gerardin, E. et al., Partially overlapping neural networks for real and imagined hand movements, *Cereb. Cortex*, 10, 1093, 2000.
113. Nair, D.G. et al., Cortical and cerebellar activity of the human brain during imagined and executed unimanual and bimanual action sequences: a functional MRI study, *Brain Res. Cogn. Brain Res.*, 15, 250, 2003.
114. de Lange, F.P., Hagoort, P., and Toni, I., Neural topography and content of movement representations, *J. Cogn. Neurosci.*, 2004.
115. Lotze, M. et al., Activation of cortical and cerebellar motor areas during executed and imagined hand movements: an fMRI study, *J. Cogn. Neurosci.*, 11, 491, 1999.
116. Hanakawa, T. et al., Functional properties of brain areas associated with motor execution and imagery, *J. Neurophysiol.*, 89, 989, 2003.
117. Sekiyama, K., Kinesthetic aspects of mental representations in the identification of left and right hands, *Percept. Psychophys.*, 32, 89, 1982.
118. Parsons, L.M., Temporal and kinematic properties of motor behavior reflected in mentally simulated action, *J. Exp. Psychol. Hum. Percept. Perform.*, 20, 709, 1994.
119. Parsons, L.M., Imagined spatial transformation of one's body, *J. Exp. Psychol. Gen.*, 116, 172, 1987.
120. Shepard, R.N. and Cooper, L.A., *Mental Images and Their Transformations*, MIT Press, Cambridge, MA, 1982.

121. Sastre-Janer, F.A. et al., Three-dimensional reconstruction of the human central sulcus reveals a morphological correlate of the hand area, *Cereb. Cortex*, 8, 641, 1998.
122. Kosslyn, S.M. and Thompson, W.L., When is early visual cortex activated during visual mental imagery? *Psychol. Bull.*, 129, 723, 2003.
123. Tanji, J. and Evarts, E.V., Anticipatory activity of motor cortex neurons in relation to direction of an intended movement, *J. Neurophysiol.*, 39, 1062, 1976.
124. Wise, S.P., Weinrich, M., and Mauritz, K.H., Motor aspects of cue-related neuronal activity in premotor cortex of the rhesus monkey, *Brain Res.*, 260, 301, 1983.
125. Fuster, J.M., *The Prefrontal Cortex: Anatomy, Physiology, and Neuropsychology of the Frontal Lobe*, Lippincott-Raven, New York, 1997.
126. Passingham, R.E., *The Frontal Lobes and Voluntary Action*, Oxford University Press, Oxford, 1993.
127. Gold, J.I. and Shadlen, M.N., Representation of a perceptual decision in developing oculomotor commands, *Nature*, 404, 390, 2000.
128. Wise, S.P. and Mauritz, K.H., Set-related neuronal activity in the premotor cortex of rhesus monkeys: effects of changes in motor set, *Proc. R. Soc. Lond. B Biol. Sci.*, 223, 331, 1985.
129. Alexander, G.E. and Crutcher, M.D., Neural representations of the target (goal) of visually guided arm movements in three motor areas of the monkey, *J. Neurophysiol.*, 64, 164, 1990.
130. Krams, M. et al., The preparation, execution and suppression of copied movements in the human brain, *Exp. Brain Res.*, 120, 386, 1998.
131. Deiber, M.P. et al., Cerebral structures participating in motor preparation in humans: a positron emission tomography study, *J. Neurophysiol.*, 75, 233, 1996.
132. Stephan, K.M. et al., Functional anatomy of the mental representation of upper extremity movements in healthy subjects, *J. Neurophysiol.*, 73, 373, 1995.
133. Kawashima, R., Roland, P.E., and O'Sullivan, B.T., Fields in human motor areas involved in preparation for reaching, actual reaching, and visuomotor learning: a positron emission tomography study, *J. Neurosci.*, 14, 3462, 1994.
134. Kornhuber, H.H. and Deecke, L., Hirnpotentialänderungen bei Willkürbewegungen und passiven Bewegungen des Menschen: Bereitschschaftspotential und reafferente Potentiale, *Pflügers Arch.*, 284, 1, 1965.
135. Godschalk, M. et al., The involvement of monkey premotor cortex neurones in preparation of visually cued arm movements, *Behav. Brain Res.*, 18, 143, 1985.
136. Vaughan, H.G., Jr., Costa, L.D., and Ritter, W., Topography of the human motor potential, *Electroencephalogr. Clin. Neurophysiol.*, 25, 1, 1968.
137. Riehle, A. and Requin, J., Monkey primary motor and premotor cortex: single-cell activity related to prior information about direction and extent of an intended movement, *J. Neurophysiol.*, 61, 534, 1989.
138. Crammond, D.J. and Kalaska, J.F., Prior information in motor and premotor cortex: activity during the delay period and effect on pre-movement activity, *J. Neurophysiol.*, 84, 986, 2000.
139. Toni, I. et al., Signal-, set- and movement-related activity in the human brain: an event-related fMRI study, *Cereb. Cortex*, 9, 35, 1999. (Published erratum appears in *Cereb. Cortex*, 9, 196, 1999.)
140. Toni, I. et al., Multiple movement representations in the human brain: an event-related fMRI study, *J. Cogn. Neurosci.*, 14, 769, 2002.
141. Thoenissen, D., Zilles, K., and Toni, I., Differential involvement of parietal and precentral regions in movement preparation and motor intention., *J. Neurosci.*, 22, 9024, 2002.

142. Wildgruber, D. et al., Sequential activation of supplementary motor area and primary motor cortex during self-paced finger movement in human evaluated by functional MRI, *Neurosci. Lett.*, 227, 161, 1997.
143. Weilke, F. et al., Time-resolved fMRI of activation patterns in M1 and SMA during complex voluntary movement, *J. Neurophysiol.*, 85, 1858, 2001.
144. Menon, R.S., Luknowsky, D.C., and Gati, J.S., Mental chronometry using latency-resolved functional MRI, *Proc. Nat. Acad. Sci. U.S.A.*, 95, 10902, 1998.
145. Formisano, E. et al., Tracking the mind's image in the brain. I. Time-resolved fMRI during visuospatial mental imagery, *Neuron*, 35, 185, 2002.
146. Richter, W. et al., Sequential activity in human motor areas during a delayed cued finger movement task studied by time-resolved fMRI, *NeuroReport*, 8, 1257, 1997.
147. Henson, R.N. et al., Detecting latency differences in event-related BOLD responses: application to words versus nonwords and initial versus repeated face presentations, *NeuroImage*, 15, 83, 2002.
148. Friston, K.J. et al., Nonlinear event-related responses in fMRI, *Magn. Reson. Med.*, 39, 41, 1998.
149. Kalaska, J.F. and Crammond, D.J., Deciding not to GO: neuronal correlates of response selection in a GO/NOGO task in primate premotor and parietal cortex, *Cereb. Cortex*, 5, 410, 1995.
150. Toni, I., Thoenissen, D., and Zilles, K., Movement preparation and motor intention, *NeuroImage*, 14, S110, 2001.
151. Riehle, A. et al., Spike synchronization and rate modulation differentially involved in motor cortical function [see comments], *Science*, 278, 1950, 1997.
152. Karni, A. et al., Functional MRI evidence for adult motor cortex plasticity during motor skill learning, *Nature*, 377, 155, 1995.
153. Shadmehr, R. and Holcomb, H.H., Neural correlates of motor memory consolidation, *Science*, 277, 821, 1997.
154. Grafton, S.T., Hazeltine, E., and Ivry, R., Functional mapping of sequence learning in normal humans, *J. Cogn. Neurosci.*, 7, 497, 1995.
155. Das, A. and Gilbert, C.D., Long-range horizontal connections and their role in cortical reorganization revealed by optical recording of cat primary visual cortex, *Nature*, 375, 780, 1995.
156. De Weerd, P. et al., Cortical mechanisms for acquisition and performance of bimanual motor sequences, *NeuroImage*, 19, 1405, 2003.
157. Muller, R.A. et al., Functional MRI of motor sequence acquisition: effects of learning stage and performance, *Brain Res. Cogn. Brain Res.*, 14, 277, 2002.
158. Sanes, J.N. and Donoghue, J.P., Plasticity and primary motor cortex, *Annu. Rev. Neurosci.*, 23, 393, 2000.
159. Eliassen, J.C., Souza, T., and Sanes, J.N., Experience-dependent activation patterns in human brain during visual-motor associative learning, *J. Neurosci.*, 23, 10540, 2003.
160. Eliassen, J.C., Souza, T., and Sanes, J.N., Human brain activation accompanying explicitly directed movement sequence learning, *Exp. Brain Res.*, 141, 269, 2001.
161. Toni, I. et al., The time course of changes during motor sequence learning: a whole-brain fMRI study, *NeuroImage*, 8, 50, 1998.
162. Toni, I. et al., Learning arbitrary visuomotor associations: temporal dynamic of brain activity, *NeuroImage*, 14, 1048, 2001.
163. Toni, I. et al., Changes of cortico-striatal effective connectivity during visuomotor learning, *Cereb. Cortex*, 12, 1040, 2002.
164. Ramnani, N. et al., Learning- and expectation-related changes in the human brain during motor learning, *J. Neurophysiol.*, 84, 3026, 2000.

165. Hazeltine, E., Grafton, S.T., and Ivry, R., Attention and stimulus characteristics determine the locus of motor-sequence encoding: a PET study, *Brain*, 120 Pt. 1, 123, 1997.
166. Kim, D.S., Duong, T.Q., and Kim, S.G., High-resolution mapping of iso-orientation columns by fMRI, *Nat. Neurosci*, 3, 164, 2000.
167. Schleicher, A. et al., A stereological approach to human cortical architecture: identification and delineation of cortical areas, *J. Chem. Neuroanat.*, 20, 31, 2000.
168. Rademacher, J. et al., Variability and asymmetry in the human precentral motor system: a cytoarchitectonic and myeloarchitectonic brain mapping study, *Brain*, 124, 2232, 2001.
169. Geyer, S. et al., Two different areas within the primary motor cortex of man, *Nature*, 382, 805, 1996.
170. Kruggel, F. et al., Analyzing the neocortical fine-structure, *Med. Image Anal.*, 7, 251, 2003.
171. Walters, N.B. et al., *In vivo* identification of human cortical areas using high-resolution MRI: an approach to cerebral structure-function correlation, *Proc. Nat. Acad. Sci. U.S.A.*, 100, 2981, 2003.

Section II

Neuronal Representations in the Motor Cortex

3 Motor Cortex Control of a Complex Peripheral Apparatus: The Neuromuscular Evolution of Individuated Finger Movements

Marc H. Schieber, Karen T. Reilly, and Catherine E. Lang

CONTENTS

ABSTRACT

Rather than acting as a somatotopic array of upper motor neurons, each controlling a single muscle that moves a single finger, neurons in the primary motor cortex (M1) act as a spatially distributed network of very diverse elements, many of which have

0-8493-1287-6/05/$0.00+$1.50

outputs that diverge to facilitate multiple muscles acting on different fingers. Moreover, some finger muscles, because of tendon interconnections and incompletely subdivided muscle bellies, exert tension simultaneously on multiple digits. Consequently, each digit does not move independently of the others, and additional muscle contractions must be used to stabilize against unintended motion. This biological control of a complex peripheral apparatus initially may appear unnecessarily complicated compared to the independent control of digits in a robotic hand, but can be understood as the result of concurrent evolution of the peripheral neuromuscular apparatus and its descending control from the motor cortex.

3.1 INTRODUCTION

Unlike a robotic hand that has been designed by human engineers, the primate hand has evolved from the pectoral fin of a primordial ancestor. What began as interconnected bony rays supporting a fin evolved into a hand with digits capable of relatively independent motion. During this evolution, the pressures of natural selection concurrently influenced both the peripheral musculoskeletal apparatus and the central mechanisms for its neural control. The resulting biological hand, which has reached its most sophisticated form in primates, especially humans, nevertheless retains many structural and functional features of the ancestral appendage. To understand how the motor cortex participates in controlling finger movements, we must appreciate certain aspects of how the peripheral apparatus of a biological hand works. Here, we will consider first the motion of the fingers themselves, then the functional organization of the muscles that move the fingers, and then how M1 controls finger movements. Because M1 plays a particularly crucial role in controlling fine, individuated finger movements, we will focus on features that affect the independence of finger movements.

3.2 THE LIMITED INDEPENDENCE OF FINGER MOVEMENTS

Modern amphibians and reptiles have forelimbs with distinct digits, but do not use these digits to grasp objects. Further along the phylogenetic scale, mammals such as rats and cats can be observed to mold the digits of the forepaw to grasp objects.[1–3] Although nonhuman primates, and especially humans, are clearly capable of more sophisticated finger movements, the vast majority of what nonhuman primates and humans do with their fingers consists simply of grasping objects. In grasping, all the digits are in motion simultaneously. Independently controlling the 15 different joints of the 5 digits presents a formidable problem for the nervous system, but analysis suggests that most control of the fingers in grasping could be simplified. Only 2 principle components — mathematical functions describing simultaneous motion of the 15 joints in fixed proportion to one another — account for most of the motion of the 15 joints.[4,5] The first principle component corresponds roughly to the simultaneous motion of all the joints in the opening and closing of all the digits. The second principle component corresponds roughly to the degree of flexion of the fingertips toward the palm or extension of the fingertips away from the palm. Together, these two principle components account for 84% of the variation in finger joint positions used by humans in grasping a wide variety of common objects. Most of the finger movements used in grasping thus could be controlled by scaling just 2 principle components, a process much simpler than independently controlling 15 joints. Whether the nervous system actually employs such a simplifying scheme to control grasping, and if so, where in the nervous system the scheme is implemented, remains unknown as yet.

Beyond grasping, the fine finger movements used in manipulating small objects, typing, or playing musical instruments are performed much less frequently. Although the fingers commonly are assumed to be moving independently during such tasks, recordings show again that these sophisticated performances entail simultaneous motion of multiple digits.[6,7] Even when specifically asked to move just one finger, both nonhuman primates and human subjects show some degree of simultaneous motion in other, noninstructed digits, whether moving the fingers isotonically, or applying forces isometrically.[8–11]

The crucial role of M1 in controlling fine, individuated movements of the fingers is evident from the common observation that such movements are the first affected and the last to recover when lesions affect M1 or its output via the corticospinal tract.[12,13] Lesions of the motor cortex, besides rendering movements weak and slow, reduce the ability to move a given body part without concurrent motion of adjacent body parts, as illustrated for the fingers in Figure 3.1.[14] From this perspective, the fingers can be hypothesized to have a fundamental level of control that produces general opening and closing of the hand for grasping.[15] This fundamental control might be accomplished by rudimentary neuromuscular structures in the periphery and driven reliably by subcortical centers in the nervous system. As evolution progressed, a capability for more sophisticated control of the fingers may have developed on top of this fundamental level. This more sophisticated control required both subdivision of the peripheral neuromuscular apparatus and evolution of a computationally more complex layer of control, in which M1 plays a major role.

3.2.1 Biomechanical Factors

The fingers of a robotic hand are mechanically independent, but the fingers of a biological hand are coupled to a measurable degree by a number of biomechanical factors. Some degree of mechanical coupling between adjacent digits is produced by the soft tissues in the web spaces between the fingers. Cutting this tissue in cadaver hands reduced the extent to which adjacent digits moved along with a passively moved digit.[16] Additional coupling is produced by interconnections between the tendons of certain muscles. In humans, the *juncturae tendinium* between the different finger tendons of *extensor digitorum communis* (EDC) are well known.[17] The tendons of *flexor digitorum profundus* (FDP) to the four different fingers also are interconnected in the palm, both by thin sheets of inelastic connective tissue and by the origins of the lumbrical muscles.[18] In macaque monkeys, these interconnections between the tendons of multitendoned muscles are more pronounced than in humans.[19] In the macaque FDP, tendon interconnections have been shown to cause

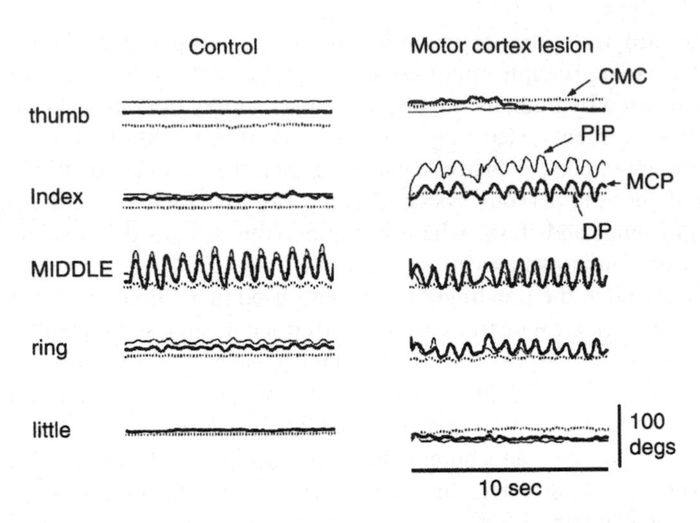

FIGURE 3.1 Loss of individuation after a motor cortex lesion. In these joint position traces, a control subject (left column) and a subject with a motor cortex lesion (right column) were instructed to move the middle finger back and forth while keeping the other fingers still. Joint position traces from the thumb are on top, followed by the index, middle, ring, and little (bottom) fingers. The thick lines show metacarpophalangeal (MCP) joint movement, the thin lines show proximal interphalangeal (PIP) joint movement, and the dotted lines show distal interphalangeal (DIP) joint movement, except for the thumb, where the dotted line shows carpometacarpal (CMC) joint movement. Joint position traces for the middle finger show that both subjects moved the middle finger as instructed. The control subject on the left made highly individuated movements of the middle finger with minimal changes in joint position of the noninstructed fingers. In contrast, the subject with a motor cortical lesion (in the contralateral precentral gyrus hand knob, extending into the white matter beneath) produced substantial changes in joint position of the index and ring fingers simultaneously with the middle finger movement.

tension exerted at one point on the proximal aponeurosis of the insertion tendon to be distributed to the distal insertions on multiple digits.[20]

Because of this biomechanical coupling of the digits, muscle activity intended to move one digit will tend to move adjacent digits as well. To move one digit more individually then, additional muscles may be activated to check the coupled motion of the adjacent digits. Such stabilizing contractions have been observed in the electromyographic (EMG) activity of finger muscles in both monkeys and humans. As a monkey flexes its little finger, for example, *extensor digiti secundi et tertii* (ED23) contracts to minimize simultaneous flexion of the index and middle fingers.[21] In humans, the portion of FDP that acts chiefly on the middle finger contracts as the subject extends either the index or the ring finger, apparently to minimize coupled extension of the middle finger (Figure 3.2).[22]

Additional requirements for stabilizing contractions result from the fact that the extrinsic finger muscles act across the wrist joint as well. When FDP and/or the *flexor digitorum superficialis* (FDS) contract, for example, they exert torque not only about the interphalangeal and metacarpophalangeal joints of the fingers, but also

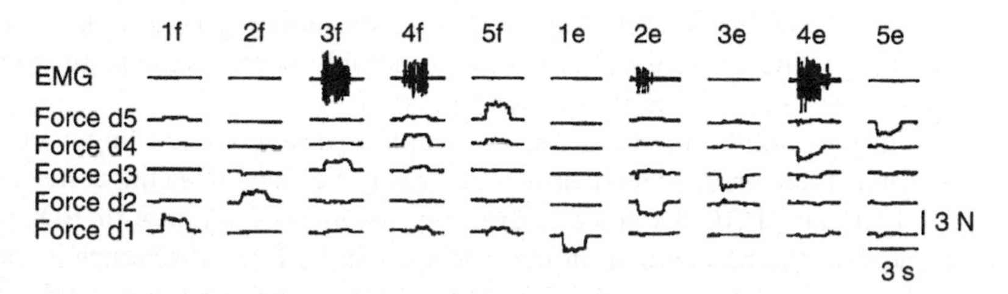

FIGURE 3.2 EMG activity in the human FDP during individuated finger movements. EMG activity from a bipolar fine-wire electrode within FDP was recorded simultaneously with the force exerted at each fingertip during individuated flexion or extension movements of each digit. Instructed movements are indicated at the top of each column by a number indicating the instructed digit (1 = thumb through 5 = little finger), and a letter indicating the instructed direction (f = flexion, e = extension). Each column shows data recorded on a single trial. Looking down each column shows that the most force was produced by the instructed digit in the instructed direction, with only small amounts of force (if any) produced in the other noninstructed digits. Looking across the ten movements shows that the amount of EMG activity recorded at the electrode varied with instructed movement. The largest EMG activity was recorded during instructed flexion of the middle finger (3f), while a smaller but still substantial amount of EMG activity was also recorded during flexion of the adjacent ring finger (4f). This region of FDP also was active during extension of the ring finger (4e), and to a lesser extent during extension of the index finger (2e). This pattern of EMG activity indicates that, in addition to acting as an agonist during middle finger flexion (3f), this region of FDP was coactivated during ring finger flexion (4f), and stabilized the middle finger against unintended extension during the instructed extension of the index (2e) or ring finger (4e).

about the wrist. This torque would flex the wrist along with the finger(s) if it were not counterbalanced by an extensor torque at the wrist. Wrist extensor muscles (e.g., *extensor carpi ulnaris* [ECU] and *extensor carpi radialis brevis* [ECRB]) indeed become active during many finger flexion movements in both monkeys and humans.[21,23] By the same token, extrinsic finger muscles can be used to produce counterbalancing torques about the wrist. In humans performing a brisk extension of the little finger, a simultaneous rise in tension in the tendon of *abductor pollicis longus* (APL) often can be palpated; presumably the contraction of APL counterbalances the wrist torque produced by the extrinsic finger muscles used to extend the little finger, EDC and *extensor digiti quinti* (EDQ).

Comparing the extrinsic finger musculature of macaque monkeys to that of humans suggests that human muscles have evolved to provide a greater degree of independence in finger movements. As noted above, when instructed to move one finger alone, lesser motion of other, noninstructed digits occurs along with that of the instructed digit in both species. Quantitatively, however, humans move their fingers more individually than macaques.[8] Part of this greater ability to individuate finger movements may result from the fact that humans have lost tendons to certain digits from multitendoned muscles. The human *extensor indicis proprius* (EIP), which extends only the index finger, is homologous to the macaque ED23, which extends both the index and middle fingers. The human EDQ, which extends only the little finger, is homologous to the macaque *extensor digiti quarti et quinti* (ED45), which extends both the ring and little fingers. (In cats, the homologous muscle extends digits 3, 4, and 5.) These species differences presumably permit humans to extend the index and little fingers more independently.

Another factor contributing to decreased coupling among human fingers is the greater mechanical separation of tendons in the extrinsic multitendoned finger muscles, particularly EDC and FDP. As noted above, the *juncturae tendinium* of EDC are less pronounced in the human than in the macaque. In FDP the difference is more dramatic. The insertion tendons of the macaque FDP all arise from a continuous aponeurotic sheet which is minimally divided as the common tendon crosses the wrist.[19] Only within the palm does the FDP tendon give rise to separate cords to each digit, including the thumb. In humans, the equivalent tendon to the thumb arises from an independent muscle, *flexor pollicis longus* (FPL), and the tendons to the other digits are separate before they cross the wrist, though tough fibrous sheets still interconnect adjacent FDP tendons within the palm. The human FDP tendon to the index finger arises from a largely separate portion of the muscle belly.[24] The greater mechanical separation of the EDC and FDP tendons contributes to the greater ability of humans to move their fingers independently.

3.2.2 Neuromuscular Compartmentalization

Mechanical separation of the FDP and EDC tendons would not result in greater independence of the fingers if all the motoneurons of these muscles still acted as a single pool. Many mammalian muscles, however, have been shown to consist of multiple neuromuscular compartments.[25,26] Each compartment consists of a distinct region of the muscle belly innervated by a primary branch of the muscle nerve. The neuromuscular compartments of a muscle can be activated differentially by the central nervous system, producing different biomechanical effects. Similarly, the multitendoned finger muscles, rather than acting as a single motoneuron pool that pulls simultaneously on all four fingers, may be subdivided to different degrees. Each subdivision then may be activated selectively by the nervous system. The extent to which multitendoned finger muscles are partially subdivided, or even fully compartmentalized, is an area of active investigation.

Of the macaque multitendoned finger muscles, FDP most clearly shows compartmentalization.[20] Four distinct neuromuscular compartments of the macaque FDP each receive their own primary nerve branch, stimulation of which produces a different distribution of tension across the five tendons. Voluntary activation during finger movements has been studied in the large radial and ulnar compartments (FDPr and FDPu) but not in the two smaller compartments. FDPr is activated during flexion of the index or middle finger but not during flexion of the ring or little finger, whereas FDPu is activated during flexion of the little or ring finger, but not during flexion of the index finger.[27] Largely because of the interconnected tendon structure described above, however, none of the four compartments exerts tension on just one digit.

Interconnected tendons are not the only factor potentially limiting neuromuscular compartmentalization. In the macaque ED45, which has quite independent tendons to the ring and little fingers, many single motor units exert tension selectively on one of the two tendons, but other single motor units exert substantial tension on both tendons.[28] When these motor units are recruited, they act on both digits simultaneously. The extent to which multitendoned motor units might be found in human muscles is under investigation. Available results indicate that most motor units in the human EDC are highly selective,[29] though some in FDP exert most of their tension on the little finger while exerting lesser tension on the ring finger.[30]

3.2.3 Central Coupling

Coupling between the fingers may not result solely from peripheral factors. Additional functional coupling may occur within the central nervous system. Many inputs to motoneurons, even the highly selective corticospinal inputs, may branch to innervate multiple motoneuron pools. Indeed, axons from single M1 neurons have been shown to branch within the cervical enlargement of the spinal cord to innervate multiple motoneuron pools,[31] and the physiological effects of such branching have been shown with spike-triggered averaging to be present in the EMG activity of multiple finger muscles.[32,33] This branching provides the motoneuron pools of different muscles, and presumably the motoneuron pools of different compartments within a multitendoned muscle, with common inputs. When these common inputs are active, concurrent activation of motoneurons of different muscles (or compartments) will be facilitated, which can be detected as short-term synchronization between motor units in different muscles.[34] As another consequence, FDP motor units recruited for flexion of one fingertip may also be recruited when slightly more flexion force is exerted at another fingertip.[35] Similarly, a region of FDP where the most EMG activity is recorded during flexion of a given finger may also be active at a lower level during flexion of adjacent fingers, as illustrated in Figure 3.2.[22]

3.2.4 Neuromuscular Evolution

How could such a complex situation have arisen, when it would have been so much simpler to have independent muscles acting on each digit? To address this question, we return to the idea that the muscles moving the fingers have evolved from simpler muscles that acted to move a pectoral fin as a whole, as illustrated in Color Figure 3.3.* Such a primordial muscle presumably had a single tendon that inserted broadly on multiple digits (Color Figure 3.3A). Motoneurons innervated muscle fibers spread through the muscle belly, which contracted as a single functional compartment. Neurons providing descending inputs synapsed extensively in the motoneuron pool, facilitating rapid, reliable contraction of the muscle.

Over time, natural selection caused the evolution of some ability to move rays of phalanges differentially (Color Figure 3.3B). The broad tendon thinned in places, permitting some degree of differential motion of its insertions on different digits. This change in the tendon could have no functional importance without changes in the muscle, however. Some motor units no longer included muscle fibers distributed so widely in the belly; their twitches therefore exerted more tension on one or the other aspect of the differentially moving tendon. Similarly, the changes in the motor

* See color insert following page 170.

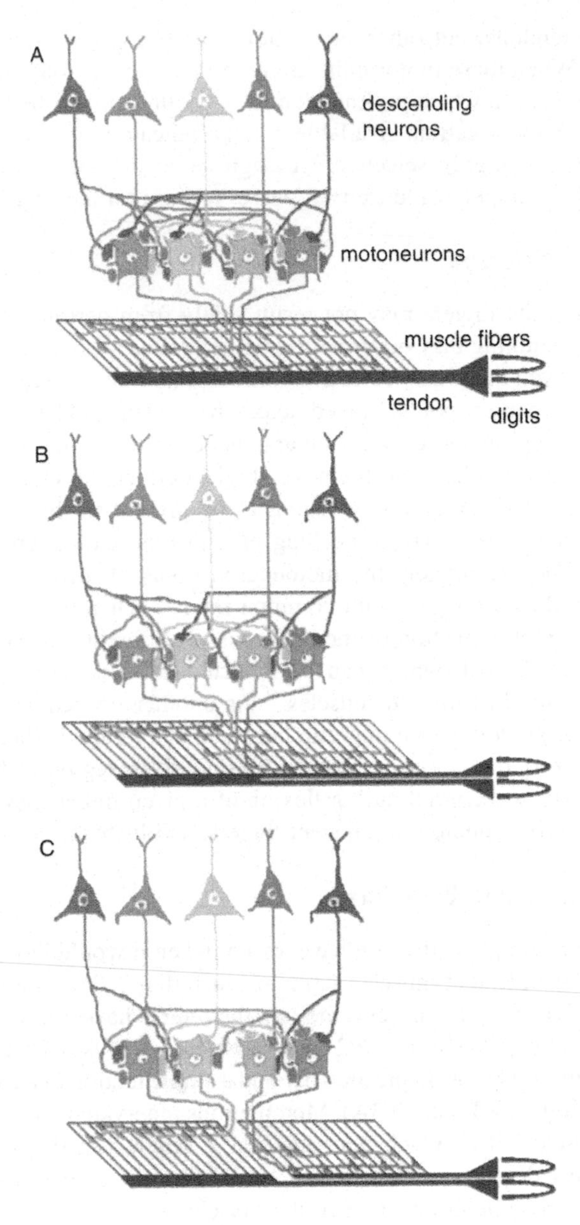

FIGURE 3.3 (see color figure) Neuromuscular evolution. A speculative scheme is illustrated through which a parent muscle (A) could become partially subdivided (B) and eventually divide into two daughter muscles (C), while still retaining some of the distributed descending neural control of the parent muscle. (A) Parent muscle. A single tendon inserts broadly on multiple digits. For simplicity only two digits are illustrated here. The four motoneurons each innervate muscle fibers distributed widely in the muscle belly, so that the four motor unit territories overlap. The motoneurons in turn are innervated by five descending neurons that each synapse widely within the motoneuron pool. Here every descending neuron innervates every motoneuron. (B) Partially subdivided muscle. The tendon has become partially divided to act differentially

units could have no effect without changes in the descending neurons. Some descending neurons no longer synapsed homogeneously throughout the motoneuron pool. Descending neurons that synapsed on populations of motoneurons exerting more tension on one aspect of the tendon thereby facilitated selective recruitment of those motoneurons, and consequently facilitated differential motion of one digit more than others. This ability to move the digits differentially to some degree conferred an evolutionary advantage. At this stage the neuromuscular system could be described as functionally subdivided, though perhaps not fully compartmentalized.

Further evolution in some cases led to complete neuromuscular compartmentalization, and then division into separate muscles (Color Figure 3.3C). As the tendon divided further, motoneurons came to innervate muscle fibers in only one subdivision or another, enabling even more differential motion of the digits. With each subdivision innervated by its own subset of motoneurons, the muscle was fully compartmentalized. Some descending input neurons came to innervate only the motoneurons of a particular compartment, allowing selective recruitment of one compartment or another. If the tendon then divided completely as well, two separate daughter muscles were formed. Nevertheless, some descending input neurons continue to innervate motoneurons of both daughter muscles, though in varying proportion. Activity of these descending neurons facilitates the activity of motoneurons in both daughter muscles when both are needed concurrently. When selective activation of only one daughter muscle is needed, which occurs less often, the more selective descending neurons can facilitate the motoneurons of that muscle, though some incompletely selective descending neurons may facilitate lesser activation of motor units in the other daughter muscle.

Of course, this evolutionary scheme is oversimplified, and many variations are possible. In some cases, complete subdivision of a tendon may precede complete compartmentalization of the muscle. The macaque ED45, for example, lies somewhere in between Color Figures 3.3B and 3.3C. Its tendons to digits 4 and 5 are

---

FIGURE 3.3 (continued) on the two digits. The motor unit territories also have become partially selective: the red and orange motoneurons innervate muscle fibers to the left, and the green and blue motoneurons innervate muscle fibers to the right, with a central region of overlap. The red and orange motoneurons thus act more strongly on one digit and the green and blue motoneurons act more strongly on the other digit. The descending inputs also have become more selective: the red and orange descending neurons no longer innervate the blue motoneuron, and the green and blue descending neurons no longer innervate the red motoneuron; hence these descending neurons can act somewhat differentially on the digits. The yellow descending neuron, however, still facilitates all four motoneurons. (C) Daughter muscles. The tendon now has divided completely in two, as has the muscle belly. The red and orange motoneurons exclusively innervate the left muscle; the green and blue motoneurons exclusively innervate the right. The descending neurons also have become more, though not completely, selective: the red descending neuron now innervates only the red and orange motoneurons, and the blue descending neuron now innervates only the green and blue motoneurons. These two descending neurons therefore selectively facilitate only the left or right daughter muscle, respectively. The orange descending neuron facilitates the left muscle more than the right; the green descending neuron facilitates the right more than the left; and the yellow descending neuron still facilitates the left and right equally.

completely separate, but some of its motor units still exert tension on both digits. Additional variations are seen in the human deep flexors. FPL has separated as a daughter muscle from the ancestral FDP, the index finger portion of FDP has its own tendon, and the belly is partially, though incompletely, separate from the middle finger portion, while the ring and little finger portions still retain a partially interconnected tendon.[24,30] We speculate further that the intrinsic hand muscles evolved from an ancestral muscle mass to become distinct daughter muscles, each acting on a different digit, while retaining some shared descending inputs that produce short-term synchronization of motor units in nearby muscles, and coactivation of muscles acting on different fingers. Evolutionary variations have made the muscles acting on a biological hand quite different from the independent motors and cables that operate a robotic hand.

3.3 MOTOR CORTEX

3.3.1 Cortical Organization and Neuromuscular Evolution

From the viewpoint of neuromuscular evolution, many features of motor cortex organization become easier to understand. Output neurons in layer V of M1 have several features of the descending neurons in the evolutionary schema described above. Many single M1 neurons have outputs that diverge to innervate multiple spinal motoneuron pools.[31,36] Spike-triggered averaging has shown that the discharges of single M1 neurons may produce effects in the motoneuron pools of several forearm and intrinsic hand muscles.[32,33] Some M1 neurons that project to wrist and finger muscles also produce effects in elbow or shoulder muscles.[37] These divergent connections from many M1 neurons to multiple muscles may be the remnants of connections to common primordial muscles that subsequently divided. The fact that divergent connections remain today suggests, however, that they are important to the present function of the motor cortex. Their importance may lie in the fact, described above, that the most frequently performed behavioral tasks, such as grasping, require the simultaneous contraction of multiple muscles acting on multiple fingers. These movements may be controlled most efficiently through M1 neurons with divergent connections to multiple muscles.

Because the output of many single M1 neurons diverges to multiple muscles (often muscles that move different digits and/or the wrist), different muscles receive inputs from intersecting sets of M1 neurons. The sets of M1 neurons that provide input to two muscles acting on the digits and wrist also are intermingled in the physical space of the cortex. Consequently, the neurons that provide input to any given muscle are spread over a relatively large cortical territory (typically a few millimeters in diameter in nonhuman primates) and the territory providing input to one muscle overlaps extensively with the territory providing input to other muscles.[38,39] This overlap limits the degree of somatotopic organization in M1.

When finger movements are made, then, active neurons are found over a relatively large M1 territory, and similar territories are activated for different finger movements. Widespread activation of the M1 hand representation during individuated finger movements has been observed in both monkeys and humans. In monkeys, microelectrode recording typically reveals a burst of the background "hash" (which presumably reflects the discharge of action potentials by numerous neurons and axons in the vicinity of the microelectrode tip) with every finger movement, no matter where within the M1 hand region the microelectrode tip is located. Single neurons likewise are observed to discharge in relation to multiple finger and wrist movements.[40] Often, a given neuron discharges in relation to movements of nonadjacent digits. The distribution of neurons active during movements of particular digits gives little if any evidence of somatotopic segregation of neurons controlling different digits. Similarly in humans, functional magnetic resonance imaging (fMRI) shows that a similar cortical territory is activated no matter which digit is moved.[41] In humans, however, subtraction of the widespread activation common to all finger movements leaves a remainder of specific activation for each digit; this remainder shows some degree of somatotopic segregation for movements of different digits.[42,43]

Widespread activation during voluntary movement, overlapping cortical territories projecting to different muscles, and single M1 neurons that project to different muscles — all are consistent with the effects of M1 lesions on movements. M1 lesions do not impair the function of particular muscles in isolation, but rather impair many functionally related muscles at the same time. This general principle applies as well within the M1 hand representation. In monkeys, injection of the gamma amino butyric acid (GABA) agonist, muscimol, at a single location in the M1 hand representation produces partial inactivation, impairing some finger movements but not others.[44,45] Which finger movements are impaired, however, has little if any relationship to the location of the injection along the central sulcus.[46] In humans, small infarcts that selectively involve the M1 hand representation are relatively uncommon (though not rare), but when such small infarcts occur they can impair the fingers differentially. Rather than producing selective impairment of different fingers in different patients, however, such infarcts impair either the radial digits (thumb and index finger) more than the ulnar digits (little, ring, and middle fingers) or vice versa.[13,47] These observations are consistent with the fMRI findings described above in suggesting that while humans have considerable overlap of M1 digit representations, humans also have steeper somatotopic gradients within the hand representation than monkeys.

3.3.2 Control of Individuated Movements without Somatotopy

If we accept the notion that movements of different fingers are not controlled simply through a somatotopic map of the hand in M1, we then must ask how M1 might control individuated finger movements. Conceivably, even if groups of functionally similar neurons were not spatially segregated in a somatotopic fashion in M1, groups of similar neurons still might control particular fingers, muscles, or muscle synergies. Neurons of different distinct functional groups could be intermingled in the physical space of M1. We have used cluster analysis to search populations of M1 neurons for such groups of functionally similar neurons.[48] For this purpose, cluster analysis has the advantage of being an objective method of searching for similar members of a population, without making assumptions as to what the function of any group might be. In three monkeys, however, cluster analysis revealed only two consistent groups of M1 neurons. A relatively large group consisted of neurons that increased discharge during most if not all finger and wrist movements; another small group decreased discharge during most movements. These two groups were found in all three monkeys, were robust against changing the method of quantifying neuronal activity or changing the clustering algorithm, and were not reproduced when the data was randomly reshuffled. In contrast, small groups of neurons that discharged during particular subsets of finger and wrist movements varied from monkey to monkey, changed when the means of quantifying neuronal activity or the clustering algorithm was changed, and appeared in randomly reshuffled data. This analysis suggests that during individuated finger and wrist movements, M1 neurons do not work as groups of functionally similar neurons. Rather, M1 neurons appear to be functionally quite diverse.

The view of M1 activity during individuated finger movements that has developed up to this point appears chaotic. Although voluntary movements of different fingers obviously can be made as desired, which finger movement is performed does not appear to be determined by where in M1 neurons are active, nor by the activity of neuronal groups controlling particular muscles, muscle synergies, digits, movements, or movement primitives. And yet the M1 neuronal populations do transmit firing rate information about which finger movement is made. Population analyses using population vector, logistic regression, and softmax approaches, all show that the discharge of M1 neurons transmits information that specifies which finger movement will be performed.[49,50] How might such a diverse population of M1 neurons generate the various patterns of concurrent activity in multiple muscles that are needed to produce specific finger movements?

3.3.3 A Corticomotoneuronal Network of Diverse Elements

A likely possibility is that the connections from M1 to motoneuron pools function as a network. The elements of the M1 layer then could be quite diverse, without categorical groups of similar neurons. M1 neurons could be diverse both in terms of the particular motoneuron pools to which they connect and in terms of their activity patterns across a set of movements. Activity of a selected subset of M1 output neurons then could facilitate activation of the correct motoneuron pools for a given movement. The population analyses described above suggest that a computer model of a fully connected neural network, in which the weights of connections between M1 neurons and motoneuron pools are adjusted by an output-optimizing algorithm, would certainly be able to reproduce output patterns of muscle activity from input patterns of a population of M1 neurons. But can the same be achieved with physiological information on which M1–muscle connections actually exist, and on the strength of those existing connections?

In general, the difficulty of obtaining such data from a real neural network has precluded direct physiological testing of the network hypothesis. The cortico-motoneuronal network provides an opportunity, however, for a first-approximation approach to this problem. As a trained monkey performs individuated finger and wrist movements, the activity of M1 neurons can be recorded simultaneously with the EMG activity of multiple muscles, each representing the activity of a large pool of spinal motoneurons. Spike-triggered averaging (SpikeTA) of rectified EMG then can provide physiological evidence of whether a functional connection exists between the M1 neuron and the motoneuron pool generating each EMG, as well as a measure of the sign and strength of any connection.

A preliminary analysis of such data collected from two monkeys in our laboratory indicates that the M1 neurons that produce SpikeTA effects in a given motoneuron pool are indeed quite diverse.[51] The firing rate modulations of M1 neurons during individuated finger movements only partially resemble the amplitude modulations of EMG activity generated by the motoneuron pools to which the M1 neurons connect (Figure 3.4). One might think that the neurons with firing rate modulations most similar to the EMG amplitude modulations have the strongest connections to the motoneuron pool, but the degree of neuron–EMG activity similarity seems to have little correlation with the strength of the neuron–EMG connection as measured by SpikeTA. Nevertheless, summing the patterns of M1 neuron activity (firing rate modulation), each weighted by the amplitude (mean percent increase [MPI]) of the SpikeTA effect of that neuron in that EMG, in some instances can produce a pattern that resembles the EMG amplitude modulation pattern. When the effect of adding in different neurons is examined more closely, some neurons clearly can be seen to contribute to increasing the similarity between the reconstructed and the actual EMG. Of course, these tend to be those neurons with the greatest individual activity pattern similarity to the actual EMG, though some are neurons with low individual similarity. Summing a selected subset of M1 neurons can produce a reconstructed EMG pattern more similar to the actual pattern than any of the individual neurons.

Many other M1 neurons, however, appear to detract from the similarity between the reconstructed and actual EMG patterns. Adding the weighted activity patterns of these neurons into the reconstructed pattern reduces its similarity to the actual EMG below that achieved by the optimal selected subset of neurons, below that of the most similar single M1 neuron, and often to nil. What might be the role played by such neurons, which produce SpikeTA effects in an EMG but show little if any correlation between the pattern of their firing rate and the modulation of the EMG? Perhaps some of the spikes of these neurons are precisely synchronized with the spikes of other neurons that have input to the motoneuron pool producing the EMG.[52–54] Such M1 neurons then might have no direct effect on the motoneuron pool,[55] or else might have effects that sum with those of other synchronized neurons and hence are not linearly related to the firing rate of the recorded neuron.

Interpretation of these observations on reconstructing EMG activity from a network of M1 neurons with output connections pruned according to connection strengths weighted by SpikeTA effects is limited, of course, by the simplifying assumptions made. In particular, this pruned network reconstruction of EMG assumes (1) that SpikeTA effects represent a constant input to motoneuron pools, the strength of which does not fluctuate during different movements or task time periods; (2) that the effects of M1 neurons on EMG activity sum linearly; and (3) that M1 neurons provide sufficient input to the motoneuron pools to create the pattern of EMG activity observed. All these assumptions are oversimplified, however. The

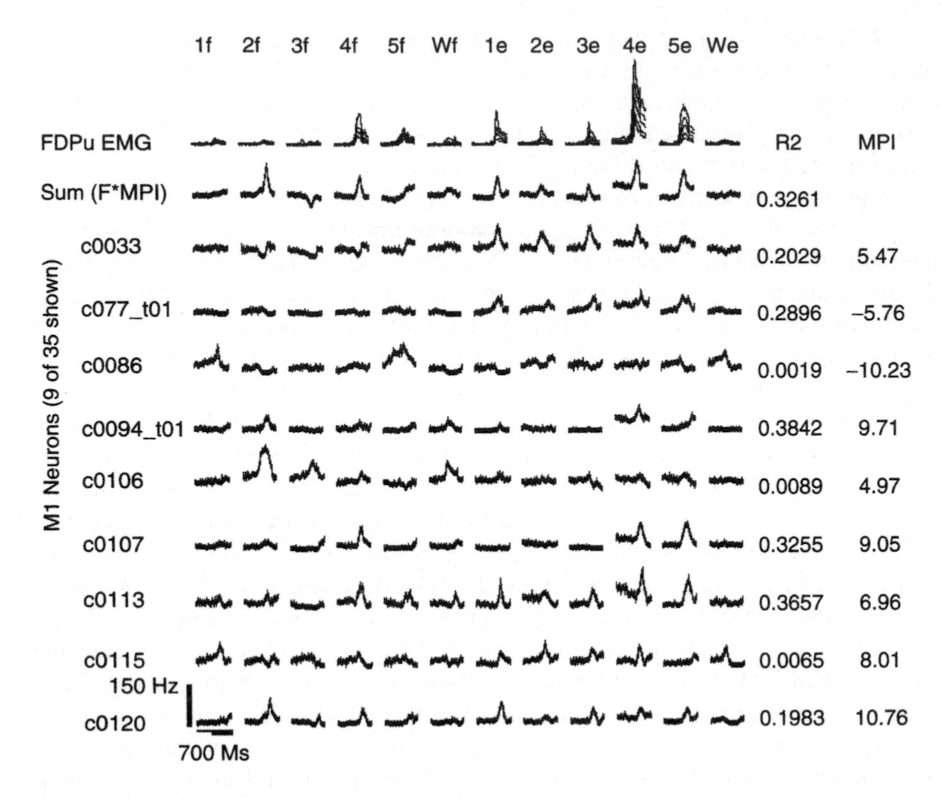

FIGURE 3.4 Reconstruction of EMG from M1 neuron activity weighted by SpikeTA effects. Thirty-five M1 neurons recorded from M1 in the same monkey all showed SpikeTA effects in the EMG activity of FDPu. The top row of traces shows the pattern of EMG activity recorded from FDPu during each of 12 individuated finger and wrist movements (abbreviated above with a number for the instructed digit and a letter for the instruction direction: 1 = thumb through 5 = little finger; W = wrist; f = flexion; and e = extension; therefore 2e = extension of index finger). For the EMG during each movement, five traces represent the mean, the mean ± SD, the maximum, and the minimum, to indicate the variability of the EMG recordings. The nine lower rows of traces show the firing rate histograms of the first 9 of the 35 M1 neurons (c0033 through c0120). Values to the right of each row represent (1) R^2 for the correlation between the 12 firing rate histograms for that neuron and the mean EMG patterns for FDPu, and (2) the mean percent increase (MPI) of the SpikeTA effect the neuron produced in FDPu EMG. Some of these M1 neurons, such as c0113, had patterns of firing rate modulation that closely resembled the modulation of FDPu EMG, while others, such as c0106, did not. In general, there was no correlation between the degree of neuron–EMG activity pattern similarity, measured by the R^2, and the strength of the neuron–EMG connection, measured by the MPI. Nevertheless, when the firing rate histograms of all 35 neurons, each weighted by its own MPI, were summed, the resulting activity pattern, (second row of traces, SUM[F*MPI]), did resemble that of FDPu. The greatest discrepancy was seen during 2f, where the sum showed considerable activity though none actually occurred in FDPu. Otherwise, the greatest activity was present during 4f and then 5f. During extensions, greatest activity was present in the sum during 4e and 5e, a moderate amount during 1e, less during 2e and 3e, and little during We.

strength of SpikeTA effects may fluctuate,[56] especially when the effects result in part from synchronized discharge of other neurons. The effects of M1 neurons on motoneuron activity and the effects of motoneuron activity on EMG do not necessarily sum linearly.[57] And during finger movements, motoneurons receive input from many other sources, including rubrospinal neurons, spinal interneurons, and sensory afferents.[58–61] The activity of M1 neurons, then, might best be viewed as combining with these other inputs to sculpt the activity of motoneurons into the patterns needed for individuated finger movements.

If output neurons distributed widely through the M1 hand representation participate in sculpting the muscle activity needed for each finger movement via a network of converging and diverging connections to the motoneuron pools, then how might the correct set of M1 neurons be brought into action for a given finger movement? Within M1, horizontally projecting axon collaterals interconnect the entire upper extremity representation.[62] For example, neurons within the digit representation (defined by intracortical microstimulation) project to the wrist, elbow, and shoulder representations, and vice versa. Presumably, this intracortical network coordinates activity throughout the M1 hand representation such that the appropriate distributed set of output neurons discharges for a particular individuated finger movement.

Adjusting this intracortical network to achieve the desired M1 output may require plasticity driven by practice and training. The weight of existing synaptic connections in M1 can be modified by long-term potentiation and depression.[63,64] New synaptic connections also may form during training at a skilled motor performance.[65] Such changes at the synaptic level may underlie the changes in stimulation maps of particular muscles or movements observed after training at a motor skill.[66,67] Such synaptic changes may help draw the necessary output neurons into the coordinated activity for particular movements. The strengthening of common inputs to the output neurons of M1 should increase the frequency with which they discharge synchronously, thereby increasing the efficiency with which M1 outputs drive spinal motoneurons. Indeed, in monkeys trained to perform individuated finger movements for a very long time, SpikeTA has provided evidence of more synchronous input to motoneuron pools.[68]

3.3.4 Evolution of the Motor Cortex

Has the same evolutionary process that produced greater biomechanical independence of the digits, and greater compartmentalization of muscles, produced corresponding changes in M1? An indication that such evolution has occurred can be gained by comparing the M1 upper extremity representation in modern rodents, monkeys, and humans. Figure 3.5 illustrates these species' differences using examples selected from the work of C. N. Woolsey and colleagues, who used similar mapping techniques in all three species.[69,70] In rodents the forelimb representation is relatively homogeneous, with little evidence of somatotopic segregation of proximal and distal parts of the limb.[71] In new world monkeys (not illustrated), representations of proximal and distal parts are intermixed, though often with a tendency for distal parts of the upper extremity to be more heavily represented posterolaterally and proximal parts to be more heavily represented anteromedially.[39,72] In old world macaques, this

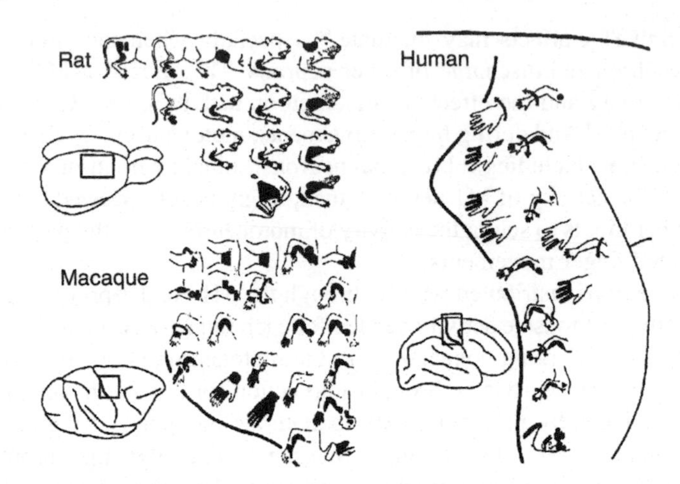

FIGURE 3.5 Phylogenetic evolution of the upper extremity representation. Individual examples are shown of the upper extremity representation as defined by electrical stimulation maps in three different species: rat, macaque monkey, and human. In each species, C. N. Woolsey and colleagues recorded the movements evoked from each stimulated point by filling a figurine of the relevant body part(s), using black to indicate the body parts that moved most prominently, and stippling or cross-hatching to indicate those that moved less prominently. Arrows indicate rotational movement. Note that the rat shows little if any separation between representation of the proximal and distal forelimb. The macaque shows a gradient with a central core of distal (fingers and wrist) representation surrounded by, and overlapping with, a horseshoe of proximal (elbow and shoulder) representation. The monkey, however, shows little if any orderly gradient of finger representations. The human shows more distinct proximodistal segregation, as well as a gradient of finger representation. Though these maps were obtained with cortical surface stimulation, more recent studies using intracortical microstimulation in animals and functional magnetic resonance imaging in humans are consistent with the same underlying features of representation (see text). The scale is different for each species, but the region shown in each stimulation map is indicated by a rectangle on the inset line drawing of the entire brain. (Modified with permission from Figures 122 (rat) and 126 (macaque) of Woolsey et al., 1952;[69] and Figure 20 (human) of Woolsey et al., 1979.[70])

proximodistal gradient is more evident still, with a posterolateral core of low threshold distal representation surrounded by, and partially overlapping with, a horseshoe-shaped zone of proximal representation.[73,74] Macaques have little if any segregation of finger representations. In humans, the proximodistal gradient of representation from medial to lateral is still more evident, and a somatotopic gradient of digit representation is demonstrable, with the thumb more heavily represented laterally and the little finger more heavily represented medially.[42,43,47] This phylogenetic trend from rodents to monkeys to humans, first for greater gradients of proximodistal representation, and then for gradients of radioulnar digit representation, suggests that the organization of M1 has evolved along with that of the peripheral neuromuscular apparatus. Paralleling the greater biomechanical and muscular independence of the digits, somatotopic gradients with stronger representation of the radial digits laterally and the ulnar digits medially are found in the human M1 hand representation.

Indeed, the degree of biomechanical and muscular independence of body parts may be related to the extent to which their M1 representations have become somatotopically segregated. In monkeys, all the digits are biomechanically coupled to a considerable degree, and have overlapping representation in M1. In humans, the thumb and index finger are somewhat independent of the other fingers and have a distinguishable gradient of representation. In both species, the hand is biomechanically coupled to the wrist and forearm, M1 representation of which overlaps that of the hand in a proximodistal gradient. Also in both species, movements of the hand are independent of the face, and the M1 representations of the hand and face also are separate.

Why has evolution of the M1 representation not proceeded to a more discrete somatotopic representation of body parts or muscles, similar to the discrete representation of the body in the primary somatosensory cortex? Because of biomechanical coupling and muscle structure, the vast majority of hand and finger movements, even the highly individuated movements used in fine motor tasks, require the simultaneous control of multiple muscles, some moving the intended digit or digits and others stabilizing the other digits and the wrist. A network of intermingled and overlapping representations may be able to accomplish such control more efficiently than a network of discrete, spatially segregated nodes. The factors that make the intermingled representation biologically more efficient are unclear, but may include a benefit of maintaining shorter interconnections with shorter conduction times.

3.4 CONCLUSIONS

Control of finger movements from the motor cortex thus appears to be achieved by a complex network of physiologically diverse cortical neurons. Neurons active during various finger movements, with outputs to various subsets of finger muscles, are intermingled with one another, such that the cortical territory representing any particular body part, muscle, or movement overlaps extensively with the territory representing any nearby body part. This system presumably has evolved to control individuated movements of a peripheral apparatus that includes incompletely subdivided muscles and biomechanical coupling among nearby body parts. Such a biological system is conceptually more complex than a robotic hand with independent digits, each driven by its own servomotor through a separate software channel. Nevertheless, a robotic hand is clumsy compared to the amazingly dextrous and flexible performance achieved by a biological hand controlled by the motor cortex.

ACKNOWLEDGMENTS

The authors thank Jennifer Gardinier and Lee Anne Schery for technical assistance, and Marsha Hayles for editorial comments. This work was supported by R01-NS27686 and R01-NS36341 from the National Institute of Neurologic Disorders and Stroke and BCS-0225611 from the National Science Foundation of the United States of America.

REFERENCES

1. Gorska, T. and Sybirska, E., Effects of pyramidal lesions on forelimb movements in the cat, *Acta Neurobiol. Exp.*, 40, 843, 1980.
2. Martin, J.H. and Ghez, C., Differential impairments in reaching and grasping produced by local inactivation within the forelimb representation of the motor cortex in the cat, *Exp. Brain Res.*, 94, 429, 1993.
3. Whishaw, I.Q. and Gorny, B., Arpeggio and fractionated digit movements used in prehension by rats, *Behav. Brain Res.*, 60, 15, 1994.
4. Santello, M., Flanders, M., and Soechting, J.F., Patterns of hand motion during grasping and the influence of sensory guidance, *J. Neurosci.*, 22, 14262, 2002.
5. Santello, M., Flanders, M., and Soechting, J.F., Postural hand synergies for tool use, *J. Neurosci.,* 18, 10105, 1998.
6. Engel, K.C., Flanders, M., and Soechting, J.F., Anticipatory and sequential motor control in piano playing, *Exp. Brain Res.,* 113, 189, 1997.
7. Soechting, J.F. and Flanders, M., Flexibility and repeatability of finger movements during typing: analysis of multiple degrees of freedom, *J. Comp. Neurosci.*, 4, 29, 1997.
8. Hager-Ross, C.K. and Schieber, M.H., Quantifying the independence of human finger movements: Comparisons of digits, hands and movement frequencies, *J. Neurosci.*, 20, 8542, 2000.
9. Reilly, K.T. and Hammond, G.R., Independence of force production by digits of the human hand, *Neurosci. Lett.*, 290, 53, 2000.
10. Schieber, M.H., Individuated finger movements of rhesus monkeys: a means of quantifying the independence of the digits, *J. Neurophysiol.,* 65, 1381, 1991.
11. Zatsiorsky, V.M., Li, Z.M., and Latash, M.L., Enslaving effects in multi-finger force production, *Exp. Brain Res.*, 131, 187, 2000.
12. Lawrence, D.G. and Kuypers, H.G., The functional organization of the motor system in the monkey. I. The effects of bilateral pyramidal lesions, *Brain*, 91, 1, 1968.
13. Schieber, M.H., Somatotopic gradients in the distributed organization of the human primary motor cortex hand area: evidence from small infarcts, *Exp. Brain Res.*, 128, 139, 1999.
14. Lang, C.E. and Schieber, M.H., Differential impairment of individuated finger movements in humans after damage to the motor cortex or the corticospinal tract, *J. Neurophysiol.*, 90, 1160, 2003.
15. Schieber, M.H., How might the motor cortex individuate movements? *Trends Neurosci.*, 13, 440, 1990.
16. von Schroeder, H.P. and Botte, M.J., The functional significance of the long extensors and juncturae tendinum in finger extension, *J. Hand Surg. Am. Vol.*,18, 641, 1993.
17. von Schroeder, H.P., Botte, M.J., and Gellman, H., Anatomy of the juncturae tendinum of the hand, *J. Hand Surg. Am. Vol.*, 15, 595, 1990.
18. Fahrer, M., Interdependent and independent actions of the fingers, in *The Hand*, Tubania, R., Ed., W.B. Saunders Company, Philadelphia, 1981, 399.
19. Serlin, D.M. and Schieber, M.H., Morphologic regions of the multitendoned extrinsic finger muscles in the monkey forearm, *Acta Anat.*, 146, 255, 1993.
20. Schieber, M.H., Gardinier, J., and Liu, J., Tension distribution to the five digits of the hand by neuromuscular compartments in the macaque flexor digitorum profundus, *J. Neurosci.,* 21, 2150, 2001.
21. Schieber, M.H., Muscular production of individuated finger movements: the roles of extrinsic finger muscles, *J. Neurosci.*, 15, 284, 1995.

22. Reilly, K.T. and Schieber, M.H., Incomplete functional subdivision of the human multitendoned finger muscle flexor digitorum profundus: an electromyographic study, *J. Neurophysiol.*, 90, 2560, 2003.
23. Beevor, C.E., *(1) The Croonian Lectures on Muscular Movements and (2) Remarks on Paralysis of the Movements of the Trunk in Hemiplegia*, The MacMillan Company, New York, 1903.
24. Segal, R.L., Catlin, P.A., Krauss, E.W., Merick, K.A., and Robilotto, J.B., Anatomical partitioning of three human forearm muscles, *Cells Tissues Organs,* 170, 183, 2002.
25. Chanaud, C.M., Pratt, C.A., and Loeb, G.E., Functionally complex muscles of the cat hindlimb. V. The roles of histochemical fiber-type regionalization and mechanical heterogeneity in differential muscle activation, *Exp. Brain Res.*, 85, 300, 1991.
26. English, A.W., Wolf, S.L., and Segal, R.L., Compartmentalization of muscles and their motor nuclei: the partitioning hypothesis, *Phys. Ther.*, 73, 857, 1993.
27. Schieber, M.H., Electromyographic evidence of two functional subdivisions in the rhesus monkey's flexor digitorum profundus, *Exp. Brain Res.*, 95, 251, 1993.
28. Schieber, M.H., Chua, M., Petit, J., and Hunt, C.C., Tension distribution of single motor units in multitendoned muscles: comparison of a homologous digit muscle in cats and monkeys, *J. Neurosci.,* 17, 1734, 1997.
29. Keen, D.A. and Fuglevand, A.J., Role of intertendinous connections in distribution of force in the human extensor digitorum muscle, *Muscle Nerve,* 28, 614, 2003.
30. Kilbreath, S.L., Gorman, R.B., Raymond, J., and Gandevia, S.C., Distribution of the forces produced by motor unit activity in the human flexor digitorum profundus, *J. Physiol. (Lond.)*, 543 Pt. 1, 289, 2002.
31. Shinoda, Y., Yokota, J., and Futami, T., Divergent projection of individual corticospinal axons to motoneurons of multiple muscles in the monkey, *Neurosci. Lett.*, 23, 7, 1981.
32. Fetz, E.E. and Cheney, P.D., Postspike facilitation of forelimb muscle activity by primate corticomotoneuronal cells, *J. Neurophysiol.*, 44, 751, 1980.
33. Buys, E.J., Lemon, R.N., Mantel, G.W., and Muir, R.B., Selective facilitation of different hand muscles by single corticospinal neurones in the conscious monkey, *J. Physiol. (Lond.)*, 381, 529, 1986 .
34. Bremner, F.D., Baker, J.R., and Stephens, J.A., Variation in the degree of synchronization exhibited by motor units lying in different finger muscles in man, *J. Physiol. (Lond.)*, 432, 381, 1991.
35. Kilbreath, S.L. and Gandevia, S.C., Limited independent flexion of the thumb and fingers in human subjects, *J. Physiol. (Lond.)*, 479, 487, 1994.
36. Shinoda,.Y., Zarzecki, P., and Asanuma, H., Spinal branching of pyramidal tract neurons in the monkey, *Exp. Brain Res.*, 34, 59, 1979.
37. McKiernan, B.J., Marcario, J.K., Karrer, J.H., and Cheney, P.D., Corticomotoneuronal postspike effects in shoulder, elbow, wrist, digit, and intrinsic hand muscles during a reach and prehension task, *J. Neurophysiol.,* 80, 1961, 1998.
38. Andersen, P., Hagan, P.J., Phillips, C.G., and Powell, T.P., Mapping by microstimulation of overlapping projections from area 4 to motor units of the baboon's hand, *Proc. Roy. Soc. Lond. B,* 188, 31, 1975.
39. Donoghue, J.P., Leibovic, S., and Sanes, J.N., Organization of the forelimb area in squirrel monkey motor cortex: representation of digit, wrist, and elbow muscles, *Exp. Brain Res.,* 89, 1, 1992.
40. Schieber, M.H. and Hibbard, L.S., How somatotopic is the motor cortex hand area? *Science*, 261, 489, 1993.

41. Sanes, J.N., Donoghue, J.P., Thangaraj, V., Edelman, R.R., and Warach, S., Shared neural substrates controlling hand movements in human motor cortex, *Science*, 268, 1775, 1995.
42. Dechent, P. and Frahm, J., Functional somatotopy of finger representations in human primary motor cortex, *Hum. Brain Mapp.*, 18, 272, 2003.
43. Kleinschmidt, A., Nitschke, M.F., and Frahm, J., Somatotopy in the human motor cortex hand area. A high-resolution functional MRI study, *Eur. J. Neurosci.*, 9, 2178, 1997.
44. Brochier, T., Boudreau, M.J., Paré, M., and Smith, A.M., The effects of muscimol inactivation of small regions of motor and somatosensory cortex on independent finger movements and force control in the precision grip, *Exp. Brain Res.*, 128, 31, 1999.
45. Kubota, K., Motor cortical muscimol injection disrupts forelimb movement in freely moving monkeys, *NeuroReport*, 7, 2379, 1996.
46. Schieber, M.H. and Poliakov, A.V., Partial inactivation of the primary motor cortex hand area: effects on individuated finger movements, *J. Neurosci.*, 18, 9038, 1998.
47. Kim, J.S., Predominant involvement of a particular group of fingers due to small, cortical infarction, *Neurology*, 56, 1677, 2001.
48. Poliakov, A.V. and Schieber, M.H., Limited functional grouping of neurons in the motor cortex hand area during individuated finger movements: a cluster analysis, *J. Neurophysiol.*, 82, 3488, 1999.
49. Georgopoulos, A.P., Pellizzer, G., Poliakov, A.V., and Schieber, M.H., Neural coding of finger and wrist movements, *J. Comp. Neurosci.*, 6, 279, 1999.
50. BenHamed, S., Schieber, M.H., and Pouget, A., Decoding M1 neuron activity during multiple finger movements, *Soc. Neurosci. Abstr.*, 27, 289.2, 2001.
51. Schieber, M.H., Reconstruction of EMG activity by a pruned network of motor cortex neurons, *Soc. Neurosci. Abstr.*, 28, 563.3, 2002.
52. Grammont, F. and Riehle, A., Precise spike synchronization in monkey motor cortex involved in preparation for movement, *Exp. Brain Res.*, 128, 118, 1999.
53. Abeles, M., Prut, Y., Bergman, H., and Vaadia, E., Synchronization in neuronal transmission and its importance for information processing, *Progr. Brain Res.*, 102, 395, 1994.
54. Vaadia, E. et al., Dynamics of neuronal interactions in monkey cortex in relation to behavioural events, *Nature*, 373, 515, 1995.
55. Baker, S.N. and Lemon, R.N., Computer simulation of post-spike facilitation in spike-triggered averages of rectified EMG, *J. Neurophysiol.*, 80, 1391, 1998.
56. Bennett, K.M. and Lemon, R.N., The influence of single monkey cortico-motoneuronal cells at different levels of activity in target muscles, *J. Physiol. (Lond.)*, 477 (Pt 2), 291, 1994.
57. Baker, S.N. and Lemon, R.N., Non-linear summation of responses in averages of rectified EMG, *J. Neurosci. Meth.*, 59, 175, 1995.
58. Mewes, K. and Cheney, P.D., Primate rubromotoneuronal cells: parametric relations and contribution to wrist movement, *J. Neurophysiol.*, 72, 14, 1994.
59. Houk, J.C., Gibson, A.R., Harvey, C.F., Kennedy, P.R., and van Kan, P.L.E., Activity of primate magnocellular red nucleus related to hand and finger movements, *Behav. Brain Res.*, 28, 201, 1988.
60. Perlmutter, S.I., Maier, M.A., and Fetz, E.E., Activity of spinal interneurons and their effects on forearm muscles during voluntary wrist movements in the monkey, *J. Neurophysiol.*, 80, 2475, 1998.

61. Flament, D., Fortier, P.A., and Fetz, E.E., Response patterns and postspike effects of peripheral afferents in dorsal root ganglia of behaving monkeys, *J. Neurophysiol.,* 67, 875, 1992.
62. Huntley, G.W. and Jones, E.G., Relationship of intrinsic connections to forelimb movement representations in monkey motor cortex: a correlative anatomic and physiological study, *J. Neurophysiol.,* 66, 390, 1991.
63. Aroniadou, V.A. and Keller, A., Mechanisms of LTP induction in rat motor cortex *in vitro, Cereb. Cortex,* 5, 353, 1995.
64. Rioult-Pedotti, M.S., Friedman, D., Hess, G., and Donoghue, J.P., Strengthening of horizontal cortical connections following skill learning, *Nat. Neurosci.,* 1, 230, 1998.
65. Kleim, J.A., Lussnig, E., Schwarz, E.R., Comery, T.A., and Greenough, W.T., Synaptogenesis and Fos expression in the motor cortex of the adult rat after motor skill learning, *J. Neurosci.,* 16, 4529, 1996.
66. Pascual-Leone, A., Nguyet, D., Cohen, L.G., Brasil-Neto, J.P., Cammarota, A., and Hallett, M., Modulation of muscle responses evoked by transcranial magnetic stimulation during the acquisition of new fine motor skills, *J. Neurophysiol.,* 74, 1037, 1995.
67. Plautz, E.J., Milliken, G.W., and Nudo, R.J., Effects of repetitive motor training on movement representations in adult squirrel monkeys: role of use versus learning, *Neurobiol. Learn. Mem.*, 74, 27, 2000.
68. Schieber, M.H., Training and synchrony in the motor system, *J. Neurosci.*, 22, 5277, 2002.
69. Woolsey, C.N., Erickson, T.C., and Gilson, W.E., Localization in somatic sensory and motor areas of human cerebral cortex as determined by direct recording of evoked potentials and electrical stimulation, *J. Neurosurg.*, 51, 476, 1979.
70. Woolsey, C.N., Settlage, P.H., Meyer, D.R., Sencer, W., Hamuy, T.P., and Travis, A.M., Patterns of localization in precentral and "supplementary" motor areas and their relation to the concept of a premotor area, *Res. Pub. Assoc. Res. Nerv. Ment. Dis.*, 30, 238, 1952.
71. Donoghue, J.P. and Sanes, J.N., Organization of adult motor cortex representation patterns following neonatal forelimb nerve injury in rats, *J. Neurosci.*, 8, 3221, 1988.
72. Gould, H.J., Cusick, C.G., Pons, T.P., and Kaas, J.H., The relationship of corpus callosum connections to electrical stimulation maps of motor, supplementary motor, and the frontal eye fields in owl monkeys, *J. Comp. Neurol.*, 247, 297, 1986.
73. Kwan, H.C., MacKay, W.A., Murphy, J.T., and Wong, Y.C., Spatial organization of precentral cortex in awake primates. II. Motor outputs, *J. Neurophysiol.*, 41, 1120, 1978.
74. Park, M.C., Belhaj-Saïf, A., Gordon, M., and Cheney, P.D., Consistent features in the forelimb representation of primary motor cortex in rhesus macaques, *J. Neurosci.,* 21, 2784, 2001.

4 Neuronal Representations of Bimanual Movements

Eilon Vaadia and Simone Cardoso de Oliveira

CONTENTS

0-8493-1287-6/05/$0.00+$1.50

4.1 INTRODUCTION

Simultaneous movements of the two arms constitute a relatively simple example of complex movements and may serve to test whether and how the brain generates unique representations of complex movements from their constituent elements, as suggested by Leyton and Sherrington: "[T]he motor cortex may be regarded as a synthetic organ for compounding … movements … from fractional movements."[1] This chapter describes studies in which we attempted to investigate how the brain assembles coordinated complex movements from their constituents, using the relatively simple example of bimanual coordination.*

To do so, we have taken a neurophysiological approach, investigating neuronal activity in behaving monkeys. The first question we ask is how the neuronal representations of unimanual movements are combined to form bimanual movements. To answer it, we compare neuronal activity during bimanual movements to the activity observed during performance of their unimanual constituents. This approach may provide preliminary evidence as to whether complex movements are coded differently from simple movements. Second, we need to define an approach to deciphering the neuronal code for complex movements; namely, how we can pinpoint which parameters of neuronal activity contain relevant information about the movement to be executed. Previous work has suggested that in the motor system, rates of neuronal populations are especially informative about the directions of upcoming movements.[2] However, a number of studies, mainly on the visual system, have suggested that temporal correlations between neuronal activities may contain information that is particularly related to the compositionality of the coded items (e.g., the coherence of moving bars[3,4,5]). Given that each arm is mainly controlled by the contralateral hemisphere, it is also likely that the temporal relationships between the hemispheres are relevant to bimanual movements.

This chapter summarizes results we have accumulated to answer the above questions, at least partially. We present evidence that bimanual representations indeed exist, both at the level of single neurons and at the level of neuronal populations (in local field potentials). We further show that population rates and dynamic interactions between the hemispheres contain information about the kind of bimanual movement to be executed.

4.2 BIMANUAL-RELATED ACTIVITY OF SINGLE NEURONS IN MOTOR CORTICAL FIELDS

One of the first efforts to resolve the first question electrophysiologically was made by Tanji et al.,[6] who trained a monkey to press buttons with the fingers of either hand separately or with both hands together. They recorded cortical neurons in the medial aspect of the frontal cortex, which was called at the time the supplementary

* The term "bimanual coordination" literally means "coordination of the two hands," yet this term has been used in the literature in studies that relate not only to the coordination of the left and right hands, but also of the left and right fingers, or of the left and right arms. This is also how we use the term in this article.

motor area (SMA*). Tanji et al.[6] found that a substantial fraction of neurons in this area were active during bimanual finger tapping and not during movements of the finger of the right or left hand separately. This finding suggests that there are some neurons that seem to be specific to bimanual movements. Their work appeared after a behavioral study by Brinkman[8] who reported bimanual deficits consecutive to SMA lesion. These and other studies (including clinical reports; for review, see Brust[9]) inspired further studies focusing on the SMA as a major candidate area for the control of bimanual coordination. Neuronal activity in SMA that is specific to bimanual movements has now been described by a number of groups using different tasks, although this specificity has been defined differently by different groups. Neuronal activity during performance of a "drawer pulling task" was tested by Wiesendanger et al.,[10] where monkeys performed naturally coordinated movements without specific training. This task involved whole arm movements, where the monkey was required to open a drawer with one hand and retrieve a raisin from it with the other. Bimanual specific SMA activity has also been described by Kermadi et al.,[11] although a different study on the same task reported that only a small percentage of neurons was exclusively activated during bimanual movements.[12]

Our group (including the authors of this chapter and Opher Donchin, Orna Steinberg, and Anna Gribova) took another approach in an attempt to capitalize on knowledge from the extensive studies of neuronal activity during arm reaching in a center-out task.[13] In what follows, we summarize a number of studies in which we used a bimanual center-out reaching task to explore neuronal representations of bimanual movements in the cortex.

4.2.1 The Bimanual Task

Macaque monkeys were trained to operate two separate manipulanda, one with each arm. The manipulanda were low weight, low friction, two-joint mechanical arms, oriented in the horizontal plane. Movement of each manipulandum produced movement of a corresponding cursor on a vertical 21" video screen. The movement of each cursor was mapped to its corresponding manipulandum movement such that each millimeter of manipulandum movement yielded one millimeter of movement of the cursor on the video display. The angular origin, 0°, was to the monkey's right, and 90° was away from the monkey for the manipulandum movement and toward the top of the screen for the display.

A trial began when the monkey aligned both cursors on 0.8 cm diameter origins, as shown in Figure 4.1 (where both cursors, left and right, are at their respective origins) and held them still for 500 msec. For each arm, one of eight peripheral target circles (0.8 cm diameter) could appear at a distance of 3 cm from the origin. This small movement amplitude was chosen to minimize postural adjustments while performing the movements. Movements taking the cursor from the origin to the target were primarily small elbow and shoulder movements. Figure 4.2 presents a few examples of trial types. In unimanual trials, only one target appeared (the upper

* SMA was later divided into SMA-proper and pre-SMA. See Reference 7.

FIGURE 4.1 The monkey sits in a primate chair holding two manipulanda and facing a video screen. Two cursors indicating the location of the manipulanda are shown on the screen (+). Each cursor appears in the corresponding origin. Possible target locations are shown as circles surrounding each origin. (Modified with permission from Reference 15.)

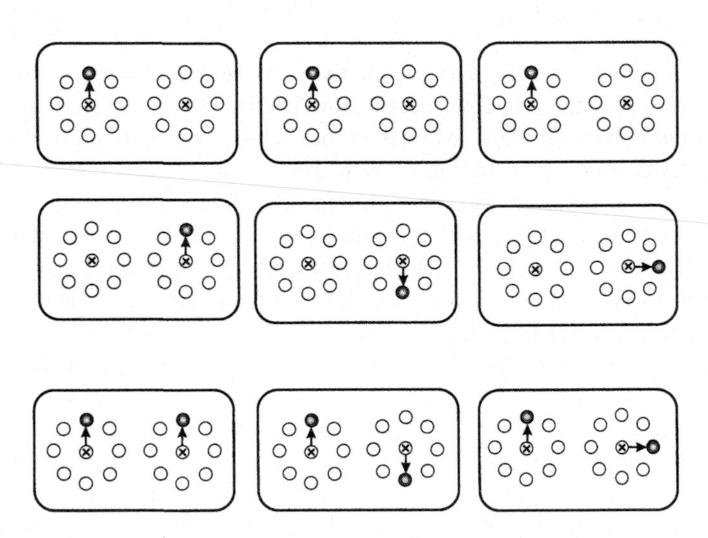

FIGURE 4.2 The behavioral task illustrated by examples of types of trials that were used in the various experiments. The empty circles are not visible to the monkey. The figure displays examples of unimanual movements (upper two rows) to 90° (up) and 270° (down) and bimanual movements (lower row).

two rows in Figure 4.2) and the monkey moved the appropriate arm to bring the corresponding cursor into the target, but did not move the other arm. If two targets appeared — signaling a bimanual trial — the monkey had to move both arms, such that the two cursors were moved into the target circles on the screen.

These structured movements made it possible to study well-controlled bimanual movements of various types. For example, parallel movements and opposite movements (lower row, Figure 4.2) were composed of unimanual movements shown in the upper rows of Figure 4.2. (Figure 4.2 shows only one direction per arm; in all cases additional directions were studied.) Other combinations, where each arm was required to move in a different direction or to cover a different distance, were also tested, as for example the movements shown in Figure 4.2 (bottom-right plot) where the arms move at 90° to each other.

4.2.2 Monkey Behavior

Neuronal activity was sampled after the monkeys were over-trained to perform bimanual trials with the two arms starting to move together and reaching the targets together quite accurately. For example, the two monkeys used for the data presented in this section initiated the bimanual movements with average interarm intervals (IAIs) of 16 to 21 msec (SD = 56 to 74 msec) and reached the targets with an average IAI of 5 to 15 msec (SD = 106 to 125 msec). These IAIs are quite short, much shorter than would be required for successful performance of the task, meaning that the monkeys tended, like humans, to synchronize their movements rather than attempting to perform two separate movements. The movements used in the tasks were small (a length of 3 cm for all movement types presented in this section and up to 6 cm in some types of movements for the experiments described in Section 4.5). The hand trajectories made to a given direction were quite similar for different movement types (but not identical; see Figure 4.6). Further, video camera observations and electromyographic (EMG) recordings failed to detect consistent variations in postural adjustments during the arm movements. EMGs of muscles on the forearm, the upper arm, the shoulders, and the back were also recorded simultaneously with neuronal activity (selected sessions). Various analyses were carried out in order to detect changes in neuronal activity that could emerge from different patterns of muscle activation. (For details about specific control measures comparing EMG during unimanual and bimanual movements see Donchin et al.[14])

4.2.3 Neuronal Recordings in MI and SMA

Single-unit activity and local field potentials were recorded from homologous sites in the two hemispheres, from the primary motor cortex (M1) and from SMA proper. (For details on recording sites see Donchin et al.[14]) The activity of 8 to 30 isolated neurons and up to eight local field potential (LFP) channels was recorded each session. The data discussed in this article were recorded from 3 monkeys and included the activity of more than 438 neurons (232 in M1 and 206 in SMA). To detect *evoked activity,* we tested the firing rate in a 500-msec period from 100 msec before movement initiation (the average activation onset across responsive units) to 400 msec after movement initiation.

The number of units whose activity varied significantly during performance of the task was high. Eighty-one percent (187/232) of the neurons recorded in M1 and 76% (157/206) of the neurons recorded in SMA were significantly modulated during performance of the task, despite the fact that no selection was made on this basis during the recording sessions.

4.2.4 Unimanual Activity of Single Cells

Figure 4.3 shows the activity of two neurons recorded from the left M1 during unimanual movements of the right and the left arm. The neuron in part A of the figure was strongly modulated during right-handed (contralateral) movements, whereas the neuron in part B was strongly modulated during left-handed (ipsilateral) movements. A simple measure for the arm preference of a single cell is the laterality index:

$$Laterality\ Index = \frac{EAcontra - EAipsi}{EAcontra + EAipsi} \tag{4.1}$$

The unit shown in Figure 4.3A had a laterality index of 0.59 (indicating contralateral preference); the neuron shown in Figure 4.3B had a laterality index of –0.77 (ipsilateral preference). In both recording areas, about one third of the neurons was activated solely during contralateral movements, whereas approximately one fifth of the neurons was only activated ipsilaterally. Analysis of the distribution of laterality indices in the two areas showed only a slight contralateral preference and a tendency for neurons in M1 to be more contralaterally activated than neurons in SMA. However, a detailed χ^2 analysis of the results (not shown) revealed no significant differences between M1 and SMA for laterality preferences.

4.2.5 Bimanual-Related Activity

Comparisons of the cells' evoked activity in unimanual and bimanual trials revealed *bimanual-related* components of activity, which are shown in Figures 4.4 and 4.5. Figure 4.4 shows the activity of one neuron recorded in the right hemisphere. It was inactive when the monkey made unimanual movements toward either 45° or 225° (middle and rightmost columns). However, the same neuron was strongly activated when the two arms moved in parallel toward 225° (row b). It was also active, but less so, in another type of bimanual trial where the two arms moved in opposite directions (row c), but did not respond at all during the other bimanual movements (rows a and d). Figure 4.5 shows another example. Here, the neuron exhibited the opposite effect: the strong responses during unimanual movements of the contralateral arm to 45° disappeared when the same arm moved in a bimanual context with the ipsilateral arm (left column, rows a and c).

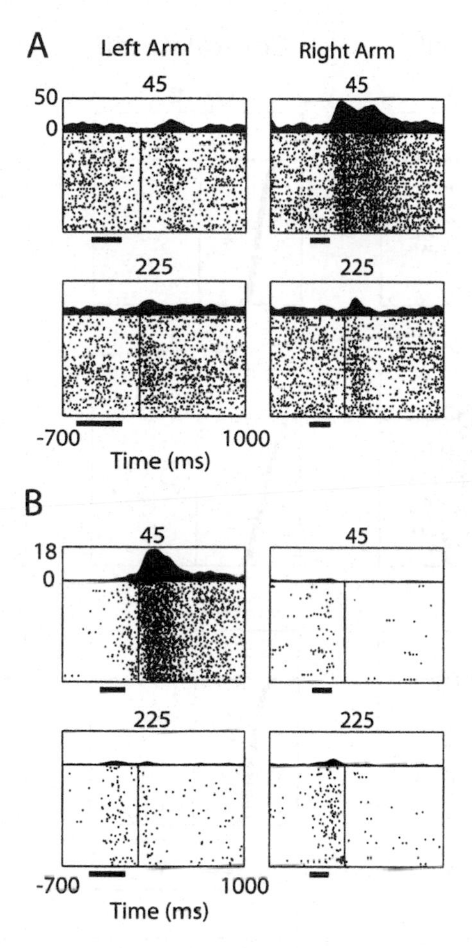

FIGURE 4.3 Activity of two neurons recorded in M1 (left hemisphere) during unimanual movements. (A) Neuron with strong contralateral preference (laterality index = 0.59; see Equation 4.1). (B) Neuron with strong ipsilateral preference (laterality index = –0.77). Each horizontal line of dots represents a trial; each action potential is represented by a dot. Trials are aligned on the beginning of movement and sorted by reaction time; the line below each plot indicates the range of target appearance times. The peri-event time histograms (PETHs) (filled black histograms above raster display) have a bin width of 2.5 msec and were smoothed using a filter with a cutoff frequency of 100 msec.

Many neurons showed bimanual-related activity that was less dramatic than the two examples above. To quantitatively compare evoked activity during bimanual movements to evoked activity during unimanual movements, it is necessary to correctly compare the evoked activity during performance of a given bimanual movement to the unimanually evoked activities. In the task used here, there were four different bimanual movements performed by the monkey — two bimanual parallel movements and two bimanual opposite movements (the four left-hand plots of Figure 4.4). We chose two types of comparisons.

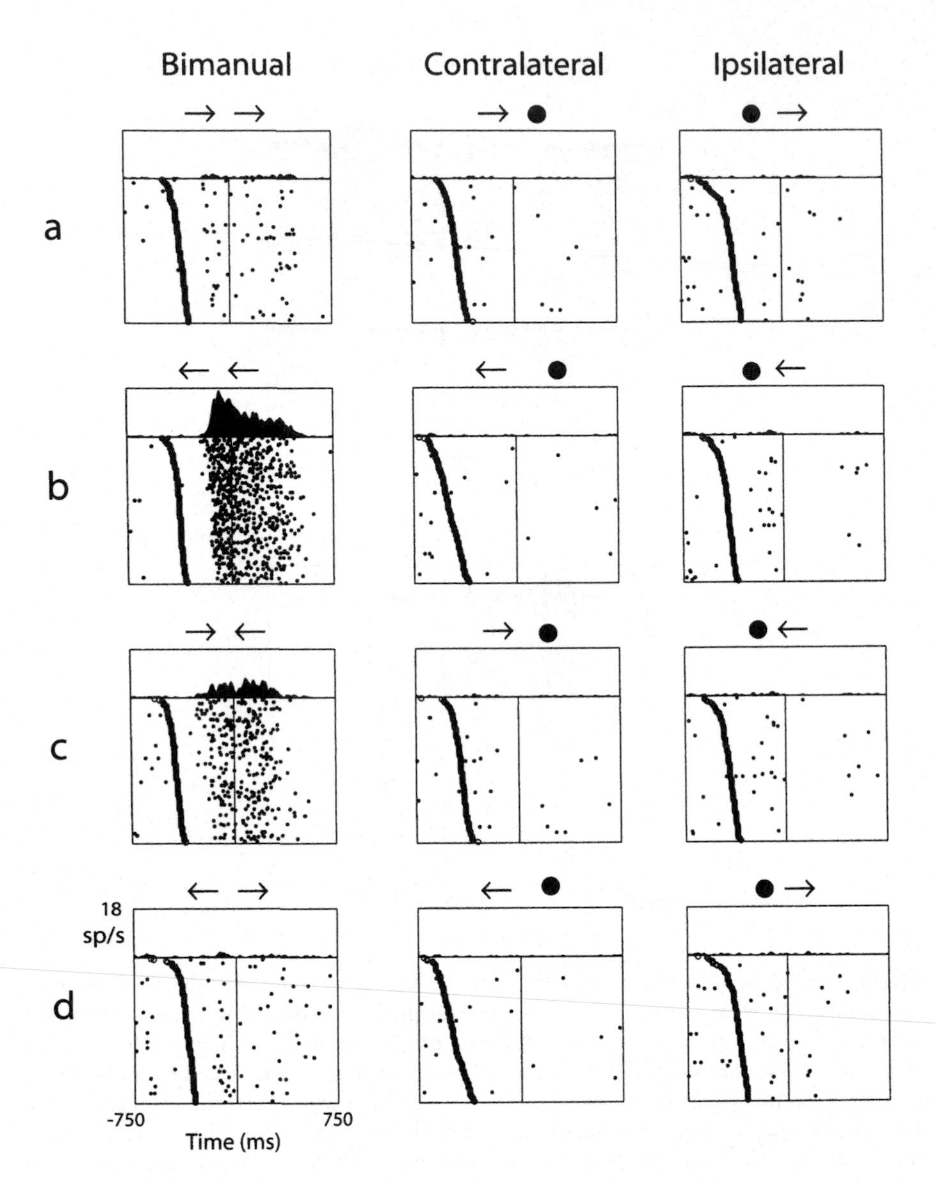

FIGURE 4.4 Bimanual-related activity of a single unit recorded in SMA (from Reference 15). Each row contains PETHs and raster displays depicting the cell activity in one type of trial. The cell only had strong activation during bimanual movements (b, left column) and no activity in unimanual trials (right column is unimanual right; middle column is unimanual left). The direction of movement of each arm is indicated by arrows or a dot if the arm does not move. Trials are aligned on the beginning of the movement (of the first arm) and sorted by reaction time. The target onset is indicated by black squares. The PETH scales are identical in all plots. The movement directions were 45° and 225°.

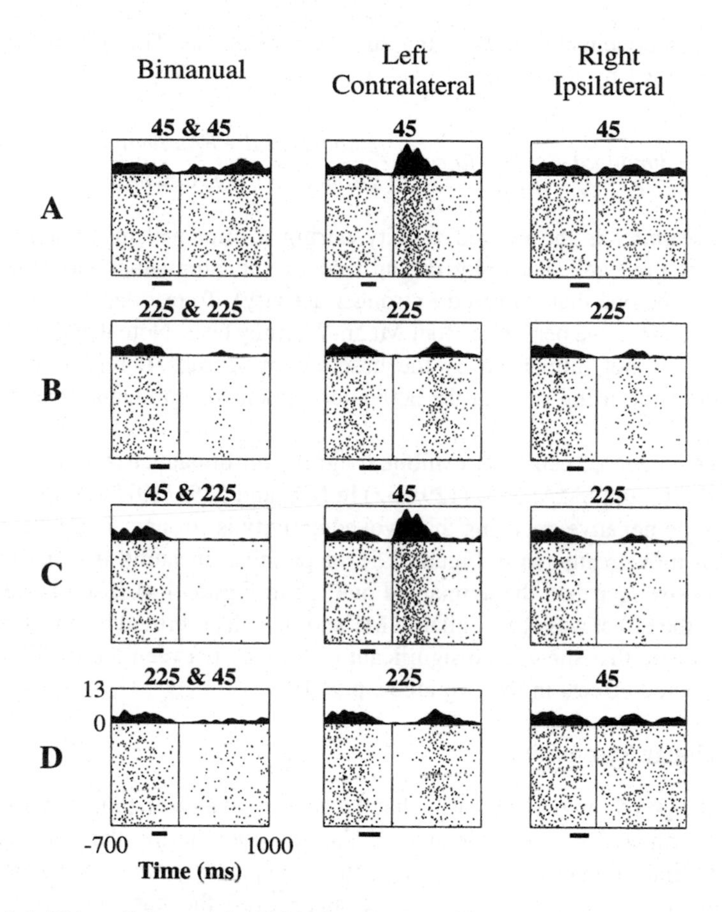

FIGURE 4.5 This cell from right SMA demonstrates bimanual related activity opposite to the activity shown in Figure 4.4. The activity evoked during unimanual contralateral movements disappears and is even suppressed (row C) during bimanual movements. The strength of the bimanual-related effect is –0.84. Format is the same as in Figure 4.4. (Reproduced with permission from Reference 14.)

4.2.5.1 Comparing Bimanual-Evoked Activity to One of the Unimanual Components

Here, the evoked activity for each of the four types of bimanual movements was compared to the activity during one of the unimanual movements that composed it. The question is which of the two unimanual movements forms the appropriate comparison. One possibility is always to compare activity during bimanual movements to activity during a unimanual contralateral movement. However, this choice disregards neurons with an ipsilateral preference in unimanual movements. If the focus is whether there is a difference between maximal activation in bimanual movements and maximal activation in unimanual movements, it is appropriate to compare neural activity during bimanual movements to the neural activity in the unimanual component that evoked the strongest response. This *bimanual-related effect* is quantified by Equation 4.2.

$$\textit{Bimanual-Related Effect} = \frac{\textit{EAbimanual} - \textit{EAunimanual}}{\textit{EAbimanual} + \textit{EAunimanual}} \tag{4.2}$$

where *EAbimanual* is the evoked activity during a bimanual movement, and *EA unimanual* is the evoked activity during one of the unimanual movements that composed it (the one that evoked the stronger activity). To test the statistical significance of this effect, we performed four Mann-Whitney tests. Note that the bimanual-related effect is not influenced by the baseline firing rate; it represents a direct comparison of the firing rates in the activation epochs of unimanual and bimanual movements.

The percentage of cells that exhibited significant bimanual-related effects was high in both M1 and SMA: 55% (129/232) in M1, and 52% (107/206) in SMA. The effect could be negative, meaning that evoked activity is stronger in unimanual than in bimanual movements (as in Figure 4.5), or positive, meaning that the bimanual activity was stronger than the unimanual one (as in Figure 4.4). The distribution of strengths of the effect was also similar in M1 and in SMA (verified by a Kolmogorov-Smirnov statistic that showed no significant difference between the distributions of bimanual-related effects in the two areas, $p > 0.1$).

4.2.5.2 Linear Summation

The second type of comparison posits that the activity in a given bimanual movement should be compared to a combination of the activities during the two unimanual movements that compose it. To conduct this comparison, we tested whether the *normalized evoked activity (NEA)* (the change from the baseline firing) during bimanual movements could be explained by a simple linear summation of the unimanual movements that compose it, which requires that the linear summation hold true for all four bimanual movements. For this purpose, the deviations from linearity in each type of bimanual movement were combined to produce a statistic that was expected to distribute like an χ^2 with 3 degrees of freedom (specifically, we calculated the sum of the squared differences between bimanual NEA and the sum of the unimanual NEAs divided by the combined variance of the bimanual and unimanual NEAs). We also tested for the possibility that NEA in bimanual movements is equal to NEA during contralateral movements, and for the third possibility that it is equal to NEA during ipsilateral movements. If all three of these null hypotheses could be rejected at $p < 0.05$, the bimanual activity of this neuron could not be accounted for by the hypothesis of linear summation. Note that our failure to correct for the multiple statistical tests effectively increases the significance level since all three null hypotheses and not just one must be rejected.

The results clearly indicate that for most of the bimanual-related neurons (~80%), the bimanual-evoked activity could not be explained by the linear summation hypothesis. In contrast, for neurons that were not bimanual-related, 60% of the neurons in M1 and 72% of the neurons in SMA failed to reject one or more of the hypotheses at this level. Their responses might be explained by a linear combination of the unimanual responses.

In an additional analysis, we fit the neuronal activity with a model that attempts to explain bimanual NEA using a more general linear combination of unimanual NEAs (see Donchin et al.[19] for details). This model again fit only a minority of the bimanual-related neurons (19 to 26% in SMA and M1, respectively). In addition, the parameters of the fit for different neurons were not clustered in any way, suggesting that there was no general rule for the combination of unimanual activities. To conclude, both analyses indicated that the majority of the bimanual-related neurons failed to be accounted for by any linear explanation of their bimanual activity.

4.2.6 Comparing Movement Kinematics and Single-Unit Activity

In all the experiments described in this section, the monkey performed short movements (3 cm) that did not require noticeable postural adjustment. Actual observation of the monkey during task performance (aided by video recordings) revealed no postural adjustments or other differences that differentiated movements during bimanual as compared to unimanual trials. Moreover, detailed analyses further demonstrated that for many of the recorded cells, the bimanual related effect was unlikely to be related to differences in kinematics or muscular activity during movements of each arm in unimanual and bimanual contexts. An example is shown in Color Figure 4.6.* The figure depicts the activity of one bimanual related cell, recorded in M1 during performance of unimanual and bimanual trials. Trials in which unimanual and bimanual movements were more similar are shown in the top displays, while trials in which the movements were less similar are shown in the lower displays. The figure demonstrates that selection of trials with similar trajectories in unimanual and bimanual conditions did not lessen the bimanual-related effect, and selection of trials with different trajectories did not increase it. Moreover, the temporal pattern of the neuron's activity was unaffected by the selection of trials.

4.3 BIMANUAL-RELATED ACTIVITY IN EVOKED LFP ACTIVATION

Besides the activity of single cells, our physiological recordings also served to measure local field potentials. The LFP is thought mainly to reflect synaptic activity in the area of the recording electrode[16] and thus may be an important tool for investigating population activity. Animal research on field potentials in the motor cortex has focused on the relationship of synchronous LFP oscillations to movements and to single-unit activity.[17–20]

More recent studies, however, have provided evidence that LFP recorded in the parietal cortex[21] and in the motor cortex[22] contains much more concrete information about behavior than was previously thought. In the study by Donchin et al.,[23] we

* See color insert following page 170.

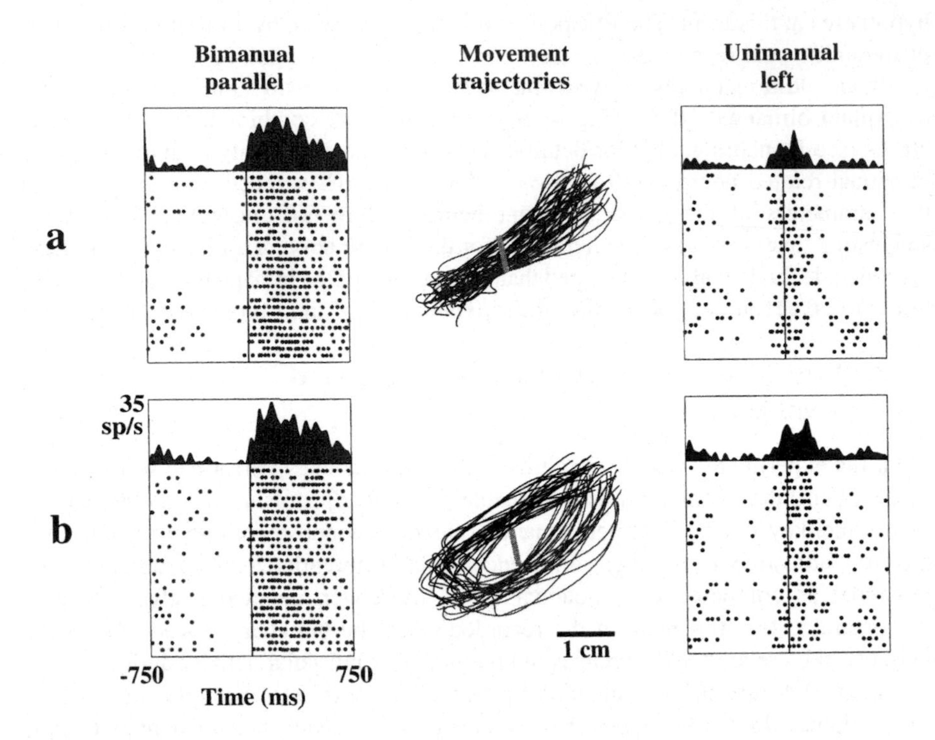

FIGURE 4.6 (see color figure) Raster displays and PETHs illustrating the activity from a right M1 cell in four conditions. The activity of the cell during bimanual parallel movements is on the left (red). The activity of the cell during unimanual left movements is on the right (blue). The middle plots show the movement paths of the left hand for bimanual parallel (red) and unimanual left (blue) movements. Row A only contains trials in which the movement path passed through a narrow band (thick green line) located between the origin and the target. Row B only contains trials that did not pass through the band. The green band was placed to maximize the difference between the trajectories in the lower display. PETHs are centered on the beginning of movement, and the scale for all PETHs is the same. The trajectories begin in the upper right and end in the lower left of the frame. Note that the cell activity in bimanual trials (in red) remains similar regardless of the precise trajectories. (Used with permission from Reference 15.)

specifically addressed the question of whether the LFP also contains information about bimanual movements.

To observe movement-evoked LFP activity, the LFP signal in repeated trials was averaged by aligning trials on the beginning of movement, producing the *movement-evoked potential* (mEP). Figure 4.7 shows examples of individual LFP traces, and the average of 100 traces from which the examples were taken. The resulting mEPs seen in the motor cortex have a characteristic shape of positive and negative deflections exemplified in the figure. The strength of an mEP may be calculated in several

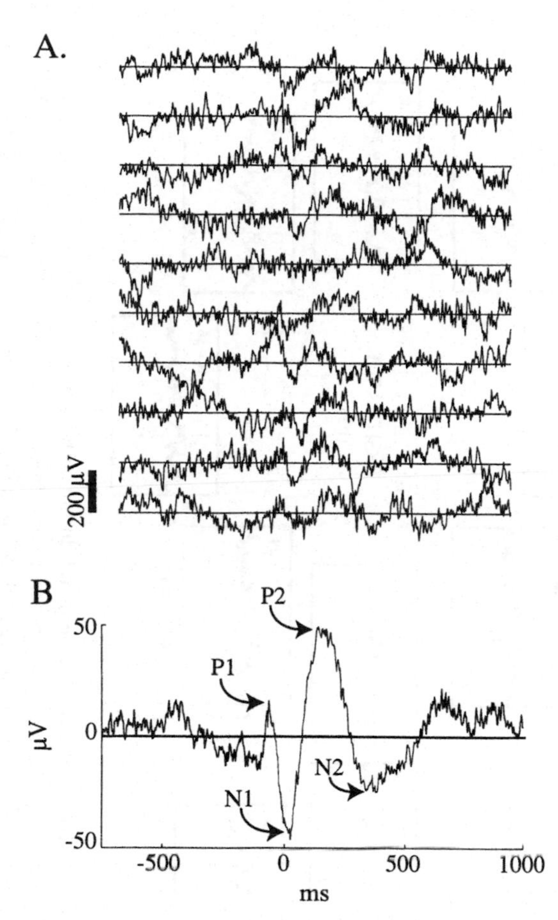

FIGURE 4.7 LFP traces and averaged LFP. (A) Ten examples of individual LFP traces selected at random from one recording site in left M1. All of these examples are taken from instances where the monkey was making unimanual right-handed movements toward 225°, and they are aligned at the beginning of movement (time 0). (B) The average of 101 LFP traces recorded during repetitions of the same movement. Identifiable peaks (N1, N2, P1, P2) are indicated. (Reproduced with permission from Reference 23.)

ways. For the purpose of this chapter we used the total root mean square (RMS) measure of the response (the square root of the integral of the squared mEP).

4.3.1 Movement-Evoked Potentials during Unimanual Movements

Figure 4.8 shows mEPs recorded by one microelectrode in M1, during performance of unimanual movements of the left (contralateral) and the right arm. Interestingly, mEPs, like single cells in the motor cortex, are directionally tuned. (See the example in Figure 4.8, which was particularly tuned for the contralateral arm.) This feature has further implications for the functional anatomy of M1 and the possible existence of clusters with correlated directional preference.[24,25] Another main feature of the

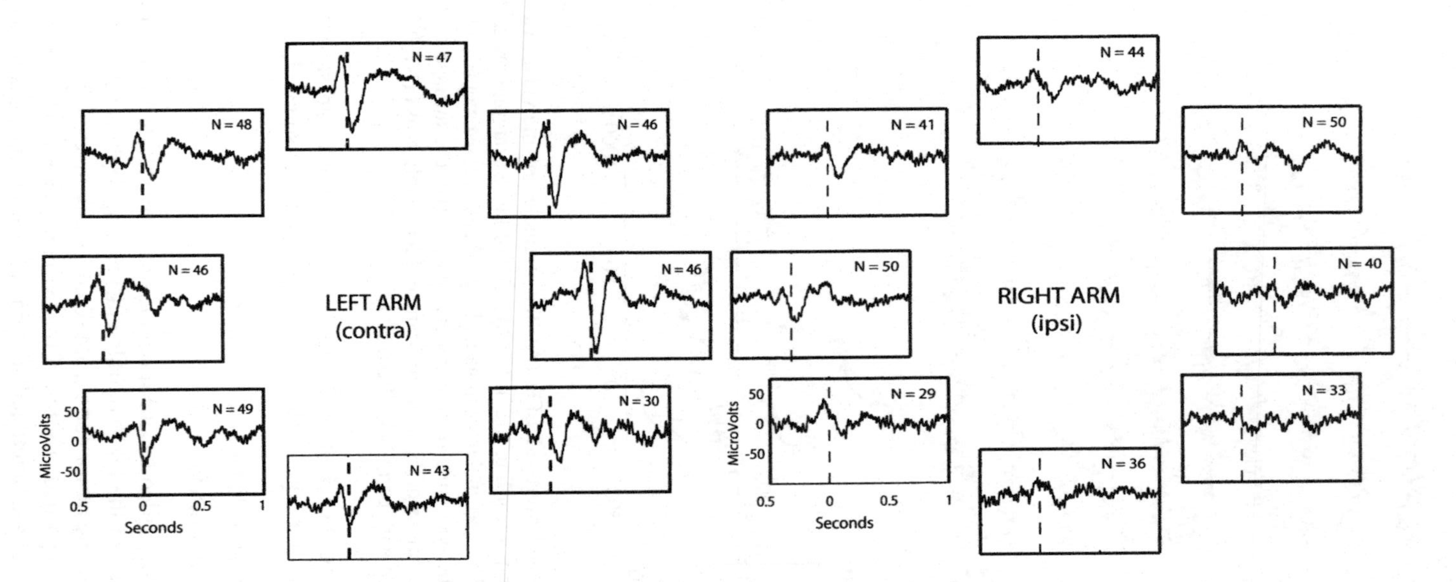

FIGURE 4.8 Movement-evoked potentials (mEPs) recorded by one microelectrode in M1 during performance of unimanual movements of the left (contralateral) and the right arm. The position of each subdisplay corresponds to the target direction for each kind of trial. The numbers in each rectangle indicate the number of trials used to generate the averaged mEP. Note the directional tuning, in particular to the contralateral arm.

mEPs differed markedly from the evoked single-unit activity. The figure shows a clear contralateral preference in the mEPs. A detailed analysis of all the recording sites in M1 indicated that for most sites, the mEP showed strong contralateral preference. As described above, this was not the case for single-unit spike activity in M1. Interestingly, recording sites in SMA did not share this property. Rather, mEPs in SMA were of similar strength for the two arms with even a slight ipsilateral preference.

4.3.2 Comparing mEPs in Different Movements

To evaluate the mEPs and compare different movement types, Donchin et al.[23] quantified the contralateral preference and the strength of the bimanual-related effect.

$$Contralateral\ preference = \frac{mEP_{contra} - mEP_{ipsi}}{\sigma_{mEP}} \tag{4.3}$$

where σ_{mEP} represents the standard deviation combined from the mEP in the two movements, the square root of a weighted average of the two variances.

The strength of the bimanual-related effect was generated using a very similar formula:

$$Bimanual\text{-}Related\ Effect = \frac{mEP_{Bimanual} - mEP_{Unimanual}}{\sigma_{mEP}} \tag{4.4}$$

where σ_{mEP} is now calculated using the variances from the evoked response during the unimanual and bimanual movements.

4.3.3 Bimanual-Related Effect in mEPs

Figure 4.9 compares mEPs recorded by one microelectrode during performance of bimanual movements and the corresponding unimanual movements that compose them. (The format is the same as for Figure 4.4.) The difference between the mEP during bimanual movements and unimanual movements was particularly evident in bimanual parallel movements to 315° where the bimanual-related effect value was 2.60 (significant at $p < 0.001$).

Figure 4.10 demonstrates that positive bimanual-related effects characterize the population. The figure shows the bimanual-related effect for all recording sites in both M1 and SMA for the full mEP. In both areas, the RMS of mEP was greater during bimanual movements than during unimanual movements for a vast majority of the recording sites.

To conclude the results of the mEP analysis, we showed that mEPs differ from single-cell activity in two major ways. First, we found a difference in the contralateral preference of M1 and SMA. Second, for nearly all recording sites, bimanual mEPs were greater than unimanual mEPs. This increase was caused mainly by an increase in the positive components of the mEP, particularly the P2 component (see Figure 4.7). This result was different from the single-unit result where the bimanual-related effect manifested as either an increase or a decrease in activity during bimanual movements.

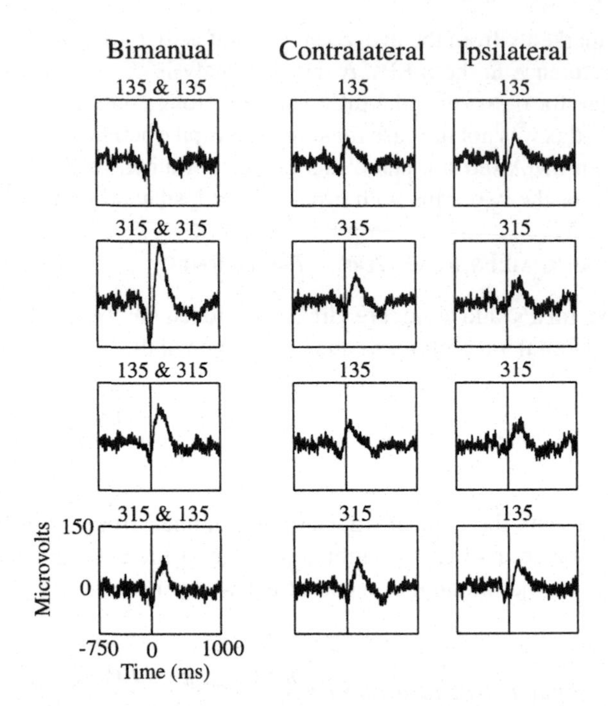

FIGURE 4.9 Example of a recording site in M1 with a bimanual-related effect. Each row shows the mEP in one bimanual movement and the two unimanual movements that comprise it. All plots are at the same scale. (Reproduced with permission from Reference 23.)

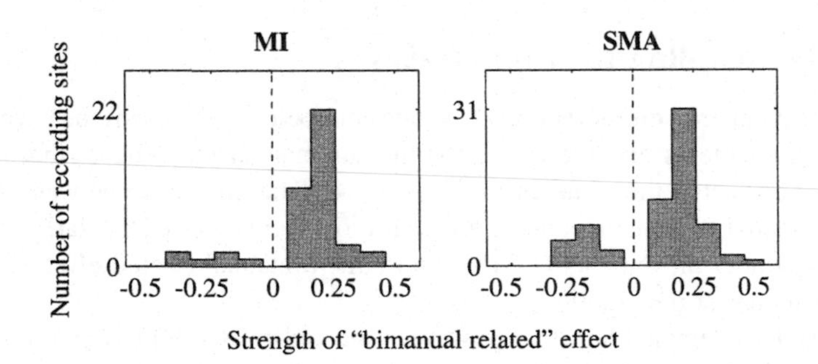

FIGURE 4.10 Distribution of the strength of bimanual-related effects in the mEPs. The histograms show the strength of the effect in the overall RMS of the mEP in M1 and SMA. Note that for almost all sites the deviation is positive.

The unidirectional nature of the bimanual-related effect in mEPs that were recorded in each of the two hemispheres supports the hypothesis that the motor cortices represent bimanual movements specifically, requiring neuronal control beyond the simultaneous production of activation represented by the two unimanual control signals. However, while lending weight to the hypothesis above, the result raises its own questions. Is there any physiological explanation for the increased LFP activation during bimanual movements? Is there any functional significance in the result?

There are three (not mutually exclusive) possibilities that provide an immediate explanation for the increased mEP during bimanual movements:

1. More neurons are active in the area of the electrode.
2. The number of neurons that send inputs (inhibitory or excitatory) to the electrode site increases during bimanual movements.
3. The synaptic activity in the area of the electrode is more synchronized.

The first possibility can be rejected because (as shown in Donchin et al.[14]) the total spike activity in both M1 and SMA does not increase during bimanual movements. The second possibility is not implausible. While for any particular neuron, maximal bimanual activation may be less than maximal unimanual activation, the sum of bimanual activation across both hemispheres could still be greater than the sum of unimanual activation. For instance, neurons in the left cortex may be more active during movements of the right hand, whereas neurons in right cortex are more active during movements of the left hand, but during bimanual movements both sets of neurons are active. Because M1 and SMA receive inputs from both the contralateral and the ipsilateral cortex, the amount of input that each cortical area receives may be greater during bimanual movements than during unimanual movements. A group investigating the neuronal response as a function of stimulus size in visual cortex found a similar result: induced oscillations in LFP increase with increased stimulus size, whereas single-unit discharge rates may increase or decrease.[26]

Whether the third possibility can also account for increased mEP size is still unclear. Work on synchronization of LFP oscillations has shown a relationship between synchronized oscillations in the LFP and synchrony in single-unit activity,[27] but this study did not find increased LFP synchrony during bimanual movements.[18] Our own study on LFP synchronization[28] revealed, in only one of the two monkeys tested, a slight transient increase of synchronization around movement onset. The major and consistent effect was a net decrease of synchronization during movements. On the other hand, it cannot be excluded that only a specific subset of neurons increased their synchronization during bimanual movements, which could account for the increased LFP size. In order to clarify this question, the circuitry of the recorded neurons should be known to the experimenter, which was not the case in the previous experiments.

In conclusion, although many questions remain regarding the interpretation of the mEP in LFPs, it seems clear that this signal does contain information about bimanual movements. The fact that the LFP shows a specific bimanual effect demonstrates that bimanual-specific signals also occur on the population level and are not confined to single neurons. In the next two sections, we will deal with the question of how the neuronal activity during bimanual movements may be read out by the system and used for the task of bimanual coordination.

4.4 REPRESENTATION OF BIMANUAL MOVEMENTS IN A POPULATION RATE CODE

It has been repeatedly suggested that single neurons in M1 are tuned to the direction of arm movements and that the activity of a population of tuned neurons faithfully predicts the direction of upcoming movements. However, the existence of bimanual-related activity means that a single neuron may be activated differently when one arm makes the very same movements as part of a unimanual movement or a bimanual movement. This was the rationale for investigating whether the population vector approach could produce reliable movement predictions for bimanual movements as well, in spite of the related bimanual effects. In the study by Steinberg et al.,[29] we tested this question by comparing the predictive quality of population vectors for unimanual and bimanual arm movements. The behavioral task was essentially similar to the bimanual task described above. Again, monkeys performed the unimanual center-out task and two classes of bimanual movements (parallel and opposite). Here, however, neuronal activity was recorded during performance of movements in all 8 directions, in all sessions.

For most cells, the directional tuning curve can be approximated by a cosine function, although the method probably overestimates tuning width.[30] In order to allow for comparison of our results with previous studies, we used the cosine approximation for directional tuning when applying the population vector approach. For the same reason, the cells were characterized in terms of *preferred direction* (PD), the direction of movement to which the cell has the strongest response, and the fit of its tuning to a cosine, estimated by the coefficient of determination (R^2). Cells with R^2 above 0.7 were defined as "directionally tuned" and the others as "non-tuned." Again, this value was selected to facilitate comparison with previous PD studies. (See for instance References 13,31,32.)

4.4.1 Directional Tuning in Unimanual and Bimanual Movements

As was expected from repeated reports of the arm area in M1, most of the sampled cells (156/212) exhibited broad symmetrical directional tuning around a preferred direction for at least one movement type. (For details, see Steinberg et al.[29]) Interestingly, about one-third of the tuned cells were directionally tuned to movements of the ipsilateral arm ($R^2 \geq 0.7$). Only a few cells (7%) were significantly tuned to all four types of movements. However, the majority of the tuned cells were tuned to more than one type (58%). An example of a cell that was tuned to all four movement types is shown in Figure 4.11.

4.4.2 Comparisons of Preferred Directions in Different Movement Types

For the population vector approach, the critical question is whether the PD of motor cortical cells changes during bimanual movements. Recently, evidence has been accumulating that directional tuning and PD may in fact change under certain conditions.[33] Comparing PDs in unimanual and bimanual movements has yielded intriguing results which are summarized in Figure 4.12 (restricted to cosine tuned cells with $R^2 > 0.7$ for the two compared movement types). The figure shows that the PDs calculated from (a) bimanual parallel, (b) bimanual opposite, or (c) ipsilateral unimanual movements were all correlated to the PDs calculated for contralateral

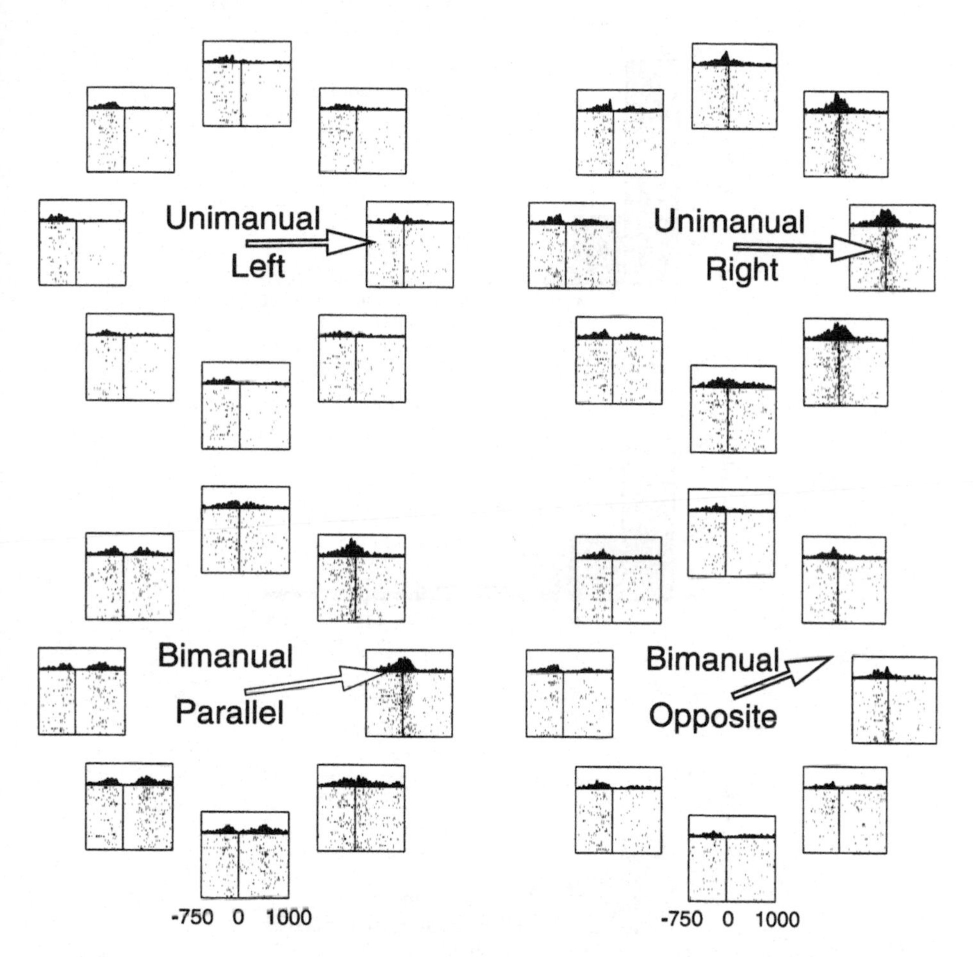

FIGURE 4.11 Activity of one cell from left M1 during performance of four different types of movement. Each quadrant of the figure shows the activity of the cell in one type of movement, in eight directions. The rasters are aligned around movement onset (time 0) in a time window of 750 msec before movement onset until 1000 msec after it. The red arrows indicate the preferred direction (PD). Their lengths are proportional to the R^2 of the cosine fit. The cosine fit of this cell with its R^2 values and directional indices is shown in Figure 4.4. Note that the PDs for all movement types are similar.

unimanual movements (the differences for all three comparisons are not uniformly distributed, Rao test, $p < 0.01$). However, the figure also illustrates that the PD of some cells can change substantially, as is most clearly seen in the comparison of the contralateral with the ipsilateral tuning (in unimanual trails).

4.4.3 Population Vectors Predict Bimanual Movements Well

In order to be able to calculate population vectors (PVs), Steinberg et al.[29] first estimated the PD of each cell as a constant, using an estimated best-fit PD taken from all four movement types. To construct separate population vectors for the two

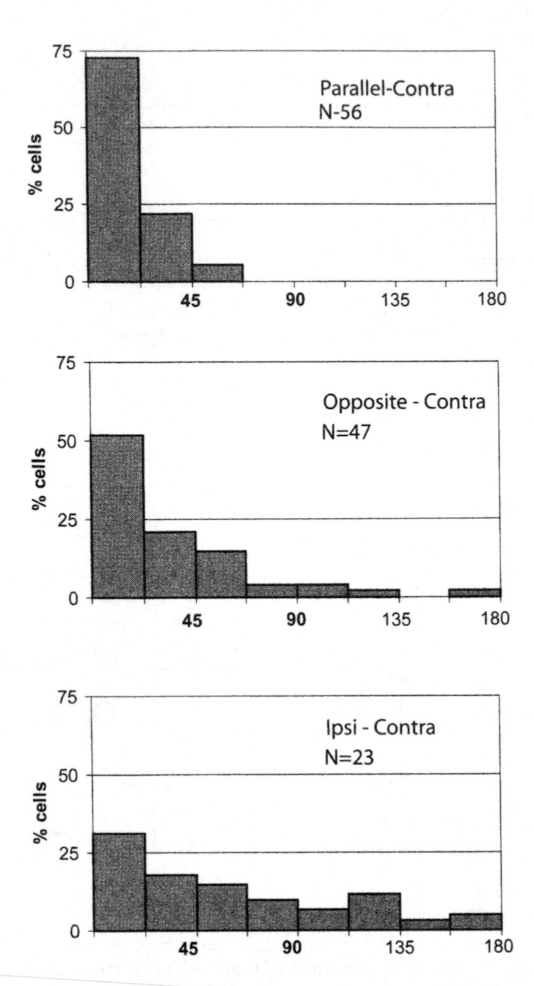

FIGURE 4.12 Comparison of PDs in different movement types. The figure shows the distributions of differences in PDs, comparing the PD during unimanual movements of the contralateral arm to (from top to bottom) bimanual parallel, bimanual opposite, and ipsilateral movements. Only cells with $R^2 \geq 0.7$ in both movement types were included in this analysis. *N* represents the number of cells included in each plot. (Reproduced with permission from Reference 29.)

arms, the population of sampled cells was divided into two subpopulations, guided by the hypothesis that bimanual movements are generated by two separate (although possibly coordinated) neuronal networks. The division into two subpopulations was motivated by two different approaches. The first natural choice was to divide the cells according to the hemisphere in which they resided. Color Figure 4.13A show PVs, for movements in 315°, where each pair was generated by the two subpopulations, one from the left hemisphere (for the right arm, in blue) and one from the right hemisphere (for the left arm, in red). The figure shows PVs for four types of movements. The two plots on the left show the prediction for unimanual movements. Note that for unimanual movements, PVs were also obtained for the non-moving arm. Although very small, these PVs did not point in random directions, but were

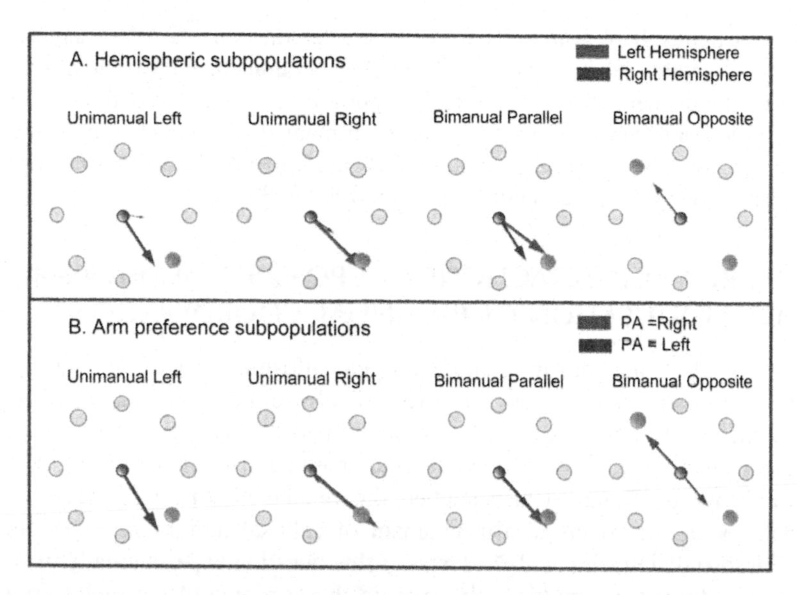

FIGURE 4.13 (see color figure) Population vectors calculated for unimanual and bimanual movements to 315°. For each movement, two PVs (colored arrows) of two neuronal subpopulations were calculated using an estimated best-fit PD. The different colors represent the PVs of the different neuronal subpopulations. (A) PVs constructed by dividing all cells into two subpopulations according to the hemisphere in which they reside. (B) PVs constructed by dividing all cells into two subpopulations according to their arm preference. (Reproduced with permission from Reference 29.)

generally aligned with the direction of the moving arm. PVs for bimanual movements are shown on the right side of the figure. Color Figure 4.13B shows predictions guided by the second approach. Here, cells were selected for each arm on the basis of their activation, under the assumption that each cell can be characterized by its "preferred arm" (PA) — i.e., the arm for which unimanual movements evoked the strongest activity — regardless of the hemisphere in which it resides.

PVs for movements in the direction of 315° generated by "PA selection" of subpopulations are shown in Color Figure 4.13B. For this specific direction, the PA-based sub-populations seem to represent the direction of simultaneous movements of the two arms somewhat better than selection by the hemispheric locations of the cells. Also, the PVs for the nonmoving arm are a little smaller in Color Figure 4.13B as compared to Color Figure 4.13A. For unimanual movements, this is an inevitable result of the reselection, but the improvement in the bimanual movements is not a trivial result. Nevertheless, when examining the PVs for all movement directions, it was impossible to demonstrate that the accuracy of PA-based PVs is higher than that of hemisphere-based PVs.

To conclude, these results show that large enough populations of neurons contain enough information to simultaneously encode for the direction of movements of the two arms in bimanual movements, despite the bimanual specific activity changes. Cells were divided into two subpopulations, either by hemisphere or by their arm preference. Even though the latter division "replaces" approximately a quarter of the cells in the contralateral hemisphere with cells from the ipsilateral one, the PVs calculated in bimanual movements from this division are not less accurate than PVs calculated when dividing by hemisphere. This result further supports the notion that both hemispheres are active and contribute to execution of both unimanual and bimanual movements.

Exactly how the two hemispheres interact and collaborate with each other was the subject of two additional studies, described below.[28,34]

4.5 NEURONAL INTERACTIONS AS A POSSIBLE MECHANISM FOR MODULATION OF BIMANUAL COORDINATION

Cortico-cortical connections through the corpus callosum are a major candidate for mediating bimanual coordination. The effect of callosotomy on the nature of bimanual performance has been repeatedly demonstrated.[35–37] However, little is known about the physiological basis of the processes mediated by the callosum. A recent modeling work by Rokni et al.[34] studied the related nonlinear bimanual effects described above and proposed a mechanism of callosal inhibition to explain this effect. Cardoso de Oliveira et al.[28] addressed this question experimentally by simultaneous recordings from multiple sites within the arm area of the motor cortex in both hemispheres. For technical and statistical reasons, studying temporal correlations between single units is problematic when firing rates are relatively low and the number of similar trials is limited (as is the case in the experiments we describe here). However, LFP correlations turned out to be quite useful, as described below.

4.5.1 Time-Averaged Correlations

The time-averaged correlation method has been used to study neuronal interactions for many years.[38,39] This measure was calculated here for LFP signals using Equation 4.5, which defines the correlation coefficient (CC) for different temporal delays (τ) in a single trial of duration T (also called the correlogram):

$$CC(\tau)=\frac{\sum_{t=1}^{T}\left(LFP1(t)-\overline{LFP1}\right)\cdot\left(LFP2(1+\tau)-\overline{LFP2}\right)}{\sqrt{\sum_{t=1}^{T}\left(LFP1(t)-\overline{LFP1}\right)^{2}\cdot\sum_{t=1}^{T}\left(LFP2(1+\tau)-\overline{LFP2}\right)^{2}}} \tag{4.5}$$

where $LFP1(t)$ and $LFP2(t+\tau)$ are values of LFPs from two electrodes, at times t and $(t+\tau)$. $\overline{LFP1}$ and $\overline{LFP2}$ are the corresponding average values of the two channels across the measurement time T (trial duration).

The analysis was performed separately for two epochs in each trial.

1. **Hold period.** An interval of 500 msec before movement onset during which the monkey held its hands stationary at the origins and waited for the target (or targets) to appear. During this period the monkey could not predict the type of movement (bimanual or unimanual) or its direction.

2. **Movement period.** The correlation during a given type of movement was calculated for a time window of 1250 msec, from 250 msec before until 1000 msec after movement onset, an interval that included movement preparation and execution.

The resulting correlogram may be affected both by similar evoked responses (similar mEPs) and by possible trial-wise interactions between the single trial signals. A typical way of distinguishing between these two features is to calculate a "shift predictor" to approximate the correlation between the averages, and then subtract it from the correlograms to estimate the "pure" trial-wise correlation. Examples for such correlograms, obtained from the hold period and from the movement period (during performance of bimanual trials) are shown in Figure 4.14. On the diagonal of each plot are the autocorrelograms of the LFPs in each of the sites. The result depicted in the figure is highly typical in three ways.

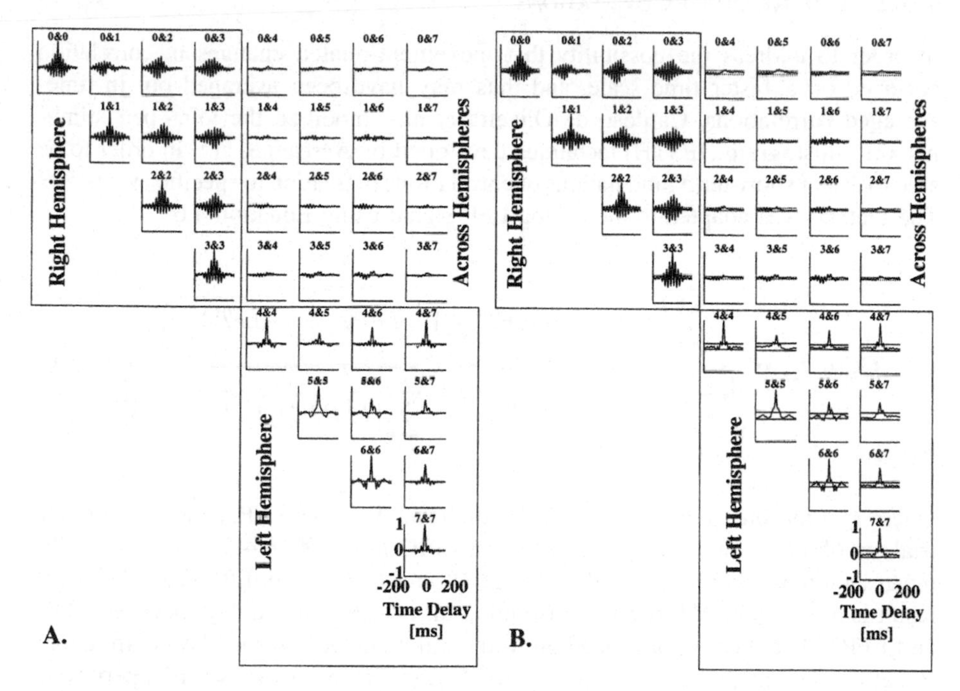

FIGURE 4.14 Time-averaged trial-by-trial cross-correlations among all simultaneously recorded LFPs from one recording session, analyzed during the *hold period* (A), and during *bimanual symmetric* movements to the front (B). Autocorrelations are shown along the diagonal. From each correlogram the shift predictor has been subtracted. Straight horizontal lines indicate a confidence limit for significant correlations based on the standard deviation of the shift predictor. Note that the correlations between the hemispheres are much smaller than those within the same hemisphere. As for the correlations within each hemisphere, note that there are clear correlation patterns. Oscillations are in the gamma range in the right hemisphere and in the alpha range in the left hemisphere. However, there are no differences between corresponding correlograms in the different behavioral conditions (A and B). (Reproduced with permission from Reference 28.)

First, the correlations between the hemispheres are significantly smaller than the correlations within the same hemisphere, suggesting that interhemispheric interactions are less intense than intrahemispheric ones. This is in agreement with EEG studies, which have reported only weak interhemispheric correlations.[40,41] Although weak, many correlations recorded between the hemispheres were statistically significant.

Second, the LFP correlations may show distinct and different correlation patterns. Figure 4.14 shows an extreme example in which all correlograms (auto and cross) in the left hemisphere show oscillatory patterns, while the pattern of correlations in the right hemisphere shows little if any oscillatory activity.

Third, in spite of the clear correlations, the vast majority of the time-averaged correlations were not influenced by behavior, such as the example in Figure 4.14, which fails to show any significant difference between the hold period and the movement period.

4.5.2 Time-Resolved Correlations

In order to address the possibility that movement-related changes in correlation occurred on a faster time scale, and thus may have been averaged out in time-averaged correlations, Cardoso de Oliveira et al.[28] modified the joint–pen–stimulus–time–histogram (JPSTH) technique developed by Aertsen et al.[42] in order to be able to detect short-term modulations of correlations in relation to specific events.[43–45] The method was adapted to the analog LFP signal using Equation 4.6:

$$CC(t_1,t_2)=\frac{\sum_{n=1}^{N}\left(LFP1_n(t_1)-\overline{LFP1(t_1)}\right)\cdot\left(LFP2_n(t_2)-\overline{LFP2(t_2)}\right)}{\sqrt{\sum_{n=1}^{N}\left(LFP1_n(t_1)-\overline{LFP1(t_1)}\right)^2\cdot\sum_{n=1}^{N}\left(LFP2_n(t_2)-\overline{LFP2(t_2)}\right)^2}} \tag{4.6}$$

where t_1 is the time bin from LFP1, t_2 is the time bin from LFP2, and n is the n-th trial out of a total of N. A bar over LFP1 or LFP2 in Equation 4.3 indicates that the mean should be taken across trials (thus, $\overline{LFP1}$ and $\overline{LFP2}$ are mEP1 and mEP2). The result is a matrix of $N \times N$ bins constituting all possible time delays between LFP1 and LFP2. The values corresponding to the simultaneous (zero-delay) correlations fall along the main diagonal of this matrix. Color Figure 4.15 shows joint peri-event time correlogram (JPETC) matrices displayed using a color-coded scale. Time progresses from the bottom-left to the top-right corner such that the value of t_1 (the time index of the first LFP) increases along the x-axis and the value of t_2 (the time index of the other LFP) increases along the y-axis. The bin-wise significance of the correlation coefficients in the JPETC was determined by testing the hypothesis that the correlation coefficient was 0, using a standard t-test. The JPETC in the figure shows an epoch starting 750 msec before movement onset and continuing to 1000 msec after movement onset. Using a time resolution of 2.5 msec, the matrix dimension size is 700 × 700 bins.

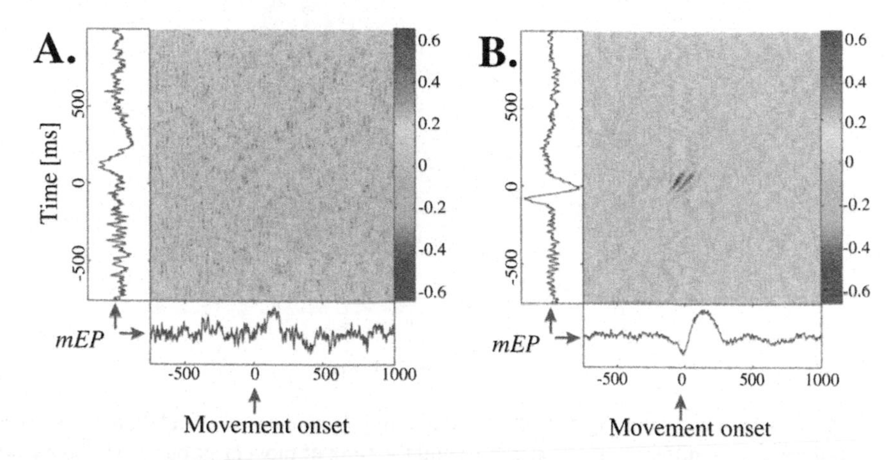

FIGURE 4.15 (see color figure) Example of joint peri-event time correlograms (JPETCs) of a pair of recording sites from different hemispheres demonstrating different correlation patterns during different movements. Each pixel in the JPETC represents the correlation coefficient between all the (single-trial) values of one local field potential (LFP) channel at the corresponding time bin of the x-axis and the values of a second LFP channel at the respective time bin of the y-axis. Correlation is expressed as a correlation coefficient (CC) and is shown in a color code, with the color scale given on the right side of the JPETC. The main diagonal depicts the correlation at delay = 0. (A) The correlation pattern during unimanual movements of the contralateral (left) arm to the front. No correlation is apparent between the two electrodes in this condition. (B) The correlation pattern for the same pair during a bimanual movement of the same amplitude. Movement directions of the two arms differ by 90°, with the left arm moving to the front (as with the unimanual condition shown in A) and the right arm moving to the right. Note the strong correlation with side peaks that arises around movement onset, and lasts for about 100 msec.

Unlike the time-averaged correlation, the JPETC revealed movement-related modulation of correlations in a majority of electrode pairs.

The example shown in Color Figure 4.15 illustrates a case of two recording sites from different hemispheres, where the LFP correlation at zero-delay (on the main diagonal) between two recording sites increased near movement onset (right-side matrix, increased correlation at just the time of movement onset). Interestingly, the same two sites did not show a similar increase when the movement tested was unimanual (left matrix). (Note that the bimanual and unimanual trials were randomly interleaved in the experiment.) This figure illustrates two main results that were consistent across the whole data sample: (1) unlike the time-averaged correlation, the JPETC method revealed movement-related modulation of correlation; (2) the modulations of correlation strength revealed by the JPETC could be movement-specific.

Color Figure 4.15 shows an example where the correlation increased during the movement period. About half (40 to 60% in different hemispheres and monkeys) of the pairs contained significant increases of correlation in relation to movements. However, in even more cases the correlation decreased. Color Figure 4.16 shows a typical example where the correlation was high during the hold period (lower left part of the diagonal in the matrix) and decreased near movement onset. Decreases

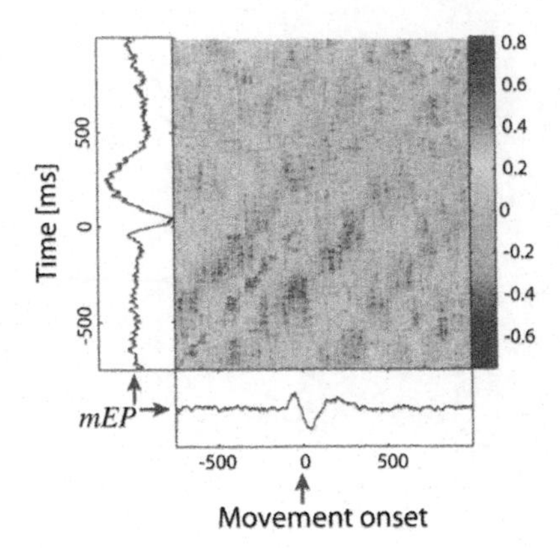

FIGURE 4.16 (see color figure) JPTEC depicting dynamics of correlation between two recording sites in different hemispheres around the time of movement onset. The movements were bimanual nonsymmetric, with the left arm moving to the left and the right arm to the front. Note that the high level of correlation before movement onset (CC ≈ 0.5) is significantly reduced near movement onset and throughout the movement duration.

of correlation during movement were detected in the majority of the diagonals tested (60–80% in different hemispheres and monkeys; note that both increases and decreases could occur in the same JPETC at different times).

In contrast to the time-averaged correlations, in which intrahemispheric correlations were stronger than interhemispheric ones, the movement-related modulations of correlations (detected by the JPETC) were as strong and as frequent for interhemispheric as for intrahemispheric sites. The changes in correlations, including correlations across hemispheres, were associated with both bimanual and unimanual movements.

Interestingly, the typical time courses of increases and decreases of correlation differed from each other. Figure 4.17A depicts, for all the JPETCs in one monkey, the total count of occurrences of significant increases (upward) and decreases (downward) of correlation as a function of time around movement onset. Note that the onset of both increases and decreases is similar at approximately 200 msec before movement, corresponding to a time when the targets had already appeared on the screen. Increases in correlation were sharply peaked around movement onset (during movement planning and initiation). In contrast, decreases in correlation were more broadly distributed and occurred preferentially during the movement. Since the decreases were more common and stronger than the increases, the net correlation change after movement initiation was a decrease of correlation, as shown in Figure 4.17B, depicting the grand average of the correlation strength.

4.5.3 Can Modulation of Interhemispheric Interactions Be Related to Bimanual Coordination?

The changes described above were found for interhemispheric as well as intrahemispheric correlations, in all movement types, with similar temporal profiles. They

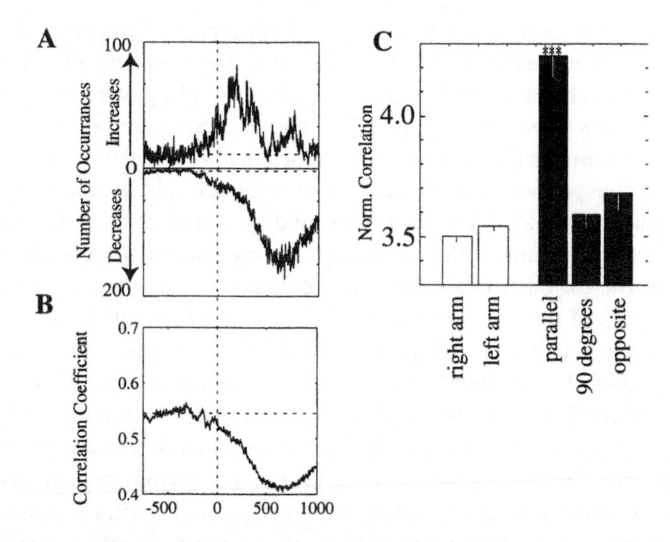

FIGURE 4.17 Movement-related modulations of correlation. Plots A and B show results from all JPETC diagonals in all movement types. (A) Rate of occurrence of increases (upward plot) and decreases (downward plot) of correlation as a function of time around movement onset. Deviations were detected by comparing each time bin in the JPETC diagonals to the hold period. The number of significantly deviating correlations in each time bin is plotted on the y-axis. The vertical dashed line at time 0 indicates movement onset. The horizontal dashed line indicates the average level of randomly occurring deviations in correlation during the hold period. (B) Grand average of the correlation at different time bins around movement onset expressed in correlation coefficients. The graph shows that the result of the increases and decreases shown in A is a net decrease in correlation during movement execution. The vertical dashed line indicates the time of movement onset. The horizontal dashed line shows the average level of correlation during the hold period. (C) Average normalized size of significant increases of correlation between hemispheres revealed within the JPETC diagonals during different unimanual (open bars) and bimanual (black bars) trial types. Note that increased correlations in bimanual symmetric movements are higher than in all other types of movements (marked by three stars; Wilcoxon rank sum test, $a < 0.001$). Lines at the end of each bar represent the standard error of the mean.

leave unanswered the intriguing question of whether these changes are related to the level of coupling between the two arms. What, if anything, characterizes neuronal interactions in relation to bimanual movements? We found that two aspects of neuronal interactions could be related to bimanual coordination.

First, decreased correlations are found with relative uniformity in all movement types, and are an oft-reported phenomenon in population activity.[17,46–49] The different time course of increased and decreased correlations could explain the behavioral finding that bimanual movements are most closely coupled at their initiation and are progressively desynchronized during movement execution.[50–52] The fact that this temporal progression of bimanual decoupling occurs both in symmetric as well as in nonsymmetric movements is consistent with our finding that decreased correlations were found for all bimanual movement types.

Second, comparing the statistics of significant increases in correlations in interhemispheric versus intrahemispheric pairs revealed that interhemispheric correlations were consistently related to the degree of bimanual coupling, whereas the intrahemispheric correlations were not. Figure 4.17C shows a comparison of the normalized average of interhemispheric correlation increases in unimanual, bimanual symmetric, and bimanual nonsymmetric movements. The figure clearly demonstrates that symmetric bimanual movements were accompanied by significantly greater increases in interhemispheric correlations than asymmetric bimanual or unimanual movements. This was not true for pairs from the same hemisphere. At the same time, we found that the velocities of the two arms were more strongly correlated with each other in symmetric than in asymmetric bimanual movements (see Cardoso de Oliveira et al.[28]). This finding suggests that interhemispheric correlations in particular contribute to interlimb coupling and aid in the production of similar movements of the two arms (bimanually symmetric movements). By the same token, interhemispheric coupling may underlie the difficulties we have in producing asymmetric movements. The significantly weaker correlation increases that we found during asymmetric movements may be the result of an active process that reduces coupling, and the residual correlations may be a neural correlate of our inability to completely decouple our arms.

The idea that interhemispheric correlations are related to bimanual coupling has also recently been supported in a study by Serrien et al.,[53] who showed that the stability of bimanual cyclic movements is related to the strength of interhemispheric correlations. This idea is also in line with the finding that split-brain patients (in which the callosal connections have been destroyed) are better than normal individuals in highly asymmetric bimanual tasks.[35,36] Thus, the findings are consistent with the view that interhemispheric correlations reflect neuronal interactions that serve to generate coordinated motor plans for the two hands. The strength of "cross-talk" between the hemispheres may determine the level of coupling between the arms. The time course of correlation changes lends further support to this notion. Correlation increases usually began before movement onset. It is thus feasible that they are related to movement programming or preparation rather than execution. This would be consistent with the observation that motor cortical, but not cerebellar, areas are activated during imaginary bimanual movements.[54] Also, interlimb cross-talk occurs even when a movement is not actually executed, such as when it is only imagined,[55] or when a limb has been amputated.[56] Like the increased correlations described here, cross-talk between two simultaneously planned movements occurs during a transient phase associated with the process of movement preparation.[57]

4.6 CONCLUSION

While we have a long way to go before we can fully understand how the coordination between simultaneous bimanual movements takes place in the brain, the work summarized here provides insights into the neuronal mechanisms involved. It demonstrates that single neurons and the population of neurons in motor cortical fields contain activity evoked specifically when bimanual movements are performed. This modulation of directionally tuned cells during bimanual movements does not interfere with the population rate-code of movement directions, since the population still faithfully predicts both components of a bimanual movement. In addition to rate modulations, our findings indicate specific temporal patterns of activity that are related to the coordination between arm movements. Transient increases of activity correlations are consistently related to the symmetry and correlation of the movements of the two arms. This relation was found solely for sites in different hemispheres. We conclude that flexible interhemispheric correlations may be a neuronal basis for achieving appropriate coordination of the individual movement components.

REFERENCES

1. Leyton, A.S.F. and Sherrington, C.S., Observations on the excitable cortex of the chimpanzee, orangutan and gorilla, *Q. J. Exp. Physiol.*, 1, 137, 1917.
2. Georgopoulos, A.P., Taira, M., and Lukashin, A., Cognitive neurophysiology of the motor cortex, *Science,* 260, 47, 1993.
3. Gray, C.M., König, P., Engel, A.K., and Singer, W., Oscillatory responses in cat visual cortex exhibit inter-columnar synchronization which reflects global stimulus properties, *Nature,* 338, 334, 1989.
4. Kreiter, A.K. and Singer, W., Oscillatory neuronal responses in the visual cortex of the awake macaque monkey, *Eur. J. Neurosci.*, 4, 369, 1992.
5. Engel, A.K., König, P., Kreiter, A., Schillen, T.B., and Singer, W., Temporal coding in the visual cortex: New vistas on integration in the nervous system, *Trends Neurosci.*, 15, 218, 1992.
6. Tanji, J., Kazuhiko, O., and Kazuko, C.S., Neuronal activity in cortical motor areas related to ipsilateral, contralateral, and bilateral digit movements of the monkey, *J. Neurophysiol.*, 60, 325, 1988.
7. Luppino, G., Matelli, M., Camarda, R., and Rizzolatti, G., Corticocortical connections of area F3 (SMA-proper) and area F6 (pre-SMA) in the macaque monkey, *J. Comp. Neurol.,* 338, 114, 1993.
8. Brinkman, C., Supplementary motor area of the monkey's cerebral cortex: short- and long-term deficits after unilateral ablation and the effects of subsequent callosal section, *J. Neurosci.*, 4, 918, 1984.
9. Brust, J.C., Lesions of the supplementary motor area, *Adv. Neurol.*, 70, 237, 1996.
10. Wiesendanger, M., Kaluzny, P., Kazennikov, O., Palmeri, A., and Perrig, S., Temporal coordination in bimanual actions, *Can. J. Physiol. Pharmacol.*, 72, 591, 1994.
11. Kermadi, I., Liu, Y., Tempini, A., Calciati, E., and Rouiller, E.M., Neuronal activity in the primate supplementary motor area and the primary motor cortex in relation to spatio-temporal bimanual coordination, *Somatosens. Mot. Res.*, 15, 287, 1998.
12. Kazennikov, O., Hyland, B., Corboz, M., Babalian, A., Rouiller, E.M., and Wiesendanger, M., Neural activity of supplementary and primary motor areas in monkeys and its relation to bimanual and unimanual movement sequences, *Neurosci.*, 89, 661, 1999.
13. Georgopoulos, A.P., Kalaska, J.F., Caminiti, R., and Massey, J.T., On the relations between the direction of two-dimensional arm movements and cell discharge in primate motor cortex, *J. Neurosci.*, 2, 1527, 1982.
14. Donchin, O., Gribova, A., Steinberg, O., Mitz, A.R., Bergman, H., and Vaadia, E., Single-unit activity related to bimanual arm movements in the primary and supplementary motor cortices, *J. Neurophysiol.*, 88, 3498, 2002.
15. Donchin, O., Gribova, A., Steinberg, O., Bergman, H., and Vaadia, E., Primary motor cortex is involved in bimanual coordination, *Nature,* 395, 274, 1998.

16. Mitzdorf, U., Properties of cortical generators of event-related potentials, *Pharmacopsychiatry,* 27, 49, 1994.
17. Sanes, J.N. and Donoghue, J.P., Oscillations in local field potentials of the primate motor cortex during voluntary movement, *Proc. Nat. Acad. Sci. U.S.A.,* 90, 4470, 1993.
18. Murthy, V.N. and Fetz, E.E., Oscillatory activity in sensorimotor cortex of awake monkeys: synchronization of local field potentials and relation to behavior, *J. Neurophysiol.,* 76, 3949, 1996.
19. Baker, S.N., Kilner, J.M., Pinches, E.M., and Lemon, R.N., The role of synchrony and oscillations in the motor output, *Exp. Brain Res.,* 128, 109, 1999.
20. Eckhorn, R., Bauer, R., Jordan, W., Brosch, M., Kruse, W., Munk, M., and Reitboeck, H.J., Coherent oscillations: a mechanism of feature linking in the visual cortex? Multiple electrode and correlation analyses in the cat, *Biol. Cybern.,* 60, 121, 1988.
21. Pesaran, B., Pezaris, J.S., Sahani, M., Mitra, P.P., and Andersen, R.A., Temporal structure in neuronal activity during working memory in macaque parietal cortex, *Nat. Neurosci.,* 5, 805, 2002.
22. Mehring, C., Rickert, J., Vaadia, E., Cardoso de Oliveira, S., Aertsen, A., and Rotter, S., Inference of hand movements from local field potentials in monkey motor cortex, *Nat. Neurosci.,* 6, 1253, 2003.
23. Donchin, O., Gribova, A., Steinberg, O., Bergman, H., Cardoso de Oliveira, S., and Vaadia, E., Local field potentials related to bimanual movements in the primary and supplementary motor cortices, *Exp. Brain Res.,* 140, 46, 2001.
24. Ben Shaul, Y., Stark, E., Asher, I., Drori, R., Nadasdy, Z., and Abeles, M., Dynamical organization of directional tuning in the primate premotor and primary motor cortex, *J. Neurophysiol.,* 89, 1136, 2003.
25. Amirikian, B. and Georgopoulos, A.P., Modular organization of directionally tuned cells in the motor cortex: is there a short-range order? *Proc. Nat. Acad. Sci. U.S.A.,* 100, 12474, 2003.
26. Bauer, R., Brosch, M., and Eckhorn, R., Different rules of spatial summation from beyond the receptive field for spike rates and oscillation amplitudes in cat visual cortex, *Brain Res.,* 669, 291, 1995.
27. Murthy, V.N. and Fetz, E.E., Synchronization of neurons during local field potential oscillations in sensorimotor cortex of awake monkeys, *J. Neurophysiol.,* 76, 3968, 1996.
28. Cardoso de Oliveira, S., Gribova, A., Donchin, O., Bergman, H., and Vaadia, E., Neural interactions between motor cortical hemispheres during bimanual and unimanual arm movements, *Eur. J. Neurosci.,* 14, 1881, 2001.
29. Steinberg, O., Donchin, O., Gribova, A., Cardosa de Oliveira, S., Bergman, H., and Vaadia, E., Neuronal populations in primary motor cortex encode bimanual arm movements, *Eur. J. Neurosci.,* 15, 1371, 2002.
30. Amirikian, B.R. and Georgopoulos, A.P., Directional tuning profiles of motor cortical cells, *Neurosci. Res.,* 36, 73, 2000.
31. Georgopoulos, A.P., Kettner, R.E., and Schwartz, A.B., Primate motor cortex and free arm movements to visual targets in three-dimensional space. II. Coding of the direction of movement by a neuronal population, *J. Neurosci.,* 8, 2928, 1988.
32. Schwartz, A.B., Motor cortical activity during drawing movements: single-unit activity during sinusoid tracing, *J. Neurophysiol.,* 68, 528, 1992.
33. Li, C.S., Padoa-Schioppa, C., and Bizzi, E., Neuronal correlates of motor performance and motor learning in the primary motor cortex of monkeys adapting to an external force field, *Neuron,* 30, 593, 2001.
34. Rokni, U., Steinberg, O., Vaadia, E., and Sompolinsky, H., Cortical representation of bimanual movements, *J. Neurosci.,* 23, 11577, 2003.

35. Eliassen, J.C., Baynes, K., and Gazzaniga, M.S., Direction information coordinated via the posterior third of the corpus callosum during bimanual movements, *Exp. Brain Res.,* 128, 573, 1999.
36. Eliassen, J.C., Baynes, K., and Gazzaniga, M.S., Anterior and posterior callosal contributions to simultaneous bimanual movements of the hands and fingers, *Brain,* 123 Pt. 12, 2501, 2000.
37. Diedrichsen, J., Hazeltine, E., Nurss, W.K., and Ivry, R.B., The role of the corpus callosum in the coupling of bimanual isometric force pulses, *J. Neurophysiol.,* 90, 2409, 2003.
38. Perkel, D.H., Gerstein, G.L., and Moore, G.P., Neuronal spike trains and stochastic point processes. II. Simultaneous spike trains, *Biophys. J.,* 7, 419, 1967.
39. Vaadia, E., Ahissar, E., Bergman, H., and Lavner, Y., Correlated activity of neurons: a neural code for higher brain functions, in *Neuronal Cooperativity,* Krüger, J., Ed., Springer-Verlag, Berlin, 1991, 249.
40. Andrew, C. and Pfurtscheller, G., Lack of bilateral coherence of post-movement central beta oscillations in the human electroencephalogram, *Neurosci. Lett.,* 273, 89, 1999.
41. Schoppenhorst, M., Brauer, F., Freund, G., and Kubicki, S., The significance of coherence estimates in determining central alpha and mu activities, *Electroencephalogr. Clin. Neurophysiol.,* 48, 25, 1980.
42. Aertsen, A.M.H.J., Gerstein, G.L., Habib, M.K., and Palm, G., Dynamics of neuronal firing correlation: modulation of "effective connectivity," *J. Neurophysiol.,* 61, 900, 1989.
43. Eggermont, J.J., Neural interaction in cat primary auditory cortex. 2. Effects of sound stimulation, *J. Neurophysiol.,* 71, 246, 1994.
44. Sillito, A.M., Jones, H.E., Gerstein, G.L., and West, D.C., Feature-linked synchronization of thalamic relay cell firing induced by feedback from the visual cortex, *Nature,* 369, 479, 1994.
45. Vaadia, E., Haalman, I., Abeles, M., Bergman, H., Prut, Y., Slovin, H., and Aertsen, A., Dynamics of neuronal interactions in monkey cortex in relation to behavioural events, *Nature,* 373, 515, 1995.
46. Baker, S.N., Olivier, E., and Lemon, R.N., Coherent oscillations in monkey motor cortex and hand muscle EMG show task-dependent modulation, *J. Physiol. Lond.,* 501, 225, 1997.
47. Donoghue, J.P., Sanes, J.N., Hatsopoulos, N.G., and Gaal, G., Neural discharge and local field potential oscillations in primate motor cortex during voluntary movements, *J. Neurophysiol.,* 79, 159, 1998.
48. MacKay, W.A. and Mendonça, A.J., Field potential oscillatory bursts in parietal cortex before and during reach, *Brain Res.,* 704, 167, 1995.
49. Pfurtscheller, G. and Lopes da Silva, F.H., Event-related EEG/MEG synchronization and desynchronization: basic principles, *Clin. Neurophysiol.,* 110, 1842, 1999.
50. Boessenkool, J.J., Nijhof, E.J., and Erkelens, C.J., Variability and correlations in bi-manual pointing movements, *Hum. Mov. Sci.,* 18, 525, 1999.
51. Fowler, B., Duck, T., Mosher, M., and Mathieson, B., The coordination of bimanual aiming movements: evidence for progressive desynchronization, *Q. J. Exp. Psychol.,* A43, 205, 1991.
52. Gribova, A., Donchin, O., Bergman, H., Vaadia, E., and Cardoso de Oliveira, S., Timing of bimanual movements in human and non-human primates in relation to neuronal activity in primary motor cortex and supplementary motor area, *Exp. Brain Res.,* 146, 322, 2002.

53. Serrien, D.J., Pogosyan, A.H., and Brown, P., Cortico-cortical coupling patterns during dual task performance, *Exp. Brain Res.*, 155, 204, 2004.
54. Nair, D.G., Purcott, K.L., Fuchs, A., Steinberg, F., and Kelso, J.A., Cortical and cerebellar activity of the human brain during imagined and executed unimanual and bimanual action sequences: a functional MRI study, *Brain Res. Cogn. Brain Res.*, 15, 250, 2003.
55. Heuer, H., Spijkers, W., Kleinsorge, T., and van der Loo, H., Period duration of physical and imaginary movement sequences affects contralateral amplitude modulation, *Q. J. Exp. Psychol.*, A51, 755, 1998.
56. Franz, E.A. and Ramachandran, V.S., Bimanual coupling in amputees with phantom limbs, *Nat. Neurosci.*, 1, 443, 1998.
57. Heuer, H., Spijkers, W., Kleinsorge, T., van der Loo, H., and Steglich, C., The time course of cross-talk during the simultaneous specification of bimanual movement amplitudes, *Exp. Brain Res.*, 118, 381, 1998.

5 What Is Coded in the Primary Motor Cortex?

James Ashe

CONTENTS

ABSTRACT

The motor cortex is the most basic of cortical motor structures and is intimately connected with the control of movement parameters. There has been a great deal of debate as to whether the motor cortex codes for the spatial aspects (kinematics) of motor output, such as direction, velocity, and position, or primarily relates to controlling muscles and forces (kinetics). Although the weight of evidence is in favor of the motor cortex controlling spatial output, the effect of limb biomechanics and forces on motor cortex activity is incontrovertable. Here, I propose (1) that the motor cortex codes for the most behaviorally relevant spatial variable, and (2) that both spatial variables and limb biomechanics are reflected in motor cortex activity.

5.1 INTRODUCTION

Voluntary movement is an elemental function. Without movement we cannot communicate, walk, feed ourselves, or interact with the environment. It is fitting, therefore, that the motor cortex was one of the first cortical areas to be explored experimentally.[1,2]

0-8493-1287-6/05/$0.00+$1.50

Unfortunately, a deep understanding of motor cortex function still eludes us. In what follows, I will deal with the motor cortex proper, i.e., Brodmann area 4, and focus on the control of reaching movements of the upper limb. Also, I will not consider in any detail issues such as how visual signals are integrated into motor behavior, the putative coordinate transformations in other motor areas necessary to convert retinal input into a reference frame meaningful for action, or how the proprioceptive and motor systems interact. I am assuming that the final command for a motor act comes from the motor cortex; this chapter will focus on the properties (characteristics) of such a command.

Our understanding of the function of the motor cortex did not progress significantly, despite the advances made in cataloguing its connectivity, from the detailed experiments of Leyton and Sherrington[3] until the early 1960s, when extracellular neural recording was introduced in studies of the motor system. The earlier studies, notably those of Evarts,[4,5] Cheney and Fetz,[6] Hepp-Reymond,[7,8] and Thach,[9] showed that there was a clear relation between neural activity and muscle force. These experiments also found a relation between cell activity and limb position, and even in those very constrained behavioral paradigms it was clear that the direction of movement was an important controlled variable quite apart from muscle force or posture.[6] Although activation of single muscles or movement at single joints appeared to correlate relatively well with neural activity in motor cortex under very restricted conditions, such explanations no longer held during more naturalistic behaviors. In the first published experiment in which motor cortex cells were studied during free arm movements to different visual targets, there was "no simple relation between the electromyographic (EMG) and single precentral neuron" activity, and "the discharge pattern of single precentral units may be temporally correlated equally well with any number of joint rotations" (p. 144).[10] At about the same time an important behavioral paper suggested that an ideal controller might more efficiently regulate movement by coding for parameters at the endpoint of the limb.[11]

5.2 DIRECTION TUNING IN MOTOR CORTEX; CONTROL OF KINEMATIC VARIABLES

In what proved to be a very influential study, Georgopoulos and colleagues[12] showed that the activity of cells in the motor cortex was best related to the direction of movement. In that experiment, monkeys made visually instructed two-dimensional movements to targets from a central starting point. The frequency of discharge of the majority of cells varied in an orderly fashion with the direction of movement (Figure 5.1); such cells were regarded as being directionally tuned, and their activity could be expressed as a sinusoidal function of the direction of movement. The preferred direction for tuned cells was that in which their neural discharge was highest. Similar results were found for movements in three dimensions;[13] the neural activity was related to the direction of movement M as follows:

$$d(M) = b_0 + b_x x + b_y y + b_z z + e \qquad (5.1)$$

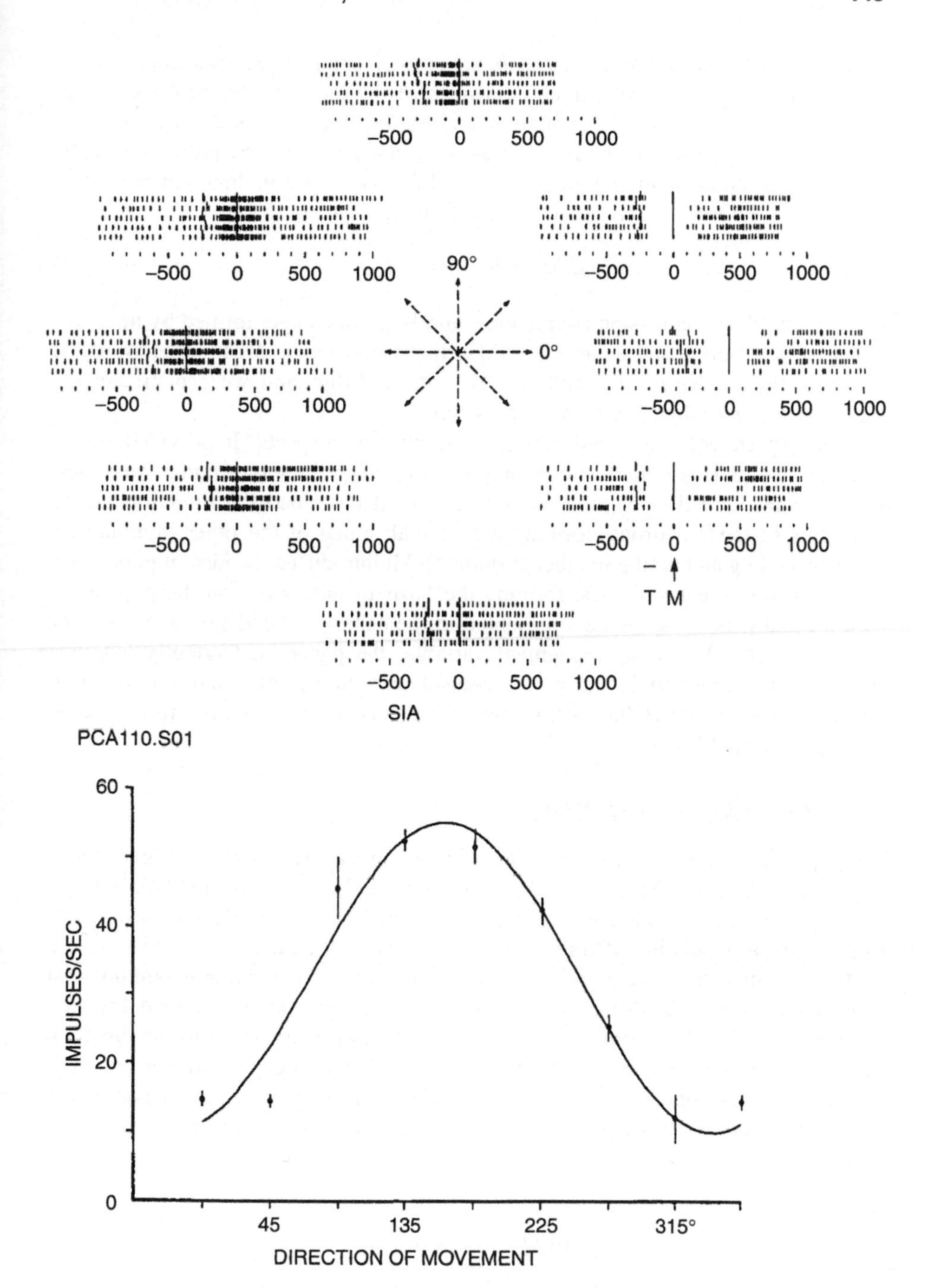

FIGURE 5.1 Variation in the frequency of discharge of a single motor cortex cell during movement in different directions. *Upper half*: Each small tick indicates an action potential; the display shows impulse activity during five repetitions (trials) of movements made in each of the eight directions indicated in the center diagram. Trials are oriented to the onset of movement *M*. *Lower half*: Direction tuning curve of the same cell; the average frequency of cell activity during the response time and movement time is plotted for each of the eight directions. (From Reference 12, Figure 4, with permission.)

where d is the frequency of discharge, b_0 is the intercept, b_x–b_z are partial regression coefficients, $[x, y, z]$ are the direction cosines of the three-dimensional movement vector, and e is an error term. Equation 5.1 implies that there is a direction C for which the discharge of the cell is the highest, which is the preferred direction of the cell. One can relate the discharge rate of a cell for movement in direction M relative to the discharge for its preferred direction C using Equation 5.2:

$$d(\boldsymbol{M}) = b_0 + k\cos \theta_{CM} \tag{5.2}$$

where b_0 and k are regression coefficients, and θ_{CM} is the angle formed by the cell's preferred direction C and a particular direction of movement M. In other words, if we know the discharge of a cell for the preferred direction we can predict the discharge for any other direction of movement.

The experiment was significant in a number of respects. It provided strong evidence that, during naturalistic behaviors, neurons in the motor cortex were best related to the direction of movement in space. It described the direction tuning properties of these neurons, forming a link with much of the other literature on direction coding and tuning in other systems.[14,15] It introduced the idea of population coding of movement variables, forming the basis of later work on the population vector. Finally, because the data presented in the paper provided strong support for the coding of a kinematic movement variable, the paper inadvertently led to a polarization of opinion in the motor control community into camps advocating kinematic and kinetic coding of movement; this debate is as vibrant now as when the paper was published.

5.3 POPULATION CODING

The population coding of movement direction was put forward first as a suggestion,[12] then as a formal hypothesis,[16,17] and finally with detailed neural data for movements in two[17] and three dimensions.[18,19] The concept underlying population coding is as follows: for a population of directionally tuned neurons, each neuron will make a vectorial contribution to the code in its preferred direction, and at a magnitude that is proportional to the angle between its preferred direction and the intended direction of movement. In other words, cells with a preferred direction close to the direction of movement make a greater contribution; those further away, a smaller one. The vector sum of the contributions of a population of cells is used to form the population vector that predicts the direction of the upcoming movement. In formal terms, the population vector can be expressed as follows:

$$P(M) = \sum_{i=1}^{N} w_i(M)C_i \tag{5.3}$$

where $P(M)$ is the neural population vector in movement direction M; C_i is the preferred direction of the ith cell in the population; and $w_i(M)$ is a weighting function that reflects the magnitude of the contribution of ith cell to the population vector for movement in direction M.

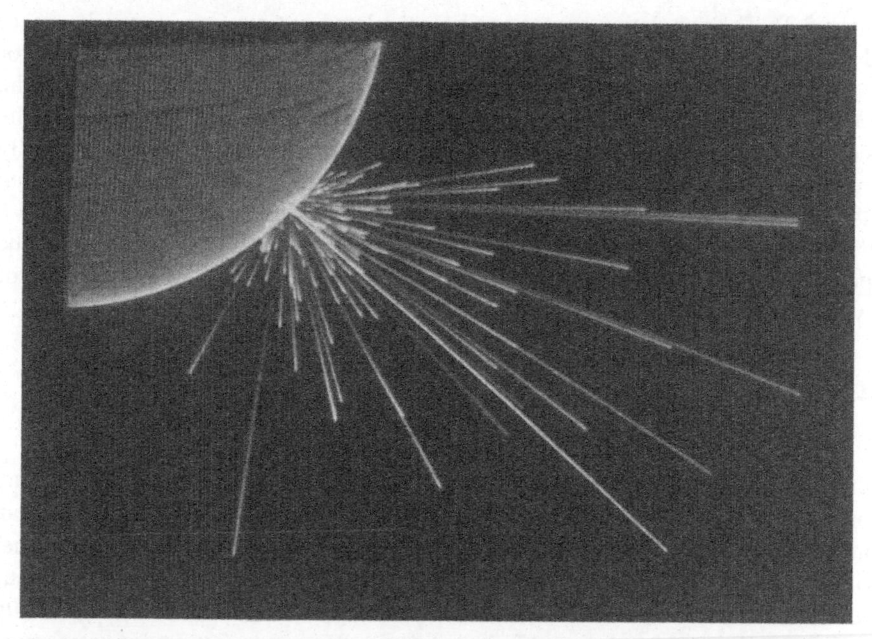

FIGURE 5.2 (see color figure) An example of the population coding of movement direction. The blue lines represent the vectorial contribution of individual cells in the population (N = 475). The actual movement direction is in yellow and the direction of the population vector is in red. (From Reference 19, Figure 1, with permission.)

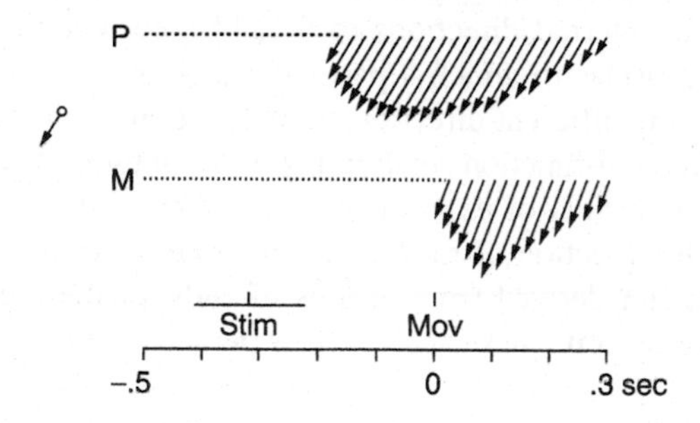

FIGURE 5.3 Evolution of the population vector in time, before and during an instructed arm movement. A time series of population (P) and movement (M) vectors is shown. The instructed movement direction is indicated by a small arrow on the far left. The population vector can be seen to increase in size and point in the direction of the upcoming movement before the movement occurs. *Stim* is onset of target instruction; *MOV* is onset of movement. (Adapted from Reference 19, Figure 4, with permission.)

The population vector is an accurate reflection of the direction of movement (Color Figure 5.2*). It can also be derived during the response time or in delay periods before movement actually begins. In these contexts, the population vector may reflect the "intention" of a population of motor cortex cells in relation to movement

* See color insert following page 170.

(Figure 5.3). The population vector can be calculated in time (e.g., every 10 msec) and therefore gives a continuous readout of the activity of cells that predicts motor behavior in advance. For the population vector analysis to accurately predict the direction of movement, the following three conditions need to be satisfied:[19] (1) the directional tuning function is one of a broad category of functions that are radially symmetric around a preferred direction; (2) the preferred directions of the tuned cells are uniformly distributed; and (3) the values of the tuning parameters k and b are randomly distributed relative to the preferred directions. The accuracy of the population vector is relatively resistant to cell loss and is a good predictor of movement direction with as little as 20 tuned cells.[20]

5.4 WHAT DO DIRECTION TUNING AND THE POPULATION VECTOR REPRESENT?

Although the directional properties of both single cells and populations of cells are strongly correlated with the direction of movement, this does not necessarily mean that the cells code for the direction of movement alone. There are many other variables, such as muscle activity, that also covary with movement direction and thus might be reflected in cell activity. It is known that single cells in motor cortex generally engage several motoneuronal pools[21–23] and thus influence the activity of many muscles often distributed across more than one joint.[24] It can be reasonably assumed that cells may engage different muscles at different strengths resulting in a set of muscles with a preferred direction. The combinatorial possibilities of different muscles, grouped together in distinct sets with different weights, would result in a very large number of potential preferred directions to which motor cortex cells might relate. Because the set of muscles to which a particular cell projects is likely to be active for movements in many different directions, the direction tuning will be quite broad. This view of the structure-function relation between motor cortex cells and muscles is consistent with the results of experiments on preferred direction and population vector. Therefore, the direction of movement in space is not the only interpretation of the population vector derived from groups of cells, or the preferred direction in individual motor cortex cells; nevertheless, it is perhaps the most parsimonious.

5.5 MUSCLES OR MOVEMENTS: THE CURRENT CONTROVERSY

The issue of what is represented in the activity of motor cortex neurons is not new. It dates back at least, and most famously, to the writings of Hughlins Jackson toward the end of the 19th century. That the topic should still be hotly debated is just as remarkable as if current geneticists and molecular biologists were battling it out over some seminal statement of Gregor Mendel. Jackson wrote, "To speak figuratively, the central nervous system [read "motor cortex"] knows nothing of muscles, it only knows movements."[25,26] Unfortunately, the statement has been interpreted literally and not "figuratively," and thus has created two polarized groups within the motor control community. One group holds that motor cortex "knows nothing of muscles," while the other, adopting an equally extreme position, believes that motor cortex knows nothing of movements.

5.6 STATEMENT OF THE PROBLEM

It is obvious what the brain must do for us to successfully make voluntary movements to a target. The target is initially represented within the visual system in retinotopic coordinate space. We assume that the representation of the target is then transformed into a coordinate frame that is relevant to the upper limb: either allocentric or "world space," or an egocentric frame that is anchored to the body. Finally, the brain commands the arm to move toward the appropriately transformed target. What form do these commands take? Does the motor cortex specify the exact activation of the muscles around the shoulder and elbow joints, so that the appropriate torques are produced to bring the arm to its target (the extreme kinetic position)? There is no disagreement that these torques have to be specified at some level for movement to occur; the issue is whether this specification takes place in the motor cortex. Alternatively, does the motor cortex plan the spatial trajectory of the movement alone (the extreme kinematic position)? Movement kinetics refers to the forces produced and their derivatives. At the lowest or most fundamental level, these are the forces produced by individual muscles. However, the level of control could equally well be that of the torques produced at specific joints or, indeed, the total force generated by the limb. Kinematics refers to the spatial variables of movement, such as position, velocity, acceleration, and direction. As is the case with kinetics, the kinematic variables can be defined for muscles, joints, or the whole limb. The class of variable, kinetic or kinematic variable, and the coordinate frame for movement control are hypothetically independent. Nevertheless, kinetic coding is much more likely in muscle or joint space. Similarly, kinematic coding is more probable in allocentric (or extrinsic) space.

5.7 THE CASE FOR KINEMATIC CONTROL

More than two decades ago, Morasso[11] demonstrated what in retrospect seems obvious: that it is computationally much less demanding to control the kinematics of the endpoint than the kinematics at the component joints during movements of the arm (Figure 5.4). There is compelling evidence from the psychophysical literature that movement is indeed planned in terms of extrinsic coordinates,[27–30] although there are other views.[31] However, though knowledge of the intrinsic geometry of the limb may be a necessary part of the planning process.[32,33] Viviani and colleagues have put forward a "vector coding" hypothesis which holds that during voluntary movement the target information delivered to the motor system is in vector format in extrinsic coordinates. Of course, the executive motor system, and particularly the motor cortex, is under no obligation to operate directly on the information in this vector format, although, again, it may be the most parsimonious approach.

The most compelling evidence in favor of the kinematic control of movement comes not from the psychophysical literature, but from direct neural recording in

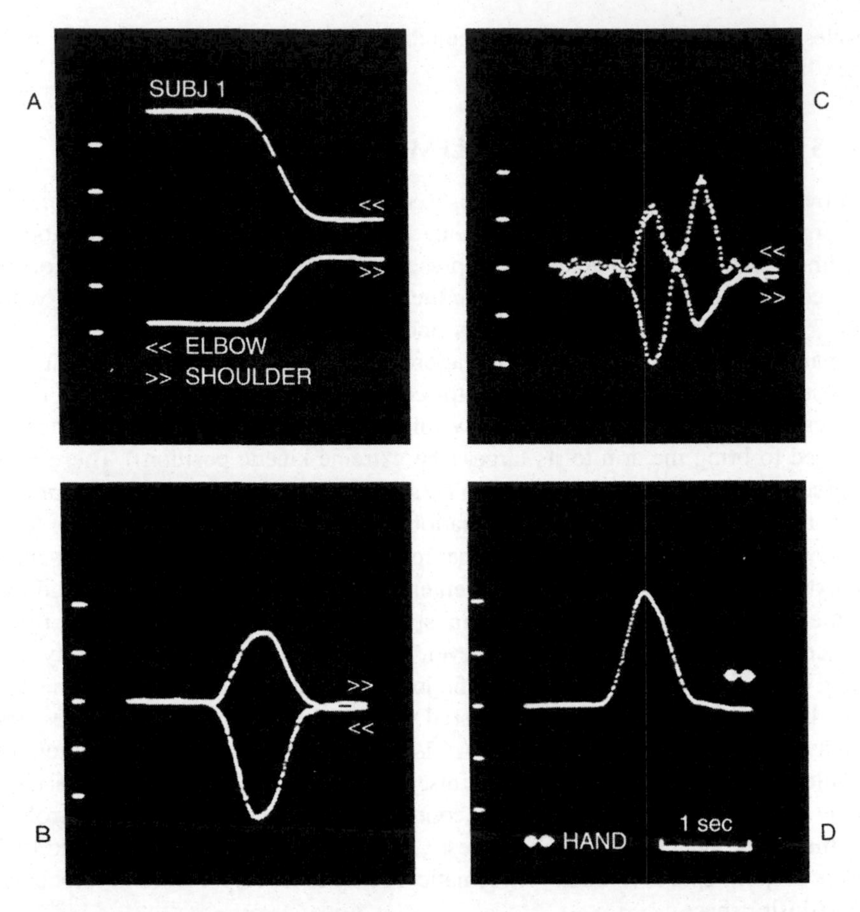

FIGURE 5.4 Records from one subject during a point-to-point movement on a two-dimensional plane. (A) Change in position at the shoulder and elbow joints. (B) Velocity at the shoulder and elbow joints. (C) Acceleration at the two joints. (D) Velocity at the hand. It is obvious that the hand velocity would be the simplest variable to control. (Adapted from Reference 11, Figure 3, with permission.)

the motor cortex during a variety of motor behaviors. As mentioned above, there is a large body of work demonstrating that cells in the motor cortex relate strongly to the direction of arm movement in space.[12,18,34–39] Because direction of movement varies along with several other movement variables, which are also potential control variables, it was necessary to dissociate these variables systematically during neural recording studies. Two sets of comprehensive dissociation experiments have been performed. Alexander and Crutcher[40,41] dissociated the direction of arm movement from the muscles used in a visually controlled task by applying loads to a one-dimensional manipulandum. Approximately one third of cells in the monkey motor cortex were related to muscle activation during the execution of movement[41] and an even smaller proportion during a preparatory period before movement began.[40] Kakei and colleagues[42] dissociated muscle activity, posture, and direction of movement in space during a two-dimensional wrist movement task. They found that about 25%

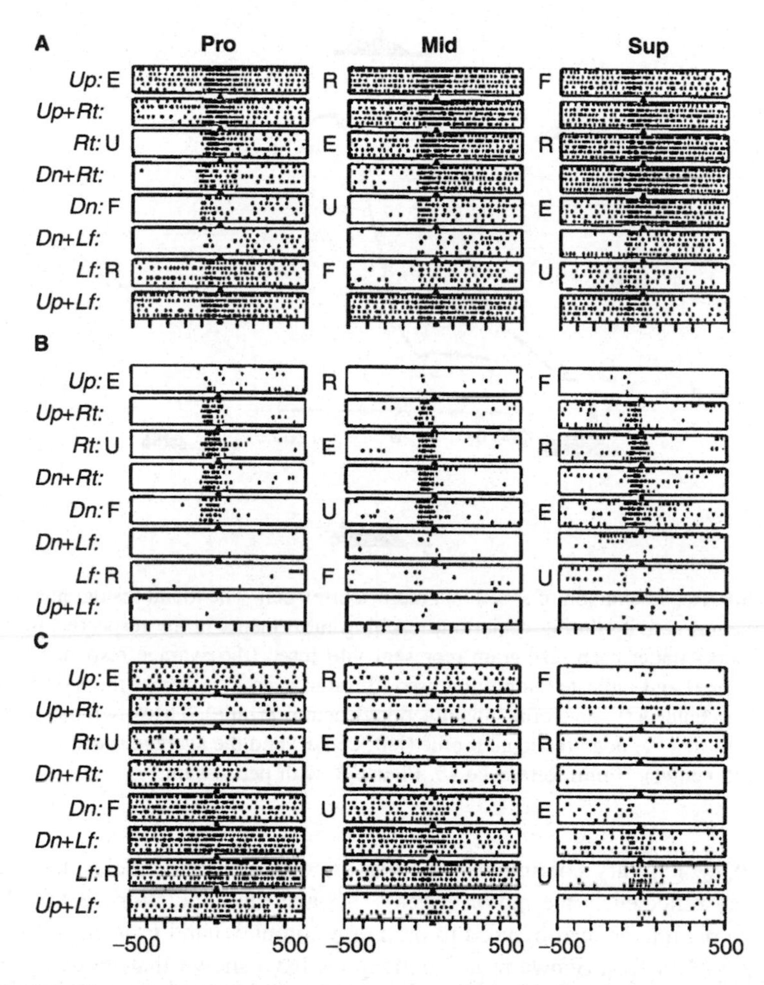

FIGURE 5.5 The records of three single neurons (A, B, C) in the motor cortex for movements in six different directions (Up, Up and Rt, etc.) in each of three different wrist postures (Pro, Mid, Sup). All the rasters are aligned to the onset of movement. The tuning of cell B does not change across the different postures; therefore, it can be categorized as "extrinsic." The activity of cell A changes consiberably for the different postures, its preferred direction changing by 79° from Pro to Sup, and can be regarded as "muscle-like." Cell C is an extrinsic-like neuron that is also influenced by wrist posture. (From Reference 42, Figure 2, with permission.)

of motor cortex cells had "muscle" properties, while approximately 50% related to direction of movement in space, and reasonably concluded that both "muscles" and "movements" were represented in the motor cortex (Figure 5.5).

The motor cortex not only seems to encode relatively static kinematic parameters such as direction during point-to-point movements, but can also reflect parameters that change continuously during straight movements such as position, velocity, and acceleration.[43] The coding principles developed for simple, straight movements to a target also hold true when applied to more complex ones. For example, direction

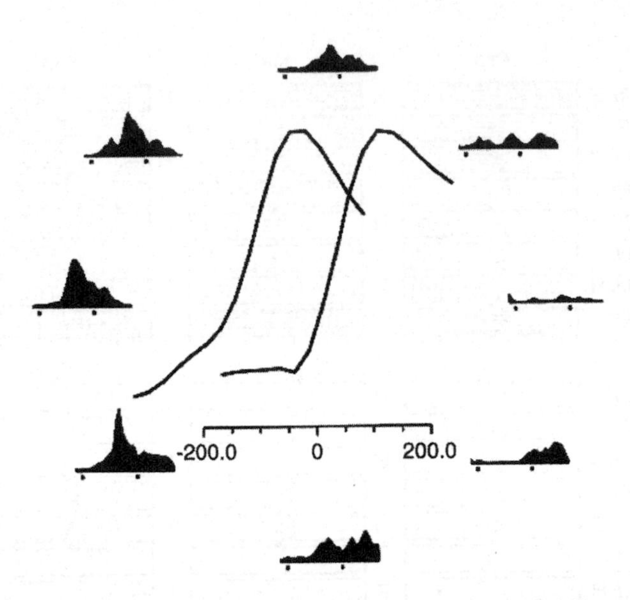

FIGURE 5.6 Representation of speed in a motor cortex cell. The radial histograms show the averaged neural activity during center-out reaching movements in the respective directions. The tick marks under each histogram represent 440 msec (the average response time plus movement time) and indicate the portion used, through averaging across the 8 directions, to generate the center-left waveform (nondirectional neural profile). There is remarkable concordance between the nondirectional profile (left, center) and the average speed of the movements (right, center). (From Reference 45, Figure 3, with permission.)

and speed, which vary continuously during drawing or spiral tracing movements, are strongly reflected in the motor cortex[38,44–46] and the population vectors derived from these parameters can be used to predict a complex hand trajectory accurately (Figure 5.6).[45] In fact, Schwartz and colleagues have shown that models that take both the direction and speed of movement into account provide more accurate descriptions of motor cortex activity than those using direction alone. Fitting neural activity to other time-varying movement parameters like EMG or joint-angle velocity resulted in a much less accurate model of the data than that obtained for the trajectory of the hand.[45] The obvious conclusion from these studies is that during drawing and other continuous movements, it is the kinematics of movement in extrinsic space that is primarily reflected in motor cortex activity.

5.8 THE CASE FOR KINETIC CONTROL

Cells in the motor cortex have prominent projections to the spinal cord, and some have monosynaptic projections to motoneurons that in turn directly control muscle activation. Muscles and their output, force, are the obvious control variable for cells in motor cortex. Evarts[4,5] was the first to show a relation between motor cortex activity and the force generated by the muscles. Since then, a large number of studies have shown relations between motor cortex and the magnitude, direction, and rate of change in force. (See References 47 and 48 for reviews.) However, with a relatively simple behavior it is possible to find a relation between almost any motor variable and motor cortex activity. Also, more than one variable may be encoded in the activity of cells.[43] The case in favor of kinetic control has rested strongly on showing that the coding of kinematic variables alone is not a complete explanation of the variations in cell activity one sees in certain motor behaviors.[49,50] These last studies showed that using different arm configurations to make movements to a set of spatial targets resulted in changes in the activity of single cells in the motor cortex, which suggested that the trajectory of the hand in space was not the only aspect of the behavior being coded, as this did not change significantly for different arm configurations. In addition, other work has demonstrated that the location of the hand, and hence the configuration of the arm, may have a systematic effect on the direction tuning of cell activity during an isometric ramp and hold task,[53] in which no actual movements were produced.

Despite the influence of arm posture on the activity of single cells in the motor cortex, the direction of the population vector, based on the activity of these cells, has been relatively resistent to changes in arm posture.[49] Nevertheless, in some circumstances, such as when three-dimensional movements are made to targets from different starting points, even the population vector appears to be modulated by limb biomechanics.[34] Further evidence for the effect of biomechanics on the population vector comes from experiments in which the biomechanics at the shoulder and elbow joints were accurately measured, showing that the nonuniformity in the distribution of population vectors was a function both of velocity and torque at the joints.[51] Taking these experiments as a whole, one cannot but conclude that the kinetic output has a distinct influence on the activity of motor cortex cells, although the effect of biomechanics on the population vector has been modest. It is obvious that arm kinematics alone cannot account for the changes in neural activity that have been observed. However, such a conclusion is quite different from stating that the motor cortex codes primarily for the kinetics of motor output.

5.9 SUCCESSIVE COORDINATE TRANSFORMATIONS

Much has been made, particularly in the psychophysical literature, of the concept of successive coordinate transformations to explain how visual targets, initially defined in retinal coordinates, can be reached by the arm, for which the coordinate frame is defined by the joints and muscles. As discussed above, there is clear evidence that movement is first specified in terms of kinematics, but the actual movement is ultimately produced by a weighted activation of groups of muscles (kinetics). The hypothesis underlying successive coordinate transformations is that different motor areas, including several sub-areas of parietal cortex, participate in the various stages of this transformation from kinematics to kinetics. The common wisdom is that the motor cortex would either be involved in the final stage of the kinematic to kinetic transformation or would implement the kinetics on instructions from a "higher" motor area such as the lateral premotor cortex or the supplementary motor area.[52] As mentioned above, the idea that the motor cortex implements kinetics alone is not tenable on the basis of current evidence. There is more evidence in favor of the motor cortex being instrumental in some kind of kinematic to kinetic transformation, though the form of such a transformation is not at all clear.[34,49,51,53]

5.10 A MODEST PROPOSAL

To paraphrase, or contort, the view of Hughlins Jackson: although the motor cortex knows of both muscles and movements, it appears to be concerned primarily with space. In other words, the motor cortex primarily codes for the most relevant spatial aspects of motor output, both in the case of movement and during behaviors that are purely isometric.[35,47,53–55] One simple experiment illustrates this point. Let us imagine that one is required to make force pulses in different target directions in the presence of opposing forces. The muscle forces exerted will not be in the direction of the targets, because one has to neutralize the opposing forces. Will the activity of motor cortex cells reflect the actual forces produced by the muscle or the resultant force (a combination of the muscle force and the opposing force), which is inevitably in the direction of the target? Using this behavioral paradigm in the monkey, it was shown that both the single cell and population activity in the motor cortex related to the resultant force, which was the most relevant spatial variable, and not to the forces produced by the muscles.[55] For more than two decades, motor control has been dominated by studies examining the relation between motor cortex cells and the spatial aspects of motor output. If motor cortex codes for the spatial aspects of behavior, in what coordinate framework does this coding occur? It is likely that the coding occurs in the coordinate frame that is most relevant for the behavior. For example, the majority of studies has used reaching movements and it is no surprise that cell activity in those cases reflects an extrinsic reference frame anchored to the hand. If instead, one were to perform the behavior using the elbow as a pointer, this would likely be the reference point. Similarly, manipulation of objects by the hands would be coded in a reference frame that might well be muscle or joint-based relative to the hands.

However, a coherent theory of coding in the motor cortex must also account for the clear effect of biomechanical factors on cell activity.[34,49–51,53] Just as gain fields have been used to explain the interaction of several different frames of reference on the activity of single neurons in the posterior parietal cortex,[56] we can perhaps use a similar framework to explain findings in motor cortex neural recordings. The composition of such a gain field is as yet uncertain. The available data suggests that if such a gain field exists, it is comprised of unequal partners, the neural activity relating to the spatial output predominating, but modulated in a systematic way dependent on the biomechanics of the limb. For motor behaviors that are performed in two dimensions — for example, the reaching movements in monkeys — one might conceptualize such a field as a plane with a relatively shallow slope. Of course, we currently lack the type of complete quantitative data that would be necessary in order to construct such putative gain fields accurately, but systematic studies of this issue are currently being conducted. We predict that the slope of such gain fields is likely to be small and that the representation of space by the motor cortex, as in the parietal cortex,[56] is likely to be distributed.

5.11 THE CHALLENGE

If such gain fields do in fact exist, then a major challenge for those concerned with the cortical control of motor behavior will be to understand how a cortical representation of space, modulated by limb biomechanics, is translated into the muscle or joint coordinate frame that will ultimately be required for implementation of the behavior. There are some intriguing possibilities. Bizzi and colleagues[57–60] have shown, in experiments in the frog and rat, that a set of "motor primitives," which could form the basis of activating specific sets of muscles during multiple joint movement, can be elicited through microsimulation of the spinal gray matter. These primitives may form the building blocks for voluntary movement by translating spatial signals from the motor cortex into appropriate muscle output. Recent data from experiments using long trains of intracortical microstimulation suggest that the motor cortex may be able to access such primitives directly.[61] In addition, other spinal interneuronal systems such as the propriospinal system in the cat[62] have been shown to be important in the patterned activation of the different muscles required for reaching. These propriospinal interneurons may participate in the integration of reaching movements at a spinal level, and may effectively translate signals from cells in the motor cortex that relate to the direction of force output of the whole limb[55] into appropriate patterns of muscle activation. Another question is how motor cortex learns to access such motor primitives. It is likely that the association between motor cortex cell activity and motor primitive "modules" at another level in the motor system is established through learning and adaptation.

Though conceptually attractive, the idea of successive coordinate transformations in frontal motor areas culminating in a muscle or joint based coding of motor output in motor cortex[63] does not have strong experimental support and should be abandoned, at least as applied to skilled movements. The search for a direct reflection of the motor periphery in the motor cortex is likely to be as futile as the quest for the representation of the single muscle.

REFERENCES

1. Fritsch, G. and Hitzig, E., Über die elektrische Erregbarkeit des Grosshirns, *Arch. Anat. Physiol. Wis. Med.*, 37, 300, 1870.
2. Ferrier, D., Experiments in the brain of monkeys, *Proc. R. Soc. Lond. (Biol.)*, 23, 409, 1875.
3. Leyton, A.S.F. and Sherrington, C.S., Observations on the excitable cortex of the chimpanzee, orangutan and gorilla, *Qu. J. Exp. Physiol.*, 11, 135, 1917.
4. Evarts, E.V., Relation of pyramidal tract to force exerted during voluntary movement, *J. Neurophysiol.*, 31, 14, 1968.
5. Evarts, E.V., Activity of pyramidal tract neurons during postural fixation, *J. Neurophysiol.*, 32, 375, 1969.
6. Cheney, P.D. and Fetz, E.E., Functional classes of primate corticomotoneuronal cells and their relation to active force, *J. Neurophysiol.*, 44, 773, 1980.
7. Smith, A.M., Hepp-Reymond, M.C., and Wyss, U.R., Relation of activity in precentral cortical neurons to force and rate of force change during isometric contractions of finger muscles, *Exp. Brain Res.*, 23, 315, 1975.

8. Hepp-Reymond, M.C., Wyss, U.R., and Anner, R., Neuronal coding of static force in the primate motor cortex, *J. Physiol.,* 74, 287, 1978.
9. Thach, W.T., Correlation of neural discharge with pattern and force of muscular activity, joint position, and direction of intended next movement in motor cortex and cerebellum, *J. Neurophysiol.,* 41, 654, 1978.
10. Murphy, J.T., Kwan, H.C., MacKay, W.A., and Wong, Y.C., Activity of primate precentral neurons during voluntary movements triggered by visual signals, *Brain Res.,* 236, 429, 1982.
11. Morasso, P., Spatial control of arm movements, *Exp. Brain Res.,* 42, 223, 1981.
12. Georgopoulos, A.P., Kalaska, J.F., Caminiti, R., and Massey, J.T., On the relations between the direction of two-dimensional arm movements and cell discharge in primate motor cortex, *J. Neurosci.,* 2, 1527, 1982.
13. Schwartz, A.B., Kettner, R.E., and Georgopoulos, A.P., Primate motor cortex and free arm movements to visual targets in three-dimensional space. I. Relations between single cell discharge and direction of movement, *J. Neurosci.,* 8, 2913, 1988.
14. Henn, V. and Cohen, B., Coding of information about rapid eye movements in the pontine reticular formation of alert monkeys, *Brain Res.,* 108, 307, 1976.
15. Dykes, R.W., Afferent fibers from mystacial vibrissae of cats and seals, *J. Neurophysiol.,* 38, 650, 1975.
16. Georgopoulos, A.P., Kalaska, J.F., Crutcher, M.D., Caminiti, R., and Massey, J.T., The representation of movement direction in the motor cortex: single cell and population studies, in *Dynamic Aspects of Neocortical Function,* Edelman, G.M., Gall, W.E., and Cowan, W.M., Eds., John Wiley & Sons, New York, 1984, 501.
17. Georgopoulos, A.P., Caminiti, R., Kalaska, J.F., and Massey, J.T., Spatial coding of movement: a hypothesis concerning the coding of movement direction by motor cortical populations, *Exp. Brain Res. Suppl.,* 7, 327, 1983.
18. Georgopoulos, A P., Schwartz, A.B., and Kettner, R.E., Neuronal population coding of movement direction, *Science,* 233, 1416, 1986.
19. Georgopoulos, A.P., Kettner, R.E., and Schwartz, A.B., Primate motor cortex and free arm movements to visual targets in three-dimensional space. II. Coding of the direction of movement by a neuronal population, *J. Neurosci.,* 8, 2928, 1988.
20. Lukashin, A.V., Amirikian, B.R., Mozhaev, V.L., Wilcox, G.L., and Georgopoulos, A.P., Modeling motor cortical operations by an attractor network of stochastic neurons, *Biol. Cybern.,* 74, 255, 1996.
21. Fetz, E.E. and Cheney, P.D., Postspike facilitation of forelimb muscle activity by primate corticomotoneuronal cells, *J. Neurophysiol.,* 44, 751, 1980.
22. Shinoda, Y., Yokota, J., and Futami, T., Divergent projection of individual corticospinal axons to motoneurons of multiple muscles in the monkey, *Neurosci. Lett.,* 23, 7, 1981.
23. Lemon, R.N., Mantel, G.W.H., and Muir, R.B., Corticospinal facilitation of hand muscles during voluntary movement in the conscious monkey, *J. Physiol. (London),* 381, 497, 1986.
24. McKiernan, B.J., Marcario, J.K., Karrer, J.H., and Cheney, P.D., Corticomotoneuronal postspike effects in shoulder, elbow, wrist, digit, and intrinsic hand muscles during a reach and prehension task, *J. Neurophysiol.,* 80, 1961, 1998.
25. Jackson, J.H., On the anatomical investigation of epilepsy and epileptiform convulsions, in *Selected Writings of John Hughlings Jackson,* Hodder & Stoughton, London, 1932, 113.
26. Jackson, J.H., On the comparative study of diseases of the nervous system, *Br. Med. J.,* 2, 355, 1889.

27. Vindras, P., Desmurget, M., Prablanc, C., and Viviani, P., Pointing errors reflect biases in the perception of the initial hand position, *J. Neurophysiol.,* 79, 3290, 1998.
28. Vindras, P. and Viviani, P., Altering the visuomotor gain. Evidence that motor plans deal with vector quantities, *Exp. Brain Res.,* 147, 280, 2002.
29. Rogosky, B.J. and Rosenbaum, D.A., Frames of reference for human perceptual-motor coordination: space-based versus joint-based adaptation, *J. Motor Behav.,* 32, 297, 2000.
30. Rosenbaum, D.A., Meulenbroek, R.G., and Vaughan, J., Planning reaching and grasping movements: theoretical premises and practical implications, *Motor Contr.,* 5, 99, 2001.
31. Soechting, J.F. and Flanders, M., Moving in three-dimensional space: frames of reference, vectors, and coordinate systems, *Annu. Rev. Neurosci.,* 15, 167, 1992.
32. Rossetti, Y., Desmurget, M., and Prablanc, C., Vectorial coding of movement: vision, proprioception, or both? *J. Neurophysiol.,* 74, 457, 1995.
33. Paillard, J. and Brouchon, M., A proprioceptive contribution to the spatial encoding of position cues for ballistic movements, *Brain Res.,* 71, 273, 1974.
34. Caminiti, R., Johnson, P.B., and Urbano, A., Making arm movements within different parts of space: Dynamic aspects in the primate motor cortex, *J. Neurosci.,* 10, 2039, 1990.
35. Kalaska, J.F., Cohen, D.A.D., Hyde, M.L., and Prud'homme, M., A comparison of movement direction-related versus load direction-related activity in primate motor cortex, using a two-dimensional reaching task, *J. Neurosci.,* 9, 2080, 1989.
36. Hatsopoulos, N.G., Ojakangas, C.L., Paninski, L., and Donoghue, J.P., Information about movement direction obtained from synchronous activity of motor cortical neurons, *Proc. Nat. Acad. Sci. U.S.A.,* 95, 15706, 1998.
37. Serruya, M.D., Hatsopoulos, N.G., Paninski, L., Fellows, M.R., and Donoghue, J.P., Instant neural control of a movement signal, *Nature,* 416, 141, 2002.
38. Schwartz, A.B., Direct cortical representation of drawing, *Science,* 265, 540, 1994.
39. Taylor, D.M., Tillery, S.I., and Schwartz, A.B., Direct cortical control of 3D neuroprosthetic devices, *Science,* 296, 1829, 2002.
40. Alexander, G.E. and Crutcher, M.D., Preparation for movement: neural representations of intended direction in three motor areas of the monkey, *J. Neurophysiol.,* 64, 133, 1990.
41. Crutcher, M.D. and Alexander, G.E., Movement-related neuronal activity selectively coding either direction or muscle pattern in three motor areas of the monkey, *J. Neurophysiol.,* 64, 151, 1990.
42. Kakei, S., Hoffman, D.S., and Strick, P.L., Muscle and movement representations in the primary motor cortex, *Science,* 285, 2136, 1999.
43. Ashe, J. and Georgopoulos, A.P., Movement parameters and neural activity in motor cortex and area 5, *Cereb. Cortex,* 4, 590, 1994.
44. Schwartz, A.B. and Moran, D.W., Motor cortical activity during drawing movements: population representation during lemniscate tracing, *J. Neurophysiol.,* 82, 2705, 1999.
45. Moran, D.W. and Schwartz, A. B., Motor cortical activity during drawing movements: population representation during spiral tracing, *J. Neurophysiol.,* 82, 2693, 1999.
46. Moran, D.W. and Schwartz, A.B., Motor cortical representation of speed and direction during reaching, *J. Neurophysiol.,* 82, 2676, 1999.
47. Ashe, J., Force and the motor cortex, *Behav. Brain Res.,* 87, 255, 1997. [Corrected and republished article originally printed in *Behav. Brain Res.,* 86, 1, 1997.]
48. Hepp-Reymond, M.-C., Functional organization of motor cortex and its participation in voluntary movements, in *Comparative Primate Biology,* Steklis, H.D. and Erwin, J., Eds., Alan R. Liss, New York, 1988, 501.

49. Scott, S.H. and Kalaska, J.F., Changes in motor cortex activity during reaching movements with similar hand paths but different arm postures, *J. Neurophysiol.*, 73, 2563, 1995. [Published errata appear in *J. Neurophysiol.*, 77, 1997, following table of contents and 77, 2856, 1997.]
50. Scott, S.H. and Kalaska, J.F., Reaching movements with similar hand paths but different arm orientations. I. Activity of individual cells in motor cortex, *J. Neurophysiol.*, 77, 826, 1997.
51. Scott, S.H., Gribble, P.L., Graham, K.M., and Cabel, D.W., Dissociation between hand motion and population vectors from neural activity in motor cortex, *Nature,* 413, 161, 2001.
52. Padoa-Schioppa, C., Li, C.S., and Bizzi, E., Neuronal correlates of kinematics-to-dynamics transformation in the supplementary motor area, *Neuron,* 36, 751, 2002.
53. Sergio, L.E. and Kalaska, J.F., Systematic changes in motor cortex cell activity with arm posture during directional isometric force generation, *J. Neurophysiol.,* 89, 212, 2003.
54. Taira, M., Boline, J., Smyrnis, N., Georgopoulos, A.P., and Ashe, J., On the relations between single cell activity in the motor cortex and the direction and magnitude of three-dimensional static isometric force, *Exp. Brain Res.,* 109, 367, 1996.
55. Georgopoulos, A.P., Ashe, J., Smyrnis, N., and Taira, M., Motor cortex and the coding of force, *Science,* 256, 1692, 1992.
56. Andersen, R.A., Essick, G.K., and Siegel, R.M., Encoding of spatial location by posterior parietal neurons, *Science,* 230, 456, 1985.
57. Bizzi, E., Mussa-Ivaldi, F., and Giszter, S., Computations underlying the execution of movement: A biological perspective, *Science,* 253, 287, 1991.
58. Giszter, S.F., Mussa-Ivaldi, F.A., and Bizzi, E., Convergent force fields organized in the frog's spinal cord, *J. Neurosci.,* 13, 467, 1993.
59. Tresch, M.C., Saltiel, P., and Bizzi, E., The construction of movement by the spinal cord, *Nat. Neurosci.,* 2, 162, 1999.
60. Bizzi, E., Tresch, M.C., Saltiel, P., and d'Avella, A., New perspectives on spinal motor systems, *Nat. Rev. Neurosci.,* 1, 101, 2000.
61. Graziano, M.S., Taylor, C.S., and Moore, T., Complex movements evoked by microstimulation of precentral cortex, *Neuron,* 34, 841, 2002.
62. Lundberg, A., Integration in a propriospinal motor centre controlling the forelimb in the cat, *Igaku Shoin, Tokyo*, 47, 1979.
63. Todorov, E., Direct cortical control of muscle activation in voluntary arm movements: a model, *Nat. Neurosci.,* 3, 391, 2000.

6 Conceptual Frameworks for Interpreting Motor Cortical Function: New Insights from a Planar Multiple-Joint Paradigm

Stephen H. Scott

CONTENTS

6.1 INTRODUCTION

Visually guided reaching is a natural motor task performed regularly by primates in order to reach for and interact with objects of interest in the environment. The well-defined goal, moving the hand to a spatial location, makes it a popular paradigm for exploring sensorimotor function.[1] In general, the problem solved by the brain is how to convert visual information about the target location, initially sensed by receptors in the retina, into motor action generated by temporal and spatial patterns of muscle activities so as to stabilize the body and move the hand to the target. This conversion of sensory to motor signals involves many cortical and subcortical regions of the CNS, and a major focus of research is to identify the role played by each of these regions.

0-8493-1287-6/05/$0.00+$1.50

The basic question posed by studies that record neural activity during behavior is this: What type of information is conveyed by the discharge pattern of individual or populations of neurons? While cells are unlikely literally to code any engineering-inspired variable, it is nonetheless valuable (and even necessary) to relate neural activity to some features of behavior reflecting sensory, cognitive, or motor aspects of the task.

How one chooses which variable to correlate depends highly on the conceptual framework used to develop the experiment. This chapter starts with the important issue of how theoretical concepts guide experimental design and data analysis.[2] Such frameworks can be explicitly defined, or in some cases, only implicitly imbedded in the experiment and analysis. I will describe two conceptual frameworks for interpreting neural activity during reaching: sensorimotor transformations and internal models. Both frameworks address the same biological problem: How does the brain control the limb to reach toward a spatial target? The key difference is that each framework focuses attention on a different aspect of the motor task and thus each leads to different experiments. The sensorimotor transformations framework has been used extensively over the past 20 years to guide neurophysiological experiments on reaching, whereas the internal models framework has only recently had an impact on experimental design.

The second half of this chapter illustrates how the notion of internal models can be used to explore the neural basis of movement. A new experimental facility is described that can sense and perturb multiple-joint planar movements and this is followed by a brief description of the mechanics of limb movement. Finally some preliminary observations are presented on neural correlates in the primary motor cortex (M1) of the mechanical properties of the limb and of external mechanical loads.

6.2 CONCEPTUAL FRAMEWORK

6.2.1 SENSORIMOTOR TRANSFORMATIONS

The most common framework for exploring the neural control of reaching has been based on the idea of coordinate frames and sensorimotor transformations.[3–5] The brain is assumed to convert visual information on target location into forelimb muscle activation patterns through intermediary coordinate frames first through various kinematic representations of movement followed by kinetic representations. One putative series of transformations is shown in Figure 6.1A, where spatial target location is sequentially converted into hand kinematics, joint kinematics, joint muscle torques, and, finally, muscle activation patterns. The use of intermediary representations to plan and control movement seems like a reasonable assumption, particularly given the ubiquitous observation that hand trajectories are relatively straight for point-to-point reaching movements.[6,7]

Based on the concept of sensorimotor transformations, it seems obvious that the key neurophysiological question is which coordinate frame is specified by the discharge patterns of individual neurons in each brain region. Over the past 20 years this framework has spawned myriad studies. As described below, some experiments

A

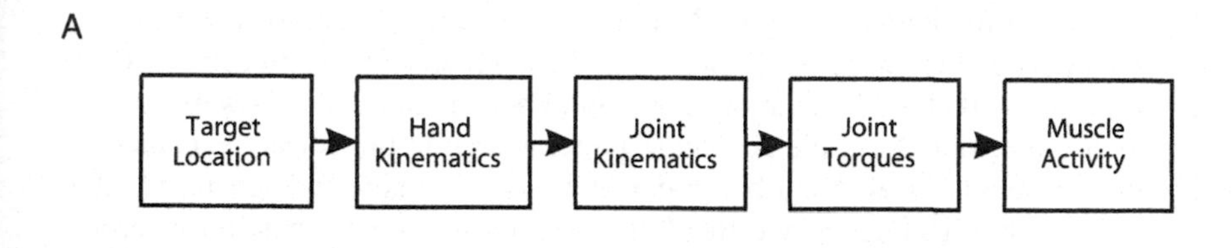

B

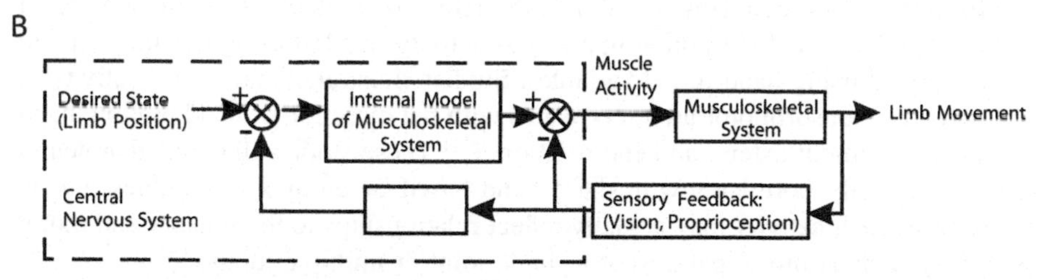

FIGURE 6.1 Two alternate frameworks for interpreting how the brain performs visually guided movements. (A) The notion of sensorimotor transformations assumes that information on spatial targets is converted into muscle activation patterns through a series of intermediary representations. This framework leads to the scientific problem of identifying how these representations are reflected in the discharge pattern of neurons in different brain regions. (B) The idea of internal models is that neural processes mimic the properties of the musculoskeletal system and physical objects in the environment. This framework leads to the scientific problem of identifying how information related to the motor periphery and physical loads is reflected in the discharge pattern of neurons.

have been designed to dissociate different variables, or levels of representation. In other cases, a specific class of variable has been chosen *a priori*, either based on the results of previous studies or simply for technical reasons.

One of the first studies to record neural activity in the motor cortex during reaching found that cell discharge was broadly tuned to the direction of hand motion.[8] This study showed that the cell discharge rate was maximal for movements in one direction, the preferred direction (PD), and that the cell's activity decreased as the angle between movement direction and the cell's PD increased. Further, the direction of hand motion could be predicted from the discharge pattern of an ensemble of neurons; this was termed the population vector hypothesis.[9,10]

A criticism often levied upon these studies has been that these hand-based correlates could be observed regardless of the type of information conveyed by individual neurons.[11–13] Variables of movement such as hand and joint motion are highly intercorrelated, such that even if neural activity coded muscle velocity, one would find significant correlations between cell discharge and hand motion. Moreover, the population vector will point in the direction of hand motion if three conditions are met: (1) neural activity is symmetrically tuned to the direction of movement; (2) the PDs of neurons are uniformly distributed; and (3) there is no coupling between a cell's PD and the magnitude of modulation during movement.[11] Any population of neurons that satisfies these conditions will predict the direction of hand motion regardless of the underlying information conveyed in its discharge patterns. A recent theoretical study by Todorov[13] reiterated this point by illustrating how a broad range of observations between hand movement and neural activity, both at the single-cell and at the population level, could be explained if cells were simply coding multidimensional muscle activation patterns. While the correct explanation of the precise details of all hand-based correlations is a matter of debate,[14–17] the article by Todorov illustrates how difficult it is to interpret the discharge of neurons with simple correlation methods.

In spite of these concerns, a school of thought was created around the population vector hypothesis and the notion that neural activity in M1 during reaching should be interpreted using hand-based variables. Studies illustrated that neural activity in M1 and other sensorimotor areas correlates with the direction of hand motion, hand velocity, movement extent, and end position.[18–21] These studies illustrate that neural activity certainly correlates with these hand-based or end-point variables, but in many cases such activity may actually reflect relationships to the sensory and motor periphery such as motor patterns at a single joint or multiple joints.[11,13]

Few believe neural activity in M1 is coding the activity of single muscles, but the hand-based framework makes a substantive leap away from the motor periphery. In the extreme, descending commands are assumed to convey only the direction of hand movement that gets converted into motor output at the spinal level.[22] The shift away from the motor periphery has been extended further to suggest that M1 may be involved in cognitive processing such as mental rotation[23,24] although such interpretations remain controversial.[25–27]

A key feature of many studies has been to dissociate explicitly different features of the task, such as sensory versus motor,[28,29] global variables (hand–target) versus joint–muscle,[30–36] or kinematic versus kinetic variables.[37–39] For example, we performed an experiment where reaching movements were performed to the same target locations but with two different arm postures: first, in a natural arm posture where the elbow tended to remain directly below the hand and shoulder, and second, in an abducted posture with the shoulder abducted and with the elbow almost at the level of the hand and shoulder.[31] This task dissociated global features of the task related to the spatial target and hand motion, which remained constant, from joint-based variables related to joint motion, joint torque, or muscle activity, which varied between arm orientations. We found that most neurons showed changes in activity either by changing their directional tuning or by modulating the overall level of activity, suggesting that neural discharge was related in some way to the motor periphery. Some cells, however, showed no changes in activity when movements were performed with the two arm orientations. Such invariances could reflect that these cells are specifying global features of the task, although it is still possible that such cells could reflect joint-based information. (See Scott and Kalaska.[31])

A cleaner dissociation between muscle- and hand-based features of movement was provided in a study where wrist movements were performed with three different forearm orientations: neutral, supinated, and pronated .[34] They found that some cells varied their directional tuning in a manner that was similar to the variation observed for muscles, whereas others showed no change in directional tuning, as would be expected if neurons reflected the spatial direction of the task. However, most of these latter spatial/hand cells still showed changes in the magnitude of activity for movements with different forearm orientations.

All these studies illustrate that primary motor cortical activity correlates to almost every imaginable task variable, including spatial target location, hand movement direction and extent, hand velocity, joint velocity, force output, and muscle activity, to name a few.[1,20,40] The obvious conclusion is that there appears to be no single unified coordinate frame in M1. This of course causes considerable problems for the population vector hypothesis, which presupposes that a global signal related to the direction of hand motion is created across the cell population. In M1, cells respond to many different variables, with some cells largely reflecting kinematic features of the task and other neurons reflecting kinetic features. Whenever force but not kinematic motion is modified, these latter cells, which modulate their activity with force output, will alter estimates of hand motion.

While most agree that neural activity in M1 reflects a mixture of different kinematic and kinetic features of movement, the notion that the brain performs a series of sensorimotor transformations to execute reaching movements assumes a certain relationship between these representations. Specifically, cells insensitive to force output are assumed to reflect a higher level representation of movement which gets converted by cortical processing into a lower level representation; cells sensitive to force output are classified as this lower level representation. Are cells that are insensitive to force output necessarily reflecting a higher level representation than cells that are sensitive to force output? This assumption would seem reasonable, if muscle activity (electromylography [EMG]) were the only feature of motor behavior controlled by the brain.

However, descending commands to the spinal cord must consider more than just muscle activity.[1,41,42] Alpha motoneurons, which innervate extrafusal muscle fibers and produce force, represent only one type of motoneuron. In each motoneuron pool, there is a large number of gamma motoneurons that innervate intrafusal fibers in muscle spindles,[43] which may be equal in proportion to alpha motoneurons in some muscles. There are even beta motoneurons innervating both intra- and extrafusal muscle fibers.[44] Another role for descending commands is to modulate and influence sensory feedback.[45] Spinal reflexes can also create various contingency plans for unexpected perturbations or errors,[46] which must also be selected or modified by descending commands. It is quite possible that up to two thirds of descending signals from the cortex to the spinal cord are related to controlling these other features of motor output. However, little is known about cortical discharge related to controlling gamma-motoneuron activity and spinal reflexes during volitional tasks since experimental paradigms, including our own, tend to focus on alpha-motoneuron activity.[47,48] It is quite possible that neurons related to these other features are relatively insensitive to variations in force output during motor behavior. Within the rubric of sensorimotor transformations, such neurons would be assumed to code a higher level representation of movement related to the kinematic features of the task when in fact they were simply involved in controlling relatively low-level but non-EMG features of the task. Furthermore, such discrete segregation between alpha-motoneuron activity and other spinal processing is highly unlikely and descending signals likely reflect a mixture of influences on spinal circuitry.

Continued support for using sensorimotor transformations as a basis for interpreting neural activity during reaching stems from the observation that hand trajectory is relatively straight during reaching, suggesting that hand trajectory may be explicitly planned or controlled by the central nervous system (CNS). However, recent theoretical studies cast doubt on even this basic assumption. It has been proposed that strategies for motor control may be optimized to minimize the effect of noise on motor performance.[49] To minimize the influence of this noise, Todorov[50] suggested that the brain may be capable of identifying and implementing optimal feedback laws for controlling body movements. Such a law optimizes feedback signals to correct movement errors based entirely on the global goal of the task. If local errors in motor performance affect the ability to attain the global goal, then motor patterns are adjusted to correct these errors. If local errors do not affect global motor performance, they are not corrected. This framework predicts several common characteristics of motor performance such as trial-to-trial variability and goal-directed corrections. More importantly, if the brain implements such laws, hand trajectory may not be planned, but may simply fall out from the optimal feedback law.

6.2.2 Internal Models

An alternate framework for interpreting limb movements has emerged in the past few years based on the notion of internal models, neural processes that mimic the mechanical properties of the limb or objects in the environment.[51–53] The notion of internal models comes from robotic control theory, and suggests that control strategies implemented by the brain and spinal cord reflect in some way the properties of the motor plant, such as the geometrical and inertial properties of the limb and the physiological properties of muscle. A simple schematic of the motor problem based on the idea of internal models is shown in Figure 6.1B.

While it is theoretically possible that the brain could generate reaching movements entirely by afferent feedback, there is ample evidence that the brain possesses some knowledge of the peripheral motor apparatus and uses it to guide action.[54–58] In particular, adaptation studies with mechanical loads illustrate that subjects quickly learn to alter motor patterns to maintain relatively straight hand paths. Unexpected removal of the load results in trajectory errors that mirror how the loads initially perturbed limb movement and illustrate that the brain has, in some way, incorporated the novel load in motor commands for movement. This adaptive change to motor output can be construed as an internal model of the mechanical load.

The conversion of visual target location into motor commands of muscle reflects an inverse internal model in that it reverses the normal causal flow from muscle activity to body motion. There is evidence that forward internal models (which mimic the normal causal flow) are also used by the brain. For example, the grip force on hand-held objects is adjusted and scaled prior to or with whole-arm movements, suggesting that knowledge of the impending limb movement is used to adjust grip force to prepare for changes in the forces generated by the object.[59,60]

An optimal feedback controller can be viewed as a special form of internal model in that it captures key mechanical features of the limb in the feedback algorithm. There are clearly many ways in which the brain may use both internal models and optimal feedback control during movement. For example, the brain could use an inverse internal model to specify a feed-forward signal to initiate movement, and then use optimal feedback laws to correct on-line errors in performance. This conceptual framework predicts that very few representations of movement may be specified by the brain to plan and control movement: high-level signals related to the global goal and relatively low-level signals related to sensory and motor features of the task. Such low-level signals may still have considerable dimensionality, reflecting the many muscles and sensory receptors that make up the somatomotor system. What would not be required are any intermediary representations reflecting hand trajectory, joint motion, or torque.

Sensorimotor transformations and internal models both provide a description of how the brain converts spatial target information into motor commands. However, each framework emphasizes different features of the problem. The notion of sensorimotor transformations, at least its application to goal-directed limb movements, has focused attention on identifying what coordinate frames best represent the discharge patterns of cells, with less emphasis on limb mechanics. The present use of internal models as a conceptual framework de-emphasizes the search for coordinate frames and focuses attention on how the physical properties of the musculoskeletal system and physical loads in the environment influence behavior. The focus of research on internal models shifts to explore how the brain represents the physical properties of the limb and environmental forces, and how these representations are learned and mapped onto brain circuitry.

6.3 AN EXPERIMENTAL PARADIGM: NEURAL CORRELATES OF INTERNAL MODELS

The notion of internal models and optimal feedback control explicitly address how the brain reflects the physical properties of the limb and the environment. Logically, an experimental paradigm to explore this problem requires precise knowledge on the physics of the limb, and the ability to manipulate these features. For pragmatic reasons, traditional reaching paradigms for studying neural activity in nonhuman primates have only monitored hand movements or applied loads directly to the hand. These reaching movements involve at least four degrees of freedom (DOF), including three at the shoulder and one at the elbow, making them technically difficult to monitor and even more of a challenge to perturb systematically, except at the hand.

In order to address how the brain controls whole-limb movements, we developed a new experimental device that can both sense and perturb multiple-joint movements.[61] The robotic device kinesiological instrument for normal and altered reaching movements (KINARM) maintains the arm in the horizontal plane, allowing planar movements of the hand from combined flexion and extension movements at the shoulder and elbow joints (Figure 6.2). Custom-made fiberglass troughs attach the forearm and upper arm each to the mechanical linkage. The lengths of the linkage are adjustable so that hinge joints on the device are aligned with the shoulder and elbow joints, allowing their motion to be paralleled by the linkage. The four-bar linkage is attached to two torque motors using timing belts, where one motor is directly

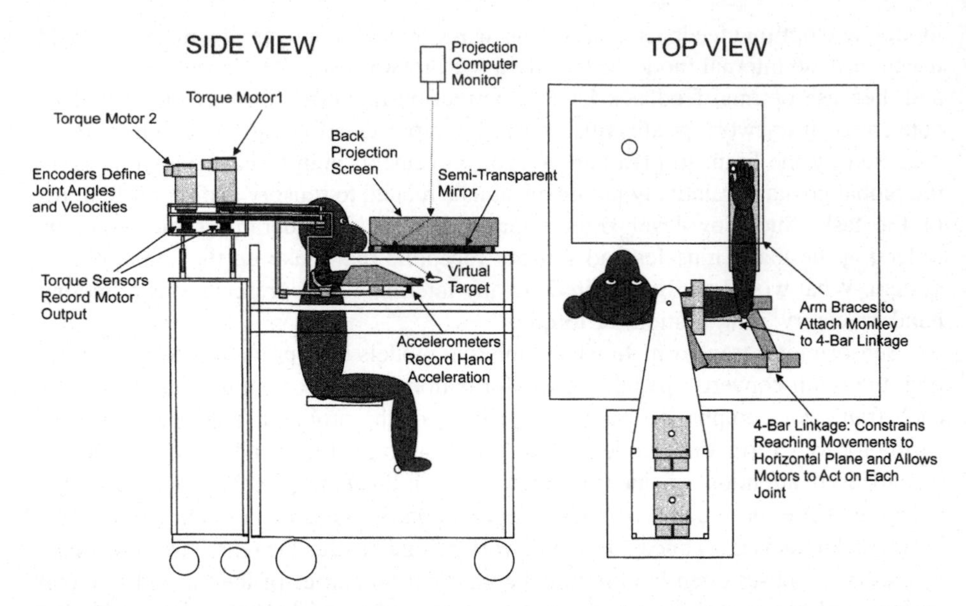

FIGURE 6.2 KINARM robotic device used to monitor and manipulate the physics of limb movement. Arm troughs attach the monkey's limb to a four-bar linkage allowing arm movements in the horizontal plane involving flexion and extension motions at the shoulder and elbow joints. Torque motors attached to the device can apply mechanical loads to each joint, independently. The computer projection system projects virtual targets onto the plane of the task. (Diagram from Reference 61, with permission.)

attached to the linkage under the upper arm and the other is indirectly attached to the forearm. These motors allow loads to be applied to the shoulder or elbow joints independently, and encoders within the motors are used to measure shoulder and elbow angles indirectly.

This paradigm is a natural extension of Ed Evarts' initial experiment, which examined neural activity in M1 while a monkey performed single DOF movements at the wrist.[37] This single-joint paradigm has been used by many researchers and has contributed substantially to our knowledge of how M1 and many other brain regions are involved in volitional motor control.[62] The present planar paradigm using KINARM extends this single-joint paradigm by exploring how two joints, each with a single DOF, are used together to create purposeful movement.

The paradigm also captures much of the behavioral richness inherent in whole-limb motor tasks where hand movements can be made to a broad range of spatial locations using a range of possible hand trajectories (trajectory redundancy). The planar limb movements include two separate joints so that well established problems of mechanical anisotropies and intersegmental dynamics influence motor performance (see below). Since joint motion is measured directly, these mechanical features of movement can be estimated using biomechanical models. Finally, motors can be used to manipulate the physics of each joint independently and to dissociate kinematic and kinetic features of movement. The one attribute of limb movements that cannot be addressed with this planar two DOF task is postural redundancy, where a given hand position can be obtained using different arm geometries.

6.4 LIMB MECHANICS

The musculoskeletal system filters and converts complex patterns of forelimb muscle activity into smooth and graceful body movements. The notion of internal models within the brain to plan and control limb movement suggests that there is information on the peripheral plant imbedded in the distributed circuitry related to sensorimotor function. Therefore, the first crucial step to examining the neural basis of this internal model is to quantify and understand the actual mechanics of limb movement.

We have trained monkeys to perform reaching movements with the right arm from a central target to 16 peripheral targets distributed uniformly on a circle, centered on the start position (center-out task). As observed in many previous studies, hand kinematics are quite simple with relatively straight hand paths and bell-shaped velocity profiles.[63] Peak hand velocity for movements in different directions was quite similar (Figure 6.3A), but it is important to note that our animals have undergone daily training for several months. Human subjects asked to perform planar reaching movements do show some asymmetries in movement velocity due to the mechanics of the limb.[64]

The remarkable point about multiple-joint movements is that, while the patterns of motion of the hand are similar for movements in different spatial directions, the underlying motion of the joints is anisotropic; that is, joint movement varies with movement direction.[63] Figure 6.3B shows peak velocity at the shoulder and elbow joints for movements in different spatial directions. Movements toward or away from the monkey require substantial excursions at the joints, whereas movements to the left or right require far less motion. This substantial variation in joint motion for different movement directions merely reflects geometry and how joint angular motion contributes to end-point motion.

Variations in peak velocity have important implications when considering how neural activity represents the properties of the musculoskeletal system. Velocity has a substantial effect on muscle force production, as described by the force-velocity curve. Peak elbow velocity for movements from the central target away from the monkey reaches 1.5 rad/sec. Brachialis, an elbow flexor, has a moment arm of 1 cm at the elbow,[65] so the angular motion at the elbow translates to approximately 1.5 cm/sec of linear shortening by the muscle. When converted into units of muscle fascicle length, one finds that peak shortening velocity reaches 0.3 L_0/sec, where L_0 is optimal fascicle length (4.3 cm).[66] The magnitude of shortening means that the force produced by the muscle is approximately half the force generated in mammalian slow twitch fibers under isometric conditions.[67] These movements were performed at rather modest speeds; the force-velocity curve would have an even greater influence in fast movements.

Joint muscular torque for a single-joint task simply equals joint angular acceleration multiplied by the moment of inertia of the moving segment. This simple coupling between motion and torque at a joint is lost when movements involve more

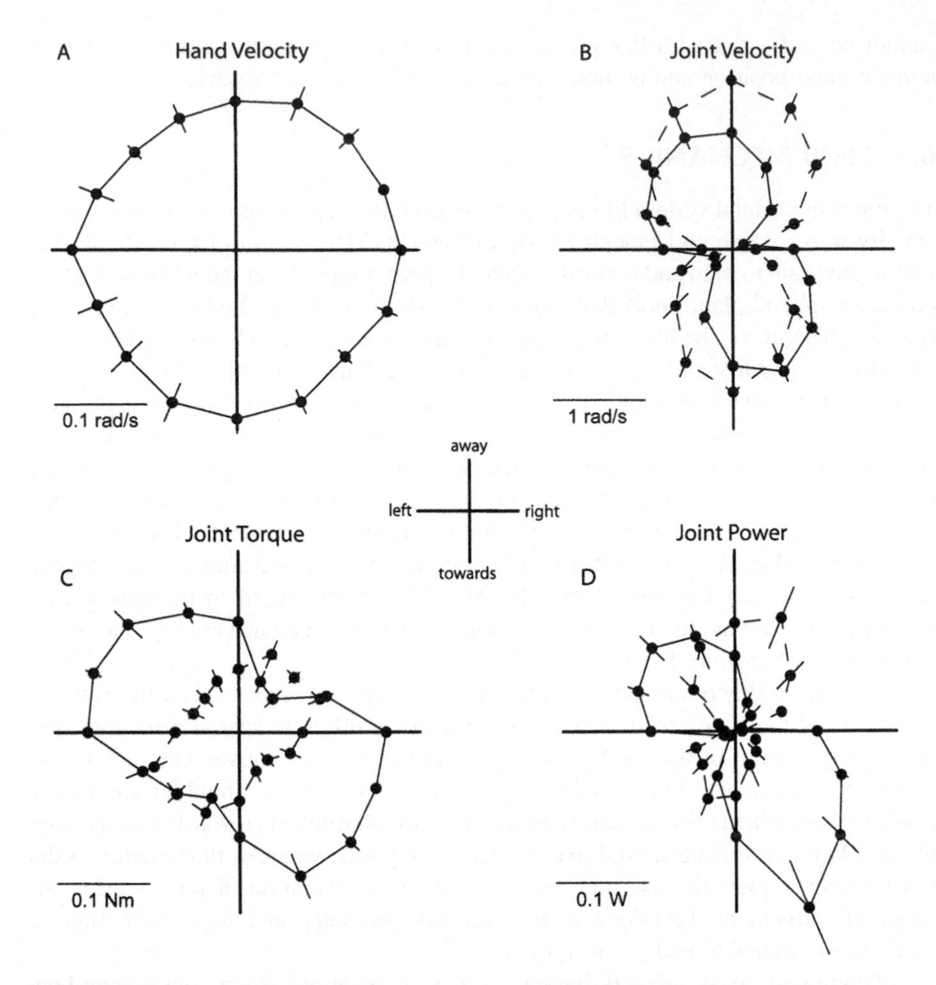

FIGURE 6.3 Kinesiology of reaching movements. (A) Polar plot of peak hand velocity where the angle denotes direction of hand motion and velocity is defined by the distance from origin. (B) Polar plot of variations in peak joint velocity for movements in different spatial directions. The solid line is shoulder velocity and the dashed line is elbow velocity. (C) Variations in peak joint torque for movements in different spatial directions. (D) Variations in peak joint power (velocity times torque) for movements in different spatial directions.

than one joint due to intersegmental dynamics.[68] Therefore, muscular torque at one joint creates motion at that joint and at other joints, dependent on limb geometrical and inertial properties, as well as contact forces or loads from the environment.

In the case of center-out reaching movements, the magnitude of muscular torque varies with movement direction such that large shoulder torques are needed for movements to the left and away from or to the right and toward the monkey, whereas elbow torques are larger for movements in the opposite quadrants (Figure 6.3C). Note that while joint velocities and torques are both anisotropic, they have very different principal axes. While angular motions are slightly greater at the elbow than at the shoulder, muscular torques at the shoulder tend to be much larger than at the elbow.

A variable that is particularly relevant for reflecting the properties of the musculoskeletal system is joint muscular power, since it is the multiplicative of the two most important features of the motor periphery: joint torque, reflecting the inertial properties of the limb, and joint angular velocity, which strongly influences muscle force output due to the force-velocity relationship of muscle. In many respects, joint power provides a very crude first approximation of the muscle activation patterns of all muscles spanning a joint. It is by no means perfect since it fails to consider both the importance of co-contraction by antagonistic muscles and the subtleties inherent in activation patterns at different joints. It also fails to capture the muscle activation associated with isometric conditions where velocity is zero.

Not surprisingly, joint power shows considerable anisotropy. Figure 6.3D illustrates that peak joint power varies strongly for movements of similar magnitude but in different spatial directions. Total peak power, the sum at both joints, is largest for movements away and to the left, and for movements toward and to the right. Peak values are much smaller for movements directly to the left or right.

6.4.1 Neural Correlates of Limb Mechanics

The description above illustrates the complex relationship between hand motion and the underlying mechanics of movement. In order to execute a reaching movement, the motor system must, in some way, compensate or consider these kinematic and kinetic anisotropies. The question we posed was whether M1 reflected these mechanical features of reaching, and thus reflected an internal model of the motor periphery.

We recorded the activity of neurons in contralateral M1 while monkeys made reaching movements from a central target to 8 or 16 peripheral targets located on the circumference of a circle, and examined the response of neurons in MI of three monkeys.[68] Figure 6.4 illustrates the discharge pattern of a cell for movements in different spatial directions. As observed in previous studies, cells were broadly tuned to the direction of movement. Peak discharge occurred when the monkey moved its hand to the right and toward itself, its PD of movement. This cell was found to be statistically tuned with a preferred direction of 328°.

We examined two possible ways in which the anisotropic properties of limb movement might influence the activity of individual neurons. First, cells with PDs associated with movements requiring greater power may show larger variations in their discharge than cells with preferred directions associated with movements requiring less power. We examined this possibility by plotting cell modulation against each cell's PD.[69] There was a wide range in the modulation of neural activity across the cell sample (26 ± 16 spikes/sec, mean ± SD). The modulation of activity for cells with PDs within ±22.5° of a given movement direction was averaged to define mean cell modulation for that direction. However, we found no correlation between this mean discharge rate and the corresponding total peak joint power ($R^2 = 0.04$).

A second possibility is that variations in the mechanics of limb motion in different spatial directions may result in a nonuniform distribution of PDs across the cell sample. Cells were divided into 16 groups (bin size = 22.5°) and plotted against movement direction in Figure 6.5A. There was a wide range in the number of neurons associated with each movement direction, from 27 to 1. One would

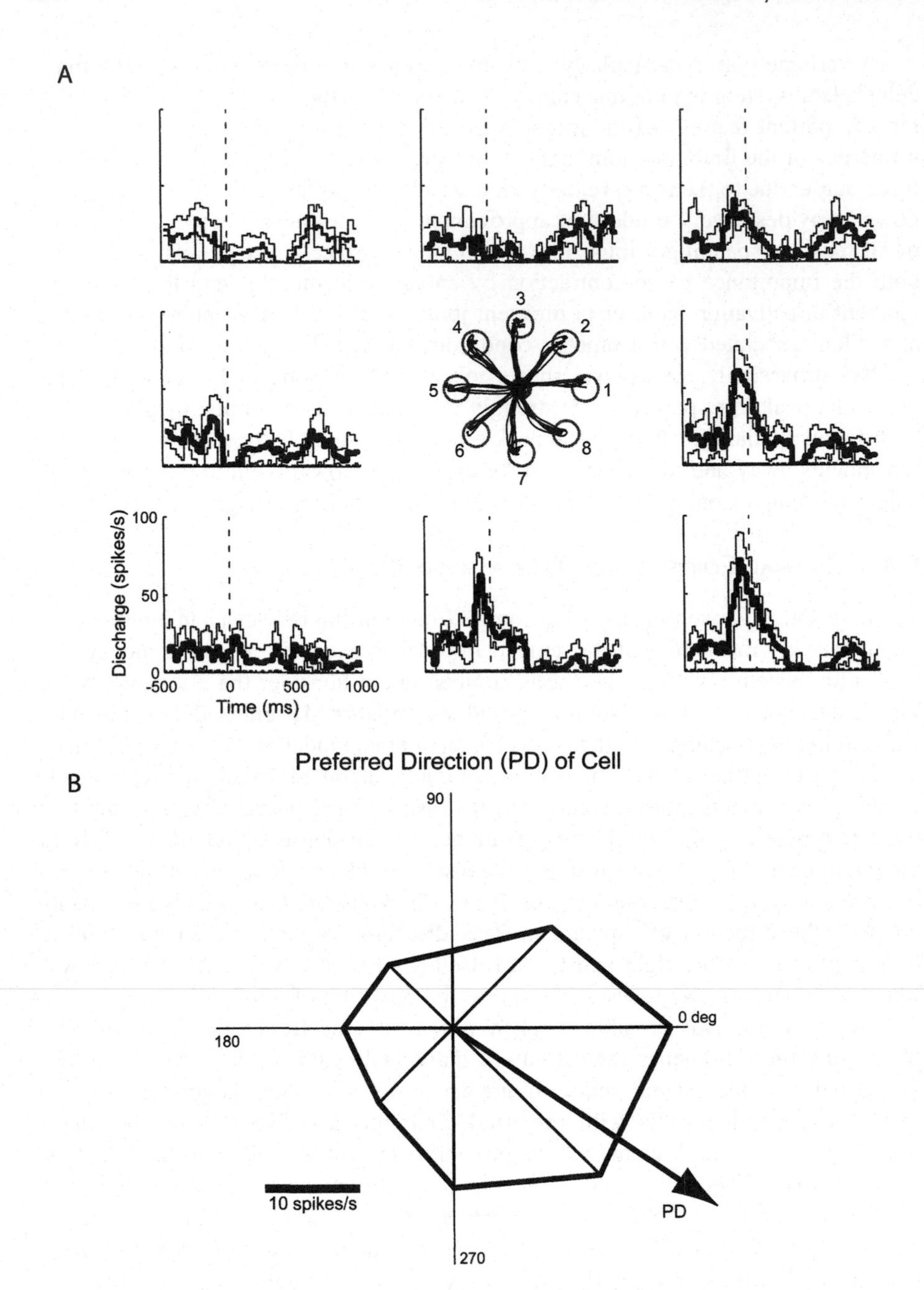

FIGURE 6.4 (A) Discharge pattern of a neuron in M1 for movements from a central target to one of eight peripheral targets (central inset panel). Cell discharge is aligned with movement onset. The thick line denotes the mean and the thin lines denote one standard deviation based on five repeat trials. (B) Cell discharge is plotted as a polar plot where the angle denotes movement direction and the distance from the origin reflects mean cell discharge (from 200 msec before to 500 msec after the onset of movement).

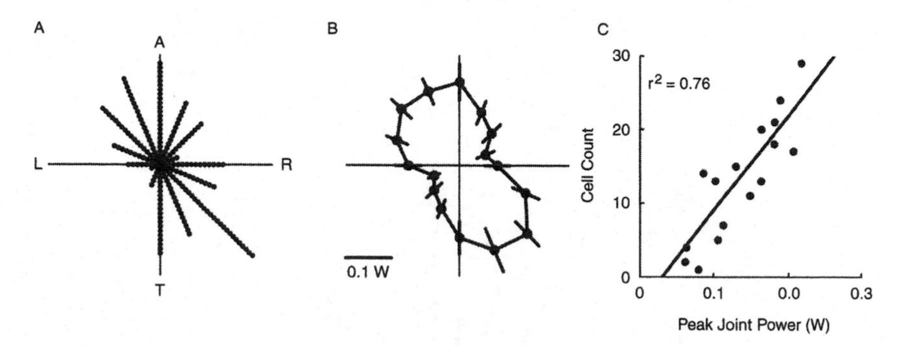

FIGURE 6.5 (A) Distribution of preferred directions (PDs) of neurons in M1. Each dot reflects the PD of an individual neuron and all dots forming a line reflect neurons with PDs in a given direction of movement (16 divisions of 22.5°). Distribution of PDs is bimodal. (B) Variation in total peak joint power at the shoulder and elbow relative to movement direction. (C) Relationship between peak joint power and the number of neurons with PDs associated with each movement direction.

identify this distribution of PDs as uniform if one only tested against a unimodal distribution, as is commonly done. However, the diagram graphically illustrates that the distribution is not uniform when compared to a bimodal distribution. The nonuniform distribution of PDs was also observed based on neural activity only during the reaction time period prior to the onset of movement, suggesting that such biases were not simply a result of afferent feedback.

The strong bias in the distribution of PDs during reaching has two profound effects on population vectors (Figure 6.6A). First, 13 of 16 population vectors did not point in the direction of hand motion. Vectors tended to be biased toward one of two directions. Second, the nonuniform distribution resulted in large variations in the magnitude of the population vector ranging from 56 to 145% of the mean vector length. This modulation in the magnitude of the population vector occurred although movements were of similar magnitude and with similar peak hand velocities. Figure 6.6B illustrates that a neural trajectory computed from the instantaneous firing rate of neurons does not predict the instantaneous direction of hand motion. The neural trajectory did not predict the direction of movement from the very beginning of hand motion.

Criticisms have been raised about these observed deviations between the population vector and movement direction.[70] One concern is that neural vectors based on neural activity for the entire reaction and movement time should not be compared to movement direction for the first half of movement since neural activity during the latter part of movement may have caused errors in the population signal. However, reanalysis of our data using the direction of movement for the entire limb movement (from movement initiation to the end of movement) still resulted in the majority of population vectors not predicting the direction of hand movement (11 of the 16 directions). We used a technique comparable to weighting function 8 in the study of Georgopoulos et al.,[10] except that cosine tuning functions were replaced with Von Mises tuning functions in order to capture the fact that many neurons are

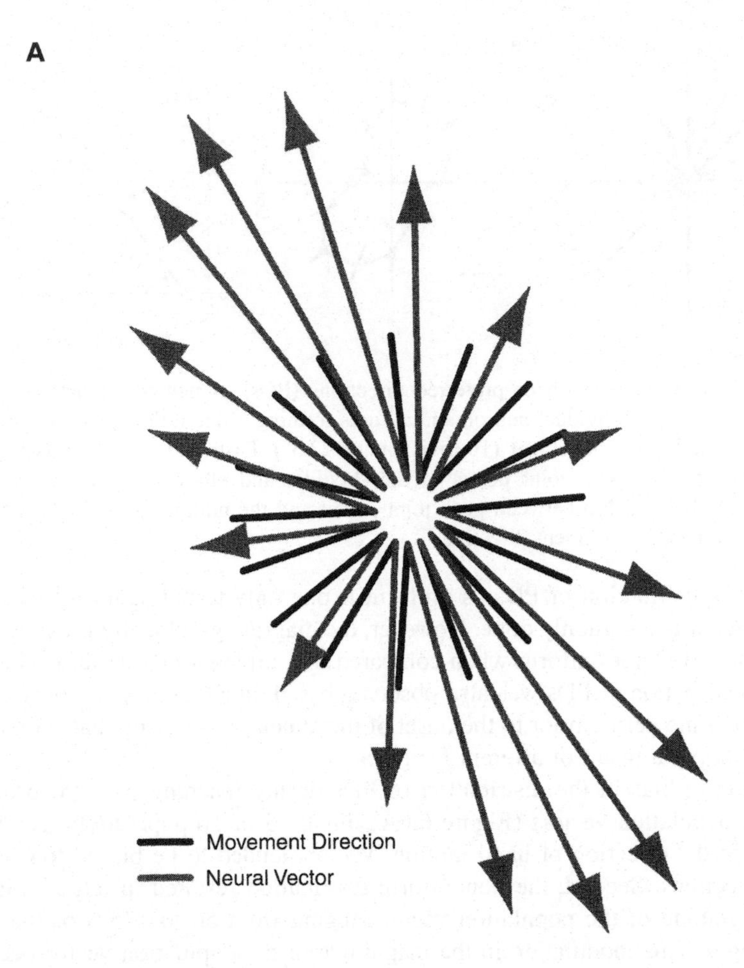

FIGURE 6.6 (A) Direction of hand movement (black lines) for 16 target directions. Population vectors for each movement are shown as grey arrows and corresponding hand path is attached to the base of the arrow. (Adapted from Reference 69.) (B) Instantaneous trajectory of the hand is shown in black for movements to the left target (25 msec intervals). The two large, dashed, light grey circles denote start (right) and target (left) spatial locations. Corresponding population vector trajectory is shown in grey. Each sequential value is added vectorally to previous data points and then scaled to match the spatial trajectories in the diagram. The size of the circle denotes a significant difference between population vector and instantaneous hand motion. Population vectors are shifted in time such that the first vector that is statistically tuned is aligned to movement onset.

more sharply tuned than a cosine function would suggest.[71] Further, all the weighting functions presented by Georgopoulos et al. fail to predict most directions of hand motion when comparing the direction of hand motion for the first 100 msec of movement based on population vectors constructed from neural activity during the reaction time period.[1]

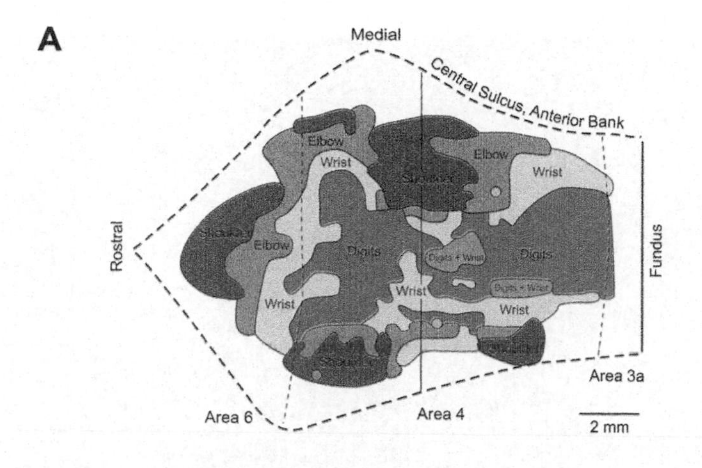

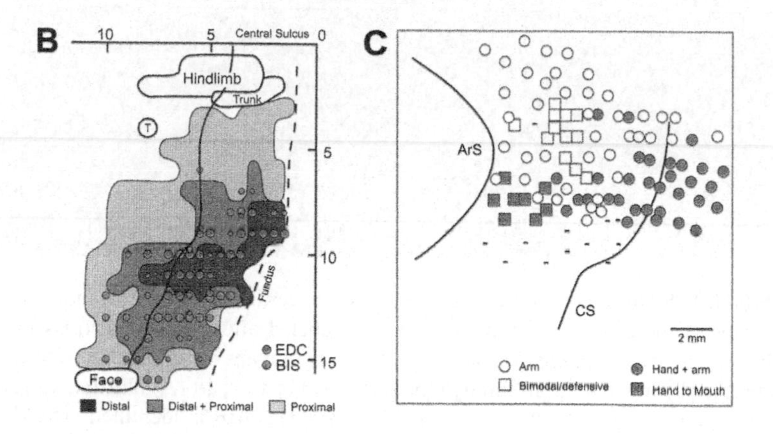

COLOR FIGURE 1.2 Intracortical stimulation maps of M1 in macaque monkeys. Note that in each map, hand movements form a central core (*red*). (A) Summary map of the movements evoked by intracortical stimulation (2–30: A) in an awake macaque monkey. (Adapted with permission from Kwan, H. C. et al., *J. Neurophysiol.*, 41, 1120, 1978. Copyright 1978 by the American Physiological Society.) (B) Summary map of muscle representation in M1 derived from stimulus-triggered averages of rectified EMG activity (15: A at 15 Hz) in an awake monkey. Sites that influenced only proximal muscles are indicated by *light shading*, those that influenced only distal muscles by *dark shading*, and those sites that influenced both proximal and distal muscles by *intermediate shading*. Sites of significant stimulus-triggered averages of rectified EMG activity for the shorthead of biceps (BIS, *blue*) and extensor digitorum communis (EDC, *red*) are indicated with size-coded dots (3, 4, 5, 6 S.D. levels above pre-trigger level baseline activity). (Adapted with permission from Park, M. C., Belhaj-Saif, A., Gordon, M., and Cheney, P. D., *J. Neurosci.*, 21, 2784, 2001. Copyright 2001 by the Society for Neuroscience.) (C) Summary of hand and arm postures produced by long train (0.5 sec), high intensity (25–150: A) intracortical stimulation in M1, the PMd, and the PMv of an awake monkey. *Arm* sites evoked postures involving the arm but without changes in the configuration of the hand. *Hand + arm* indicates sites where stimulation evoked postures involving both the hand and arm. *Hand to mouth* indicates sites that evoked grasp-like movements of the hand which was brought to the mouth. *Bimodal/defensive* indicates sites where neurons received visual input and stimulation moved the arm into a defensive posture. See text for further explanation. (Adapted with permission from Graziano, M. S., Taylor, C. S., and Moore, T., *Neuron*, 34, 841–51, 2002. Copyright 2002 by Cell Press.)

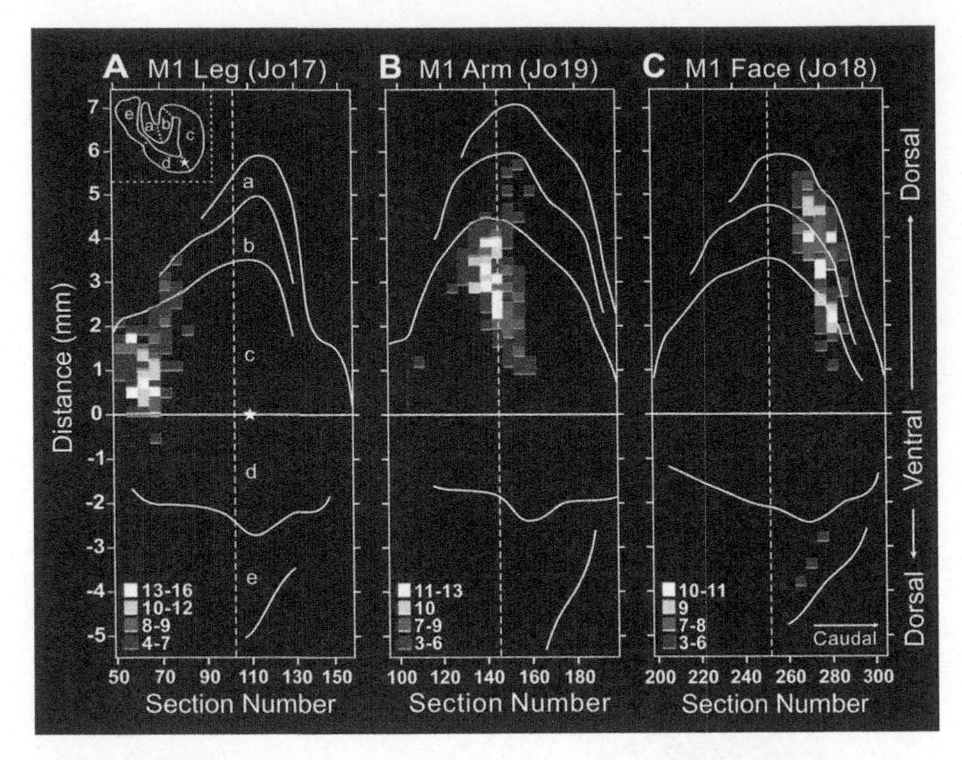

COLOR FIGURE 1.10 Somatotopic organization of dentate output channels to M1. Unfolded maps of the dentate illustrate the neurons labeled after HSV1 injections into the (A) leg, (B) arm, and (C) face representations of M1. These maps of the dentate were created by unfolding serial coronal sections through the nucleus. Inset in part (A) illustrates a coronal section of the dentate where each segment in the unfolded map is identified. The *dashed vertical line* indicates the rostro-caudal center of the nucleus. (Adapted with permission from Dum, R. P. and Strick, P. L., *J. Neurophysiol.*, 89, 634, 2003. Copyright 2003 by the American Physiological Society.)

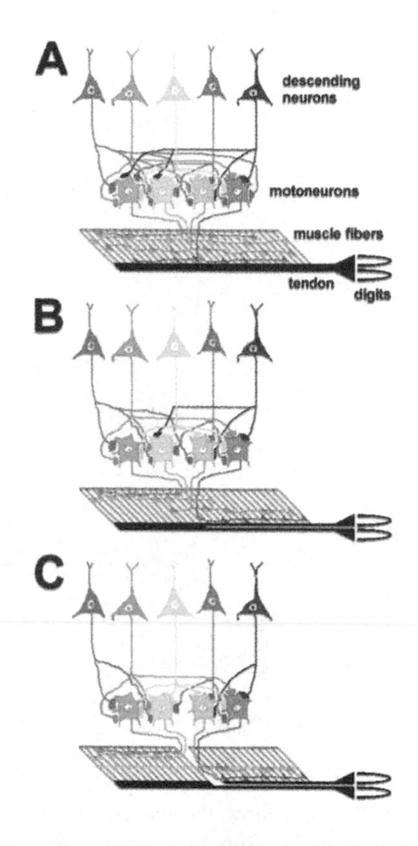

COLOR FIGURE 3.3 Neuromuscular evolution. A speculative scheme is illustrated through which a parent muscle (A) could become partially subdivided (B) and eventually divide into two daughter muscles (C), while still retaining some of the distributed descending neural control of the parent muscle. A. Parent Muscle. A single tendon inserts broadly on multiple digits. For simplicity only two digits are illustrated here. The four motoneurons each innervate muscle fibers distributed widely in the muscle belly, so that the four motor unit territories overlap. The motoneurons in turn are innervated by five descending neurons that each synapse widely within the motoneuron pool. Here every descending neuron innervates every motoneuron. B. Partially Subdivided Muscle. The tendon has become partially divided to act differentially on the two digits. The motor unit territories also have become partially selective: the red and orange motoneurons innervate muscle fibers to the left, the green and blue motoneurons muscle fibers to the right, with a central region of overlap. The red and orange motoneurons thus act more strongly on one digit and the green and blue motoneurons act more strongly on the other digit. The descending inputs also have become more selective: the red and orange descending neurons no longer innervate the blue motoneuron; the green and blue motoneurons no longer innervate the red motoneuron; hence these descending neurons can act somewhat differentially on the digits. The yellow descending neuron, however, still facilitates all four motoneurons. C. Daughter Muscles. The tendon now has divided completely in two, as has the muscle belly. The red and orange motoneurons exclusively innervate the left muscle; the green and blue motoneurons exclusively innervate the right. The descending neurons also have become more, though not completely, selective: the red descending neuron now innervates only the red and orange motoneurons, and the blue descending neuron now innervates only the green and blue motoneurons. These two descending neurons therefore selectively facilitate only the left or right daughter muscle, respectively. The orange descending neuron facilitates the left muscle more than the right, the green descending neuron facilitates the right more than the left, and the yellow descending neuron still facilitates the left and right equally.

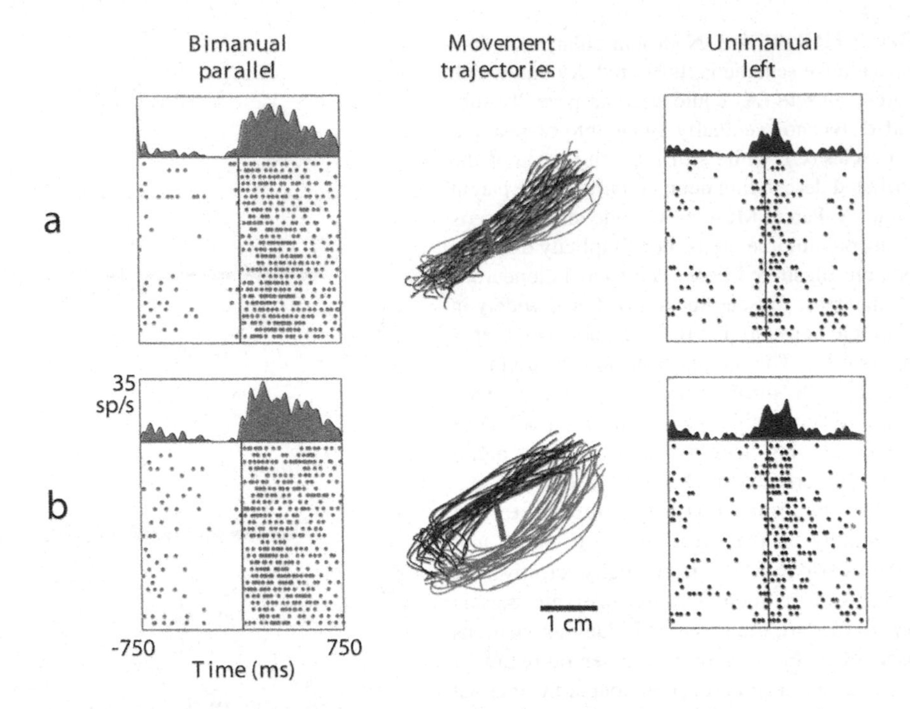

COLOR FIGURE 4.6 Raster displays and PETHs illustrating the activity from a right MI cell in four conditions. The activity of the cell during bimanual parallel movements is on the left (red). The activity of the cell during unimanual left movements is on the right (blue). The middle plots show the movement paths of the left hand for bimanual parallel (red) and unimanual left (blue) movements. Row A only contains trials in which the movement path passed through a narrow band (thick green line) located between the origin and the target. Row B only contains trials that did not pass through the band. The green band was placed to maximize the difference between the trajectories in the lower display. PETHs are centered on the beginning of movement, and the scale for all PETHs is the same. The trajectories begin in the upper right and end in the lower left of the frame. Note that the cell activity in bimanual trials (in red) remains similar regardless of the precise trajectories.

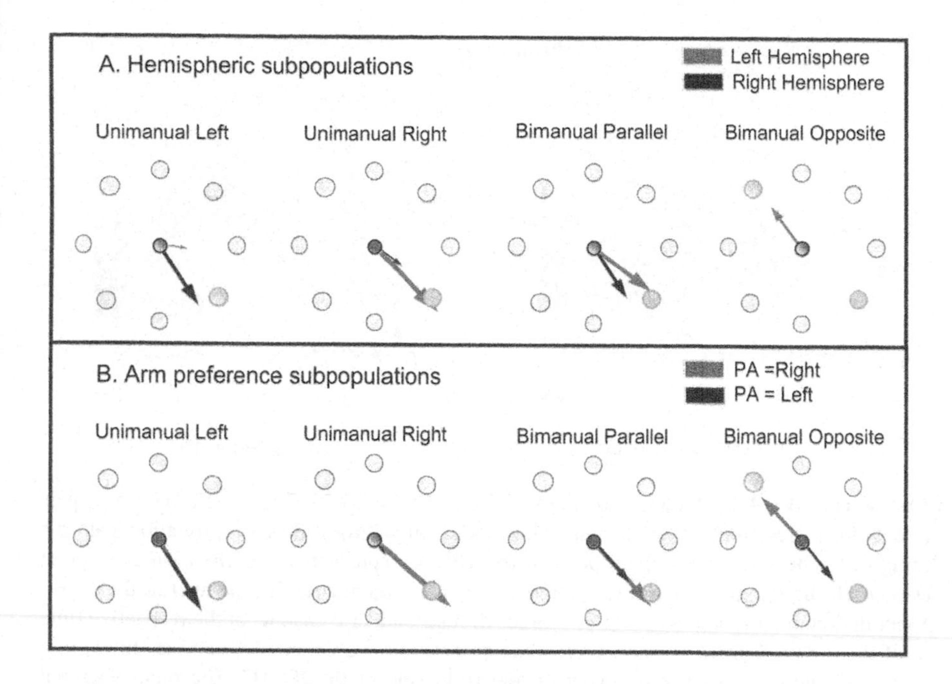

COLOR FIGURE 4.13 Population vectors calculated for unimanual and bimanual movements to 315°. For each movement, two PVs (colored arrows) of two neuronal subpopulations were calculated using an estimated best-fit PD. The different colors represent the PVs of the different neuronal subpopulations. A. PVs constructed by dividing all cells into two subpopulations according to the hemisphere in which they reside. B. PVs constructed by dividing all cells into two subpopulations according to their arm preference. (Reproduced with permission from Steinberg, O., Donchin, O., Gribova, A., Cardosa de Oliveira, S., Bergman, H., and Vaadia, E., Neuronal populations in primary motor cortex encode bimanual arm movements, *Eur. J. Neurosci.,* 15, 1371, 2002.)

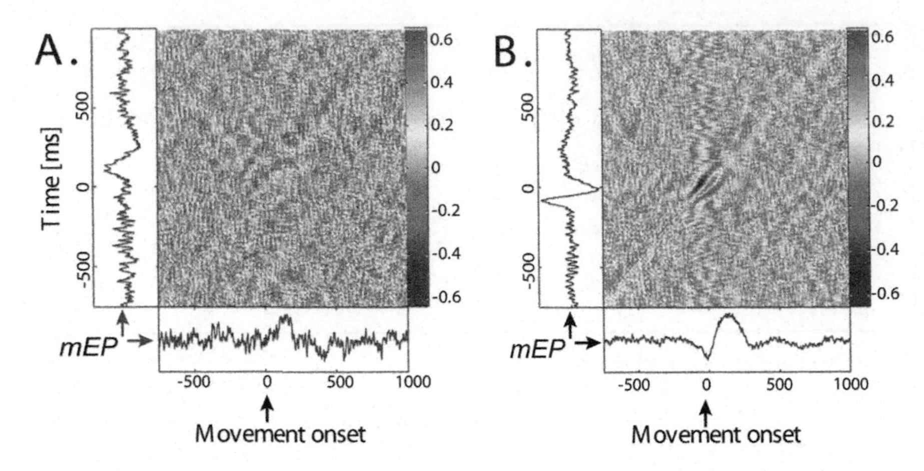

COLOR FIGURE 4.15 Example of joint peri-event time correlograms (JPETCs) of a pair of recording sites from different hemispheres demonstrating different correlation patterns during different movements. Each pixel in the JPETC represents the correlation coefficient between all the (single-trial) values of one local field potential (LFP) channel at the corresponding time bin of the x-axis and the values of a second LFP channel at the respective time bin of the y-axis. Correlation is expressed as correlation coefficient (CC) and is shown in a color code, with the color scale given on the right side of the JPETC. The main diagonal depicts the correlation at delay = 0. A. The correlation pattern during unimanual movements of the contralateral (left) arm to the front. No correlation is apparent between the two electrodes in this condition. B. The correlation pattern for the same pair during a bimanual movement of the same amplitude. Movement directions of the two arms differ by 90 degrees, with the left arm moving to the front (like in the unimanual condition shown in A) and the right arm moving to the right. Note the strong correlation with side peaks that arises around movement onset, and lasts for about 100 msec.

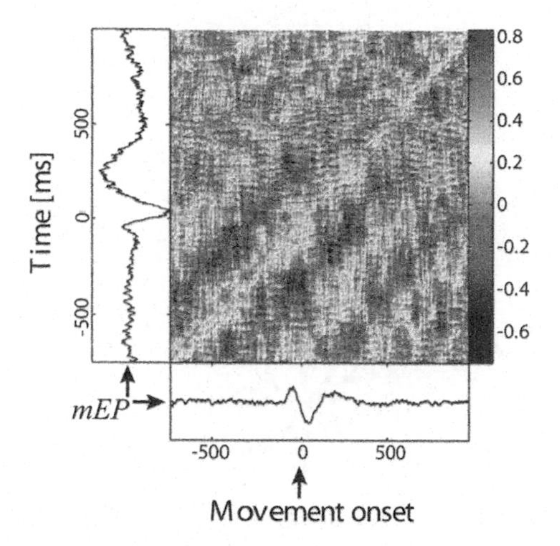

COLOR FIGURE 4.16 JPTEC depicting dynamics of correlation between two recording sites in different hemispheres around the time of movement onset. The movements were bimanual non-symmetric with the left arm moving to the left, and the right arm to the front). Note that the high level of correlation before movement onset (CC ≈ 0.5) is significantly reduced near movement onset and throughout the movement duration.

COLOR FIGURE 5.2 An example of the population coding of movement direction. The blue lines represent the vectorial contribution of individual cells in the population (N = 475). The actual movement direction is in yellow and the direction of the population vector is in red. (From Georgopoulos, A.P., Kettner, R.E., and Schwartz, A.B., Primate motor cortex and free arm movements to visual targets in three-dimensional space. II. Coding of the direction of movement by a neuronal population, *J. Neurosci.*, 8, 2928, 1988, Figure 1, with permission.)

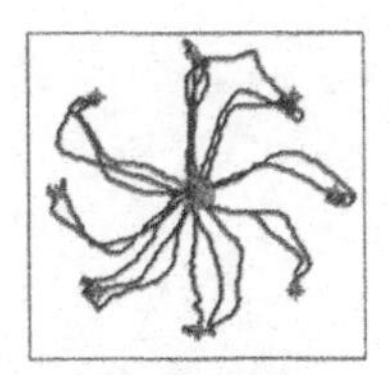

Early Force

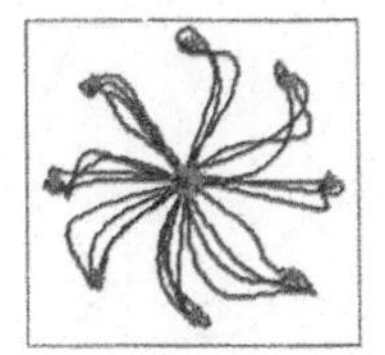

Early Washout

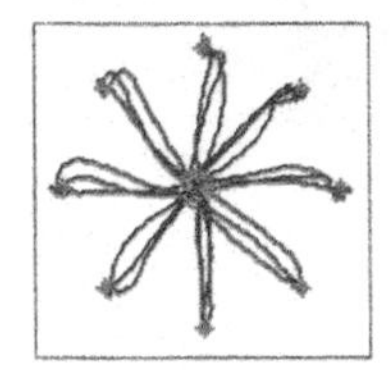

Late Baseline

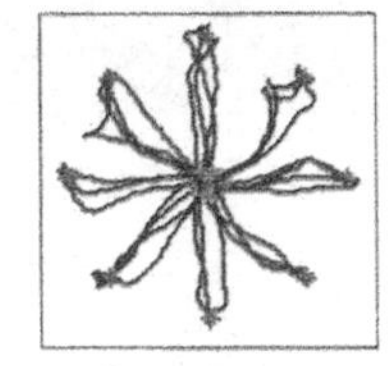

Late Force

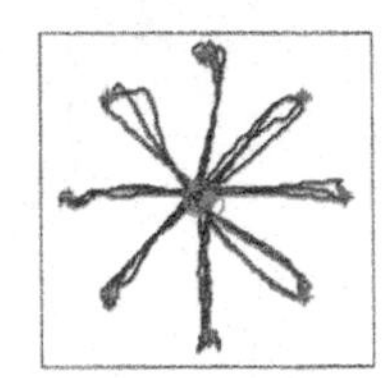

Late Washout

COLOR FIGURE 12.6 Psychophysics of motor learning. Data are shown from a representative experimental session. (A) Trajectories in real space. The trajectories are roughly straight when the movements are not perturbed (baseline). When a counterclockwise (CCW) force field is turned on, trajectories are deviated at first (early force). After the perturbing force is turned off, the first movements show an aftereffect, inasmuch as they are deviated in the clockwise direction (early washout). Within a few trials, however, the monkey readapts to the unperturbed condition, and trajectories become straight again (late washout). (From Gandolfo, F., Li, C., Benda, B.J., Padoa-Schioppa, C., and Bizzi, E., Cortical correlates of learning in monkeys adapting to a new dynamical environment, *Proc. Natl. Acad. Sci. USA,* 97, 2259, 2000, with permission.)

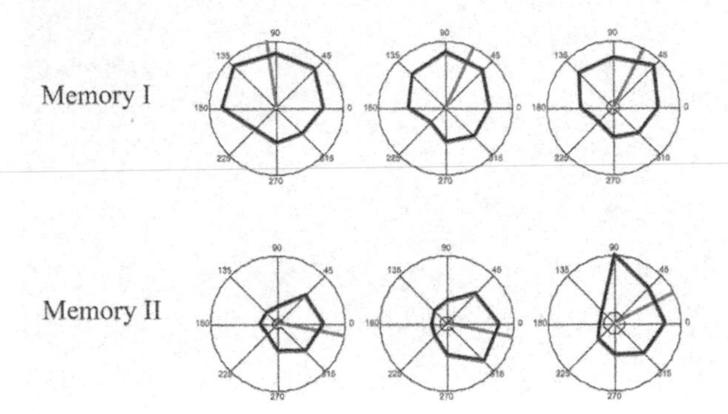

COLOR FIGURE 12.7 The tuning curves are plotted in polar coordinates. For each cell, the three plots represent the movement-related activity in the Baseline (left), in the Force epoch (center), and in the Washout (right). In each plot, the circle in dashed line represents the average activity during the center hold time window, when the monkey holds the manipulandum inside the center square and waits for instructions. Examples of memory I and memory II cells, in terms of the modulation of the Pd. All cells were recorded with a clockwise force field. (From Li, C.S., Padoa-Schioppa, C., and Bizzi, E., Neuronal correlates of motor performance and motor learning in the primary motor cortex of monkeys adapting to an external force field, *Neuron,* 30, 593, 2001, with permission.)

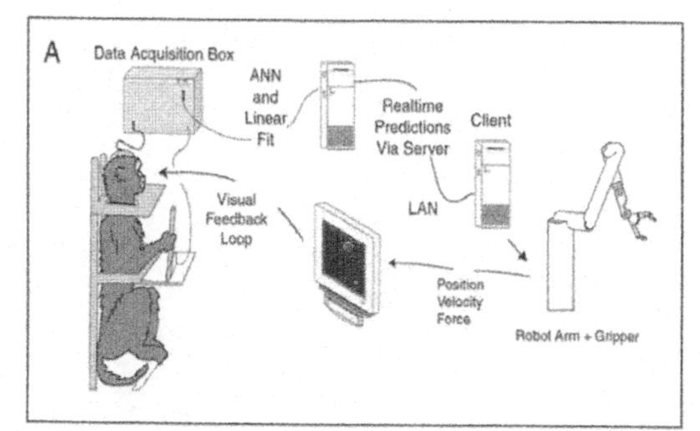

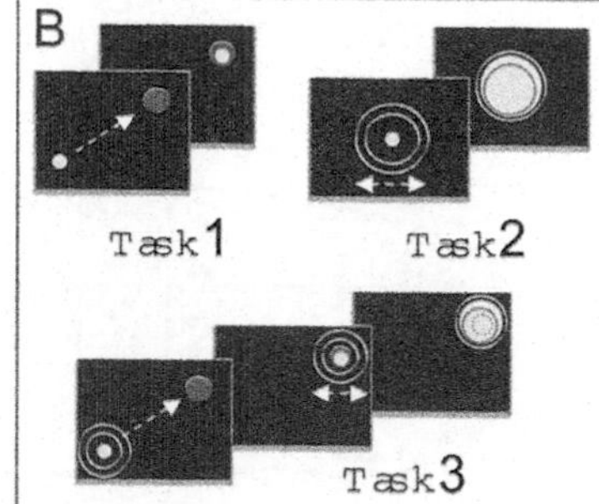

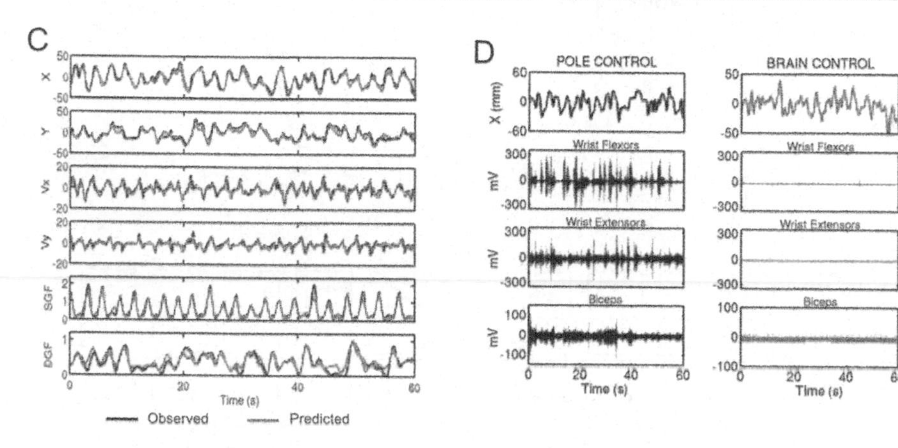

COLOR FIGURE 13.2 (A) Experimental setup and control loops, consisting of a data acquisition system, a computer running multiple linear models in real time, a robot arm equipped with a gripper, and a monkey visual display. The pole was equipped with a gripping force transducer. Robot position was translated into cursor position on the screen, and feedback of the gripping force was provided by changing the cursor size. (B) Schematics of three behavioral tasks. In task 1 the monkey's goal was to move the cursor to a visual target (green) that appeared at random locations on the screen. In task 2 the pole was stationary, and the monkey had to grasp a virtual object by developing a particular gripping force instructed by 2 red circles displayed on the screen. Task 3 was a combination of tasks 1 and 2. The monkey had to move the cursor to the target and then develop a gripping force necessary to grasp a virtual object. (C) Motor parameters (blue) and their prediction using linear models (red). From top to bottom, hand position (HPX, HPY) and velocity (HVx, HVy) during execution of task 1, and gripping force (GF) during execution of tasks 2 and 1. (D) Surface EMGs of arm muscles recorded in task 1 for pole control (left) and brain control without arm movements (right). Top plots show X-coordinate of the cursor; plots below display EMGs of wrist flexors, wrist extensors, and biceps. EMG modulations were absent in brain control. (Extracted from Reference 10.)

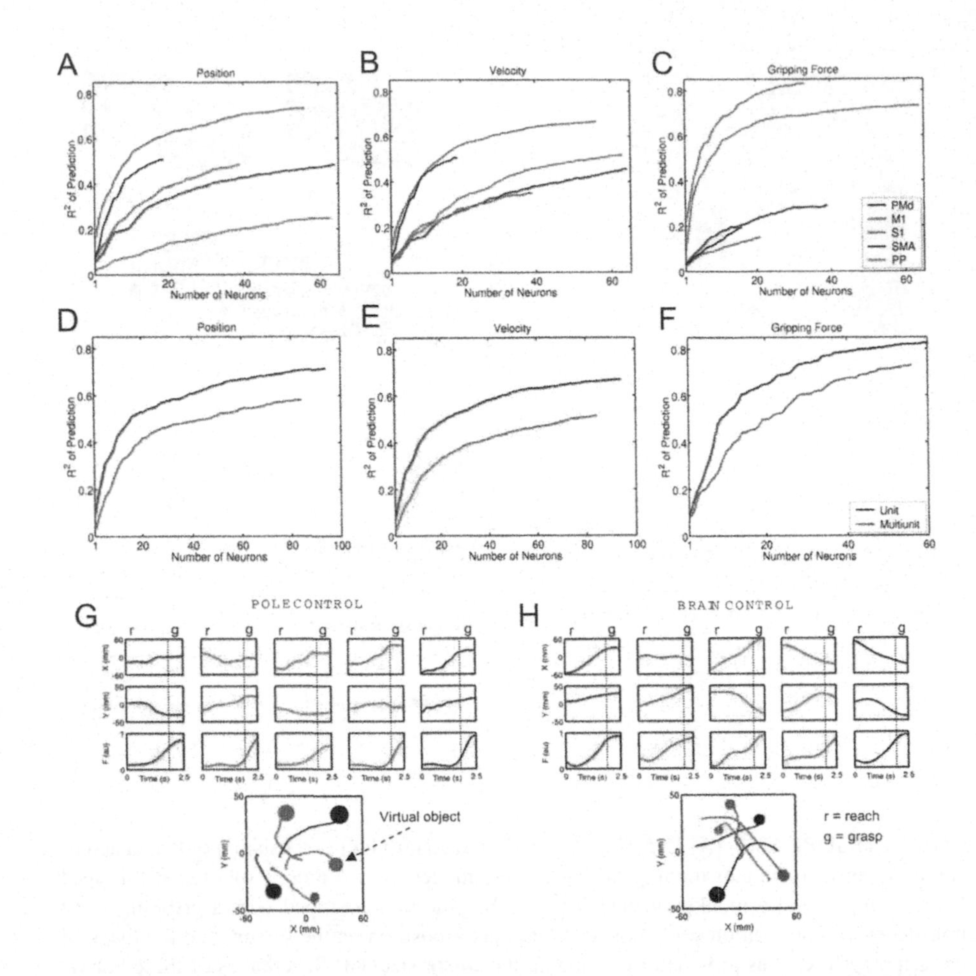

COLOR FIGURE 13.3 (A–F) Contribution of different cortical areas to model predictions of hand position, velocity (task 1), and force (task 2). For each area, neuronal dropping curves represent average prediction accuracy (R^2) as a function of the number of neurons needed to attain it. Contributions of each cortical area vary for different parameters. Typically more than 30 randomly sampled neurons were required for an acceptable level of prediction. (G–I) Comparison of the contribution of single units (blue) and multiple units (red) to predictions of HP, HV, and GF. Single units and multiple units were taken from all cortical areas. Single units' contribution exceeded that of multiple units by ~20%. (G, H) Representative robot trajectories and gripping force profiles in an advanced stage of training in task 3 during both pole and brain control. The bottom graphs show trajectories and the amount of the gripping force developed during grasping of each virtual object. The dotted vertical lines in the panels indicate the end of reach. Note that during both modes of BMI operation, the patterns of reaching and grasping movements (displacement followed by force increase) were preserved. (Extracted from Reference 10.)

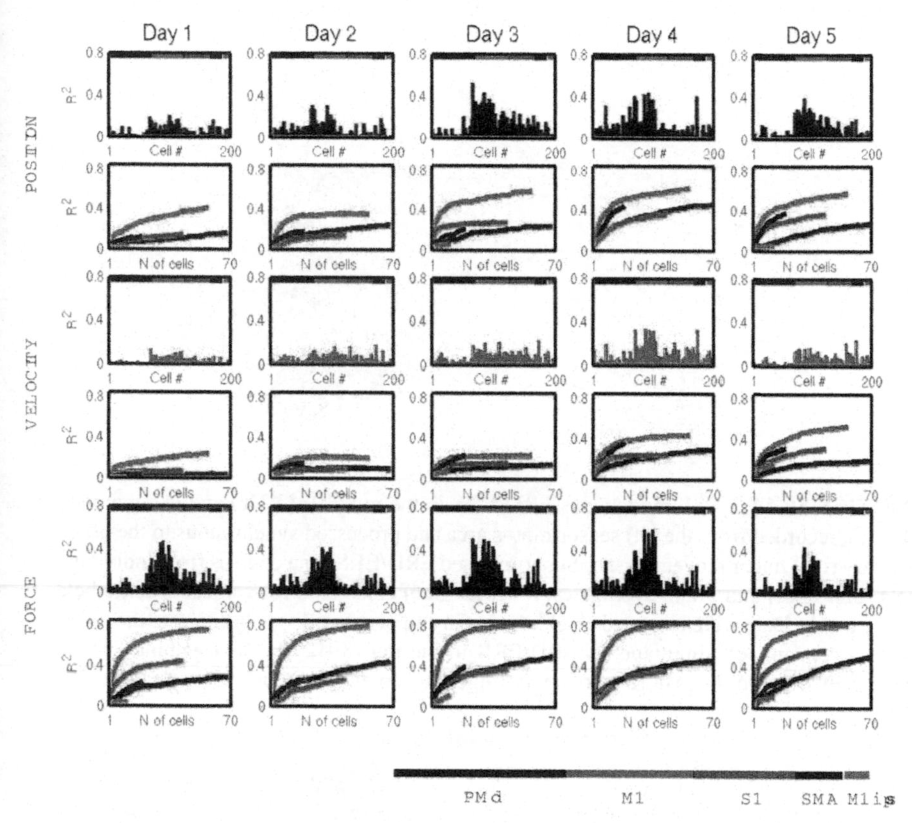

COLOR FIGURE 13.4 Variability in contributions of individual neurons and cortical areas to the representation of multiple motor parameters (from top to bottom: hand position, hand velocity, and gripping force). Note the clear increase of accuracy in predictions for individual neurons and cortical areas during the 5 day period. During the same period, a high degree of variability in both neuronal and real contributions was observed. The color bar indicates the sample size for each cortical area. (Extracted from Reference 45.)

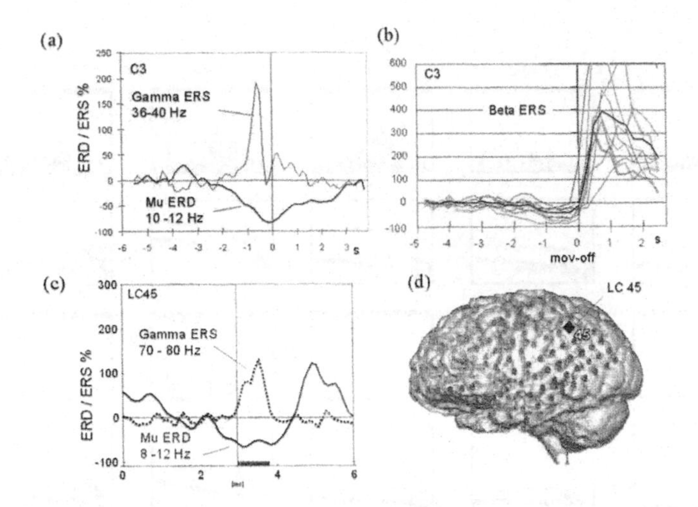

COLOR FIGURE 14.2 a) Simultaneous ERD/ERS in the mu (10–12 Hz) and gamma band (36–40 Hz), recorded from the left sensorimotor area and processed synchronous to the offset of voluntary right finger movements. b) Superimposed ERD/ERS time courses from individual subjects (thin lines) and grand average curve (thick line) calculated for subject-specific beta frequency bands. Data were recorded on electrode C3 and processed synchronous to the offset of wrist movements. c) Simultaneous ERD/ERS in the mu (8–12 Hz) and gamma (70–80 Hz) band in ECoG recordings during voluntary finger movement. d) ECoG electrode locations.

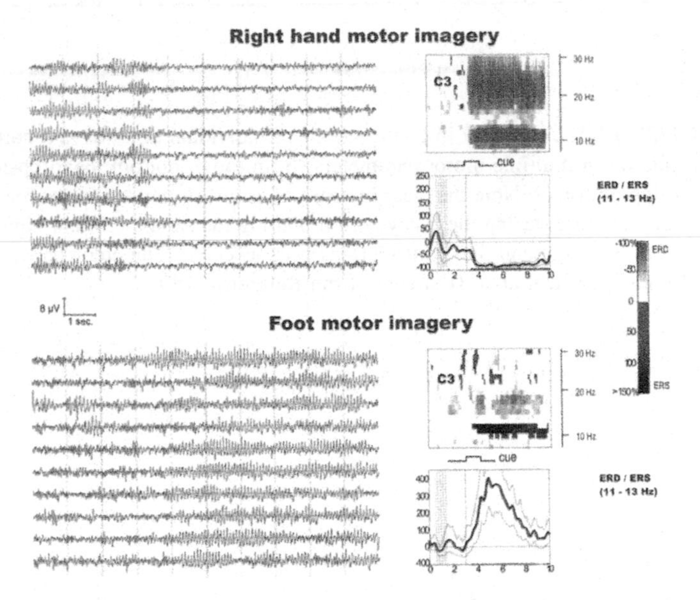

COLOR FIGURE 14.4 Left side: examples of single EEG trials recorded from electrode position C3 during right hand (upper panel) and foot (lower panel) motor imagery. Right side: ERD/ERS time frequency maps and time curves of the frequency band 11–13 Hz recorded from electrode position C3 during right hand (upper panels) vs. foot (lower panels) motor imagery. Onset of cue presentation at second 3.

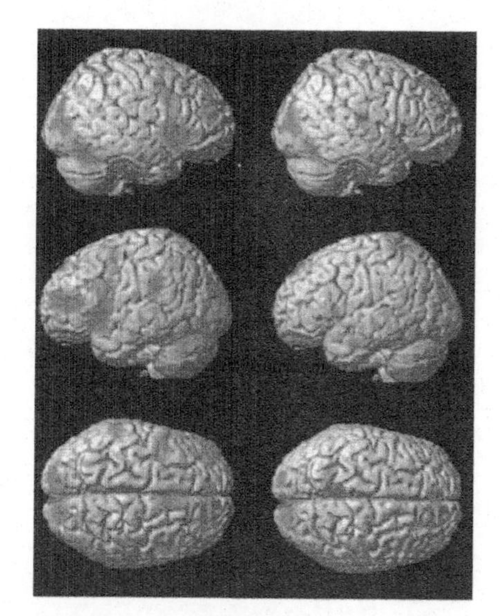

COLOR FIGURE 14.5 Averaged functional magnetic resonance images (fMRI) of 10 subjects during self-regulation of cortical negativity (left) and positivity (right). Red symbolizes increase, green decrease of BOLD-response.

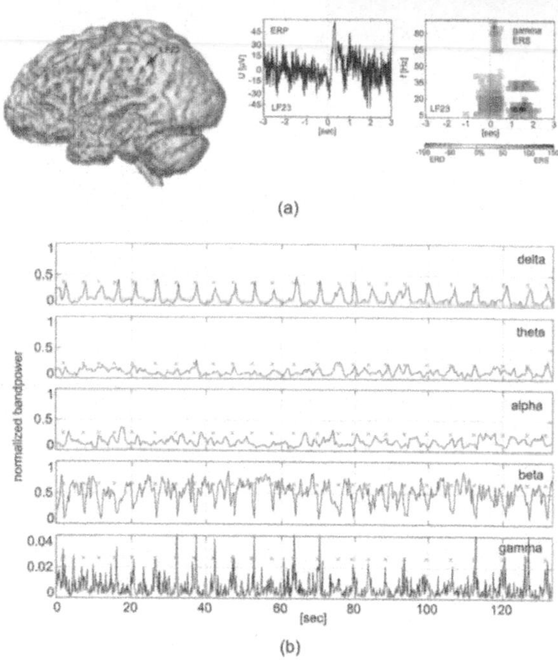

COLOR FIGURE 14.15 ERP, ERD/ERS map, and short-time Fourier time course of one exemplary ECoG channel. (a) ERP template calculated from 23 trials. This template was used for the CCTM method. The ERD/ERS maps represent averaged oscillatory activity in a frequency range from 5 to 100 Hz. ERD is colored in red, and ERS is colored in blue. Movement onset is indicated by the vertical dash–dotted line. (b) The short-time Fourier time courses show ongoing normalized bandpower of movement-related patterns in the delta (< 3.5 Hz), beta (12.5–30 Hz), and gamma (70–90 Hz) band. Theta (3.5–7.5 Hz) and alpha (7.5–12.5 Hz) bands do not show distinct peaks around movement-onsets indicated by the crosses. (Modified from Graimann, B., Movement-related patterns in ECoG and EEG: visualization and detection, TU Graz, Austria: Ph.D. thesis, 2002, with permission.)

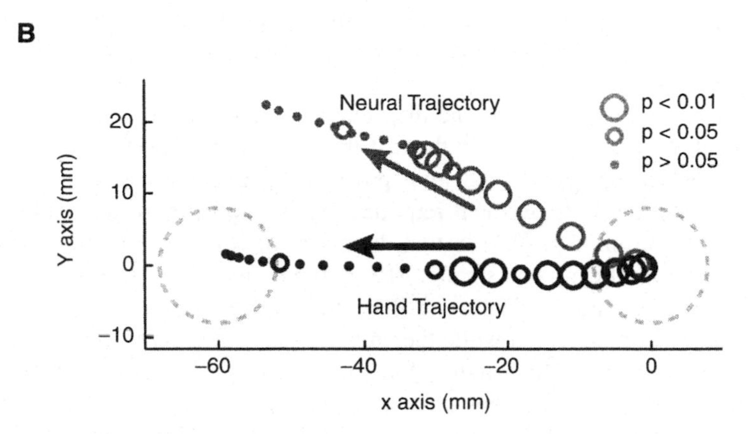

FIGURE 6.6 (continued)

A second criticism is that population vectors for 16 movement directions cannot be predicted when most cells were recorded in only 8 directions. As stated by Georgopoulos,[70] "Eight points are insufficient to estimate accurately intermediate points for an intensely curved tuning function." This is a surprising concern since practitioners of the population vector method often replace the actual discharge pattern of cells with cosine functions, thus removing any "intensely curved" components of a cells' tuning function.[10,72] Further, it has been shown that correlations between the direction of population vectors and movement direction are similar whether the actual discharge rate of the cell or fitted cosine tuning functions are used in constructing the population signal.[10] We found similar errors in predicting movement direction whether the population vector was constructed from the actual discharge of neurons or based on Von Mises functions.[1,69]

Interestingly, Georgopoulos[70] did not raise any criticisms regarding our observed bimodal distribution of PDs, the key reason why population vectors failed to predict movement direction. Some functions described by Georgopoulos et al.[10] can compensate for a unimodal bias in the distribution, but none can compensate for a bimodal distribution.[1]

Bimodal distributions have been noted previously for hand movements in the horizontal plane but with the arm in two different arm postures: one with the arm roughly in a horizontal plane and the other with the limb oriented vertically.[31] The distribution of preferred directions tends to be more skewed when the arm is maintained in the horizontal plane as predicted by mathematical models that assume that neural activity reflects features of the motor periphery. Even the distribution of PDs shown in Figure 12 of Schwartz et al.[73] appears to have a bimodal distribution with a greater number of neurons having PDs oriented away from and toward the monkey. While the article states that PDs were uniformly distributed, it is not stated whether the distribution was tested against both a unimodal and a bimodal distribution.

The present study illustrates that nonhuman primates are capable of generating movements of the hand to spatial targets even though population vectors constructed from neural activity in M1 do not point in the direction of hand movement. This certainly does not disprove that some neural activity in M1 may convey information related to the hand.[28,34]

Of particular interest is why the distribution is not uniform. Figures 6.5B and 6.5C illustrate that the variations in the distribution of PDs of neurons in M1 appear to parallel the anisotropy in total joint power for movements in different spatial directions: directions of movement requiring the greatest power also had a greater proportion of neurons with PDs in that direction. In contrast, correlations of joint velocity were significant but smaller (R^2 = 0.54) and there was essentially no correlation between the distribution of PDs and muscle torque.

It is important to note that while the present study suggests that limb mechanics has a strong influence on the activity of neurons in M1, it does not mean that neural activity at the single cell or population level is explicitly coding joint power. M1 appears to reflect many different features of movement. The covariation between the distribution of PDs and joint power can be viewed as a reflection of the internal model of the motor periphery used to guide and control limb movement. A high correlation was found, since power captures two key elements of the peripheral motor apparatus: torque and velocity. Neural activity in M1 is influenced sufficiently by these features of the motor periphery that it biases the activity of many neurons to be preferentially active for movements in one of two spatial directions. The reduction from three to one DOF of motion at the shoulder also likely plays a role in the bias in the distribution of PDs.[1,31]

6.5 NEURAL CORRELATES OF MECHANIC LOADS

Not only can primates compensate their motor patterns for mechanical loads, they do it almost effortlessly for many types of loads and under many behavioral contexts, such as picking up a shell as we swim under water, shooting a puck while wearing protective equipment, or even juggling while balancing on a unicycle. Each object or environment creates forces with different temporal and spatial features such as constant, bias forces (i.e., gravitational force), viscous forces (swimming), and accelerative forces (moving a mass). How does the brain represent the wide range of mechanical loads encountered in our daily lives?

Two qualitatively distinct hypotheses have been proposed to explain how internal models for different loads are implemented by the brain.[74] One possibility is that internal models for different loads are represented within a single controller that encapsulates all possible loads. A second possibility is a more modular scheme in which multiple controllers coexist, each suitable for one context or a small set of contexts. These two hypotheses suggest striking differences as to how individual neurons in regions of the brain will respond to loads; either a cell responds to all mechanical loads (the former), or it responds only to a subset of loads (the latter).

We addressed this issue by exploring the response of neurons in M1 during reaching with and without velocity-dependent (viscous) loads applied to the shoulder or elbow joints.[39] Loads applied only to the shoulder (viscous shoulder [VS]) or only to the elbow (viscous elbow [VE]) allowed us to examine whether mechanically independent loads are represented by separate populations of neurons or distributed across a single neural population. A third load condition, where viscous loads were applied to joints simultaneously (viscous both [VB]), allowed us to examine how mechanically dependent loads with common features or characteristics are represented neurally. We found that many cells changed their activity for one, two, and in some cases all three load conditions as compared to their activity during unloaded reaching. The representation of VS and VE loads were not completely independent, but demonstrated at least a partial overlap across the cell population in M1. Of the 51 cells that responded to either loading condition, 27 were sensitive only to VE, 9 were sensitive only to VS, and 15 showed significant changes in discharge for both VS and VE ($p < 0.05$, analysis of variance [ANOVA]). There was a highly consistent relationship between how a cell responds in VS and VE. Cells that increase discharge for VS also tend to increase discharge for VE, while decreases in discharge for VS are likewise associated with discharge decreases in VE.

Perhaps the most important observation was that there was considerable overlap in the representations of VB and either VS or VE. Almost all neurons that changed their activity for VB as compared to unloaded movements, also showed significant changes of activity related to VE or VS. We found that almost all cells showed similar signs of change across all three load conditions. If a neuron increased its discharge for a given loading condition, its response to any other load condition would also be an increase in discharge. If a neuron decreased its discharge for a load condition, responses to other loads would also tend to be a reduction in discharge.

With regard to whether the brain uses a single internal model or multiple internal models for different mechanical contexts, the present results illustrate that neural activity in M1 appears to reflect a single internal model for both these single- and multiple-joint loads. However, other regions of the brain, such as the cerebellum, may use separate internal models for these different contexts. Further, because only velocity-dependent loads were used in this study, it is quite possible that neural representations for different types of mechanical loads (i.e., viscous versus elastic) may be treated separately.[75]

One of the key differences between the present study and previous studies is that loads were applied at different parts of the motor apparatus: shoulder versus elbow. This mechanical segregation allowed us to illustrate that load-related activity for some neurons was limited to loads at only one of the two joints, whereas other neurons responded to loads applied to either joint. These results suggest that there is some separation, but not a complete separation, in neurons responding to loads at different joints, reflecting a course somatotopic map within M1.[76,77] We are presently developing cortical maps of neurons related to the shoulder and elbow joints to observe if there is any variation in their distribution within the cortex.

The present data on neural responses for single- and multiple-joint loads allow us to ask how information related to different joints is integrated together. We tested two possible models, one in which load-related activity related to each joint is linearly summed across all joints. However, we found this model consistently overestimated the response of neurons to multiple-joint loads. We examined a second model that assumed that the response of a neuron reflected vector summation of its response to loads at each joint. This vector summation model assumes that activity related to each joint can be treated as orthogonal vectors and that multiple-joint loads reflect the vector sum of these single-joint loads. Our data illustrated that the response of neurons tended to follow this simple rule. We are presently assessing whether this integrative feature of multiple-joint loads reflects an inherent feature of cortical processing or simply parallels the activity of shoulder and elbow muscles for these movement-dependent loads.

Another of our recent studies examined the response of neurons to constant-magnitude (bias) loads applied to the shoulder or elbow as the monkey maintained its hand at a central target.[78] The response of many neurons paralleled our results on viscous loads applied during reaching: some neurons responded to loads at only one of the two joints, whereas others responded to loads at both joints. Load-sensitive cells again responded to both multiple-joint loads and at least one of the two single-joint loads so that there was no segregation between neural responses to single- and multiple-joint loading conditions. Further, the response of neurons to multiple-joint loads again could be predicted using a vector summation model from the response of neurons to single-joint loads.

A key feature of both of these studies was that we could load the shoulder and elbow joints independently. It seems reasonable to assume that these single-joint loads would selectively influence the response of muscles that span that joint. We were mistaken. Many muscles that only spanned one of the two joints modified their activity for loads applied to the other joint. For example, brachioradialis, an elbow flexor muscle, increased its activity when the monkey generated either an elbow flexor *or* a shoulder extensor muscular torque (Figure 6.7). The greatest activity level was observed when the monkey generated an elbow flexor and a shoulder extensor torque simultaneously. At first, this seems paradoxical, but it simply reflects the action of biarticular muscles that span both joints. Changes in a biarticular muscle's activity for loads applied at one joint necessarily create torque at the other joint. As a result, the activity of muscles spanning this second joint must change to compensate for the change in activity of the biarticular muscles.[79,80]

This coupling of muscle activity at one joint to the mechanical requirements of another joint obfuscates any simple mapping between torque at a joint and the activity of muscles spanning that joint. This has important implications with regard to the response of neurons during single- and multiple-joint loads. While the response of single-joint muscles was almost always greater for loads applied to the spanned as compared to the nonspanned joint, its effect cannot be discounted. Therefore, one cannot assume that neurons that changed their activity for loads applied to both joints are necessarily related to controlling muscles at both joints.

This example underlines the inherent complexity of the peripheral motor apparatus. Our description earlier illustrated that joint torque does not match joint motion for multiple-joint movements due to intersegmental dynamics. The present observations on EMG activity related to mechanical loads illustrates that muscle activity does not match joint torque at a given joint. Therefore, all three levels of description — motion, torque, and muscle activity — provide unique, complementary information on limb motor function. Our ongoing studies are continuing to explore limb mechanics including using simulations to better understand the relationship between muscle activity and motor performance.

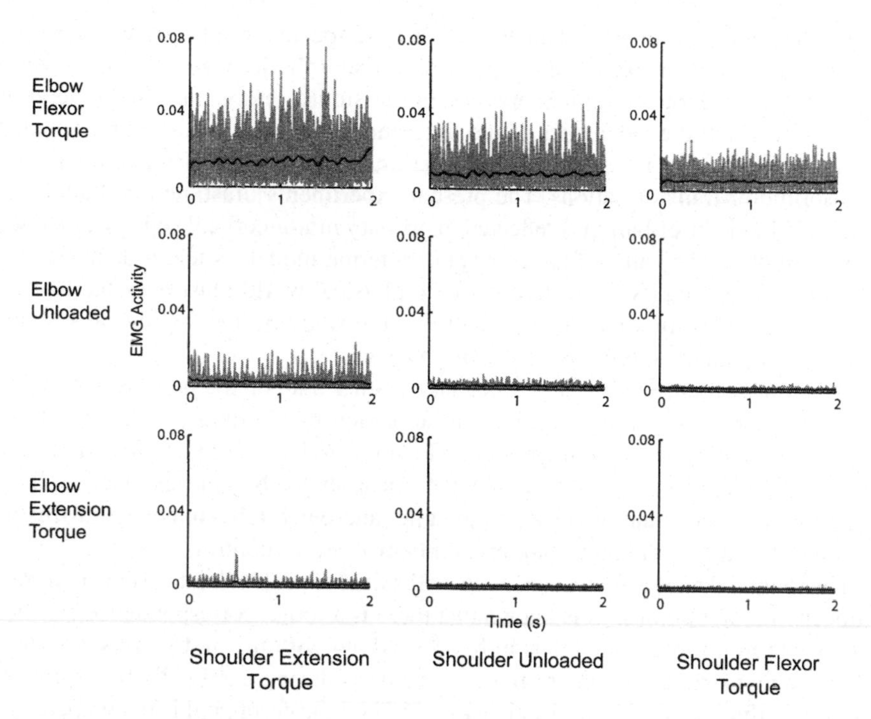

FIGURE 6.7 Activity of brachioradialis, an elbow flexor muscle, when the monkey maintains a constant hand position, but with different constant loads (0.11 Nm) applied to the shoulder or elbow joints. Nine different loading conditions were examined, generating flexor, null, or extensor muscular torque at each joint.[78] The central panel shows the activity of the muscle when no loads were applied to the joints (the solid line is the mean of five repeat trials). Muscle activity increases when the monkey generates an elbow flexor torque. However, its magnitude also varies with shoulder muscle torque such that it increases when the monkey generates a shoulder extensor torque. Therefore, brachioradialis muscle activity varies with shoulder muscle torque even though this muscle does not span the shoulder joint.

6.6 SUMMARY AND CONCLUSIONS

The goal of this chapter was twofold. The first goal was to describe two conceptual frameworks, sensorimotor transformations and internal models, for interpreting how the brain controls visual-guided reaching. This comparison was presented because it helps to explain how conceptual frameworks, whether implicitly or explicitly defined, strongly influence the design, analysis, and interpretation of experimental data. What seems like a logical experiment from one perspective can be irrelevant from another. My recent experiments have been designed and interpreted based on the concept of internal models, where the brain mimics or reflects the physical properties of the limb and the environment. This concept has been very influential for human studies on motor performance and learning and appears to be ideal, at this time, for exploring the neurophysiological basis of movement in nonhuman primates.

The second goal of this chapter was to describe the results from our recent studies using a planar experimental paradigm. Our robotic device can both sense and perturb limb motor function, and our initial studies have illustrated several of the ways in which the mechanics of the limb and of physical loads are represented in M1. It is important to realize that the present results do not disprove the notion of sensorimotor transformations. The present experiments illustrate that both kinematic and kinetic information is reflected in primary motor cortical activity, as shown by other studies. The value of the concept of internal models is that it demonstrates that body motion and its interaction with the physical world must obey the laws of Newtonian physics. In effect, motor control is the study of how biological systems consider and manage these basic laws of physics.

We initially focused on well-learned tasks rather than on the process of learning such tasks. This was largely a pragmatic approach to the question of how stable neural states are represented in the brain. However, M1 is clearly involved in motor learning and adaptive control.[81–84] KINARM can apply loads at the shoulder or elbow joint based on almost any variable imaginable, and our future studies will focus on exploring the role of the motor cortex in adaptive motor control.

The robotic device also appears to be well suited for exploring the role of afferent feedback in motor control. It is known that there is a rough correspondence between sensory and motor representations in M1.[85,86] Along with the motor tasks described above, we regularly record the response of neurons during passive limb movements and to perturbations during postural tasks.[87,88] The long-term goal is to compare and contrast the sensory and motor responses of individual neurons in order to better understand how afferent feedback contributes to motor function.

ACKNOWLEDGMENTS

I would like to thank Kirsten M. Graham, Kimberly D. Moore, and Jon Swaine for technical assistance. Stephen H. Scott is supported by a CIHR Grant (MOP-13462) and a CIHR Investigator Award.

REFERENCES

1. Scott, S.H., The role of primary motor cortex in goal-directed movements: insights from neurophysiological studies on non-human primates. *Curr. Opin. Neurobiol.*, 13, 671, 2003.
2. Kuhn, T.S., *The Structure of Scientific Revolutions*, 2nd edition, University of Chicago Press, Chicago, 1970.
3. Saltzman, E., Levels of sensorimotor representation, *J. Math. Psychol.*, 20, 91, 1979.
4. Soechting, J.F. and Flanders, M., Moving in three-dimensional space: frames of reference, vectors, and coordinate systems, *Annu. Rev. Neurosci.*, 15, 167, 1992.
5. Kalaska, J.F. et al., Cortical control of reaching movements, *Curr. Opin. Neurobiol.*, 7, 849, 1997.
6. Morasso, P., Spatial control of arm movements, *Exp. Brain Res.*, 42, 223, 1981.

7. Sergio, L.E. and Scott, S.H., Hand and joint paths during reaching movements with and without vision, *Exp. Brain Res.*, 122, 157, 1998.
8. Georgopoulos, A.P. et al., On the relations between the direction of two-dimensional arm movements and cell discharge in primate motor cortex, *J. Neurosci.*, 2, 1527, 1982.
9. Georgopoulos, A.P. et al., Interruption of motor cortical discharge subserving aimed arm movements, *Exp. Brain Res.*, 49, 327, 1983.
10. Georgopoulos, A.P., Kettner, R.E., and Schwartz, A.B., Primate motor cortex and free arm movements to visual targets in three-dimensional space. II. Coding of the direction of movement by a neuronal population, *J. Neurosci.*, 8, 2928, 1988.
11. Mussa-Ivaldi, F.A., Do neurons in the motor cortex encode movement direction? An alternative hypothesis, *Neurosci. Lett.*, 91, 106, 1988.
12. Sanger, T.D., Theoretical considerations for analysis of population coding in motor cortex, *Neural Comp.*, 6, 29, 1994.
13. Todorov, E., Direct cortical control of muscle activation in voluntary arm movements: a model, *Nat. Neurosci.*, 3, 391, 2000.
14. Moran, D.W. and Schwartz, A.B., Letter to the editor: "One motor cortex, two different views," *Nat. Neurosci.*, 3, 964, 2000.
15. Georgopoulos, A.P. and Ashe, J., Letter to the editor: "One motor cortex, two different views," *Nat. Neurosci.*, 3, 963, 2000.
16. Todorov, E., Reply in "One motor cortex, two different views," *Nat. Neurosci.*, 3, 963, 2000.
17. Scott, S.H., Reply in "One motor cortex, two different views," *Nat. Neurosci.*, 3, 964, 2000.
18. Schwartz, A.B., Motor cortical activity during drawing movements: single-unit activity during sinusoid tracing, *J. Neurophysiol.*, 68, 528, 1992.
19. Fu, Q.G. et al., Temporal encoding of movement kinematics in the discharge of primate primary motor and premotor neurons, *J. Neurophysiol.*, 73, 836, 1995.
20. Johnson, M.T., Mason, C.R., and Ebner, T.J., Central processes for the multiparametric control of arm movements in primates, *Curr. Opin. Neurobiol.*, 11, 684, 2001.
21. Poppelle, R. and Bosco G., Sophisticated spinal contributions to motor control, *Trends Neurosci.*, 26, 269, 2003.
22. Georgopoulos, A.P., On the translation of directional motor cortical commands to activation of muscles via spinal interneuronal systems, *Cogn. Brain Res.*, 3, 151, 1996.
23. Georgopoulos, A.P. et al., Mental rotation of the neuronal population vector, *Science*, 243, 234, 1989.
24. Lurito, J.T., Georgakopoulos, T., and Georgopoulos, A.P., Cognitive spatial-motor processes. 7. The making of movements at an angle from a stimulus direction: studies of motor cortical activity at the single cell and population levels, *Exp. Brain Res.*, 87, 562, 1991.
25. Whitney, C.S., Reggia, J., and Cho, S., Does rotation of neuronal population vectors equal mental rotation? *Connection Sci.*, 9, 253, 1997.
26. Cisek, P. and Scott, S.H., An alternative interpretation of population vector rotation in macaque motor cortex, *Neurosci. Lett.*, 272, 1, 1999.
27. Moody, S.L. and Wise, S.P., Connectionist contributions to population coding in the motor cortex, *Progr. Brain Res.*, 130, 245, 2001.
28. Shen, L. and Alexander, G.E., Neural correlates of a spatial sensory-to-motor transformation in primary motor cortex, *J. Neurophysiol.*, 77, 1171, 1997.
29. Zhang, J. et al., Dynamics of single neuron activity in monkey primary motor cortex related to sensorimotor transformation, *J. Neurosci.*, 17, 2227, 1997.

30. Caminiti, R., Johnson, P.B., and Urbano, A., Making arm movements within different parts of space: dynamic aspects in the primate motor cortex, *J. Neurosci.,* 10, 2039, 1990.
31. Scott, S.H. and Kalaska, J.F., Reaching movements with similar hand paths but different arm orientations: I. Activity of individual cells in motor cortex, *J. Neurophysiol.*, 77, 826, 1997.
32. Sergio, L.E. and Kalaska, J.F., Systematic changes in directional tuning of motor cortex cell activity with hand location while generating static isometric forces in constant spatial directions, *J. Neurophysiol.*, 78, 1170, 1997.
33. Sergio, L.E. and Kalaska, J.F., Systematic changes in motor cortex cell activity with arm posture during directional isometric force generation, *J. Neurophysiol.*, 89, 212, 2003.
34. Kakei, S., Hoffman, D.S., and Strick, P.L., Muscle and movement representations in the primary motor cortex, *Science*, 285, 2136, 1999.
35. Scott, S.H., Sergio, L.E., and Kalaska, J.F., Reaching movements with similar hand paths but different arm orientations. II. Activity of individual cells in dorsal premotor cortex and parietal area 5, *J. Neurophysiol.*, 78, 2413, 1997.
36. Kakei, S., Hoffman, D.S., and Strick, P.L., Direction of action is represented in the ventral premotor cortex, *Nat. Neurosci.*, 4, 1020, 2001.
37. Evarts, E.V., Representation of movements and muscles by pyramidal tract neurons of the precentral motor cortex, in *Neurophysiological Basis of Normal and Abnormal Motor Activities*; Yahr, M.D. and Purpura, D.P., Eds., Raven Press, New York, 1967.
38. Kalaska, J.F. et al., Comparison of movement direction-related versus load direction-related activity in primate motor cortex, using a two-dimensional reaching task, *J. Neurosci.*, 9, 2080, 1989.
39. Gribble, P.L. and Scott, S.H., Overlap of internal models in motor cortex for mechanical loads during reaching, *Nature,* 417, 938, 2002.
40. Fetz, E.E., Are movement parameters recognizably coded in the activity of single neurons? *Behav. Brain Sci.*, 15, 679, 1992.
41. Lundberg, A., To what extent are brain commands for movements mediated by spinal interneurons? *Behav. Brain Sci.*, 15, 775, 1992.
42. Loeb, G.E., Brown, I.E., and Cheng, E.J., A hierarchical foundation for models of sensorimotor control, *Exp. Brain Res.*, 126, 1, 1999.
43. Burke, R.E. et al., Anatomy of medial gastrocnemius and soleus motor nuclei in cat spinal cord, *J. Neurophysiol.*, 40, 667, 1977.
44. Barker, E. et al., Types of intra- and extrafusal muscle fibre innervated by dynamic skeletofusimotor axons in cat peroneus brevis and tennissimus muscles as determined by the glycogen depletion method, *J. Physiol. Lond.*, 266, 713, 1977.
45. Seki, K., Perlmutter, S.I., and Fetz, E.E., Sensory input to primate spinal cord is presynaptically inhibited during voluntary movement, *Nat. Neurosci.*, 6, 1309–1316, 2003.
46. Cole, K.J. and Abbs, J.H., Kinematic and electromyographic responses to pertubation of a rapid grasp, *J. Neurophysiol.*, 57, 1498, 1987.
47. Conrad, B. et al., Cortical load compensation during voluntary elbow movements, *Brain Res.*, 71, 507, 1974.
48. Evarts, E. and Tanji, J., Reflex and intended responses in motor cortex pyramidal tract neurons of monkey, *J. Neurophysiol.*, 39, 1069, 1976.
49. Harris, C.M. and Wolpert, D.M., Signal dependent noise determines motor planning, *Nature*, 394, 780, 1998.

50. Todorov, E. and Jordan, M.I., Optimal feedback control as a theory of motor coordination, *Nat. Neurosci.*, 5, 1226, 2002.
51. Kawato, M., Furukawa, K., and Suzuki, R.A., Hierarchical neural-network model for control and learning of voluntary movement, *Biol. Cybern.*, 57, 169, 1987.
52. Miall, R.C. and Wolpert, D.M., Forward models for physiological motor control, *Neural Netw.*, 9, 1265, 1996.
53. Scott, S.H. and Norman, K.E., Computational approaches to motor control and their potential role for interpreting motor dysfunction, *Curr. Opin. Neurol.*, 16, 693, 2003.
54. Shadmehr, R. and Mussa-Ivaldi, F.A., Rapid adaptation to coriolis force perturbations of arm trajectory, *J. Neurosci.*, 14, 3208, 1994.
55. Lackner, J.R. and DiZio, P., Rapid adaptation to coriolis force perturbations of arm trajectory, *J. Neurophysiol.*, 72, 299, 1994.
56. Sainburg, R.L., Ghez, C., and Kalakanis, D., Intersegmental dynamics are controlled by sequential anticipatory, error correction, and postural mechanisms, *J. Neurophysiol.*, 81, 1045, 1999.
57. Wolpert, D.M., Ghahramani, Z., and Flangan, J.R., Perspectives and problems in motor learning, *Trends Cogn. Sci.*, 5, 487, 2001.
58. Wolpert, D.M. and Ghahramani, Z., Computational principles of movement neuroscience, *Nat. Neurosci.*, 3, 1212, 2000.
59. Flanagan, J.R. and Wing, A.M., The role of internal models in motion planning and control: evidence from grip force adjustments during movements of hand-held loads, *J. Neurosci.*, 17, 1519, 1997.
60. Westling, G. and Johansson, R.S., Factors influencing the force control during precision grip, *Exp. Brain Res.*, 53, 277, 1984.
61. Scott, S.H., Apparatus for measuring and perturbing shoulder and elbow joint positions and torques during reaching, *J. Neurosci. Meth.*, 89, 119, 1999.
62. Porter, R. and Lemon, R., *Corticospinal Function and Voluntary Movement*, Clarendon Press, Oxford, 1993.
63. Graham, K.M. et al., Kinematics and kinetics of multi-joint reaching in non-human primates, *J. Neurophysiol.*, 89, 2667, 2003.
64. Favilla, M. et al., Trajectory control in targeted force impulses. VII. Independent setting of amplitude and direction in response preparation, *Exp. Brain Res.*, 79, 530, 1990.
65. Graham, K.M. and Scott, S.H., Morphometry of macaca mulatta forelimb. III. Moment are of shoulder and elbow muscles, *J. Morphol.*, 255, 301, 2003.
66. Cheng, E.J. and Scott, S.H., Morphometry of *Macaca mulatta* forelimb. I. Shoulder and elbow muscles and segment inertial parameters, *J. Morphol.*, 245, 206, 2000.
67. Scott, S.H., Brown, I.E., and Loeb, G.E., Mechanics of feline soleus: I. Effect of fascicle length and velocity on force output, *J. Musc. Res. Cell Motil.*, 17, 207, 1996.
68. Hollerbach, J.M. and Flash, T., Dynamic interactions between limb segments during planar arm movement, *Biol. Cybern.*, 44, 67, 1982.
69. Scott, S.H. et al., Dissociation between hand motion and population vectors from neural activity in motor cortex, *Nature*, 413, 161, 2001.
70. Georgopoulos, A.P., Cognitive motor control: spatial and temporal aspects, *Curr. Opin. Neurobiol.*, 12, 678, 2002.
71. Amirikian, B. and Georgopoulos, A.P., Directional tuning profiles of motor cortical cells, *Neurosci. Res.*, 36, 73, 2000.
72. Georgopoulos, A.P., Schwartz, A.B., and Kettner, R.E., Neuronal population coding of movement direction, *Science*, 233, 1416, 1986.

73. Schwartz, A.B., Kettner, R.E., and Georgopoulos, A.P., Primate motor cortex and free arm movements to visual targets in three-dimensional space. I. Relations between single cell discharge and direction of movement, *J. Neurosci.*, 8, 2913, 1988.
74. Wolpert, D.M. and Kawato, M., Multiple paired forward and inverse models for motor control, *Neural Netw.,* 11, 1317, 1998.
75. Tong, C., Wolpert, D.M., and Flanagan, J.R., Kinematics and dynamics are not represented independently in motor working memory: evidence from an interference study, *J. Neurosci.* 22, 1108, 2002.
76. Sanes, J.N. and Schieber, M.H., Orderly somatotopy in primary motor cortex: does it exist? *NeuroImage*, 13, 968, 2001.
77. Park, M.C. et al., Consistent features in the forelimb representation of primary motor cortex in rhesus macaques, *J. Neurosci.*, 21, 2784, 2001.
78. Cabel, D.W., Cisek, P., and Scott, S.H., Neural activity in primary motor cortex related to mechanical loads applied to the shoulder and elbow during a postural task, *J. Neurophysiol.*, 86, 2102, 2001.
79. Buchanan, T.S., Rovai, G.P., and Rymer, W.Z., Strategies for muscle activation during isometric torque generation at the human elbow, *J. Neurophysiol.*, 62, 1202, 1989.
80. van Zuylen, E.J. et al., Coordination and inhomogeneous activation of human arm muscles during isometric torques, *J. Neurophysiol.*, 60, 1523, 1988.
81. Classen, J. et al., Rapid plasticity of human cortical movement representation induced by practice, *J. Neurophysiol.*, 79, 1117, 1998.
82. Li, C.S.R., Padoa-Schioppa, C., and Bizzi, E., Neuronal correlates of motor performance and motor learning in the primary motor cortex of monkeys adapting to an external force field, *Neuron*, 30, 593, 2001.
83. Sanes, J.N. and Donoghue, J.P., Plasticity and primary motor cortex, *Annu. Rev. Neurosci.*, 23, 393, 2000.
84. Paz, R. et al., Preparatory activity in motor cortex reflects learning of local visuomotor skills, *Nat. Neurosci.*, 6, 882, 2003.
85. Murphy, J.T., Wong, Y.C., and Kwan, H.C., Sequential activation of neurons in primate motor cortex during unrestrained forelimb movement, *J. Neurophysiol.*, 53, 435, 1985.
86. Scott, S.H., Comparison of onset time and magnitude of activity for proximal arm muscles and motor cortical cells prior to reaching movements, *J. Neurophysiol.*, 77, 1016, 1997.
87. Singh, K. and Scott, S.H., Neural circuitry influences learning and generalization of novel loads during reaching in humans, *Soc. Neurosci. Abstr.*, 28, 269.9, 2002.
88. Korbel, T.K. and Scott, S.H., Neural activity in primary motor cortex related to multi-joint pertubations during a postural task, *Soc. Neurosci. Abstr.,* 28, 61.10, 2002.

7 Wheels of Motion: Oscillatory Potentials in the Motor Cortex

William A. MacKay

CONTENTS

ABSTRACT

From their earliest recognition, the oscillatory electroencephalogram (EEG) signals in the sensorimotor cortex have been associated with stasis: a lack of movement, static postures, and possibly physiological tremor. It is now established that 10-, 20-, and 40-Hz motor cortical oscillations are associated with constant, sustained muscle contractions, again a static condition. Sigma band oscillations of about 14 Hz may be indicative of maintained active suppression of a motor response. The dynamic phase at the onset of an intended movement is preceded by a marked drop in oscillatory power, but not all frequencies are suppressed. Fast gamma oscillations coincide with movement onset. Moreover, there is increasing evidence that oscillatory potentials, even of low frequencies (4–12 Hz), may be linked to dynamic episodes of movement. Although their overall power is reduced, these oscillations

0-8493-1287-6/05/$0.00+$1.50

appear to exert motor executive or preparatory functions. Most surprisingly, the 8-Hz cortical oscillation — the neurogenic component of physiological tremor — is emerging as a major factor in shaping the pulsatile dynamic microstructure of movement, and possibly in coordinating diverse actions performed together.

7.1 INTRODUCTION

Within 10 years of Hans Berger's discovery of the EEG and brain rhythms, many features of oscillatory potentials in the sensorimotor cortex were established. For example, what might be called the Kornmüller doctrine appeared in 1932. Based on a thorough study of rabbit cortical activity, Kornmüller postulated that alpha rhythms dominate in granular cortex, and beta rhythms in agranular cortex.[1] In the rolandic area, this concept has been repeatedly supported to this day.[2,3] On the other hand, the functional significance of brain rhythms in motor control remains a topic of much debate.

Extracellular field potentials are generated by neuronal dipoles created within elongated dendritic fields, aligned in parallel arrays. Cortical pyramidal cells with their long apical dendrites are the classic example of dipole generators. The current sink is the site of net depolarization, and the source is the site of normal membrane polarity or of hyperpolarization. Oscillatory potentials are generated by a combination of mechanisms. Many cortical neurons have pacemaker-like membrane properties such that they can produce oscillatory potentials at a variety of frequencies; generally, the higher the depolarization, the faster the frequency.[4] For the oscillations to be stabilized and sustained, however, a resonant circuit needs to be recruited. Inevitably such circuits involve inhibitory interneurons to reinforce the excitation–inhibition alternation.[5,6] Furthermore, the circuits entrain components both within the cerebral cortex and the associated parts of the thalamus to create a thalamocortical network.[5]

The coupling of oscillators sharing a common frequency may be found not just between the cortex and thalamus, but among a set of cortical areas, and between the cortex and the spinal cord. (The cerebral cortex and cerebellum and the cortex and striatum may also have coupled oscillations.) Essentially these are pulse-coupled oscillators, linked by a quasiperiodic train (or rhythmic burst) of action potentials. Linked oscillators, with close to the same frequency, invariably synchronize over time.[7] That is, there is no phase delay between them. Therefore, oscillations offer a useful strategy for circumventing, or at least offsetting, conduction delays within the nervous system, but the price to pay is a preparatory, oscillatory lead-up time.

Brain rhythms can be monitored with noninvasive EEG or magnetoencephalographic (MEG) recording, electrocorticograms (ECoG) recorded with subdural grids of electrodes placed on the surface of the brain, or local field potentials (LFPs) recorded with microwires or microelectrodes within the brain tissue. They can also be seen in the firing patterns of single neurons, but not as clearly and reliably. In sensorimotor cortex, mu and beta rhythms, which may be prominent prior to a movement, are commonly suppressed at the onset of a movement. These changes in oscillatory potentials are time-locked to the event, but unlike event-related potentials, they are not usually phase-locked to the event.[8] Therefore they cannot be extracted by averaging, but require frequency analysis for detection. Such phenomena are manifested as frequency-specific changes of the ongoing electromagnetic activity of the brain. A decrease in power in a given frequency band is thought to be due to a decrease in synchrony of the underlying neuronal population, and therefore is commonly termed an event-related desynchronization (ERD). Conversely an increase in power in a frequency band is termed an event-related synchronization (ERS) of the neuronal population.[8]

The major frequency bands of cortical oscillation considered here are theta (4–8 Hz), mu (or sensorimotor alpha, 8–12 Hz), sigma (12–15 Hz), beta (15–30 Hz), and gamma (>30 Hz). Because of the well-established ERD of both mu and beta rhythms at the time of movement onset, and their reappearance when movement stops, they are commonly equated with "inhibition" or "deactivation" of the motor cortex.[8,9] However, it is not sufficient to equate an oscillation with inhibition or any other single process, when it is, in fact, cyclic. Even if the inhibitory duty cycle is more than 50%, there is a regular period of depolarization following the inhibition. Neglecting this dichotomy can severely limit the usefulness of many ideas about the function of rhythms in sensorimotor cortex.

The appearance of a strong LFP oscillation in a cortical area does have an influence on trains of action potentials. As Figure 7.1 shows, action potentials tend to occur near the negative peaks of oscillatory field potentials, corresponding to the

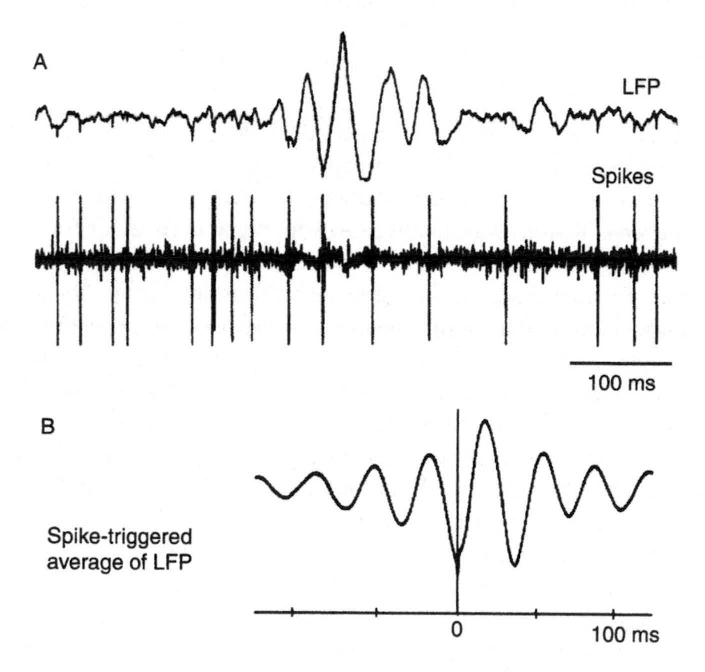

FIGURE 7.1 Relationship of cortical LFP oscillation to single-unit discharge pattern. (A) The extracellularly recorded spike train shows more regular discharge during the LFP oscillation. Both the spikes and LFP were recorded simultaneously with the same electrode. (B) For a similar unit to that shown in A, and only using spikes occurring during an oscillatory episode, spike-triggered averaging of the LFP shows that spiking occurs preferentially at the time of the negative peak in the LFP oscillation. (Adapted from Reference 10, with permission.)

depolarization phase of local pyramidal cells.[10–12] In effect, the oscillation tends to regularize the spike train into a quasiperiodic pattern,[10] which would be useful for some purposes but not others. Most importantly, pyramidal tract (PT) neurons are entrained to the oscillatory LFPs in the motor cortex,[12,13] thereby conveying the oscillation to spinal motoneurons.

As a result of these findings in the past 10 years, consensus seems to be settling on the hypothesis that motor cortical rhythms accompany intervals of stationary sensorimotor processing. The duration of an oscillatory potential would correspond to an episode of relatively stable activity in the neuronal territory exhibiting it. Shifts in power among frequency bands, or in correlation strength among oscillatory potentials in different regions, indicate functional transitions within the motor system from one state to another. During stationary (postural) states, cortical oscillations provide an economic way of driving motor units. Partial synchronization of the discharge of corticomotoneuronal cells would allow them to recruit motor units while maintaining as low a firing rate as possible.[12,14,15] During intervals of overt movement and dynamic fluxes in joint torque, it is generally observed that cortical oscillations below 50 Hz are blocked or desynchronized.[8,9,16] As will be seen below, this is not universally true. It is possible that motor cortical oscillations may provide an economical means of driving motor units or spinal interneurons during dynamic as well as stationary phases of motor control.

7.2 THETA BAND

Lukatch and colleagues,[17] working in neocortical slices, showed that it is possible to elicit realistic theta rhythm (4–8 Hz) by applying carbachol, a cholinergic agonist, and bicuculline, a gamma-amino butyric acid ($GABA_A$) antagonist. The circuitry for generating theta oscillations was localized to superficial cortical layers. A prominent current sink was found in layers 2 to 3, with the source in layer 5. Rhythmic activity required intact glutamatergic transmission as well as inhibition. By analogy with the hippocampus, the inhibitory interneurons in the network probably have a slow spiking frequency, and terminate on distal dendrites of pyramidal neurons.[18] Thalamic neurons also commonly have oscillatory membrane potentials at 6 Hz.[5] Therefore, a thalamocortical feedback loop is likely involved in sustaining theta oscillations.

Although the circuit to generate it may be present, theta rhythm is not conspicuous in motor cortical recordings. Relationships of theta oscillations to movement typically involve neighboring regions. For example, in cats, a widespread 5-Hz LFP oscillation with foci in somatosensory (S1) and in the visual cortex correlates to general disinterest in the environment and drowsiness.[19] However, theta oscillations can occur simultaneously with faster rhythms, in the context of a movement. Popivanov et al.[20] recorded scalp EEG in subjects who moved a manipulandum with the right hand to direct a light beam at a target. The authors developed an autoregressive model of the linear dynamics of EEG, and found that abrupt short transients of model coefficients occured during the movement preparatory period, related to dynamic changes. These transients marked temporal nonstationarities in the alpha, beta, and gamma bands that seemed to reflect boundaries separating successive phases of motor preparation. When the EEG was averaged with respect to the first

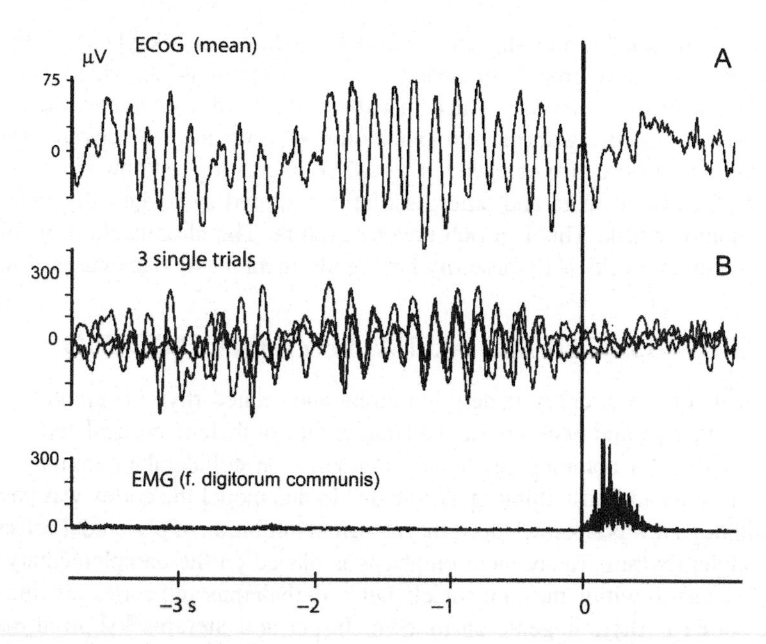

FIGURE 7.2 Preparatory 7-Hz rhythm recorded in human inferior parietal cortex. (A) Mean ECoG of 30 trials of hand grip aligned on EMG onset. (B) Subset of three successive trials showing superimposed ECoG waveforms and surface EMG (rectified) of the flexor digitorum communis muscle. (Adapted from Reference 21, with permission.)

transient, it coincided with increased theta activity (3–7 Hz) and increased gamma activity (35–40 Hz), which continued in sporadic bursts during the movement. These theta plus gamma oscillations were observed mainly over the supplementary motor area (SMA), premotor and parietal cortex. The authors interpreted the oscillatory bursts as markers of the succession of dynamic stages in the production of movement.[20]

Recording with subdural electrodes directly from the cortex of epileptic patients undergoing surgical intervention, Turak and colleagues[21] found a remarkable 6–9 Hz oscillation in the inferior parietal and superior temporal area immediately preceding movement onset. The movement was the grasping of a joystick, self-paced about every 10 sec. As shown in Figure 7.2, the rhythm was clearly preparatory to movement because it was phase-locked to the onset of the electromyogram (EMG) onset, but rapidly disappeared when movement started. Some leads showing this activity were in the rolandic area, or over the cingulate motor area. The patients were on reduced antiepileptic medication, and had no motor deficits.[21] Nonetheless, this rhythm may have been influenced by medication. The theta rhythm was obviously not driving movement, but it could have been influencing a premotor network involved in setting up the motor performance.

Because the theta rhythm was phase-locked to EMG onset, there is a possibility that it was used as a temporal reference for preparatory activity, analogous to the use of theta rhythm in hippocampus for coding spatial position.[22] Again with ECoG recording in humans, Caplan and coworkers[23] have found extensive theta oscillations both within rolandic cortex and far beyond it, as subjects performed a virtual "taxi cab" navigation task by pressing arrows and other keys on a keyboard. The theta activity was much more prominent during virtual movement on the videoscreen than while "standing still." They concluded that the theta rhythm facilitates the sensorimotor integration occuring during dynamic motor performance.[23] Since other rhythms (beta and gamma) were also observed, it would be interesting to know whether phase relationships between theta and other oscillations altered systematically in relation to the visuomotor task. This is a rich area to explore. The classification of rolandic theta, however, is murky. Are these rhythms really theta, or in some cases slow mu?

7.3 MECHANISMS OF MU (ALPHA) RHYTHM GENERATION

Early models of the circuitry underlying alpha and related rhythms emphasized the thalamus. Andersen and Eccles postulated that groups of thalamocortical neurons were synchronized into a common rhythm by mutual axon collateral inhibition (via the reticular nucleus) and postinhibitory rebound.[24] In this model the cortex was passively driven, although it was acknowledged that recurrent inhibition of pyramidal cells would enhance alpha rhythm. Today more emphasis is placed on the complementary oscillatory mechanisms within the cortex itself, but both thalamus and cortex are important.

The cortex is rhythmogenic on its own. Jasper and Stefanis[25] showed that one to two shocks of intralaminar thalamus were sufficient to elicit a spindle of "recruiting responses," at about 10 Hz, in cat motor cortex (Figure 7.3). The surface negative peak of the spindle oscillation corresponded with a wave of depolarization and spiking in pyramidal tract neurons. They suggested that the rhythm of oscillatory waves was determined by inhibitory phasing of alternating excitatory and inhibitory postsynaptic potentials (PSPs), facilitated by postinhibitory rebound hyperexcitability. Since antidromic activation of PT cells also could elicit the oscillation (although not quite as well as thalamic stimulation), recurrent collaterals of pyramidal cells probably contributed to the rhythmogenic feedback loop. Above all, synchronization of a population of neurons into a common oscillatory cycle is accomplished by

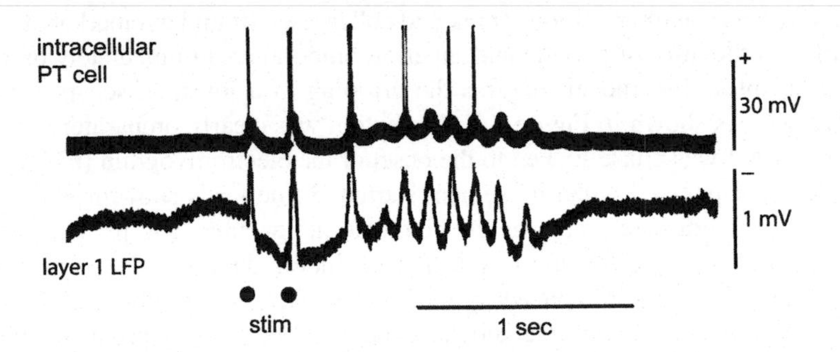

FIGURE 7.3 Motor cortical 10-Hz oscillation elicited by thalamic stimulation. A double shock stimulus to the intralaminar thalamus, marked by two solid circles, triggers a subsequent spindle oscillation in the cat motor cortex. Simultaneous recordings were made of the membrane potential of a PT cell, and of the LFP in overlying cortical layer 1. Note that the PT cell is regularly depolarized in synchrony with the negative peak of each LFP cycle, and occasionally discharges. (Adapted from Reference 25, with permission.)

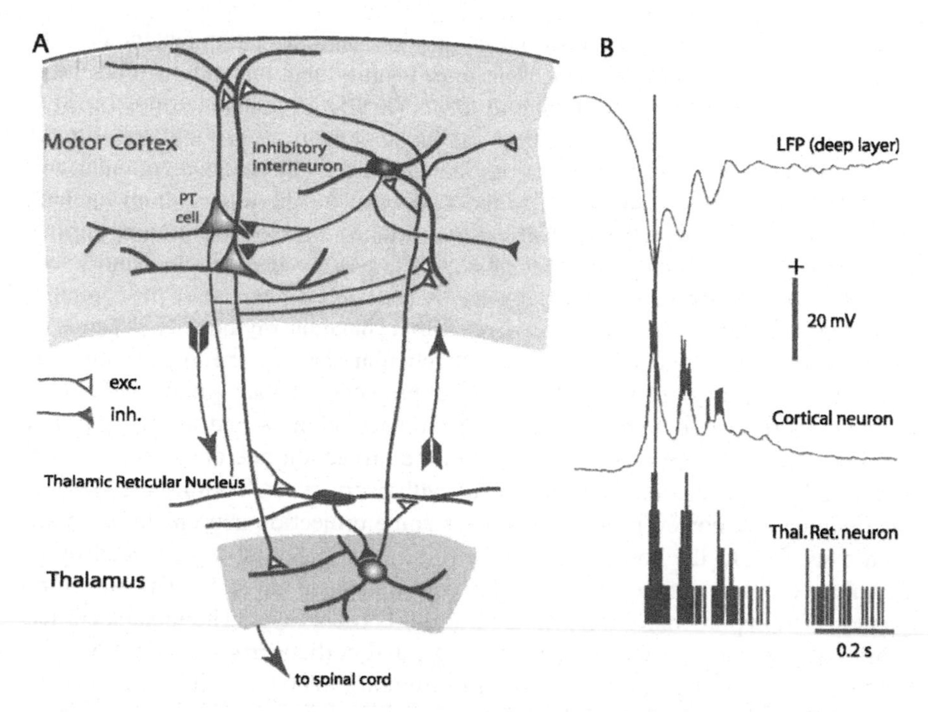

FIGURE 7.4 Schematic diagram of basic thalamocortical oscillatory circuit. (A) Inhibitory interneurons, both in the cortex and thalamus, are central to rhythmogenic mechanisms. A critical mass of synchronous discharge is recruited via gap junctions linking large numbers of inhibitory interneurons (not shown). Pyramidal neurons discharge during postinhibitory rebound excitation. (B) Mean record of ten spontaneous spindle oscillations in the cat motor cortex. Simultaneous LFP in deep layers of area 4, intracellular potential of an unidentified cortical neuron, and histogram of extracellular spikes from a thalamic reticular neuron. All are temporally aligned on the first LFP negative peak. Pyramidal cell discharge is generally the same as the "cortical neuron." (Part B adapted from Reference 29, with permission.)

electrical coupling of inhibitory interneurons within both the cortex[6] and the thalamic reticular nucleus.[26]

Jackson and coworkers,[27] studying the motor cortex in awake behaving monkeys, have confirmed that PT stimulation evokes an 8-Hz oscillation. Since the earliest effect following the stimulus was a suppression of firing, inhibitory feedback was a key mechanism, as always. The same stimulus also reset ongoing beta rhythm (20 Hz), suggesting a close relationship in the circuitry underlying both oscillatory frequencies. These effects were expressed only during a static motor episode, namely a maintained precision grip. PT stimulation during dynamic phases elicited much smaller oscillatory responses.[27] PT cells may not be an essential component of the rhythmogenic circuit (Figure 7.4A), but they are closely connected to it, and readily entrained by it.

Intrinsic membrane properties of both thalamic and neocortical neurons allow them to oscillate within different frequency spectra, usually in the range of 1–20 Hz.[5] Specific ion channels are required (a combination of sodium, calcium, and potassium conductances) other than the ones necessary to generate action potentials. The reticular thalamic neurons oscillate even more readily than thalamic neurons themselves.[5] They deliver powerful inhibitory PSPs (IPSPs) to thalamocortical neurons, which produce rebound excitation. Both the thalamic neurons give collaterals to the reticular thalamic nucleus on their way to the cortex, and the cortical pyramidal cells also give collaterals as they return to the thalamus. So the entire circuit includes recurrent inhibitory loops within both the thalamus and cortex, reinforcing intrinsically oscillatory membrane potentials, plus coupling of the thalamus and cortex such that they synchronize in the same rhythm.[28] The general layout of the circuit is illustrated in Figure 7.4A. Modeling reveals that thalamic rhythmicity depends on cortical excitatory input being stronger to reticular thalamic neurons (which are themselves inhibitory) than to thalamocortical projection cells.[28] Although the reticular neurons discharge with every cycle of the oscillation — and thereby generate IPSPs in thalamocortical cells — spiking in the projection neurons, during postinhibitory rebound excitation, does not occur with every cycle.[29] Within the synchronized population, however, there are always some projection neurons firing in any given cycle.[28,29] In the cortex, the 10-Hz oscillatory potential is generated by an equivalent dipole centered on pyramidal cell somata in layer 5.[5] The waveform reverses with depth from the top to the bottom cortical layer. The thalamocortical circuit initially entrains a small cortical focus, and corticocortical connections distribute the synchronization up to 4 mm in different directions. Inhibitory synapses in the circuit contain both $GABA_A$ and $GABA_B$ receptors.[28] However, diazepam — an enhancer of $GABA_A$ inhibition — has little effect on mu activity, so it appears that mu rhythm is not dependent on $GABA_A$ receptors.[30]

Gastaut, recording directly from the cortical surface in epileptic patients, was the first to point out the distinctive shape of the 10-Hz oscillation in motor and premotor cortex.[31] He originally named it the "rythme en arceau," or wicket rhythm (but later classified it as mu rhythm) with a range of 7–11 Hz and mean of about 9 Hz. It had alternating sharp positive peaks and rounded negative waves, as shown in Figure 7.5A. From the start, Gastaut recognized that the mu rhythm was closely allied with beta rhythm; he considered mu an amplified form of beta at half the frequency.[31] The asymmetric, 7–11 cycle/sec wicket rhythm tended to have a 20 Hz ripple superimposed on it.

The arched shape with the 20-Hz ripple is clearly retained in MEG records (Figure 7.5B). In MEG, the frequency spectrum has peaks at 10 and 21 Hz.[32] Tiihonen and colleagues found that the sharp transients in the mu waveform could be explained by a dipole model with an origin in S1.[32] However, the slow mu component was not explained by a dipole model. Moreover, higher mu frequencies were not exact harmonics of lower ones; spindles could appear at different moments of time at the 10- and 21-Hz component frequencies. Therefore, the characteristic comb shape of the mu rhythm might be composed of two superposed rhythms with different generators. It is notable that mu and beta rhythms do not engage exactly the same cortical territories. Recording subdural ECoG in humans, Ohara and coworkers observed corticomuscular coherence during isometric contraction of wrist extensor muscles.[33] Within M1, there were different locations for 17- and 8-Hz peak coherences.

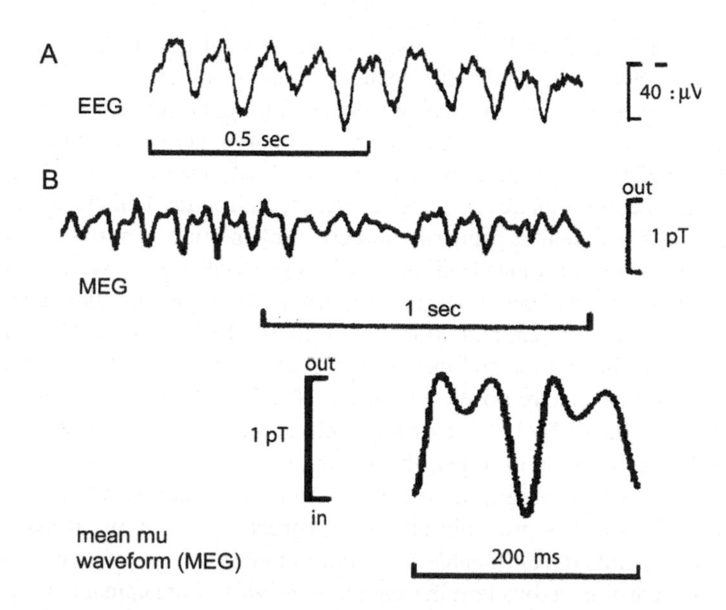

FIGURE 7.5 Asymmetric cycle of mu rhythm in humans. (A) Sample bipolar EEG recording over the hand sensorimotor cortex. (Adapted from Reference 49, with permission.) (B) Sample MEG recording over human sensorimotor cortex. At the bottom, the mean mu cyclic waveform is shown aligned on the inward peak. (Adapted from Reference 3, with permission.)

Whatever the exact circuit underlying cortical mu rhythm, it can readily entrain the discharge of pyramidal cells projecting elsewhere. Some large layer 5 projection pyramidal cells are rhythmically bursting at burst frequencies of 5–15 Hz depending on the level of depolarization.[34] Given these intrinsic pacemaker properties, they would be recruited by the local network oscillation. Nevertheless, spontaneous PT cell discharge is only modulated and synchronized by oscillatory potentials; there is much activity that occurs independently of the 10-Hz rhythm.[25] Moreover, if PT discharge is not advantageous during mu rhythm, then (presumably) PT cells can be independently inhibited. The inhibition would be necessitated rather than produced by the mu rhythm.

7.4 MU (ALPHA) BAND AND PHYSIOLOGICAL TREMOR

Recording the scalp EEG from humans, Jasper and Andrews were the first to study mu rhythm in the sensorimotor area thoroughly.[35] They noted an association between 10 Hz physiological tremor (registered in the index finger) and motor cortical mu rhythm. They also noted a secondary tremor frequency of about 25 Hz, which they suggested might be related to beta rhythm. However, they were quick to emphasize an important caveat; the relationship between cortical oscillations and tremor movements was sporadic. For example, it was lost when subjects were nervous and apprehensive: the tremor persisted but the mu rhythm virtually disappeared. They concluded that the tremor was driven by "continuous rhythmic subcortical discharges" which could momentarily entrain the activity of cortical cells "if the cortex was not sufficiently activated."[35] In this view the cortical mu oscillation is clearly state dependent, but is only indirectly related to motor output.

Mechanical properties are recognized as the major determinant of physiological tremor; nevertheless there is a central neurogenic component in about one third of the normal population, observed as synchronized EMG bursts mainly in the 7–13 Hz band.[36] The mean frequency of physiological tremor in the hand is about 7.7 Hz. The mechanical component is primarily influenced by inertia and stiffness; it changes according to the body part and loading conditions.[37] The neurogenic component is largely unaffected by changes in inertia or stiffness. In finger muscles, EMG oscillations occur at 8–12, 20–25, and 40 Hz, but only the 8–12 and 20–25 Hz rhythms were observed in the tremor and tremor–EMG coherence.[38] Adding a load to the finger decreased the relative power within the 20–25 Hz EMG band, increased the relative power of the 40-Hz band, but had no effect on the 8–12 Hz EMG frequency band, identifying the latter as a neurogenic component.[38]

The neurogenic component has a central origin, but despite the reservations of Jasper and Andrews, it is probably driven by cortical mu rhythm, at least in part. Subjects with X-linked Kallmann's syndrome exhibit mirror movements due to branching corticospinal axons terminating on both sides of the spinal cord in homologous motor nuclei. These subjects show significant coherence between left and right index finger tremor during sustained extension.[37] Normal controls show no such coherence. Furthermore, the EMG of the extensor indicis muscle was bilaterally coherent and there was significant bilateral cross-correlation of motor units, which is never seen in normals. Coherence was strongest at about 7–8 Hz, but smaller peaks were found up to 40 Hz.[37] This is strong evidence that the neurogenic component of physiological tremor involves the motor cortex and corticospinal tract.

Raethjen and coworkers[39] did epicortical recording from a grid over sensorimotor cortex. Sites showing corticomuscular coherence in the 6–15 Hz range were localized to the motor cortex, and were somatotopic. Moreover, the frequency of coherence remained stable with added inertial load. Coherence by itself is not convincing evidence of a causal connection between two oscillations. It is really necessary to show a fixed phase relationship. In this case, the phase spectrum between the motor ECoG and EMG showed a constant delay between cortex and muscle; the delay was 16 msec for deltoid,[39] a value that is very similar to that obtained by transcranial magnetic stimulation (TMS) of the motor cortex.

Stationary rats show a fine 9-Hz tremor of the jaw and/or vibrissae, not to be confused with the large 7-Hz vibrissal sweep of exploratory sniffing.[40] The tremor-related EMG bursts were found to be synchronous with bursts of multiple-unit activity in the ventrobasal complex of the thalamus and with spikes in the EEG of sensorimotor cortex.[40] The neural activity was not due to reafference; rhythmicity continued when local anesthetic injected into the face abolished sensory responses of thalamic units. The tremor was abolished by ablating the anterior half of the cortex bilaterally. Semba and Komisaruk postulate that the function of synchrony between mu rhythm and vibrissal tremor may be to increase the gain of sensory input, facilitating reception of sensory input.[40] However, this could only be true in one phase of the tremor cycle.

Recording MEG in humans performing sustained wrist movements, Marsden and coworkers observed corticomuscular coherence only during isometric postural contractions, not phasic movements.[41] This included a relationship between mu rhythm and the 10-Hz EMG component (physiological tremor) during isometric contraction of wrist extensor muscles in some subjects. Both 10- and 20-Hz components in the MEG were coherent with EMG and showed the same dipoles to estimate the source in sensorimotor cortex. Again the close relationship between mu and beta rhythms in the rolandic cortex was confirmed. Other studies have also noted corticomuscular coherence at 10 Hz, in some subjects, during steady muscle contraction.[42,43] Mu rhythm, however, may have a much more than tremorigenic importance.

An MEG study by Gross and coworkers has found a correlation between 6–9 Hz oscillations in the motor cortex and the pulsatile subcomponents of slow finger movements.[44] The latter are manifested as regular modulations of velocity and EMG at a frequency of about 8 Hz. During slow and continuous index finger flexion–extension, EMG was significantly coherent with the contralateral sensorimotor cortex at a peak frequency of 7 Hz. Analysis of the direction of coupling between the coherent oscillations indicated that motor cortex led muscle (efferent drive), but muscle led somatosensory cortex (afferent drive).[44] A cerebellar cortical region was also found to be coherent with EMG. Further analysis of coherences among brain regions identified a ring of zones contributing to the 6–9 Hz activity in the motor cortex, all showing significant phase synchronization. The direction of coupling ran from the cerebellum to the thalamus to the "premotor" cortex (which looks like it might actually be prefrontal) to the motor cortex and back to the cerebellum.[44] If, indeed, the microstructure of movement — interleaved agonist–antagonist bursts — is directly associated with a synchronous mu network in the cerebrocerebellar circuit, then mu rhythm in the motor system assumes fundamental importance. Indeed, it has already been shown that movements performed together, combined eye and hand tracking of a visual target, are coherent at 8 Hz.[45] In other words, a central 8-Hz oscillation may be used to fuse a functional synergy.

7.5 MU RHYTHM ERD

Gastaut was the first to report that rolandic mu rhythm was blocked by voluntary movement.[31] There was, however, a subsequent rebound of mu activity. Gastaut also noted that mu oscillations were actually much rarer than beta rhythms in the rolandic cortex. Since then, mu ERD with movement onset or somatosensory stimulation has been universally seen in EEG, MEG, and ECoG recorded in humans.[8]

The mu rhythm ERD occurs in both the somatosensory and the motor cortex, and is roughly centered on the representation of the body part moved or stimulated.[46] In other words, a moving arm does not block mu in the face area, and vice versa. However, there are conflicting findings; indeed the ease of finding mu ERD virtually proves that it is far more widespread than the sensorimotor zone being activated. The study by Crone and colleagues[47] sheds some light on this issue. Using a subdural recording grid, they found a difference between the early and late phases of a motor response. In the early phase, mu ERD occurred in a diffuse spatial pattern that was bilateral and not somatotopically specific. During late phases, ERD usually became more focused and somatotopically specific.[47]

Mu ERD starts 0.5–2 sec before movement, as a subject prepares to move.[3,8,46,47] The onset of mu ERD is the same whether a movement is performed slowly or briskly, but ERD magnitude is relatively greater for brisk movements.[48] Mu is also blocked by passive movements of the same body part, and by ipsilateral active movement, although less so by the latter.[46] The mu rebound appears 0.5–2.5 sec after a movement.[3,9] The onset of the rebound is related to the duration of muscle activity; it is slower for long-duration (i.e., slow) movements.[47] Tactile stimulation of the hand also blocks mu rhythm in sensorimotor cortex.[2,49]

Recording ECoG with a subdural grid, Toro and colleagues found that the power changes associated with mu ERD were widely distributed over the rolandic cortex.[50] The multiple-joint movement performed by the patients involved moving a manipulandum to position a videomonitor cursor on target. Movement amplitude influenced the magnitude, duration, and extent of the spatial distribution of mu power changes. The authors concluded that the overall changes in mu activity reflect shifts in the functional state of neuronal ensembles involved in the initiation and execution of motor tasks.[50]

There are reports, however, of sensorimotor 10-Hz oscillations not being suppressed by voluntary movement. In the MEG study of Tiihonen et al.,[32] a 10-Hz oscillation was suppressed during fist clenching only in some subjects. Using EEG recording, Andrew and Pfurtscheller found two alpha band rhythms in the rolandic area.[51] The 12-Hz band showed typical ERD during finger movement, and was not bilaterally coherent. They equated this with classic mu rhythm. A more localized 9-Hz rhythm did not show ERD, and was bilaterally coherent. Both rhythms appeared to be generated within the rolandic area; they were not volume conducted.[51] Two components in the alpha band are very commonly observed. It is confusing, however, because the lower frequency is the one that corresponds to the original 9-Hz definition of mu rhythm, not the 12-Hz component.[31,49] A factor that may be involved here is cortical plasticity. Strens et al. found that coherence between the sensorimotor hand area and the premotor (or prefrontal) zone in front of it showed two peaks at 9 and 13 Hz which were both increased by repetitive TMS of hand motor cortex (TMS at 1 Hz, 90% threshold, 1500 pulses).[52] The coherence was also increased bilaterally between the two hand motor areas. The effect lasted up to 25 minutes after stimulation.[52] Clearly, coherence among cortical areas is labile, and highly subject to whatever the subject has just been doing.

7.6 MU ERS DURING MOVEMENT

Although the power in mu oscillations drops before and during movements, nonetheless there is an increase in mu coherence between motor cortex and frontal areas, and between the two motor cortices, during this same interval.[53] After EMG onset, the increase in coherence extends to more posterior sites. Both before and after EMG onset, phase coherence showed a lead of anterior areas on more posterior regions. This suggests that mu oscillations are still functioning in the processes of motor preparation and execution, and that mu ERD simply reflects a much more focused and controlled expression of the oscillation. In a report on LFPs in the motor cortex of monkeys performing a maintained precision grip, Jackson and coworkers presented a spectrogram (their Figure 2) that indicated a modest relative increase in mu power before and at the time of movement onset.[27] This may in fact have been due to a touch evoked potential with a 10-Hz waveform (A. Jackson, personal communication). A brief mu oscillation can arise from more than one mechanism, so caution is necessary. Nevertheless, the gross picture of mu ERD may mask a more complex pattern at the level of functional clusters of neurons.

In an analysis of human EEG–EMG phase coherence, Feige, Aertsen, and Kristeva-Feige also found corticomuscular synchronization in the 2–14 Hz range during brief finger movements.[54] Peak phase coherence was actually in the theta band (5 Hz). They suggested that low frequency corticomuscular synchronization represented a "functional state of the oscillatory network related to the pulse movement execution." Current density mapping localized the phase coherence to the hand area of the motor cortex and the premotor cortex, and also to the region overlying the supplementary motor area.[54] Note that the increased *coherence* could coincide with a decline in mu *power*. In other words, although the extent of the oscillating network may be reduced, oscillatory activity may still have a motor function.

Similarly, McKeown and Radtke found distinct EEG–EMG coupling at 10 Hz (and it would appear also at about 6 Hz) using independent component analysis (ICA).[43] This is a novel approach, based on the premises that no one scalp site will contain all the signal related to any given muscle activity, and conversely that any one scalp site will be related to synergies that can encompass several muscles. During dynamic arm movements, alternating elbow extension–wrist pronation to flexion–supination, the ICA of the EMGs demonstrated tonic and phasic EMG ICs, each with unique coupling to the EEG.[43]

In spite of movement-related mu ERD, Ohara and coworkers found increased partial coherence between the motor cortex and the supplementary motor area (SMA), at mu frequencies, prior to and during a brisk, voluntary finger extension.[16] The data were from ECoG recordings.

These observations of mu synchronization between motor cortex and other areas during the dynamic phase of movement necessitate reinterpretation of the concept of mu desynchronization during movement. An important factor that is invariably ignored is the degree to which mu ERD is related to arousal or attention rather than to movement per se. When subjects tap their index fingers continuously, like an automaton, without thinking about each movement, mu rhythm can be quite prominent.[55] Indeed, in some subjects it is phase-locked to the finger tap, as was the case for the subject shown in Figure 7.6A. A different subject, shown in Figure 7.6B, rhythmically flexed his index finger. Again, a spindle of mu rhythm was synchronized to the onset of flexor EMG activity. In both subjects, the maximal 10-Hz synchronization occured about 200 msec prior to movement onset. The cyclic depolarization–hyperpolarization sequence of the oscillation would provide an ideal mechanism to efficiently synchronize a motor volley. During most of the oscillation the corticospinal neurons must, of necessity, be carefully inhibited. When it comes time to move, the inhibition is lifted, and both corticospinal neurons, and subsequently motor units, will be activated in a fixed phase relationship to the oscillation. Given that

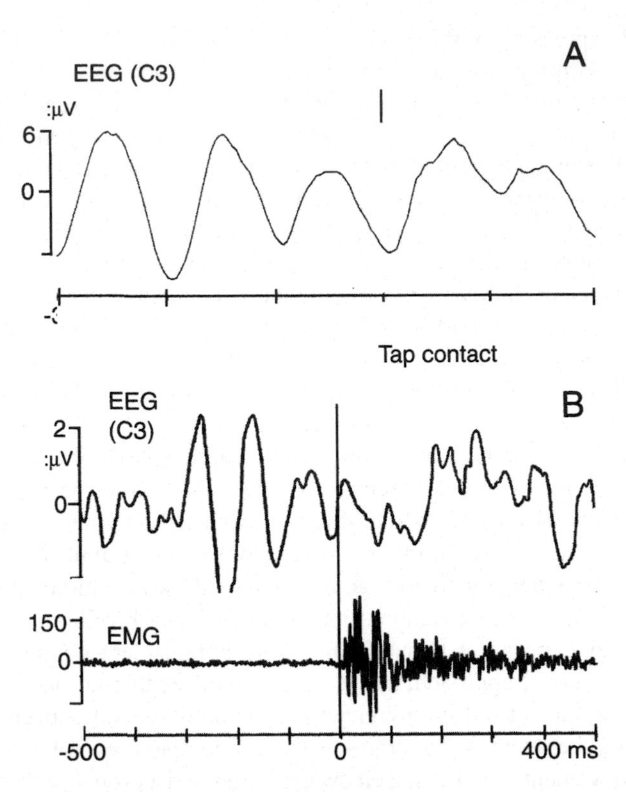

FIGURE 7.6 Preparatory 10-Hz rhythm synchronized to movement onset. Both human EEG recordings from C3. (A) Subject performed continuous rhythmic tapping at a self-paced rate of 1.27 Hz. (The subject was a 20-year-old highly trained pianist.) Mean EEG for 50 taps, aligned on moment of tap contact. (B) Subject performed continuous rhythmic finger flexion at a self-paced rate of 1.56 Hz; mean EEG and EMG for 40 movements, aligned on EMG onset (*flexor digitorum superficialis* muscle). (Unpublished data of W. MacKay and S. Makhamra.[55])

any surge of arousal or sensory activity is highly likely to destabilize the synchronous network generating mu rhythm, many interpretations of ERD are possible. It may actually signal the scrambling of a highly efficient generator of motor pulses, in favor of a desynchronized network that requires greater energy input to do the same job.

7.7 SIGMA BAND

The 12–15 c/s range spans the transition between mu and beta rhythms, and most often is ignored, being considered neither one nor the other but a fuzzy mix of both. Some studies, however, indicate otherwise; that it is, in fact, a functionally important rhythm in its own right. It has even been called the "sensorimotor rhythm."[56] Roth et al. related this oscillatory frequency in sensorimotor cortex strictly to the development of inhibitory behavior, e.g., when a cat suppressed bar-pressing, or expressly delayed a response.[56] Similarly, in cats operantly trained to enhance 12–14 c/s sensorimotor cortical activity, the occurence of this rhythm was associated with

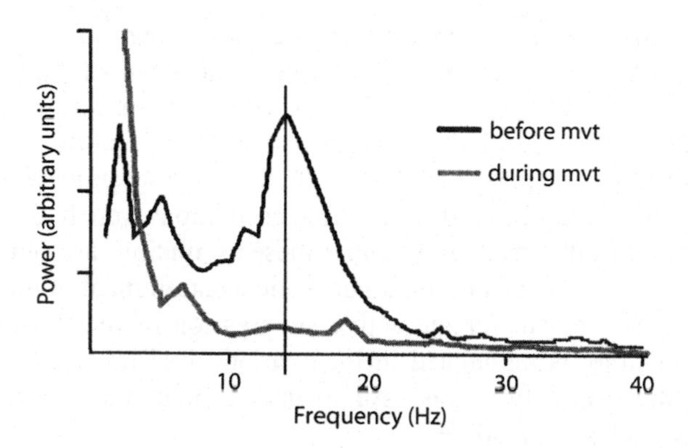

FIGURE 7.7 Preparatory 14-Hz rhythm in the monkey motor cortex. Mean LFP power spectrum for 40 trials, computed for the 1 s period prior to an arm movement (black line), and the 0.5-sec period following onset of the pointing movement (gray line). The LFP was recorded in a task-related zone of the arm representation in the motor cortex. (Unpublished data of S. Roux, W. MacKay, and A. Riehle.[58])

behavioral immobility, a depression of somatic motor activity and a general shift toward parasympathetic activity (including a drop in heart rate).[57]

When a cat is in a position of expectancy, waiting for an unseen mouse to appear at a hole, a rhythm of 14 Hz is observed in the forelimb zone of S1.[19] This is a state of attentive immobility. There are similar oscillations in ventrobasal (VP) thalamus, but the rhythmic cells are not sensory relay cells, and the relay cells do not oscillate. There was strong coherence between thalamic and S1 oscillations at a peak frequency of about 16–17 Hz.[19]

Furthermore, in the monkey motor cortex, we have observed a 14-Hz oscillation during a preparatory period (initiated by pushing a button), that ceased at the time of the response signal to move.[58] The difference in the LFP frequency spectrum before and after movement onset is shown in Figure 7.7. The monkey was trained to recognize a specific time interval of waiting, during which the prepared response was actively suppressed. Therefore, this sensorimotor rhythm does fit the description of active inhibitory behavior. Note that many neurons in motor and premotor cortex display "preparatory" activity with a time course that parallels this 14-Hz oscillation.[59] It remains to be determined, however, to what degree preparatory unitary discharge tends to be synchronized to the ongoing sigma rhythm.

A similar 15-Hz oscillation was seen in the prefrontal cortex in monkeys waiting for a visual stimulus.[60] Trials were initiated by depressing a lever, which is similar to the previous study; but the importance of this detail is unknown. In the prestimulus period and lasting to about 90 msec after a visual response signal, 15-Hz oscillations appeared at three prefrontal sites and were coherent among these sites, but not with any other sites (in motor cortex or the temporal lobe). Within this preparatory network, 15-Hz power and coherence were highly correlated to the amplitude and latency of early visual evoked potential components in visual association areas, and to response time.[60]

Human EEG also shows an increase in the 12–17 Hz range during the preparatory period of a reaction time task (i.e., between the instruction and "go" cues).[61] The increase in sigma activity is more prominent over the motor cortex than the somatosensory cortex.[61] Sigma oscillations are not restricted to the cortex. Courtemanche, Fujii, and Graybiel have reported LFP oscillations in the striatum of monkeys, with a frequency centered around 14–15 Hz.[62] These oscillations were highly synchronous across large regions of the striatum. Whether these oscillations are synchronous with cortical oscillations is an open question, but cortical entrainment of the basal ganglia could provide a mechanism for the active suppression of movement. The broad, background synchrony is modulated in local striatal foci involved with a specific movement. In the oculomotor zone, for example, small foci pop in and out of synchrony as saccades are made.[62]

The frequency range of sigma rhythm is significant in another respect. Grosse and Brown observed a 14-Hz component in the EMG of proximal arm muscles, such as deltoid and biceps, only during an acoustic startle reflex, not during similar voluntary movements.[63] This component was weak in a distal muscle such as first dorsal interosseous, and is not a peak in corticomuscular coherence. Grosse and Brown suggest that it is a sign of reticulospinal activation, associated with bilateral EMG coherence in homologous proximal muscles (deltoid and biceps) but not in distal muscles. This is a finding that has many implications. Does a cortical oscillation at 14 Hz essentially put direct cortical motor control on hold in deference to the brainstem?

7.8 BETA BAND ERD AND ERS

Beta oscillations probably involve the same basic circuit outlined in Figure 7.4A, but with a different population of inhibitory interneurons, possibly targeting $GABA_A$ receptors. Administration of diazepam greatly increases the power of 20-Hz oscillations in sensorimotor cortex, but has little effect on mu rhythm power.[30] Parallel to mu rhythm, however, beta rhythms have been reported to reverse polarity below a cortical depth of about 0.8 mm.[64]

Berger was the first to identify a beta rhythm (18–22 Hz) as the characteristic frequency of the motor cortex.[65] This was confirmed in a thorough study by Jasper and Penfield who did bipolar recording from the cortical surface in epilepsy patients.[66] In the resting individual, the dominant frequency in the motor cortex was about 25 Hz, but premotor areas registered a slower beta rhythm (17–22 Hz). Beta in the motor cortex was blocked by voluntary movement, namely clenching the contralateral fist, or by somatosensory stimulation.[66] The beta ERD only lasted about 1 sec; even though fist clenching was maintained, the beta oscillation resumed as before. It was blocked again when the subject relaxed. Also about 1 sec after relaxation there could be a brief burst of mu rhythm at about half the beta frequency (12 Hz). For Jasper and Penfield, the return of beta during a sustained contraction represented "a state of equilibrium of activity permitting again a synchronization of unit discharge."[66] Conversely, they concluded that precentral beta ERD was "closely related to the mechanisms of attention or readiness to respond." They did not see a sharp separation of beta and mu rhythms at the central sulcus. Although alpha rhythm was prominent throughout the parietal lobe in a resting individual, without significant beta or other frequencies, the postcentral gyrus showed a mix of alpha and beta frequencies.[66]

The Jasper and Penfield findings have been repeatedly confirmed ever since, but details have been added. For example, beta ERD is the same during voluntary muscle contraction or relaxation, but the rebound ERS following relaxation is much stronger, with a sharper onset, than the gradual return of beta power during a sustained contraction.[67]

The frequency of the beta rebound differs for a hand or foot movement.[68] After hand movement (or electrical stimulation), the rebound frequency at C3 averaged 17.4 Hz; for the foot at Cz, it was 21.5 Hz.[68] In this case, the different frequencies suggest a specific strategy of separation, perhaps so that they do not accidentally entrain one another. There are similar differences in beta rebound frequency in different motor areas.[69] Following index finger dorsal flexion, the rebound frequency was 18.9 Hz in hand motor cortex, and 25.5 Hz in the midline over SMA.[69] It is fascinating that these frequencies are separated, suggesting that at this moment the two cortical areas are not interacting functionally.

Beta ERD is identical for both slow and brisk finger movements prior to movement onset, but differs afterward.[70] The recovery of beta is earlier for brisk movements than for slow ones. Moreover, beta ERD is widespread, extending well beyond the representation of the finger being moved.[70] It seems to peak in postcentral cortex. The focus of beta recovery, however, is more localized and different than the ERD; it centers on the hand zone in M1.[70]

Both the beta ERD and the rebound are bilateral, although the movement or somatosensory stimulus is unilateral.[71,72] During the rebound, there is no coherence between the beta oscillations on the two sides.[72] It is interesting that, at rest, beta oscillations of about 21 Hz occur independently in the sensorimotor hand area, on one side or the other. But when they occur prominently on both sides simultaneously, they are phase-locked with a near-zero phase lag.[73] Interhemispheric coherence may play a role in the coordination of bimanual movements. In monkeys, beta oscillations can occur simultaneously in the left and right motor cortex, and often synchronize during bimanual manipulations.[64] However, synchronization occurred as often and strongly for unimanual manipulations.[64] Note that beta activity was observed during dynamic hand movement.

Similarly, Serrien and Brown had subjects perform bimanual in-phase or anti-phase cyclic movements, and measured coherence between C3 and C4 in both the mu (10 Hz) and beta (20 Hz) bands.[74] In spite of the dynamic nature of the task, coherence was seen in both bands. For the in-phase movement, no significant change in coherence was observed as the cycle period changed. For the anti-phase movement, coherence at 10 Hz stayed the same, but for 20 Hz, bilateral coherence declined markedly as cycle rate increased (and performance deteriorated).[74] No phase analysis was reported between C3 and C4; it is important to know if it differs for the two tasks. Even so, the study demonstrates that both mu and beta rhythms are still functioning during cyclic hand movements.

Overall mu and beta band power may decrease during cyclic voluntary movements, but this does not preclude relative power increases at the time of EMG onset. In subjects performing cyclic, auditory-cued thumb movement at 1/s, a very brief ERD occured in the 16–20 Hz band at EMG onset.[75] At a fast rate of 4/s a small ERS occured instead.[75] We have also observed a beta ERS during cyclic finger tapping at the subjects' preferred frequency, but the timing of the ERS relative to EMG activity was very variable among subjects.[55] It could be either during EMG activity or afterward. As with mu rhythm, beta ERD at the time of movement onset coincides with an increase in 18–22 Hz coherence among frontal and parietal areas.[53] It therefore appears to be performing a coordinative motor function.

7.9 BETA BAND CORTICOMUSCULAR COHERENCE

Since the first demonstration of synchronization between LFP motor cortical beta oscillations and EMG activity by Murthy and Fetz in 1992,[76] this area of research has expanded to include MEG, EEG, and ECoG studies in humans. Not surprisingly, there are some discrepancies between what is seen with the "gross" methods of EEG and MEG compared to the finer scale of ECoG and especially LFPs.

The typical observation in all studies is that corticomuscular coherence around 20 Hz occurs during maintained muscle contraction of weak to moderate strength (steady posture) but not during the dynamic phase of movement.[12-15,77–82] This is illustrated in Figure 7.8. Brown postulates that beta oscillations coincide with a stable state — "a free running mode of motor cortex that may maintain stable motor output with a minimum of effort."[14] Baker and coworkers have provided the pivotal evidence that the corticospinal tract links the cortical and spinal oscillations (rather than both being driven in parallel by a brainstem oscillator).[12] In the monkey motor cortex, the discharge of identified PT neurons was phase-locked to 20 Hz LFP oscillations, as the monkey maintained a steady precision grip. The cortical oscillation was coherent with the rectified EMG of contralateral hand and arm muscles (Figure 7.8).[12] Significant coherence between LFPs and identified PT neurons occured in three frequency bands, 10–14, 17–31, and 34–44 Hz.[13] Corticomuscular coherence for the adductor pollicis muscle was largely expressed in the beta band (at about 20 Hz), was not expressed at all at 10 Hz, and exhibited a small peak in the 35–40 Hz range.[13]

Further evidence that the corticospinal tract mediates the beta band coherence was found in a mirror movement subject (a probable case of Kallmann's syndrome). During an intended unilateral hand grip, coherent EMG oscillations were observed in muscles of both hands at 20–22 Hz. Moreover, the motor cortex contralateral to the intended movement was coupled to the muscles of both hands at 20–25 Hz.[83]

Corticomuscular coherence in the mu band is seen in only about 25% of subjects,[79] but is ubiquitous in the beta band. This may be due to preferred firing rates of spinal neurons. For example, monkey spinal interneurons have a basal discharge rate of 14/s.[84] During generation of static torques, they fire at 19/s for wrist flexion, 24/s for extension. The firing is regular with periodic features in the autocorrelogram.[84] Therefore, synchronous rhythmic firing of corticospinal connections could

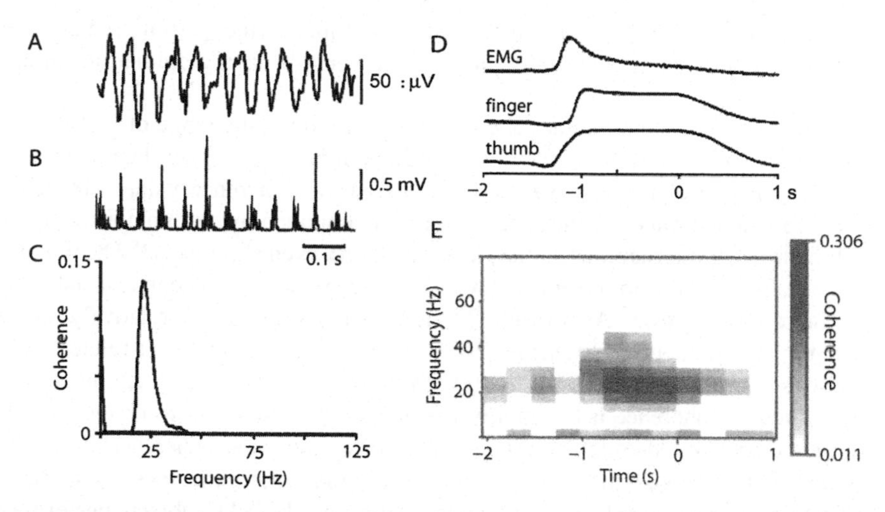

FIGURE 7.8 Corticomuscular coherence in the beta frequency band. LFP in the hand area of the monkey primary motor cortex (A) was recorded simultaneously with rectified EMG from the adductor pollicis muscle (B), as the monkey performed precision grips sustained for over 1 sec (D). In D, the mean time course of finger and thumb displacements producing the grip is shown, along with the mean rectified EMG. During the period of maintained grip, the EMG exhibited distinct oscillatory bursts that were coherent with LFP oscillations at a frequency of about 25 Hz (C). The coherence spectrogram in E (mean of 274 trials), shows that the corticomuscular coherence was largely confined to the duration of constant muscle contraction, not involving movement initiation. (Adapted from Reference 12, with permission.)

provide an efficient means of modulating these cells. It is remarkable that the range of reported frequencies of corticomuscular coherence, 18–24 Hz,[12,42,74–83] is virtually the same as the range for spinal interneuron mean firing rates.

By itself, coherence between motor cortical and EMG oscillations is never sufficient to prove a causal link. A consistent phase relationship needs to be shown. If single motor unit spikes are recorded, this is neatly done by spike-triggered averaging of the cortical oscillation.[77,82] To measure the time lag between the cortical and EMG oscillations, most commonly the phase spectrum is computed for all the coherent frequencies in the cortical and muscle signals. If there is a constant conduction delay between cortex and muscle that is responsible for the frequency coherence, then the phase delay will progressively increase for successively higher frequencies. By fitting a line to this linear trend, the conduction delay can be estimated.[82] Many other methods have also been applied to extract the delay between cortex and muscle, including use of the Hilbert transform[82] and ICA.[43] Published values of the delay time, by whatever method, sometimes agree with those measured using TMS, but are often much shorter.[42] Some values may be off because the band of coherent frequencies is too narrow to make an accurate linear fit of phase lag, but there is a basic fact here to consider. The spinal cord has rhythmogenic capabilities of its own.[84,85] The descending discharge from motor cortex may function as an entraining signal rather than a driving one. Two connected oscillators with similar frequencies will inevitably become synchronous in time.[7] Is the quest to find a delay equal to the conduction time actually missing the point of having oscillations in the first place?

Furthermore, corticomuscular coherence is generally in the range of 0.05–0.1.[13] It can go up to 0.2, but even in invasive recordings it is rarely higher than that. (See Figure 7.8 for one of those maximal moments.) Low coherence values are to be expected. Sustained muscle contractions can be maintained by autonomous activity of motoneurons or interneurons within the spinal cord, even in humans.[85] The plateau potentials observed in motoneurons, giving rise to membrane potential bistability, play a large part in this.[85] As a result, cortical synaptic input is a relatively small contributor to motoneuron discharge. Even when cortical, movement-related beta oscillations increase substantially, for example when diazepam is administered, corticomuscular coherence is essentially unchanged.[30] Changes in force level of an isometric contraction also do not change beta corticomuscular coherence.[79]

Mima and colleagues have shown that corticomuscular coherence is probably not due to reafferent signals from the contracting muscle.[42] As subjects performed thumb and little finger apposition, vibration of the abductor pollicis brevis muscle tendon at 100 Hz had no significant effect on coherence (in either the mu or beta band). Similarly, functional deafferentation by ischemia failed to change corticomuscular coherence.[67,86] One may conclude that movement-related cortical oscillations reflect motor rather than sensory activity. Moreover, the location of peak beta corticomuscular coherence generally corresponds to the appropriate muscle representation in motor cortex, as determined by TMS.[42]

The beta rebound after a completed movement can also give rise to corticomuscular synchrony. After a finger flexion–extension, corticomuscular phase coherence was seen at about 23 Hz, lasting 1–2 sec, with a concomitant increase in EMG.[54] (The subject had to reposition his finger exactly where it started.) Using EEG current density analysis, the cortical site of synchronization was localized to a broad region in the motor and premotor cortex. Feige and coworkers concluded that beta synchronization between multiple cortical areas and muscle reflects a transition of the motor network into a new equilibrium state.[54]

There is great variation among individuals in the strength of corticomuscular coherence,[87] and it waxes and wanes over time.[85] Some of the variation may be due to the exact task performed. Coherence is much stronger for an auxotonic task than for an isometric one.[29,81,87] Moreover, inadvertent and uncontrolled movement of the contralateral hand in some protocols would certainly affect corticomuscular coherence on both sides.[88] But the most important factor may be training and usage of the muscle studied in given individuals.[89]

Finally, beta corticomuscular coherence does not totally disappear during dynamic movement. Marsden et al., recording ECoG with a subdural grid in patients, found that corticomuscular coherence around 20 Hz was "by no means abolished on movement, and at some sites even increased during movement."[90] Similarly, Murthy and Fetz observed beta LFP oscillations in monkey sensorimotor cortex — oscillations synchronous with modulation in both flexor and extensor muscles — that occurred most often during exploratory arm and hand movements.[64,76]

7.10 MECHANISMS FOR GAMMA OSCILLATION GENERATION

The gamma band generally includes all frequencies over 30 Hz. Most of the work done on gamma oscillatory mechanisms is based in the hippocampus. But there are lots of neurons throughout the neocortex that can sustain high-frequency oscillations; all they need is sufficient depolarization.[4,34] Szabadics et al. postulated at least two spatially segregated networks of inhibitory interneurons for high frequency synchronization, one directed at pyramidal cell dendrites, the other at pyramidal cell somata.[91] They identified a population of cortical interneurons in layers 2 to 3 of rat somatosensory cortex with a dendritic target preference. These interneurons form a network, interacting via gap junctions and GABAergic synapses, that is capable of engaging coherent activity. The network can be activated by local pyramidal cells at beta and gamma frequencies.[91]

However, others attribute gamma generation only to the perisomatic network of fast-spiking basket cells, interconnected both synaptically and electrically by gap junctions (which are necessary for synchrony).[18] The gamma-generating basket cells probably contain parvalbumin.[92] Pyramidal cells are rhythmically inhibited by the basket cells, and discharge by postinhibitory rebound between the IPSPs. At least in the hippocampus, an electrically connected plexus of pyramidal cell axon collaterals is important to sustain gamma oscillations.[93] Individual pyramidal neurons are unlikely to faithfully fire in every cycle, but as long as a few in the population discharge each time, then the spikes are "shared" via the axon collateral plexus and the basket cells receive sustaining feedback at the correct frequency.[18,93]

A difficulty with the basket cell mechanism is that it does not explain the lack of polarity reversal with cortical depth that is sometimes reported for gamma oscillations in neocortex.[94] Tallon-Baudry, Bertrand, and Pernier suggest an alternative model consisting of a ring-shaped distribution of dipoles oriented horizontally within the cortex.[94] Such a source would generate a field potential that does not reverse through cortical depth, and would generate no magnetic field, or only a weak one, at the surface. Gamma rhythms are not seen as consistently in MEG as in EEG or in LFPs.[3] The geometry of the proposed ring of dipoles would correspond to horizontally oriented dendritic fields, but as yet there is no physical substrate to support the model. Furthermore, it seems that at least some gamma rhythms do exhibit phase reversals at the bottom of layer 4.[64,95]

Although it is often assumed that gamma rhythms are localized to the cortex — and cortical networks on their own can generate gamma oscillations[95] — the thalamus is also capable of sustaining them.[4] The same basic circuit outlined in Figure 7.4A is probably involved, but with a different population of interneurons (fast-spiking) at the cortical level.

7.11 GAMMA BAND ERS

An increase in gamma power is most dramatically observed in a cat preparing to catch a visible mouse.[96] Oscillations of 35 Hz appear in the motor cortex, in the parietal association cortex, and in the related parts of the thalamus. The gamma oscillations stop at the instant of the pounce.[96] It is very possible that the oscillations are providing tonic drive to limb motor units. In the cat's crouched posture the muscles are certainly isometrically active. In part, these gamma oscillations may be dependent on the dopaminergic system because they are suppressed after lesions of the ventral tegmental area.[5]

Gamma oscillations have been elicited by stimulation of reticular activating systems.[97] In other words, they are associated with high levels of arousal.[4,5] During rat exploratory whisking, a burst of gamma oscillation (30–35 Hz) in S1 has been observed to precede the onset of whisking.[98] The mean lead time was 268 msec, and the oscillation ceased very soon after the onset of whisking. The gamma rhythm may reflect anticipatory activities in the barrel cortex for the subsequent sensory input from the whiskers.[98]

Nonetheless, gamma oscillations in the motor cortex are directly linked to muscle activity, generally vigorous muscle activity. Gamma band corticomuscular coherence requires strong isometric contractions.[14,15,79] Increased attention also increases gamma activity in the motor cortex,[61] but does not increase corticomuscular coherence.[15] EMG frequencies in the 30–60 Hz range are often called the "Piper band," after Hans Piper, the first to observe motor unit activity in the gamma range.[14] Mima and coworkers found that gamma band power in motor cortex only increased for isometric movement at 80% maximal force.[79] The gamma oscillation was partly correlated with muscle Piper rhythm (coherence around 35 Hz) and had an anterior focus in the frontal lobe, probably in the premotor cortex.[79]

Most importantly, gamma band corticomuscular coherence is not restricted to sustained isometric contractions, but is also observable during dynamic phases of movement.[14,15] This is most frequently documented in invasive studies. Marsden and colleagues, recording ECoG from the sensorimotor cortex, found that gamma band corticomuscular coherence was actually more frequently seen during phasic muscle contractions than during sustained contractions.[90] Salenius et al. reported 40 Hz corticomuscular coherence during fast and especially slow finger movement in one subject.[99] Although EEG and MEG recording can show gamma ERS at the onset of movement, in the frequency band 36–40 Hz,[8,99] ECoG recording consistently reveals gamma ERS up in the 60–100 Hz range.[16,90,100] According to Crone et al. it is high frequency gamma (85–95 Hz) that shows ERS restricted to the onset of movement.[100] As illustrated in Figure 7.9, ERS at 40–50 Hz developed gradually after movement onset and was sustained throughout a movement. The temporal evolution of gamma oscillations fits with the fact that motor unit firing rates are only elevated above 50/sec at the onset of a movement, but it does raise concerns about EMG contamination of the ECoG signal. The fit is almost too close for comfort. However, high gamma ERS was found to be much more focused to somatotopically relevant areas than was the case for mu or beta ERD; it was also strictly contralateral.[100] It therefore appears to be truly a brain signal and not muscle.

Donoghue and colleagues, recording LFPs in monkey motor cortex, found that gamma bursts could occur immediately prior to movement onset.[11] However, both beta and gamma oscillations ceased with movement onset in trained tasks. Fast

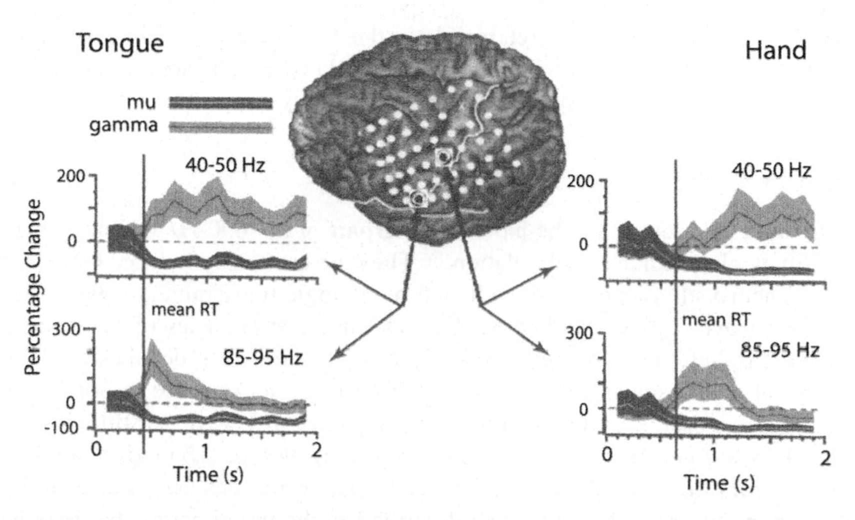

FIGURE 7.9 High frequency gamma ERS coincident with mu ERD. Opposite changes in mu and gamma bands of motor cortical ECoG during sustained tongue protrusion (left side) and fist clenching (right side). Sites of ECoG recordings for the two tasks are indicated. Mean response time for the movement (mean RT) is marked by a vertical line. Note that the ERS time course for low and high frequency gamma bands is quite different. (Adapted from Reference 100, with permission.)

oscillations always reappeared quickly upon the transition from quiet sitting to resumption of task performance, indicating an association with task engagement. Moreover, during untrained movements, gamma oscillations (at about 35 Hz) often appeared during the movement itself, but they were erratic and not reliably correlated with elevated cortical discharge rates.[11] This corroborates the findings of Murthy and Fetz.[64] Similarly, in human ECoG recordings, an increase of gamma activity (around 40 Hz) was observed during performance of visuomotor tasks such as threading pieces of tubing or moving a cursor to a visual target.[101] There is clearly a big difference between repeated, overtrained movements and spontaneous ones. Paradoxically, it appears that extensive training may increase synchronization among corticospinal neurons.[89] Does this mean that oscillatory network entrainment of corticospinal neurons is not needed during a trained movement because the neurons are already adapted for synchrony? Possibly cortical oscillations preferentially assist the assembly of an untrained movement.

Many LFP, ECoG, and even EEG studies have noted that relatively large cortical territories can be phase-linked, or at least coherent, during episodes of oscillations.[11,16,64,90,101,102] Synchronous areas stretch from the postcentral to the premotor cortex and SMA, often separated by patches that are not synchronous.[90] In visuomotor tasks they can encompass sensorimotor and occipital areas.[102] Not only gamma, but beta and mu oscillations are synchronous over an extended network. For example, increased M1–SMA and S1–SMA coherence in the 10–20 Hz band occurs for 1 sec before and after movement onset.[16] The phase lag in each case is near zero. Also, there is a significant increase in coherence between M1 and S1 around the time of movement onset, mainly in the 1–25 Hz band.[16] The concept that common oscillatory frequencies link functionally related areas at appropriate times[103] seems to be increasingly supported within the motor system.

7.12 I-WAVES

Indirect waves (I-waves) are the motor counterpart of the 600-Hz wavelets evoked in area 3b by electrical nerve stimulation.[104] They are artificially induced by the high degree of neuronal synchronization elicited by a single transcranial magnetic stimulus. The I-waves follow the direct waves (D-waves) at latencies of 1.3 msec. As the name implies, the D-wave is associated with the direct depolarization and discharge of corticospinal cells. Exactly what cortical elements are responsible for the ensuing fast oscillation that produces the sequence of I-waves is still a mystery. Perhaps they are generated by the synchronous activation of "chattering cells" that fire in intraburst frequencies of up to 800 Hz.[34] Chattering cells are a subset of layer 2 to 3 pyramidal cells. They were first described in the visual cortex, but have also been observed in areas 3b and 4 in cats.[4] Assuming that they are present in the motor cortex of humans, a 600–700 impulse/sec synchronous burst in a population of chattering cells would bombard layer 5 corticospinal cells, entraining their collective discharge to the same rhythm. Clearly this defines the upper limit for oscillations in the motor cortex. The very high frequency of rhythmogenesis indicates fast intraneuronal mechanisms only, with no feedback loops. It is probably of limited physiological interest because under normal circumstances, the population of neurons generating I-waves would not be so synchronized.

7.13 CONCLUSIONS AND DISCUSSION

The evidence is decisive that many motor cortical rhythms, at approximately 10, 20, 40, and even 60–100 Hz, can and do partially entrain the discharge of corticospinal cells and thereby influence motor unit activity. The cortical oscillations parallel the general trend of motor unit firing rates, fastest at the start of a movement, then slower. Accordingly, high frequency gamma oscillations coincide with movement onset, then are replaced by slower gamma or beta oscillations to sustain the muscle contraction. As many investigators have suggested, these synchronized rhythms are adapted to provide the most energetically efficient motor drive to motor units.[12–15] Although they may decline markedly in power, mu and beta rhythms do not disappear during the initial or dynamic phase of movement. Beta, mu, and even theta corticomuscular coherences have been observed during the dynamic phase of a movement.[43,54,90] Whatever the frequency, corticomuscular coherence is spatially focused and somatotopically relevant.[79] Moreover, single cortical sites can be coherent with more than one muscle,[12,43] suggesting a relationship between the oscillation and a functional synergy.

The driving of motor units is likely a minor function of motor cortical oscillations. The same entraining power that is weakly expressed in motor unit discharge can be exerted within and among cortical (and other brain) areas. Again the oscillation provides the most efficient signaling link among a local population of neurons and among related areas in a network. Synchrony within the network is a very useful property to recruit the requisite neurons and build up the power necessary to launch a motor action. Motor preparation is a state when oscillations are commonly observed.[11,21,56,60,61,96,105,106] However, for most of the duration of such oscillations, corticospinal neurons must be inhibited, to prevent their premature recruitment. The inhibition is not so much *by* the oscillatory process, but *because* of it. This interpretation presupposes two things: (1) oscillations can proceed in the motor cortex without the participation of corticospinal cells, and (2) the interneurons that tonically suppress corticospinal activity are different from those that generate the oscillation. Both of these suppositions remain to be explicitly demonstrated.

Preparatory unit activity is common in the premotor cortex, but is also found in a population of motor cortical cells.[57] Periodic synchronized volleys of excitatory PSPs (EPSPs) would provide an efficient means of driving the preparatory activity. The situation is analogous to corticomotoneuronal cells driving motor units during a sustained muscle contraction, except that it is a population of cortical neurons being kept in readiness. Although it is very likely that preparatory unit activity is synchronous with LFP oscillations, again, this has never been explicitly shown to be true. From LFP recording in the monkey motor cortex, it is clear that a wide range of oscillation frequencies can be involved in movement preparation.[11,56] Many areas can also be involved.[11,16,97,105] Although some areas manifest oscillations right up to the time of movement onset (Figures 7.2, 7.6, and 7.7),[11] this is not always the case. In areas 5 and 7 of the parietal lobe, beta oscillations are prominent early in the preparatory period but then fade away before the time to initiate movement.[105] This raises the possibility that oscillations in one locus may spawn oscillations elsewhere, and dissipate once they have successfully transplanted.

What happens at the transition point of movement onset to dampen the preparatory oscillation? At this time, the inhibition of corticospinal cells would be lifted and a wave of excitatory input would briskly ignite a descending motor volley. Two factors could then contribute to the degradation of the oscillation. One is the volley of corticospinal activity itself, which via collaterals will send a prolonged surge of activity into the oscillating network at least partly out of phase. Second, a volley of reafferent activity will promptly arrive from the moving limb. Both collateral activation and sensory input can elicit or reset cortical rhythms.[2,25,27] However, highly synchronous pulses are required to do this; the less coherent afferent volleys associated with spinal reflexes simply extinguish motor cortical oscillations.[35,49]

Phase-locking of theta or mu oscillations to movement onset (Figures 7.2 and 7.6) suggests use of the cycle period as a temporal reference frame, which may be used to coordinate elements of the developing motor signal. Place coding in the hippocampus makes use of phase relationships with theta oscillations.[22] If it is used in one place in the central nervous system (CNS), it is assuredly used elsewhere, and the motor system would be a logical fit. A 100- to 200-msec period would allow sufficient time for simple couplings among muscle representations in a synergy, such as the agonist–antagonist sequence in a ballistic action. The slowly accumulating evidence for an 8-Hz intermittency in motor signals, and its link to theta–mu oscillatory activity in motor cortex, suggests that it may be a microstructural unit of motor organization.[43] The nervous system can rely on limb viscoelastic properties to ultimately smooth and dampen a succession of pulsatile submovements. Both static posture and dynamic movement can be controlled by intermittent signals, but the data reviewed here strongly suggest that different "pulse periods" may be involved in each case. It is possible, however, that the different frequencies are usually embedded in a common 8–10 Hz oscillation. For example, the ubiquitous 20-Hz oscillation associated with sustained muscle contraction is usually accompanied by significant mu activity although the latter may not give rise to corticomuscular coherence.[30,82,88]

Very commonly in human subjects, mu and beta oscillations are suppressed in the EEG or MEG of sensorimotor cortex, 1 sec or more before movement onset, apparently precluding any association with motor preparation.[3,8,15] The state of arousal, or degree of attention, may be the critical issue. When attention is focused on the movement being executed, lower frequency oscillations may be lost simply as a result of increased depolarization.[4,61] Given the proclivity of cortical neurons to oscillate, it may be postulated that they will do so preferentially. The only thing normally preventing oscillation in a given area would be a surfeit of incoherent input.

Cortical oscillations do not appear to be necessary for movement.[10] What is different, therefore, about a movement where oscillations are not observed and one where they are observable? It may simply be a matter of neuronal efficiency; the movement is energetically easier to produce with the aid of oscillations. The oscillatory wheels of motion roll like a well-oiled machine. What is implied by this argument is that there are at least two modes of motor control, the subconscious and the conscious. Only the former makes maximum use of oscillations for efficiency. With conscious interference, excess neuronal activity is interjected into motor networks, still getting the job done but without the aid of oscillation because it has been dampened.

As these speculations reveal, when it comes to understanding the functional link between motor cortical oscillations and motor behavior, we have hardly scratched the surface. To proceed much further, it may be helpful to reverse the traditional approach of using behavioral markers as temporal reference points, and let the brainwaves be the guide. Shifts in frequency, relative changes in band power, and shifts in the pattern of synchrony between different areas, all could be treated as events related to the dynamics underlying motor control. Some of this is already being done for purposes of brain–machine interfacing, as described in Chapter 14. Ultimately, the study of oscillations in the motor cortex, and in all the areas functionally tied to it, may be the key to unlocking the mechanisms of the elusive motor program.

Two papers of great relevance to this chapter appeared in the spring of 2004. The first, by Hughes et al. (*Neuron*, 42, 253, 2004) presented compelling evidence that both theta and alpha rhythms are induced by metabotropic glutamate receptor-1a activation in the thalamus. Strong activation induced alpha frequencies, whereas moderate activation induced oscillation in the theta range. As many have suggested, the theta-alpha band should perhaps be viewed as a continuum. Secondly, an infomax ICA of EEG recorded during a visuomotor task by Makeig et al. (*PloS Biology*, 2(6), 747, 2004) confirmed that a frontocentral theta power increase occurs at the time of movement (button press in this case). This was previously observed in References 21, 43, and 54.

REFERENCES

1. Kornmüller, A.E., Architektonische Lokalisation bioelektrischer Erscheinungen auf der Grosshirnrinde, *J. Psychol. Neurol.*, 44, 447, 1932.
2. Cheyne, D. et al., Neuromagnetic imaging of cortical oscillations accompanying tactile stimulation, *Cogn. Brain Res.*, 17, 599, 2003.
3. Salmelin, R. and Hari, R., Spatiotemporal characteristics of sensorimotor neuromagnetic rhythms related to thumb movement, *Neuroscience*, 60, 537, 1994.
4. Steriade, M., Synchronized activities of coupled oscillators in the cerebral cortex and thalamus at different levels of vigilance, *Cereb. Cortex*, 7, 583, 1997.
5. Steriade, M. et al., Basic mechanisms of cerebral rhythmic activities, *Electroenceph. Clin. Neurophysiol.*, 76, 481, 1990.
6. Beierlein, M., Gibson, J.R., and Connors, B.W., A network of electrically coupled interneurons drives synchronized inhibition in neocortex, *Nat. Neurosci.*, 3, 904, 2000.
7. Mirollo, R.E. and Strogatz, S.H., Synchronization of pulse-coupled biological oscillators, *SIAM J. Appl. Math.*, 50, 1645, 1990.
8. Pfurtscheller, G. and Lopes da Silva, F.H., Event-related EEG/MEG synchronization and desynchronization: basic principles, *Clin. Neurophysiol.*, 110, 1842, 1999.
9. Pfurtscheller, G., Event-related synchronization (ERS): an electrophysiological correlate of cortical areas at rest, *Electroenceph. Clin. Neurophysiol.*, 83, 62, 1992.
10. Murthy, V.N. and Fetz, E.E., Synchronization of neurons during local field potential oscillations in sensorimotor cortex of awake monkeys, *J. Neurophysiol.*, 76, 3968, 1996.
11. Donoghue, J.P. et al., Neural discharge and local field potential oscillations in primate motor cortex during voluntary movements, *J. Neurophysiol.*, 79, 159, 1998.
12. Baker, S.N. et al., The role of synchrony and oscillations in the motor output, *Exp. Brain Res.*, 128, 109, 1999.
13. Baker, S.N., Pinches, E.M., and Lemon, R.N., Synchronization in monkey motor cortex during a precision grip task. II. Effect of oscillatory activity on corticospinal output, *J. Neurophysiol.*, 89, 1941, 2003.
14. Brown, P., Cortical drives to human muscle: the Piper and related rhythms, *Prog. Neurobiol.*, 60, 97, 2000.
15. Hari, R. and Salenius, S., Rhythmical corticomotor communication, *NeuroReport*, 10, R1, 1999.
16. Ohara, S. et al., Increased synchronization of cortical oscillatory activities between human supplementary motor and primary sensorimotor areas during voluntary movements, *J. Neurosci.*, 21, 9377, 2001.
17. Lukatch, H.S. and MacIver, M.B., Physiology, pharmacology, and topography of cholinergic neocortical oscillations *in vitro*, *J. Neurophysiol.*, 77, 2427, 1997.
18. Whittington, M.A. and Traub, R.D., Inhibitory interneurons and network oscillations *in vitro*, *Trends Neurosci.*, 26, 676, 2003.
19. Rougeul-Buser, A. and Buser, P., Rhythms in the alpha band in cats and their behavioural correlates, *Int. J. Psychophysiol.*, 26, 191, 1997.
20. Popivanov, D., Mineva, A., and Krekule, I., EEG patterns in theta and gamma frequency range and their probable relation to human voluntary movement organization, *Neurosci. Lett.*, 267, 5, 1999.
21. Turak, B. et al., Parieto-temporal rhythms in the 6–9 Hz band recorded in epileptic patients with depth electrodes in a self-paced movement protocol, *Clin. Neurophysiol.*, 112, 2069, 2001.

22. Huxter, J., Burgess, N., and O'Keefe, J., Independent rate and temporal coding in hippocampal pyramidal cells, *Nature*, 425, 828, 2003.
23. Caplan, J.B. et al., Human oscillation related to sensorimotor integration and spatial learning, *J. Neurosci.*, 23, 4726, 2003.
24. Andersen, P. and Eccles, J.C., Inhibitory phasing of neuronal discharge, *Nature*, 196, 645, 1962.
25. Jasper, H. and Stefanis, C., Intracellular oscillatory rhythms in pyramidal tract neurones in the cat, *Electroenceph. Clin. Neurophysiol.*, 18, 541, 1965.
26. Long, M.A., Landisman, C.E., and Connors, B.W., Small clusters of electrically coupled neurons generate synchronous rhythms in the thalamic reticular nucleus, *J. Neurosci.*, 24, 341, 2004.
27. Jackson, A. et al., Rhythm generation in monkey motor cortex explored using pyramidal tract stimulation, *J. Physiol.*, 541, 685, 2002.
28. Destexhe, A., Contreras, D., and Steriade, M., Mechanisms underlying the synchronizing action of corticothalamic feedback through inhibition of thalamic relay cells, *J. Neurophysiol.*, 79, 999, 1998.
29. Contreras, D. and Steriade, M., Spindle oscillations in cats: the role of corticothalamic feedback in a thalamically generated rhythm, *J. Physiol.*, 490, 159, 1996.
30. Baker, M.R. and Baker, S.N., The effect of diazepam on motor cortical oscillations and corticomuscular coherence studied in man, *J. Physiol.*, 546, 931, 2003.
31. Gastaut, H., Étude électrocorticographique de la réactivité des rythmes rolandiques, *Revue Neurol.*, 87, 176, 1952.
32. Tiihonen, J., Kajola, M., and Hari, R., Magnetic mu rhythm in man, *Neuroscience*, 32, 793, 1989.
33. Ohara, S. et al., Electrocorticogram-electromyogram coherence during isometric contraction of hand muscle in human, *Clin. Neurophysiol.*, 111, 2014, 2000.
34. Connors, B.M. and Amitai, Y., Making waves in the neocortex, *Neuron*, 18, 347, 1997.
35. Jasper, H. and Andrews, H.L., Brain potentials and voluntary muscle activity in man, *J. Neurophysiol.*, 1, 87, 1938.
36. Raethjen, J. et al., Determinants of physiologic tremor in a large normal population, *Clin. Neurophysiol.*, 111, 1825, 2000.
37. Mayston, M.J. et al., Physiological tremor in human subjects with X-linked Kallmann's syndrome and mirror movements, *J. Physiol.*, 530, 551, 2001.
38. Vaillancourt, D.E. and Newell, K.M., Amplitude changes in the 8–12, 20–25, and 40 Hz oscillations in finger tremor, *Clin. Neurophysiol.*, 111, 1792, 2000.
39. Raethjen, J. et al., Corticomuscular coherence in the 6–15 Hz band: is the cortex involved in the generation of physiologic tremor? *Exp. Brain Res.*, 142, 32, 2002.
40. Semba, K. and Komisaruk, B.R., Neural substrates of two different rhythmical vibrissal movements in the rat, *Neuroscience,* 12, 761, 1984.
41. Marsden, J.F., Brown, P., and Salenius, S., Involvement of the sensorimotor cortex in physiological force and action tremor, *NeuroReport*, 12, 1937, 2001.
42. Mima, T. et al., Electroencephalographic measurement of motor cortex control of muscle activity in humans, *Clin. Neurophysiol.*, 111, 326, 2000.
43. McKeown, M.J. and Radtke, R., Phasic and tonic coupling between EEG and EMG demonstrated with independent component analysis, *J. Clin. Neurophysiol.*, 18, 45, 2001.
44. Gross, J. et al., The neural basis of intermittent motor control in humans, *Proc. Nat. Acad. Sci. U.S.A.*, 99, 2299, 2002.
45. McAuley, J.H., et al., Common 3 and 10 Hz oscillations modulate human eye and finger movements while they simultaneously track a visual target, *J. Physiol.*, 515, 905, 1999.

46. Arroyo, S. et al., Functional significance of the mu rhythm of human cortex: an electrophysiologic study with subdural electrodes, *Electroenceph. Clin. Neurophysiol.*, 87, 76, 1993.
47. Crone, N.E. et al., Functional mapping of human sensorimotor cortex with electrocorticographic spectral analysis. I. Alpha and beta event-related desynchronization, *Brain*, 121, 2271, 1998.
48. Stancák, A. and Pfurtscheller, G., Mu-rhythm changes in brisk and slow self-paced finger movements, *NeuroReport,* 7, 1161, 1996.
49. Chatrian, G.E., Petersen, M.C., and Lazarte, J.A., The blocking of the rolandic wicket rhythm and some central changes related to movement, *Electroenceph. Clin. Neurophysiol.*, 11, 497, 1959.
50. Toro, C. et al., 8–12 Hz rhythmic oscillations in human motor cortex during two-dimensional arm movement: evidence for representation of kinematic parameters, *Electroenceph. Clin. Neurophysiol.*, 93, 390, 1994.
51. Andrew, C. and Pfurtscheller, G., On the existence of different alpha band rhythms in the hand area of man, *Neurosci. Lett.*, 222, 103, 1997.
52. Strens, L.H.A. et al., The effects of subthreshold 1 Hz repetitive TMS on cortico-cortical and interhemispheric coherence, *Clin. Neurophysiol.*, 113, 1279, 2002.
53. Leocani, L. et al., Event-related coherence and event-related desynchronization/synchronization in the 10 Hz and 20 Hz EEG during self-paced movements, *Electroenceph. Clin. Neurophysiol.*, 104, 199, 1997.
54. Feige, B., Aertsen, A., and Kristeva-Feige, R., Dynamic synchronization between multiple cortical motor areas and muscle activity in phasic voluntary movements, *J. Neurophysiol.*, 84, 2622, 2000.
55. MacKay, W.A. and Makhamra, S.R., unpublished data.
56. Roth, S.R., Sterman, M.B., and Clemente, C.D., Comparison of EEG correlates of reinforcement, internal inhibition and sleep, *Electroenceph. Clin. Neurophysiol.*, 23, 509, 1967.
57. Chase, M.H. and Harper, R.M., Somatomotor and visceromotor correlates of operantly conditioned 12-14 c/sec sensorimotor cortical activity, *Electroenceph. Clin. Neurophysiol.* 31, 85, 1971.
58. Roux, S., MacKay, W.A., and Riehle, A., unpublished data.
59. Riehle, A. and Requin, J., Monkey primary motor and premotor cortex: single-cell activity related to prior information about direction and extent of an intended movement, *J. Neurophysiol.*, 61, 534, 1989.
60. Liang, H. et al., Synchronized activity in prefrontal cortex during anticipation of visuomotor processing, *NeuroReport*, 13, 2011, 2002.
61. Nashmi, R., Mendonça, A.J., and MacKay, W.A., EEG rhythms of the sensorimotor region during hand movements, *Electroenceph. Clin. Neurophysiol.*, 91, 457, 1994.
62. Courtemanche, R., Fujii, N., and Graybiel, A.M., Synchronous, focally modulated-band oscillations characterize local field potential activity in the striatum of awake behaving monkeys, *J. Neurosci.*, 23, 11741, 2003.
63. Grosse, P. and Brown, P., Acoustic startle evokes bilaterally synchronous oscillatory EMG activity in the healthy human, *J. Neurophysiol.*, 90, 1654, 2003.
64. Murthy, V.N. and Fetz, E.E., Oscillatory activity in sensorimotor cortex of awake monkeys: synchronization of local field potentials and relation to behavior, *J. Neurophysiol.*, 76, 3949, 1996.
65. Berger, H., Über das Elektrenkephalogramm des Menschen. III. *Arch. Psychiatrie*, 94, 16, 1931.

66. Jasper, H., and Penfield, W., Electrocorticograms in man: effect of voluntary movement upon the electrical activity of the precentral gyrus, *Arch. Psychiatrie und Z. Neurologie*, 182, 163, 1949.
67. Toma, K. et al., Desynchronization and synchronization of central 20-Hz rhythms associated with voluntary muscle relaxation: a magnetoencephalographic study, *Exp. Brain Res.*, 134, 417, 2000.
68. Neuper, C. and Pfurtscheller, G., Evidence for distinct beta resonance frequencies in human EEG related to specific sensorimotor cortical areas, *Clin. Neurophysiol.*, 112, 2084, 2001.
69. Pfurtscheller, G. et al., Early onset of post-movement beta electroencephalogram synchronization in the supplementary motor area during self-paced finger movement in man, *Neurosci. Lett.*, 339, 111, 2003.
70. Stancák, A. and Pfurtscheller, G., Desynchronization and recovery of rhythms during brisk and slow self-paced finger movements in man, *Neurosci. Lett.*, 196, 21, 1995.
71. Alegre, M. et al., Beta electroencephalograph changes during passive movements: sensory afferences contribute to beta event-related desynchronization in humans, *Neurosci. Lett.*, 331, 29, 2002.
72. Andrew, C. and Pfurtscheller, G., Lack of bilateral coherence of post-movement central beta oscillations in the human electroencephalogram, *Neurosci. Lett.*, 273, 89, 1999.
73. Nikouline, V.V. et al., Interhemispheric phase synchrony and amplitude correlation of spontaneous beta oscillations in human subjects: a magnetoencephalographic study, *NeuroReport*, 12, 2487, 2001.
74. Serrien, D.J. and Brown, P., The functional role of interhemispheric synchronization in the control of bimanual timing tasks, *Exp. Brain Res.*, 147, 268, 2002.
75. Toma, K. et al., Movement rate effect on activation and functional coupling of motor cortical areas, *J. Neurophysiol.*, 88, 3377, 2002.
76. Murthy, V.N. and Fetz, E.E., Coherent 25- to 35-Hz oscillations in the sensorimotor cortex of awake behaving monkeys, *Proc. Nat. Acad. Sci. U.S.A.*, 89, 5670, 1992.
77. Salenius, S. et al., Cortical control of human motoneuron firing during isometric contraction, *J. Neurophysiol.*, 77, 3401, 1997.
78. Halliday, D.M. et al., Using electroencephalography to study functional coupling between cortical activity and electromyograms during voluntary contractions in humans, *Neurosci. Lett.*, 241, 5, 1998.
79. Mima, T. et al., Force level modulates human cortical oscillatory activities, *Neurosci. Lett.*, 275, 77, 1999.
80. Mima, T. and Hallett, M., Electroencephalographic analysis of cortico-muscular coherence: reference effect, volume conduction and generator mechanism, *Clin. Neurophysiol.*, 110, 1892, 1999.
81. Kilner, J.M. et al., Human cortical muscle coherence is directly related to specific motor parameters, *J. Neurosci.*, 20, 8838, 2000.
82. Gross, J. et al., Cortico-muscular synchronization during isometric muscle contraction in humans as revealed by magnetoencephalography, *J. Physiol.*, 527, 623, 2000.
83. Pohja, M. Salenius, S., and Hari, R., Cortico-muscular coupling in a human subject with mirror movements—a magnetoencephalographic study, *Neurosci. Lett.*, 327, 185, 2002.
84. Prut, Y. and Perlmutter, S.I., Firing properties of spinal interneurons during voluntary movement. I. State-dependent regularity of firing, *J. Neurosci.*, 23, 9600, 2003.

85. Nozaki, D., et al., Sustained muscle contractions maintained by autonomous neuronal activity within the human spinal cord, *J. Neurophysiol.*, 90, 2090, 2003.
86. Pohja, M. and Salenius, S., Modulation of cortex-muscle oscillatory interaction by ischaemia-induced deafferentation, *NeuroReport*, 14, 321, 2003.
87. Salenius, S. and Hari, R., Synchronous cortical oscillatory activity during motor action, *Curr. Opin. Neurobiol.*, 13, 678, 2003.
88. Kilner, J.M. et al., Task-dependent modulations of cortical oscillatory activity in human subjects during a bimanual precision grip task, *NeuroImage*, 18, 67, 2003.
89. Schieber, M., Training and synchrony in the motor system, *J. Neurosci.*, 22, 5277, 2002.
90. Marsden, J.F. et al., Organization of cortical activities related to movement in humans, *J. Neurosci.*, 20, 2307, 2000.
91. Szabadics, J., Lorincz, A., and Tamás, G., and frequency synchronization by dendritic GABAergic synapses and gap junctions in a network of cortical interneurons, *J. Neurosci.*, 21, 5824, 2001.
92. Freund, T., Rhythm and mood in perisomatic inhibition, *Trends Neurosci.*, 26, 489, 2003.
93. Traub, R.D. et al., GABA-enhanced collective behavior in neuronal axons underlies persistent gamma-frequency oscillations, *Proc. Nat. Acad. Sci. U.S.A.*, 100, 11047, 2003.
94. Tallon-Baudry, C., Bertrand, O., and Pernier, J., A ring-shaped distribution of dipoles as a source model of induced gamma-band activity, *Clin. Neurophysiol.*, 110, 660, 1999.
95. Buhl, E.H., Tamás, G., and Fisahn, A., Cholinergic activation and tonic excitation induce persistent gamma oscillations in mouse somatosensory cortex *in vitro*, *J. Physiol.*, 513, 117, 1998.
96. Bouyer, J.J., Montaron, M.F. and Rougeul, A., Fast fronto-parietal rhythms during combined focused attentive behaviour and immobility in cat cortical and thalamic localization, *Electroenceph. Clin. Neurophysiol.*, 51, 244, 1981.
97. Munk, M.H.-J. et al., Role of reticular activation in the modulation of intracortical synchronization, *Science*, 272, 271, 1996.
98. Hamada, Y., Miyashita, E., and Tanaka, H., Gamma-band oscillations in the "barrel cortex" precede rat's exploratory whisking, *Neuroscience*, 88, 667, 1999.
99. Salenius, S. et al., Human cortical 40 Hz rhythm is closely related to EMG rhythmicity, *Neurosci. Lett.*, 213, 75, 1996.
100. Crone, N.E. et al., Functional mapping of human sensorimotor cortex with electrocorticographic spectral analysis. II. Event-related synchronization in the gamma band, *Brain*, 121, 2301, 1998.
101. Aoki, F. et al., Changes in power and coherence of brain activity in human sensorimotor cortex during performance of visuomotor tasks, *BioSystems*, 63, 89, 2001.
102. Classen, J. et al., Integrative visuomotor behavior is associated with interregionally coherent oscillations in the human brain, *J. Neurophysiol.*, 79, 1567, 1998.
103. Singer, W., Neuronal synchrony: a versatile code for the definition of relations? *Neuron*, 24, 49, 1999.
104. Paulus, W., Fast oscillations in the sensorimotor cortex, *NeuroReport*, 10 (8), iii, 1999.
105. MacKay, W.A. and Mendonça, A.M., Field potential oscillatory bursts in parietal cortex before and during reach, *Brain Research*, 704, 167, 1995.
106. MacKay, W.A., Synchronized neuronal oscillations and their role in motor processes, *Trends Cogn. Sci.*, 1, 176, 1997.

8 Preparation for Action: One of the Key Functions of the Motor Cortex

Alexa Riehle

CONTENTS

ABSTRACT

One of the most striking processes involved in motor behavior is preparation for action. It is considered to be based on central processes, which are responsible for the maximally efficient organization of motor performance. A strong argument in favor of such an efficiency hypothesis of preparatory processes is the fact that providing prior information about movement parameters or removing time uncertainty about when to move significantly shortens reaction time. In this chapter, I will briefly summarize the behavioral effects of prior information and then describe some underlying neuronal correlates encountered in motor cortical areas of behaving

0-8493-1287-6/05/$0.00+$1.50

monkeys. The types of changes in neuronal activity and their selectivity during preparation will be portrayed and compared with other cortical areas that are involved in motor behavior. Furthermore, by linking motor cortical activity directly to behavioral performance, the trial-by-trial correlation between single neuron firing rate and reaction time revealed strong task-related cortical dynamics. Finally, the cooperative interplay among neurons, expressed by precise synchronization of their action potentials, will be illustrated and compared with changes in the firing rate of the same neurons. New concepts, including the notion of coordinated ensemble activity, and their functional implication during movement preparation will be discussed.

8.1 INTRODUCTION

Human motor behavior is remarkably accurate and appropriate even though the properties of our own body as well as those of the objects with which we interact vary over time. To adjust appropriately, the motor system has to assess the context in which it acts, including the properties of objects in the surrounding world and the prevailing environmental conditions. Since we often face problems that need to be solved immediately, the most essential processes underlying interactive behavior and performed in an interactive way include attention, intention, estimation of temporal and spatial constraints, anticipation, motivation, judgment, decision-making, and movement preparation. To perform all these processes, the brain continuously needs to monitor the external world, read out important information, input the desired information, retrieve related information from memory, manipulate and integrate all types of information, select the appropriate (motor) response, and then output the information necessary for initiating the response to particular brain areas. It is also needed to suppress unnecessary output to inappropriate brain areas and to inhibit inappropriate actions in order to perform spatially and temporally coordinated actions. It would be too long-winded to go into the details of all these processes and the related concepts here. For instance, many conceptual discussions about the linkage between attention, intention, and preparation within the framework of information processing operations are presented in the literature.[1–3] Since both selective attention and movement preparation can be viewed as covering internally triggered selective processes, they are closely related and possibly not separable.

One of the most fascinating processes involved in motor behavior is movement preparation. It is based on central processes responsible for the maximally efficient organization of motor performance. (For a review, see Reference 4.) A strong argument in favor of the efficiency hypothesis for preparatory processes is the fact that providing prior information about movement parameters or removing time uncertainty about when to move significantly shortens reaction time. In order for motor performance to be efficiently organized, both contextual and sensory information have to be assembled and integrated to shape the motor output. The notion of uncertainty, which is related to the manipulation of contextual information, is at the core of preparatory processes. The best-suited paradigm for studying such processes is the so-called "preparation paradigm." In this paradigm, two signals are presented successively to the subject in each trial: the first, the preparatory signal, provides prior information about what to do after occurrence of the second, the response

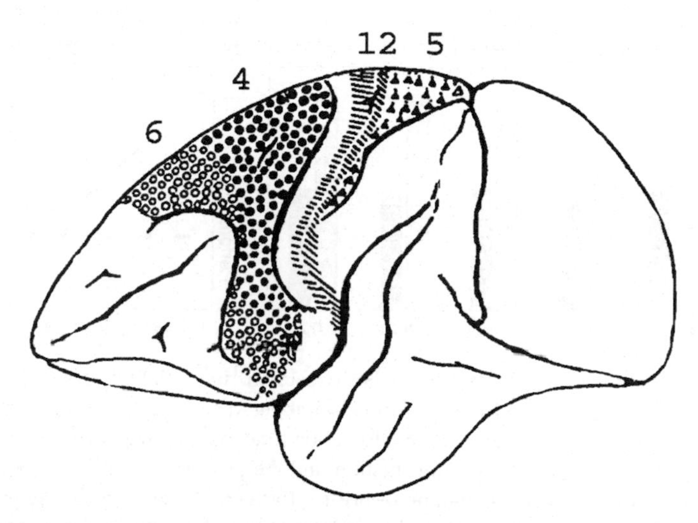

FIGURE 8.1 A lateral view of a monkey brain, with the anterior part at the left and the posterior at the right. This picture, modified after the seminal anatomical work of Korbinian Brodmann, which appeared in 1909,[5] shows, among others, the location of the primary motor cortex, area 4 (filled circles), just in front of the central sulcus (the curved line between numbers 4 and 1) and the premotor cortex (area 6, empty circles). Furthermore, posterior to the central sulcus, the somatosensory cortex is located (areas 1 and 2, stripes) as well as parietal area 5 (triangles). Neuronal activity presented in this chapter was mainly recorded in the primary motor cortex and the dorsal premotor cortex.

signal, or about when to do it. By means of such prior information, the context in which the subject is placed can be experimentally manipulated. The subject knows with more or less precision both *what* to do and *when* to initiate the requested movement, and has to adjust movement preparation accordingly.

Requin and colleagues[4] reviewed the topic of movement preparation in great detail. The focus here will be restricted to a description of the neuronal correlates of movement preparation obtained mainly in motor cortical areas such as the hand areas of the primary motor cortex and the dorsal premotor cortex (Figure 8.1) by using the preparation paradigm. In the following, I briefly summarize the behavioral effects of providing prior information. Then, the types of neuronal activity and the selectivity encountered during the preparation paradigm will be described, and its respective percentages will be compared with other cortical areas, which are involved in motor behavior. Furthermore, the direct trial-by-trial correlation between neuronal activity and behavior will be discussed. And finally, the cooperative interplay among neurons within a population will be illustrated and compared with changes in the firing rate of the same neurons in the population.

8.2 THE PREPARATION PARADIGM AND MOTOR BEHAVIOR

Reduction of uncertainty is one of the basics for understanding the mechanisms underlying preparation for action. In this context, (un)certainty is equivalent to

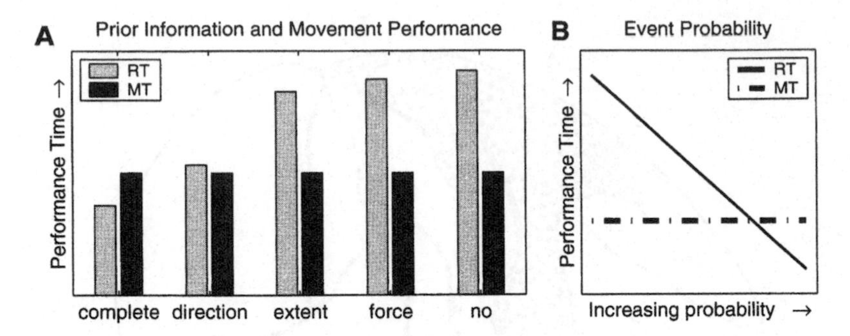

FIGURE 8.2 Schematic representations of both reaction times (RT) and movement times (MT) are shown as a function of prior information about various movement parameters (A) and the probability of signal occurrence (B). Reaction time, but not movement time, is clearly affected by both the content of prior information and the probability for the response signal to occur. In other words, the manipulation of prior information intervenes during preparation of movement, but not during its execution. (Data were schematically summarized from References 4,6–19.)

information about the required motor response. A modified preparation paradigm, the precueing paradigm, was introduced for allowing selective manipulation of prior information.[6,7] Two main categories of information may be manipulated by the preparatory signal. On the one hand, providing prior information about the spatial or kinematic parameters of the movement — e.g., direction, extent, or force — reduces uncertainty such that it leads to a significant reduction in reaction time.[6–16] Important insights in preparatory processes were gained by comparing the reduction in reaction time in relation to prior information about various movement parameters with the condition in which no information is provided, inducing the longest reaction times. It has been shown, both in human and monkey, that providing complete information, thus entirely removing uncertainty, shortens reaction time more than providing partial prior information. Since reaction time shortening is directly related to information, the most interesting condition is providing partial information and comparing information about different single movement parameters. It has been shown in various experiments[6,8–11,13,16] that information about movement direction shortens reaction time more than information about extent or information about force (Figure 8.2A). Given that the difference in reaction time is attributed to the processing time(s) for the uncued parameter(s), this would indicate that processing directional information takes longer than processing information about extent or force. Furthermore, one might infer that these processing operations may be serially performed if processing times are additive with the number of uncued parameters. Finally, when the reaction time difference associated with one parameter occurs only when another parameter is simultaneously precued, one could infer that the latter parameter is necessarily processed before the former.[4,7,15] However, most of the reaction time studies have failed to support unequivocally a hierarchy between planning operations or to decide clearly between a serial and a parallel organization of movement planning. (See References 4,8 for reviews.)

On the other hand, manipulating temporal aspects of the task by systematically varying the duration of the preparatory period has been shown to alter the preparatory state of the subject efficiently.[17,18] When presenting a finite number of durations of the preparatory period at random, but with equal probability, reaction time decreases with increasing duration (Figure 8.2B). (For a review see Reference 4.) Indeed, as time goes on during the trial and the response signal is not presented at the first possible moment, the probability for its occurrence increases with each next possible moment.[19] Interestingly, in both cases — manipulating prior information and manipulating the probability of signal occurrence — movement time remains at a stable level regardless of the nature or the amount of information (Figure 8.2). This consistently obtained result has been interpreted as a strong argument in favor of the hypothesis that providing prior information intervenes during processes of movement preparation or planning, but not during those of movement execution. The use of the preparatory paradigm thus makes it possible, first, to dissociate in time movement planning from its execution and, second, to study selectively preparatory processes by comparing data obtained in various conditions of partial prior information.

8.3 NEURONAL CORRELATES OF PREPARATORY PROCESSES

In many studies, in which only one movement parameter was precued, substantial proportions of neurons were found in various cortical (and subcortical) areas that changed their activity when prior information about direction was provided. (See References 1,20,21 for reviews.) In order to compare preparatory neuronal activity in various conditions of prior information, the precueing paradigm used in human subjects[6–8] was adapted to monkeys.[10–12] Using this paradigm, it was not only possible to identify selective processing operations, but also to make inferences about potentially different preparatory processes. Considering the functional meaning of changes in neuronal activity with respect to the behavioral features of the task, we proposed three criteria for tagging such activity changes as preparatory:[22] First, activity changes related to preparation are expected to appear within the preparatory period, i.e., the interval between the (instructive) preparatory signal and the (imperative) response signal. Second, changes in neuronal activity during the preparatory period should be selectively related to specific prior information. They must be viewed as an important step in establishing a functional, preparatory meaning of these changes. Indeed, the systematic manipulation of prior information induces parameter-specific reductions in reaction time (see Section 8.2). Third, preparatory changes in activity should be predictive for motor performance, for instance reaction time. Examples follow for each of the three criteria.

8.3.1 Types of Neurons Encountered by Using the Preparation Paradigm

Classification of neuronal activity is common to almost all studies in which a collection of data is presented that are recorded during a particular behavioral task.

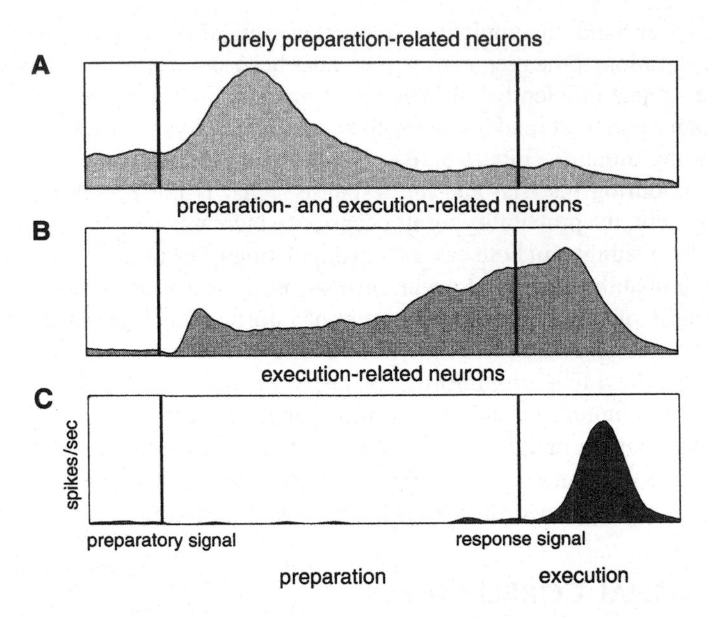

FIGURE 8.3 Three main types of neurons encountered during the preparation paradigm. Type I: purely preparation-related neurons (light gray); type II: preparation- and execution-related neurons (dark gray); type III: execution-related neurons (black). The first vertical lines correspond to the preparatory signal and the second ones to the response signal, the delay between them being usually 1 to 2 seconds.

However, classification criteria vary widely from study to study and are usually closely related to the specific question the study is dealing with. One of the simplest means of classification is to look for the temporal characteristics of changes in activity, i.e., the moment during the task when these changes occur. In Figure 8.3, three main types of neurons are presented, which were systematically encountered during the preparation paradigm. The definition for belonging to one or the other type concerns only the temporal appearance of the changes in activity in respect to the behavioral events. For instance, neurons of the first type, the so-called *purely preparation-related neurons* (Figure 8.3A), changed their activity only and exclusively in relation to movement preparation, that is, in relation to the meaning of the preparatory signal, and not at all in relation to the execution of the requested movement. Its transient character in Figure 8.3A is only an example and not a necessary condition for belonging to this type of neurons; it corresponds to the "signal-related" neurons described by Weinrich et al.[21] (See Figure 8.9A). However, there were as many purely preparation-related neurons, which were tonically activated during the preparatory period, defined as "set-related" neurons by Weinrich et al.[21] (See Figure 8.5A). In the same sense, neurons of the third type, the *purely execution-related neurons* (Figure 8.3C), changed their activity exclusively in relation to movement execution, i.e., after occurrence of the response signal, and did not modulate their activity during the preparatory period. Neurons of the most common type shared both properties by modifying their activity in relation to both movement *preparation and execution* (Figure 8.3B). For all three types, the shape

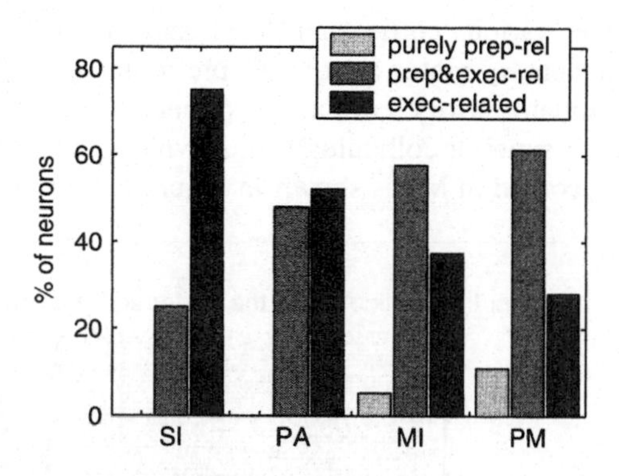

FIGURE 8.4 Distributions of the three main types of neurons, presented in Figure 8.3, in four cortical areas (see Figure 8.1). S1: area 1 and 2 of the somatosensory cortex; PA: area 5 of the posterior parietal cortex; M1: primary motor cortex; PM: dorsal premotor cortex. Type I: purely preparation-related neurons (light gray); type II: preparation- and execution-related neurons (dark gray); type III: execution-related neurons (black). (Data were summarized from References 10–12,22.)

of activity modulation did, of course, strongly vary from neuron to neuron, from phasic to tonic, including different onset and offset latencies, or increasing or decreasing in activity. What is essential here is the temporal relation either to preparatory processes, or to the executive processes, or to both. However, the attribution to one or the other type of neurons is not a clear-cut property; there is a gradual shift from preparation to execution.

In a series of experiments,[10–12,22] we compared neuronal activity recorded in four cortical areas — hand area of primary motor cortex (M1), dorsal premotor cortex (PM), area 5 of the posterior parietal cortex (PA), and areas 1 and 2 of the somatosensory cortex (S1) — during the execution of wrist extension and flexion movements, by manipulating partial information about various movement parameters. In Figure 8.4, the distribution of these three above-mentioned types of neurons is presented for each of the four cortical areas. All three types of neurons were recorded only in M1 and PM, and purely preparation-related neurons were extremely rare, having a higher percentage in PM than in M1. However, preparation-related activity in combination with execution-related activity was very common in all four cortical areas, although with different proportions. The fact that the highest percentage of purely execution-related neurons was recorded in S1 is mainly due to their definition. It relates to the fact that changes in activity occurred, by definition, after the response signal, but it does not indicate whether neuronal activity was related to movement initiation, the corollary discharge, or the sensory input related to movement execution.

8.3.2 Neuronal Representation of Movement Features during Preparation

Considering the second criterion for labeling changes in neuronal activity as preparatory, we studied their selectivity in various brain structures. Far from being a privileged property of motor cortical areas, selective preparatory processes are largely distributed over various cortical and even subcortical areas (PM,[1,10–12,20–30] M1,[10–12,22,31–36] supplementary motor area,[31,37,38] prefrontal cortex,[35,39–41] frontal eye fields,[42,43] primary somatosensory cortex,[11,12] parietal cortex,[11,12,44–47] basal ganglia,[31,48,49] cerebellum,[50] superior colliculus[51,52]). A typical example of such selective preparatory activity recorded in M1 is shown in Figure 8.5. Prior information about

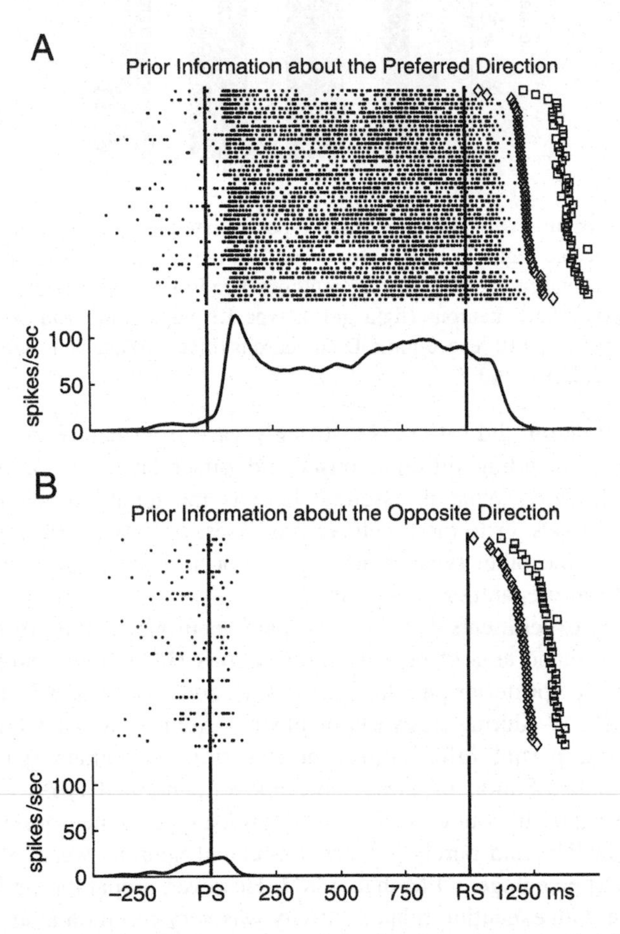

FIGURE 8.5 Directionally selective, preparation-related neuron recorded in the primary motor cortex. In (A), the preparatory signal (PS) provided prior information about the neuron's preferred direction, whereas in (B), information about the opposite direction was provided. At the top of each subfigure, a raster display of the neuron's activity is shown in which each dot corresponds to an action potential, and each line to a behavioral trial. The first vertical lines correspond to the occurrence of the PS and the second ones to the occurrence of the response signal (RS). The time between the two signals was 1 second. Trials were rank-ordered off-line according to increasing reaction time. Reaction time is defined as the time between the occurrence of the response signal and movement onset (diamonds). Squares correspond to movement end, defining movement time as the time between movement onset and offset. Below each raster display, a histogram indicates the mean discharge rate, calculated over all trials, in spikes per second. (A. Riehle, A. Bastian, and F. Grammont, unpublished data.)

the neuron's preferred direction was precued in Figure 8.5A and about the opposite to the preferred direction in Figure 8.5B. It can clearly be seen that the neuron discharged vigorously during the preparatory period when its preferred direction was precued, but was inhibited during preparation of the opposite movement.

In the series of experiments mentioned above,[10–12,22] we compared selective processing operations related to three movement parameters by manipulating prior information about two of them in each experiment within the whole series. The precued parameters were direction and extent,[10,22] direction and (frictional) force,[11] and extent and (frictional) force.[12] In each experiment, four conditions of prior information were presented to the animal at random and with equal probability: the condition of complete information, i.e., information about both manipulated parameters defining the movement entirely; two conditions of partial information, i.e., only information about one of the two parameters was provided, while the other remained to be specified by the response signal; and the condition of no information, i.e., a parametrically noninformative precue indicated solely the start of the preparatory period and both parameters remained to be specified after occurrence of the response signal. Each of the 4 conditions was combined with each of the 4 possible movements, for instance 2 directions of 2 extents each; hence, 16 types of trials were presented in each experiment.

The comparison of preparatory activities in several cortical areas shows that most of the neurons exhibited nonselective preparatory changes in activity (Figure 8.6A). Whatever prior information was presented by the preparatory signal, the neuron consistently changed its activity during the preparatory period. A neuron was labeled as selective when it changed its preparatory activity in one of the conditions of partial prior information in relation to information content — selective in respect to extension and flexion (Figure 8.5), in respect to large and small extent,

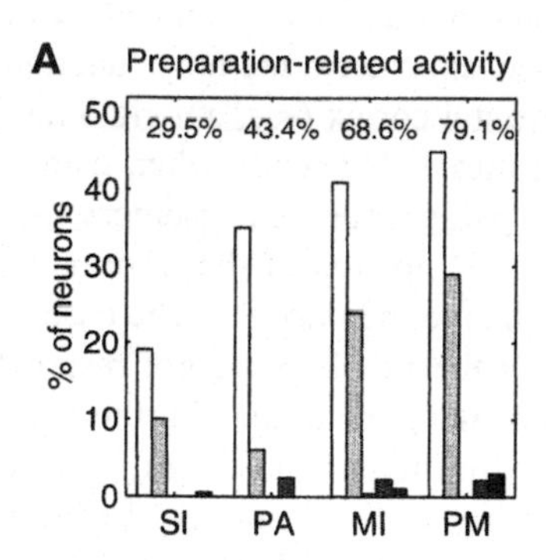

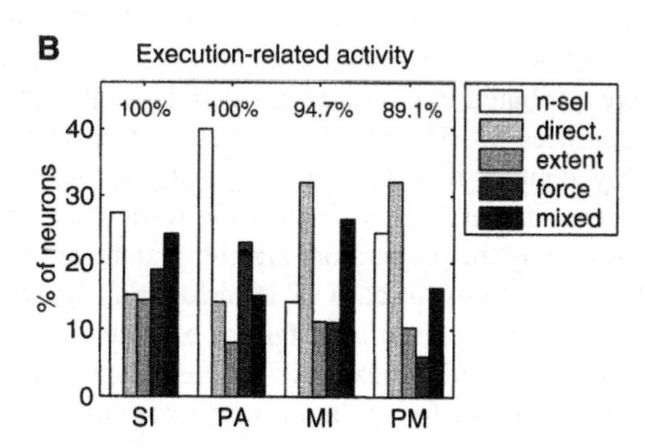

FIGURE 8.6 Distributions of preparation-related (A) and execution-related (B) activity changes encountered in four cortical areas. S1: area 1 and 2 of the somatosensory cortex; PA: area 5 of the posterior parietal cortex; M1: primary motor cortex; PM: dorsal premotor cortex. Gray levels from white to black: nonselective, direction-related, extent-related, force-related, and "mixed" changes in activity. For each cortical area the percentages of both preparation- and execution-related neurons are indicated, irrespective of whether they were selective or not. Note that one neuron could belong to both types of changes in activity. (Data were summarized from References 10–12,22.)

or in respect to weak and strong force. It was a consistent finding that many more changes in preparatory activity were selective in relation to information about movement direction than in relation to force or extent. The small percentage of "mixed" preparatory activities is due to the fact that a number of neurons changed their activity selectively in relation to one parameter only when the other parameter was known as well. For instance, in the direction–extent experiment, in which information about two directions and two extents was manipulated, differences in activity in relation to large or small extent were only obtained when movement direction was known as well — that is, in the condition of complete information. Similar results were obtained by Kurata,[27] who provided during the preparatory period, successively but in random order, two pieces of information about either direction or extent. Extent-related changes in activity were only detected when information about movement direction was available.

The high percentage of nonselective changes in activity could partly be explained by the fact that movements in only two directions were performed; that is, some directionally selective neurons might be missed because their preferred direction was perpendicular to the two opposite movement directions. Furthermore, one has to keep in mind that the condition of prior information called "no information" might be misleading. There is always some information available about the task constraints. In each experiment, the third, unmanipulated parameter remained constantly known. For instance, in the experiment in which the parameters force and extent were manipulated, the movements had to be executed in only one movement direction.[11] In this case, the permanent certainty about movement direction may have contributed to the general nonselective preparatory activity, such that the percentage of nonselective preparation-related neurons was higher in this experiment than in the others (43% versus 31% and 26% for direction–extent and direction–force, respectively).

The interpretation of delay-related neuronal activity has not been exclusively related to preparatory processes. Indeed, interpretation was mainly related to the brain structure in which it was recorded. For instance, short-term memory functions were attributed to delay-related activity in the prefrontal cortex (see, among others, References 53–55) but also in posterior parietal cortex.[56] However, when using a delayed motor task it is not always easy to clearly separate short-term memory from movement preparation, even if the response signal did not repeat the information provided by the preparatory signal[56] and the animal had to memorize it. The absence of delay activity in case of an error trial, in which the monkey did not respond, might be interpreted as a failure of both memory and preparation. Furthermore, Mountcastle and colleagues[57] proposed, specifically for the parietal cortex, a "command" function for initiating motor activity on the basis of a synthesis of sensory information. This was subsequently challenged (for a review see Reference 58) and the debate focused on whether this increase was due to an attentional facilitation of sensory processes or a preparatory facilitation of motor processes. The impossibility of unequivocally demonstrating the sensory versus motor function of this "enhancement" phenomenon stressed the difficulty in delimiting the boundary between perception and action, which is exactly as one would expect for an interfacing neural system responsible for making connections between perception and action representations. (See also the discussion of Boussaoud et al.[3] for frontal cortical areas.) The extension of preparation-related neurons into the somatosensory cortex, whatever their selectivity in relation to prior information, may be linked to setting the gain of somatosensory input pathways.[59] The large extension of preparatory activity into the postcentral areas, i.e., the parietal and somatosensory cortex (PA and S1), adds to the emerging notion that a wide distribution across the cortical structures of neuronal networks is responsible for the building of a motor act.[4,60]

For comparison, the percentages of execution-related selectivity are plotted in Figure 8.6B for the same sample of neurons. It is interesting to note that during execution many more selective activity changes were encountered than nonselective ones, whereas the percentages of selective neurons in relation to individual movement parameters did not vary as strongly in relation to both movement parameter and cortical area as they did during preparation. Furthermore, the number of "mixed" neurons (black bars) increased significantly compared to preparation. The fact that during preparation virtually all selective neurons changed their activity in relation to only one movement parameter — and not to a combination of parameters ("mixed") — suggests that movement preparation seems to be performed by rather segregated neuronal networks, each of which is responsible for processing information about that single movement parameter only. Conversely, the high number of "mixed" neurons present during execution suggests that common output networks, which represent the whole movement rather than single movement parameters, may be used. Finally, the fact that about two-thirds of primary motor cortical neurons changed their activity in relation to prior information (see percentages indicated in Figure 8.6 for each cortical area) demonstrates clearly the strong involvement of this area in preparatory processes. Hence, preparation for action is one of the key functions of motor cortical structures, including the primary motor and the premotor cortex.

8.3.3 Preparatory Activity: A Predictive Value for Performance Speed

In the framework of the preparation paradigm, it has been shown that in an identical behavioral condition (for instance, a condition in which prior information indicates that a pointing movement has to be made to a particular target), reaction time varies from trial to trial. One possible explanation for this might be that the level of attention or some other more general arousal effect spontaneously modulated the internal state of the subject, leading to changes in reaction time. The observation of delay-related neuronal activity in such a behavioral condition revealed in a high percentage of cortical neurons a statistically significant trial-by-trial correlation between neuronal firing rate and reaction time; the higher the firing rate, the shorter reaction time (see Figure 8.7). In other words, the trial-by-trial activity of individual motor cortical neurons, at the end of the preparatory period and before movement execution, reliably predicts movement performance, as expressed by reaction time.

The correlation between preparatory activity and reaction time has been shown to be statistically significant in almost 40% of primary motor cortical neurons, provided they exhibited some level of activity during the preparatory period. The same type of statistically significant correlation was also found in 27 to 35% of

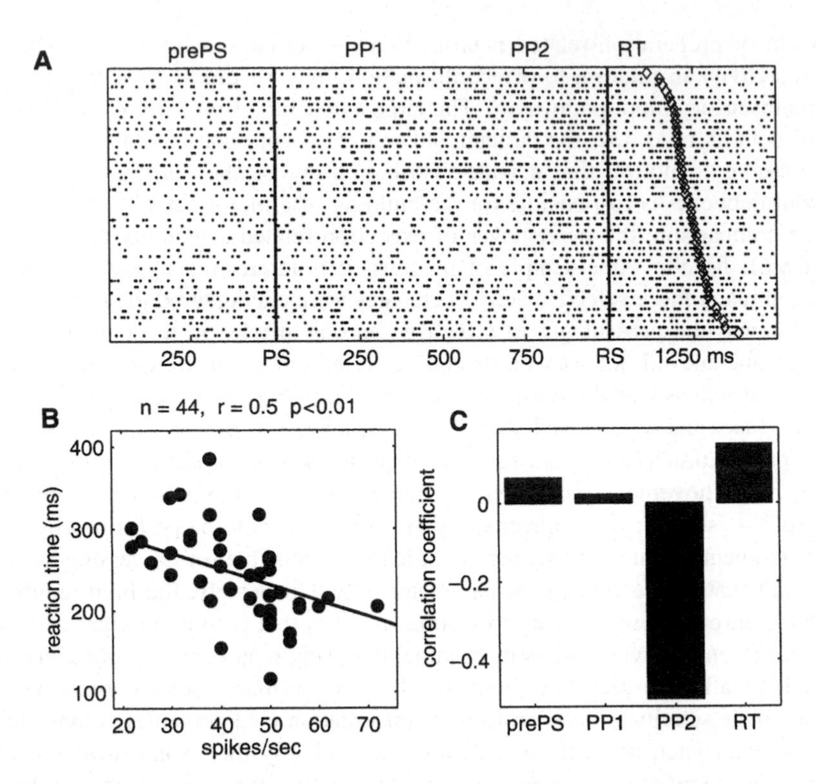

FIGURE 8.7 A typical example of a trial-by-trial correlation between the preparatory activity of a motor cortical neuron and reaction time. (A) The raster display is shown. For details see Figure 8.5. (B) The linear regression between the trial-by-trial firing rate during PP2, i.e., the last 500 msec before the response signal (RS), and reaction time is shown. The correlation coefficient r, which was highly significant, is indicated, and the regression line is drawn. Correlation indicates that the higher the firing rate at the end of the preparatory period, the shorter the reaction time. (C) Correlation coefficients are plotted, calculated in various periods during the trial: the 500 msec before the preparatory signal (prePS), during which, by definition, the animal did not know in which direction the upcoming movement has to be performed ($r = 0.064$); the first 500 msec after the preparatory signal (PP1, $r = 0.022$); the last 500 msec before the response signal (PP2, $r = -0.5$, $p < 0.01$, $df = 43$; see [B]), and finally during reaction time (RT, $r = 0.15$). During all periods, apart from PP2, no significant relationship between neuronal activity and reaction time can be seen. (A. Riehle, unpublished data.)

neurons within other cortical areas (Figure 8.8A). The percentages of neurons whose trial-by-trial firing rate was significantly correlated with reaction time depended on the behavioral condition in which the correlation was calculated. More neurons were significantly correlated in conditions in which movement direction was precued than in conditions of prior information about extent or force[11,12,22] (Figure 8.8B). However, the selective correlation did not depend on the selectivity of the preparation-related changes in activity. The trial-by-trial activity of a nonselective neuron could be selectively correlated with reaction time in only one condition of prior information

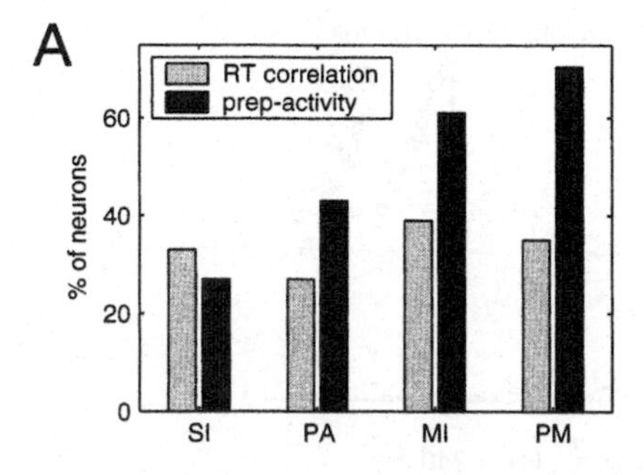

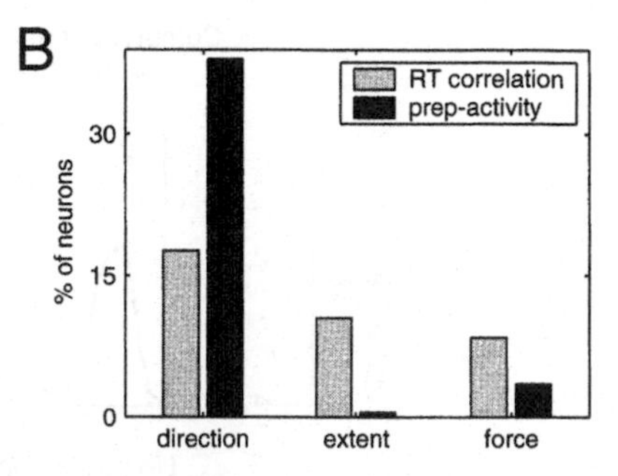

FIGURE 8.8 Distributions of preparation-related neurons (black) and neurons whose preparatory activities were significantly correlated with reaction time (gray), both as a function of cortical areas (A) and prior information (B). S1: area 1 and 2 of the somatosensory cortex; PA: area 5 of the posterior parietal cortex; M1: primary motor cortex; PM: dorsal premotor cortex. (Data were summarized from References 10–12,22.)

by exhibiting the same mean activities in all conditions.[12,22] This suggests that trial-by-trial variability of cortical activity, which is involved in inducing variability at the behavioral output, is independent of processing prior information about distinct movement parameters by means of mean changes in activity. A further argument in favor of this hypothesis is that the neurons that were significantly correlated with reaction time were more uniformly distributed over cortical areas than were the neurons that changed their mean activity in relation to movement preparation (Figure 8.8A). The same is true for the specific relation to prior information about single movement parameters (Figure 8.8B). Many more neurons changed their activity in relation to movement direction than were significantly correlated with reaction time in the same behavioral condition. However, many more neurons were correlated with trial-by-trial performance in conditions of prior information about extent or force than changed selectively their mean discharge rate during the preparatory period in relation to these parameters.

Two important results are in agreement with the behavioral results, which show that reaction time reduction was largest in conditions of prior information about direction (see Figure 8.2). First, many more neurons changed their mean preparatory activity exclusively in relation to movement direction rather than to any other precued parameter and, second, the trial-by-trial preparatory activity of many more neurons was significantly correlated with reaction time in the condition of information about direction than in other conditions of prior information.

8.4 TWO CONCEPTS OF SELECTIVE PREPARATORY PROCESSES: PREPROCESSING AND PRESETTING

In 1985, Requin[2] proposed two concepts of preparatory processes that might be responsible for the reduction in reaction time when providing prior information. (See also Reference 4.) Each process may intervene at different steps along the sensorimotor transformation, going from processing the information contained in

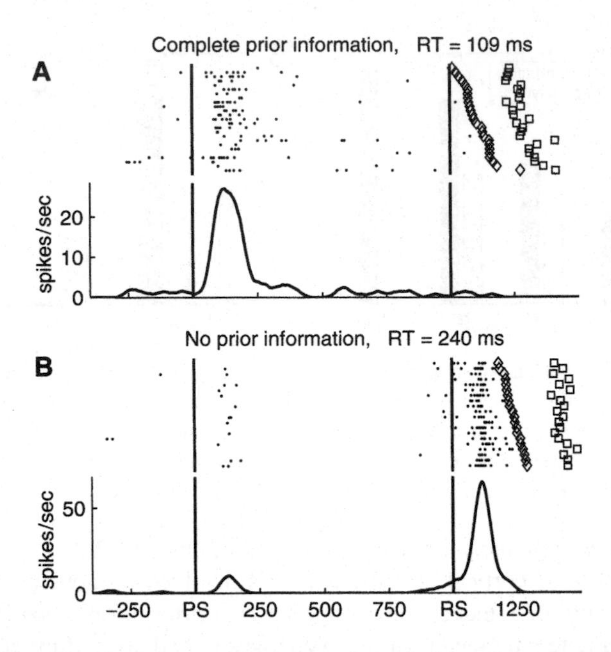

FIGURE 8.9 Example of a preprocessing neuron recorded in primary motor cortex. In (A), the preparatory signal (PS) provided complete prior information about the forthcoming movement, whereas in (B), no information was provided. Trials were rank-ordered off-line according to increasing reaction time. Reaction time (RT) is defined as the time between the occurrence of the response signal (RS) and movement onset (diamonds). Squares correspond to movement end, defining movement time as the time between movement onset and offset. The time between the two signals was 1 second. For details, see Figure 8.5. (A. Riehle, A. Bastian, F. Grammont, unpublished data.)

the stimulus to execution of the requested movement. In the first concept, the *preprocessing* view of motor preparation, *to prepare is to process in advance*. Some of the processes, which are triggered by the imperative response signal in a condition when no prior information is provided, would be triggered by the preparatory signal as long as it contains any necessary information about the requested movement. Information processing then takes place during the delay between the preparatory and the response signal and not during reaction time. In other words, what has been done in response to the preparatory signal no longer has to be done when the response signal is presented. This leads to a shorter reaction time. In the second concept, the *presetting* view of preparation, *to prepare is to facilitate movement initiation.* This means that preparatory processes, induced by the preparatory signal, accelerate processes that will be executed after the response signal, and therefore reduce reaction time. Here, the effect of preparation would result from processes induced by the preparatory signal that are different from those induced by the response signal.

Figure 8.9 shows an example of a preprocessing neuron recorded in the primary motor cortex. In Figure 8.9A, the preparatory signal provided complete prior information about the forthcoming movement. The neuron increased its activity phasically

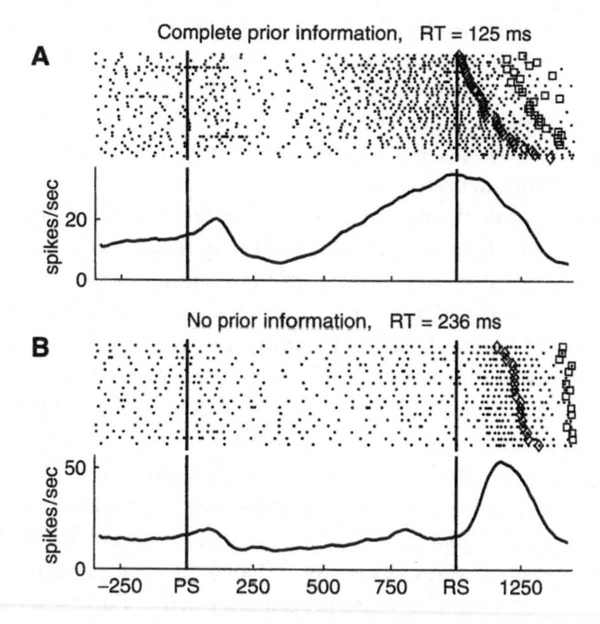

FIGURE 8.10 Example of a presetting neuron recording in the primary motor cortex. In (A), the preparatory signal (PS) provided complete prior information about the forthcoming movement, whereas in (B), no information was provided. Trials were rank-ordered off-line according to increasing reaction time. Reaction time (RT) is defined as the time between the occurrence of the response signal (RS) and movement onset (diamonds). Squares correspond to movement end, defining movement time as the time between movement onset and offset. The time between the two signals was 1 second. For details, see Figure 8.5. (A. Riehle, A. Bastian, F. Grammont, unpublished data.)

after signal presentation and then remained silent during the rest of the trial. In this condition, the monkey could anticipate movement initiation, and thus mean reaction time was very short (109 msec). In Figure 8.9B, however, the noninformative preparatory signal induced virtually no change in activity in that same neuron, and it was only after the response signal that it discharged vigorously with a constant latency.[61] Here, movement parameters could only be specified after the response signal provided the necessary information. The movement could not be anticipated, and thus reaction time was much longer (240 msec). The key property of preprocessing neurons is that the change in activity induced by the response signal depends upon prior information provided by the preparatory signal.[10,23] Note that the large majority of the purely preparation-related neurons presented in Figure 8.4 are preprocessing neurons.

In Figure 8.10, an example of a presetting neuron, recorded in the primary motor cortex, is shown. When the preparatory signal provided complete information about the forthcoming movement, the discharge frequency of this neuron increased progressively during the second half of the preparatory period and peaked in temporal relation to movement onset. However, when no information was provided in advance, the neuron did not change its activity, and it was only after the response signal that it vigorously increased its activity, in close relation to movement onset. In contrast to preprocessing neurons, presetting neurons discharge in a similar way after the occurrence of the response signal in both conditions of prior information.[10] Note that a large majority of the preparation- and execution-related neurons presented in Figure 8.4 are presetting neurons.

The timing of the peak changes in activity is in favor of the hypothesis presented above, according to which each type of neuron may intervene at different moments during sensorimotor transformation. First, in the majority of the preprocessing neurons, the discharge was time-locked to the response signal in the condition of no prior information (see Figure 8.9B). This indicates a rather early involvement in sensorimotor transformation, closely related to processing the information contained in the signal rather than to movement execution. On the contrary, most of the presetting neurons changed their activity time-locked to movement onset (Figure 8.10B), reflecting a relatively direct link to movement initiation. Second, the number of presetting neurons whose preparatory activities were significantly correlated with reaction time, thus clearly modulating the moment of movement initiation, was much higher than that of preprocessing neurons.[22] Third, the comparison of the strength of the trial-by-trial correlation between preparatory activity and reaction time in the condition of prior information about the neuron's preferred direction revealed a higher mean correlation coefficient in presetting neurons than in preprocessing neurons,[22] thus indicating once more that presetting neurons were much more strongly involved in movement initiation than were preprocessing neurons.

8.5 CORTICAL POPULATION REPRESENTATION OF MOVEMENT PARAMETERS DURING PREPARATION

The representation of movement direction in motor cortical areas has been studied by systematically relating the change in neuronal activity to the experimentally manipulated parameter. When gradually manipulating movement direction, it has been shown that the selectivity of neuronal discharge is rather broad, i.e., the activation of a neuron is largest for its preferred movement direction and gradually decreases as the movement trajectory diverges from that direction (Figure 8.11). As a consequence, it has been suggested that large populations of neurons should be active when any single motor act is executed. This has led to the notion of *population coding*, in which all activated neurons are assumed to contribute to the specification of individual values of movement parameters.[62,63] Using the technique of the population vector,[64] in which movement direction is computed as a circular mean from discharges of individual neurons, the representation of movement direction was observed in entire populations of motor cortical neurons, not only during movement execution,[64] but also during movement preparation in a delayed multidirectional pointing task.[65]

The continuous and gradual specification of movement parameters has formed the basis of a model of motor preparation.[66,67] The *dynamic field model* consists of an activation field defined over relevant movement parameters — for instance,

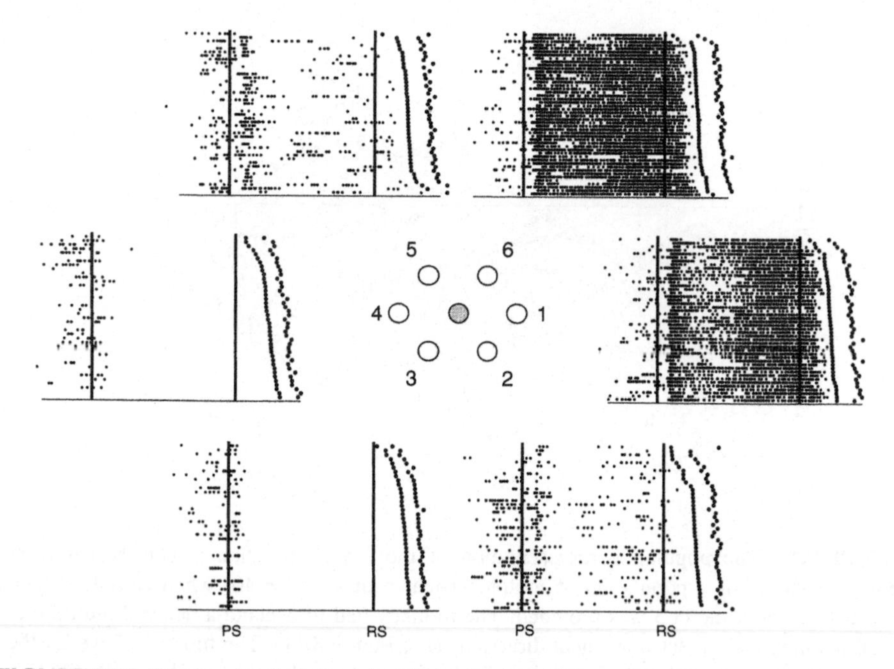

FIGURE 8.11 Discharge of a directionally selective, preparation-related neuron during a delayed multidirectional pointing task including six movement directions (see inset). The preparatory signal (PS) provided complete prior information about the forthcoming movement. Trials were rank-ordered off-line according to increasing reaction time, the time between the response signal (RS) and movement onset (first range of large dots). The second range corresponds to movement end. The time between the two signals was 1 second. (For details see Reference 32.) (A. Riehle, A. Bastian, F. Grammont, unpublished data.)

movement direction. Information about movement direction is represented as peaks of activation localized at those sites of the field that are mapped onto the indicated movement directions. The field evolves in time under the influence of two types of input. First, prior information preshapes the field by preactivating cued sites. Second, command input leads to the development of a single full-fledged peak of activation representing the final motor choice. This model predicts a large set of reaction time effects based on the fact that more highly preactivated field sites reach thresholds faster than less highly preactivated sites. Bastian et al.[32,68] (see also Reference 69) observed experimentally the preshaping of the population activity in the motor cortex. Based on the directional tuning curves of individual motor cortical neurons, determined during reaction time, distributions of the population activation were constructed, which were then extended into the preparatory period (Figure 8.12). The authors found that these distributions were preshaped by prior information, with a peak of activation centered over the precued movement direction. This peak sharpened as the response signal approached. Wider ranges of precued movement directions were represented by broader distributions, and the peak shifted to the requested movement direction as soon as it was specified by the response signal.[32,68]

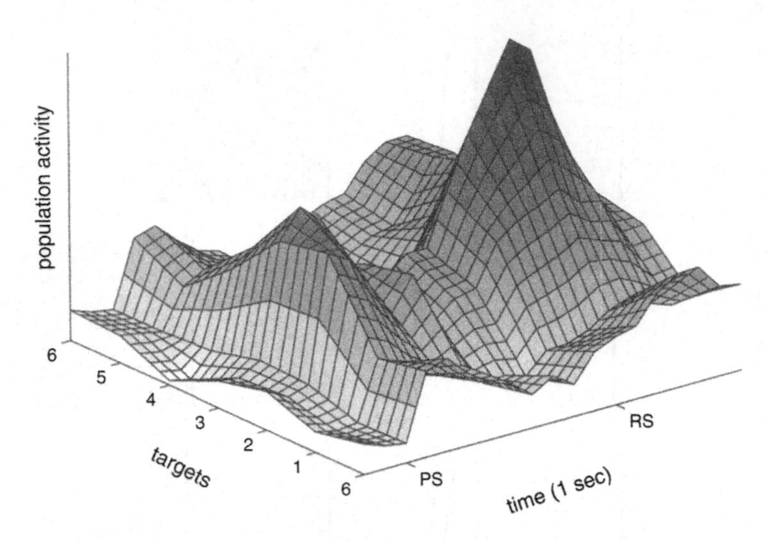

FIGURE 8.12 The population representation of movement direction is constructed from neural responses of a population of motor cortical neurons (n = 40) when complete prior information about target 3 was provided. The monkey had to execute a delayed multidirectional pointing task in six movement directions (see Figure 8.11). The time windows for the computation of the population distribution are 100 msec. Note that the population distribution is preshaped in response to the preparatory signal (PS). Location and width of activation reflect precisely prior information as early as it is provided. The activation peak is localized over the precued target during the preparatory period and the distribution increases in activation and sharpens subsequent to the response signal (RS). Time runs along the x-axis, targets (i.e., movement directions) along the y-axis, and the amplitude of the population activation along the z-axis. (Data from Reference 32.)

Thus, the shape of these distributions of population activity and their temporal evolution during the preparatory period up to movement initiation can be observed and related to prior information and movement performance.

These findings extend our knowledge about neuronal mechanisms underlying motor preparation. In particular, the concept of specific preactivation of distributions of population activation defined over the relevant movement parameter space appears to be a powerful one, which accounts for how partial prior information is integrated with new sensory information during movement preparation. In the dynamic field theory,[67,68] this is the concept of "preshaping." Because neurophysiological data were collected within task settings in which a limited set of movements was relevant, motor representations were potentially always preshaped (see Section 3.2), so that observed patterns of activity must be seen in relation to such prestructuring. The preshape concept may be relevant quite generally for examining context-dependent processes underlying behavior. For instance, the method of constructing distributions of population activity with a preshaping approach might be useful for getting further insight into processes that underly decision-making — as, for instance, in Go–NoGo tasks,[70,71] stimulus–response-compatibility tasks,[72,73] and categorization tasks.[74]

8.6 COOPERATIVITY IN CORTICAL NETWORKS RELATED TO COGNITIVE PREPARATORY PROCESSES

It is well accepted that sensorimotor functions including preparatory processes are based on conjoint processing in neuronal networks, which are widely distributed over various brain structures. However, it is much less clear how these networks organize dynamically in space and time to cope with momentary computational demands. The concept emerged that some computational processes in the brain could also rely on the relative timing of spike discharges among neurons within such functional groups,[75–80] commonly called *cell assemblies*.[81] An essential ingredient of coordinated ensemble activity is its flexibility and dynamic nature. In other words, neurons may participate in different cell assemblies at different times, depending on stimulus context and behavioral demands. To test whether such a temporal scheme is actually implemented in the central nervous system, it is necessary to observe the activities of many neurons simultaneously, and to analyze these activities for signs of temporal coordination. One type of temporal coordination consists of coincident spiking activities. Indeed, it has been argued that the synaptic influence of multiple neurons converging onto others is much stronger if they fire in coincidence,[82,83] making synchrony of firing ideally suited to raise the saliency of responses and to express relations among neurons with high temporal precision.

If cell assemblies are involved in cortical information processing, they should be activated in systematic relation to the behavioral task. We simultaneously recorded the activity of a small sample of individual neurons in the motor cortex of monkeys performing a delayed multidirectional pointing task.[88,89] (For statistics see References 84,86,87.) A surprising result revealed that many neurons, which were classified on the basis of their firing rate as being functionally involved in different processes, for instance one related to movement preparation and another related to execution (see Figure 8.3), significantly synchronized their spiking activity[88] (Figure 8.13).

Interestingly, they did not continually synchronize their activity during the whole task, but they were transitorily "connected" by synchrony during the transition from preparation to execution. The classification of the two neurons forming the pair based on their firing rate would by no means allow one to describe the functional link between them, which can only be detected by means of the synchronization pattern. Transient synchronization of spiking activity in ensembles of coactive neurons may help to strengthen the effectiveness within such groups and thereby help, for instance, to increase performance speed, complementary to the already described increase in firing rate (see Section 8.3.3). Indeed, it has been demonstrated that both the strength of synchronous activity[19] and the temporal precision of statistically significant synchrony increased toward the end of the preparatory period.[88] Moreover, highly time-resolved cross-correlation studies have shown that neurons strongly synchronized their activity at the end of the preparatory period in trials with short reaction times, but not at all, or with much less temporal precision, in trials with long reaction times[90] (Figure 8.14).

In the preceding sections, only spatial and kinematic aspects of movement preparation were taken into account. The problem of time uncertainty, however,

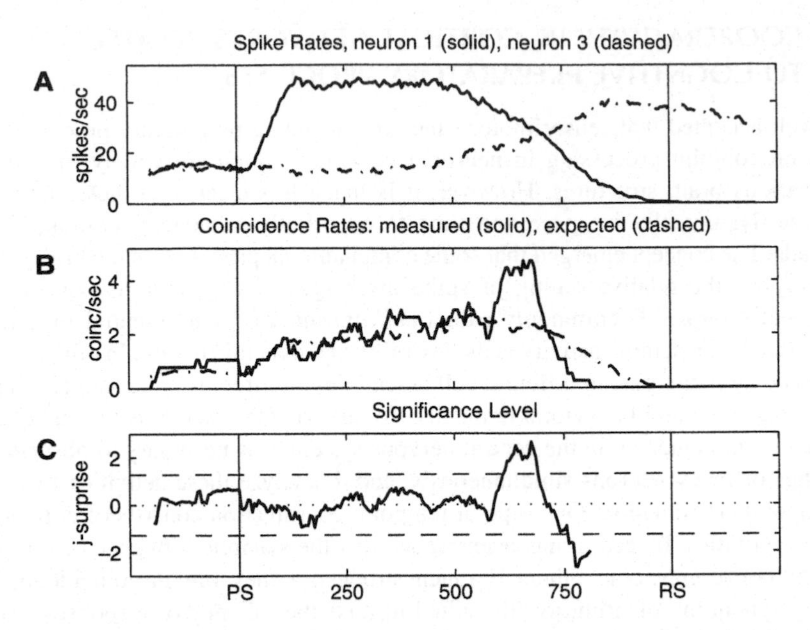

FIGURE 8.13 Dynamic changes of synchronous spiking activity of a pair of neurons recorded in the monkey primary motor cortex during a delayed pointing task. For calculation, a sliding window of 100 msec was shifted along the spike trains in 5 msec steps. The allowed coincidence width was 1 msec. (For calculation and statistics see Reference 84.) Time is running along the x-axis and is indicated in milliseconds. PS: preparatory signal; RS: response signal. (A) Firing rate of the two neurons in spikes/second. Neuron 1 may be classified as preparation-related, neuron 3 as execution-related (see Figure 8.3). (B) Measured (solid) and expected (dashed) coincidence rates are shown in coincidences/second. Expectancy was calculated by taking into account the instantaneous firing rates of each neuron.[84] (C) For each sliding window, the statistical significance (joint-surprise value[85]) was calculated for the difference between measured and expected coincidence rates. The result of each window was placed in its center. Whenever the significance value exceeded the threshold (upper dashed line, $p = 0.05$), this defined an epoch in which significantly more coincidences occurred than would be expected by chance. Coincidences within such an epoch are called "unitary events."[86,87] Occasionally, this value dropped below the lower dashed line, thus indicating epochs in which significantly fewer coincidences occurred ($p < 0.05$) than would be expected by chance. The significance level at zero corresponds to chance level. It can clearly be seen that only during a short epoch the activities of the neurons were significantly correlated. (Data from Reference 88.)

relates also to preparatory processes that are activated when a subject anticipates a behavioral demand. Removing time uncertainty by increasing the probability of signal occurrence significantly shortens reaction time[4] (see Section 8.2). At the neuronal level, it has been shown that at the moment when a signal was expected (but did not occur), motor cortical neurons significantly synchronized their discharges without necessarily changing their firing rate[19] (Figure 8.15). Furthermore, neurons correlated their spiking activity more strongly when subjects were highly motivated than when they were unmotivated.[19] In other words, groups of neurons may correlate their activity in relation to internal states of alertness at some moments,

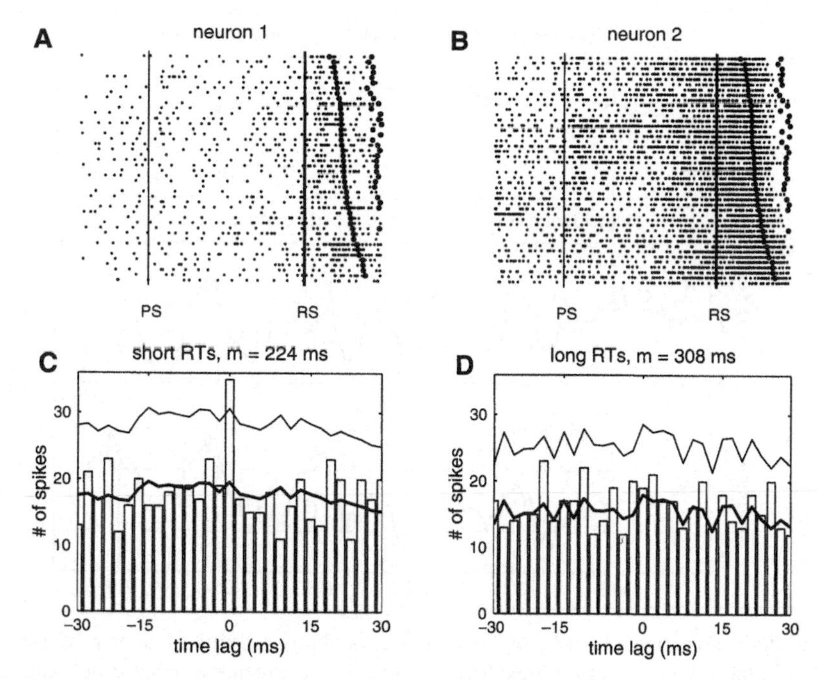

FIGURE 8.14 Cross-correlation histograms (CCHs) between the activities of two motor cortical neurons recorded simultaneously during the execution of a delayed pointing task. In (A) and (B), the spiking activities of the two neurons are shown, trials being rank-ordered according to increasing reaction times (first ranges of large dots). PS: preparatory signal; RS: response signal. The delay between PS and RS was 1 second. In (C) and (D), CCHs were calculated during the period between 500 msec before to 100 msec after RS. Binwidth: 2 msec. Thick line: shift predictor; thin line: 99% confidence limit.[91,92] In (C), the CCH was calculated for the upper half of trials including short reaction times (mean RT = 224 msec), and in (D) from the second set of trials comprising the long reaction times (mean RT = 308 msec). A distinct peak of 2 msec width can clearly be seen with zero time-lag in relation to short reaction times, but not in relation to long reaction times. (From Reference 90, with permission.)

but fire independently at others, without necessarily changing their discharge frequency. It is interesting to note that the modulation of correlated firing can by no means be predicted by simply inspecting the firing rate of individual neurons (Figure 8.15B).

The modulation of the discharge rate and the modulation of neuronal cooperativity in terms of synchronizing the occurrences of individual spikes precisely in time (in the millisecond range) suggest that the brain uses different strategies in different contextual situations. In order to deal with internal and purely cognitive processes such as expecting an event, increasing motivation, or modifying an internal state, neurons may preferentially synchronize their spike occurrences without necessarily changing their firing rates. In contrast, when processing external, behaviorally relevant events such as the appearance of a signal providing prior information or cueing the execution of the requested movement, neurons may preferentially modulate their firing rates.[19] Thus, both a temporal code (e.g., the precise synchronization of spikes)

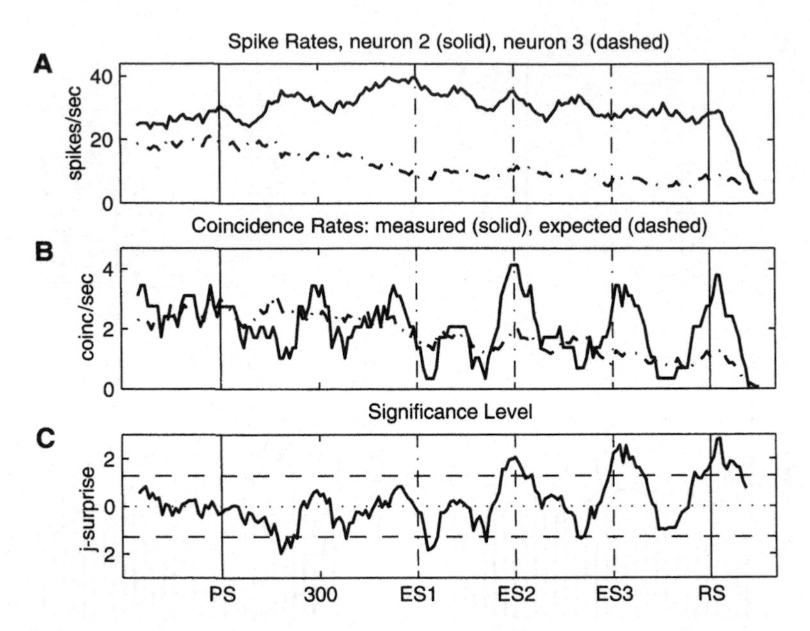

FIGURE 8.15 Dynamic changes of synchronous spiking activity of a pair of neurons recorded in the primary motor cortex. The monkey had to execute a delayed pointing task, in which 4 delay durations (600, 900, 1200, and 1500 msec) were presented at random with equal probability. Thus, as time goes on during the trial, the probability for the response signal (RS) to occur increased with each possible time step (see Section 8.2 for more details). In other words, signal expectancy increased with increasing time. All trials with the longest duration were pooled in this figure. PS: preparatory signal; ES1-3: expected (response) signals. For calculation, a sliding window of 100 msec was shifted along the spike trains in 5-msec steps. The allowed coincidence width was 2 msec. (For calculation and statistics see Reference 84.) Time is running along the x-axis. (A) Firing rate of the two neurons in spikes/second. (B) Measured (solid) and expected (dashed) coincidence rates are shown in coincidences/second. (C) Statistical significance for the difference between measured and expected coincidence rates. For more details, see Figure 8.13. (Data from Reference 19.)

and a rate code (the modulation of the discharge frequency) may serve different and complementary functions, acting in conjunction at some times and independently at others, depending on the behavioral context. The combination of both strategies allows extraction of much more information from a single pattern of neuronal activity, and thus increases the dynamics, the flexibility, and the representational strength of a distributed system such as the cerebral cortex. During preparation, in the motor cortex particularly, abrupt changes in firing rate (transient bursts) are probably deliberately kept to a minimum, if not totally prevented. This could be to prevent accidental activation of downstream motor nuclei. Most often the changes in firing rate are gradual until after occurrence of the response signal. Therefore, during preparation, phasic signalling at a precise time would be preferentially mediated by a temporal code such as transitory spike synchronization, in order to indicate internal events or to modify the internal state.

8.7 OUTLOOK

During the last decades, our understanding of higher brain functions involved in movement preparation advanced tremendously and a large body of knowledge about neuronal correlates and fundamental mechanisms was accumulated. However, the pendulum swung from information processing models introduced by cognitive psychology, including hierarchically ordered stages or processing operations, to dynamical models for the study of context-dependent processes underlying motor behavior. Dynamical neuronal representations imply large populations of neurons, where interactions both within a cortical area and between areas play an important role. Our current understanding of the neuronal correlates of sensorimotor transformations do not give a definitive answer to the question of whether there is an ordered group of different neuronal populations, which are recruited sequentially from sensory via associative to purely movement-related neurons,[70,93] or whether, alternatively, the preparatory processes take place within a single network whose firing pattern evolves in time, reflecting these different subprocesses. For instance, "complex" neurons were discovered, whose activities were neither entirely target-dependent nor entirely movement-dependent, by using an experimental design in which target location could be dissociated from actual limb movement direction.[94] A very similar type of neuron was found to contribute to the stimulus–response association in a stimulus–response compatibility task.[72] These findings suggest that single neurons are involved in movement preparation at various moments in time and in various roles. The time course, parametric dependence, and correlation with behavior of the representations carried by these populations may be observable by applying, for instance, methods such as constructing distributions of population activity[68,69] to different neuronal subpopulations, whose elements are selected on the basis of their response properties. This may make it possible to discover the functional role of these different groups in the processes of movement specification. Furthermore, the time courses of the interaction within neuronal networks in terms of synchrony of their spiking activities and of the mean firing rate of the same neurons appeared to be very different. Indeed, there was a clear tendency for synchrony to precede firing rate, but there was no simple parallel shifting in time of these two measures.[89,95] This makes it unlikely that the two coding schemes are tightly coupled by any kind of stereotyped transformation; they seem to obey rather different dynamics, suggesting that the coherent activation of cell assemblies may trigger the increase in firing rate in large groups of neurons. The functional relationship between synchrony and firing rate involved in preparatory processes remains to be established. For instance, the preparatory coherent activation of cell assemblies, by way of synchrony, may generate the increase in firing rate in large cortical networks, which in turn communicate with the periphery for initiating the movement.[95]

ACKNOWLEDGMENTS

I wish to thank Michel Bonnet, Driss Boussaoud, Guillaume Masson, Bill MacKay, and Sébastien Roux for many helpful discussions and comments. The work on this chapter was supported in part by the CNRS and the French government (ACI Cognitique: Invariants and Variability).

REFERENCES

1. Evarts, E.V., Shinoda, Y., and Wise, S.P., *Neurophysiological Approaches to Higher Brain Functions,* John Wiley & Sons, New York, 1984.
2. Requin, J., Looking forward to moving soon: ante factum selective processes in motor control, in *Attention and Performance XI,* Posner, M. and Marin, O., Eds., Lawrence Erlbaum Ass., Hillsdale, NJ, 1985, 147.
3. Boussaoud, D., Di Pellegrino, G., and Wise, S.P., Frontal lobe mechanisms subserving vision-for-action versus vision-for-perception, *Behav. Brain Res.,* 72, 1, 1996.
4. Requin, J., Brener, J., and Ring, C., Preparation for action, in *Handbook of Cognitive Psychophysiology: Central and Autonomous Nervous System Approaches,* Jennings, R.R. and Coles, M.G.H., Eds., John Wiley & Sons, New York, 1991, 357.
5. Brodmann, K. *Vergleichende Lokalisationslehre der Grosshirnrinde in ihren Prinzipien dargestellt aufgrund des Zellenbaues*, Barth, Leipzig, 1909.
6. Rosenbaum, D.A., Human movement initiation: specification of arm, direction, and extent, *J. Exp. Psychol. Gen.*, 109, 444, 1980.
7. Rosenbaum, D.A., The movement precuing technique: assumptions, applications, and extensions, in *Memory and Control of Action*, Magill, R.A., Ed., North-Holland Publishing Company, Amsterdam, 1983, 274.
8. Lépine, D., Glencross, D., and Requin, J., Some experimental evidence for and against a parametric conception of movement programming, *J. Exp. Psychol. Hum. Percept. Perf.,* 15, 347, 1989.
9. MacKay, W.A. and Bonnet, M., CNV, stretch reflex and reaction time correlates of preparation for movement direction and force, *Electroencephal. Clin. Neurophysiol.,* 76, 47, 1990.
10. Riehle, A. and Requin, J., Monkey primary motor and premotor cortex: single-cell activity related to prior information about direction and extent of an intended movement, *J. Neurophysiol.,* 61, 534, 1989.
11. Riehle, A. and Requin, J., Neuronal correlates of the specification of movement direction and force in four cortical areas of the monkey, *Behav. Brain Res.,* 70, 1, 1995.
12. Riehle, A., MacKay, W.A., and Requin, J., Are extent and force independent movement parameters? Preparation- and movement-related neuronal activity in the monkey cortex, *Exp. Brain Res.,* 99, 56, 1994.
13. Bonnet, M., Requin, J., and Stelmach, G.E., Specification of direction and extent in motor programming, *Bull. Psychon. Soc.,* 19, 31, 1982.
14. Vidal, F., Bonnet, M., and Macar, F., Programming response duration in a precueing reaction time paradigm, *J. Mot. Behav.,* 23, 226, 1991.
15. Zelaznik, H.N. and Hahn, R., Reaction time methods in the study of motor programming: the precuing of hand, digit, and duration, *J. Mot. Behav.,* 17, 190, 1985.
16. Larish, D.D. and Frekany, G.A., Planning and preparing expected and unexpected movements: reexamining the relationships of arm, direction, and extent of movement, *J. Mot. Behav.,* 17, 168, 1985.
17. Bertelson, P. and Boons, J.P., Time uncertainty and choice reaction time, *Nature,* 187, 131, 1960.
18. Bertelson, P., The time course of preparation, *Q. J. Exp. Psychol.,* 19, 272, 1967.

19. Riehle, A., Grün, S., Diesmann, M., and Aertsen, A., Spike synchronization and rate modulation differentially involved in motor cortical function, *Science,* 278, 1950, 1997.
20. Wise, S.P., The nonprimary motor cortex and its role in the cerebral control of movement, in *Dynamic Aspects of Neocortical Function*, Edelman, G., Gall, W.E., and Cowan, W.H., Eds., John Wiley & Sons, New York, 1984, 525.
21. Weinrich, M., Wise, S.P., and Mauritz, K.H., A neurophysiological study of the premotor cortex in the rhesus monkey, *Brain,* 107, 385, 1984.
22. Riehle, A. and Requin, J., The predictive value for performance speed of preparatory changes in activity of the monkey motor and premotor cortex, *Behav. Brain Res.,* 53, 35, 1993.
23. Crammond, D.J. and Kalaska, J.F., Prior information in motor and premotor cortex: activity during the delay period and effect on pre-movement activity, *J. Neurophysiol.,* 84, 986, 2000.
24. Godschalk, M. and Lemon, R.N., Involvement of monkey premotor cortex in the preparation of arm movements, *Exp. Brain Res. Suppl.,* 7, 114, 1983.
25. Godschalk, M., Lemon, R.N., Kuypers, H.G.J.M., and Van der Steen, J., The involvement of monkey premotor cortex neurones in preparation of visually cued arm movements, *Behav. Brain Res.,* 18, 143, 1985.
26. Kubota, K. and Hamada, I. Visual tracking and neuron activity in the post-arcuate area in monkeys, *J. Physiol. Paris,* 74, 297, 1978.
27. Kurata, K., Premotor cortex of monkeys: set- and movement-related activity reflecting amplitude and direction of wrist movements, *J. Neurophysiol.,* 69, 187, 1993.
28. Weinrich, M. and Wise, S.P., The premotor cortex of the monkey, *J. Neurosci.,* 2, 1329, 1982.
29. Wise, S.P., The primate premotor cortex: past, present, and preparatory, *Annu. Rev. Neurosci.,* 8, 1, 1985.
30. Wise, S.P. and Mauritz, K.H., Set-related activity in the premotor cortex of rhesus monkeys: effects of changes in motor set, *Proc. Roy. Soc. Lond. B,* 223, 331, 1985.
31. Alexander, G.E. and Crutcher, M.S., Preparation for movement: neural representations of intended direction in three motor areas in the monkey, *J. Neurophysiol.,* 64, 133, 1990.
32. Bastian, A., Riehle, A., Erlhagen, W., and Schöner, G., Prior information preshapes the population representation of movement direction in motor cortex, *NeuroReport,* 9, 315, 1998.
33. Georgopoulos, A.P., Crutcher, M.D., and Schwartz, A.B., Cognitive spatial-motor processes. 3. Motor cortical prediction of movement direction during an instructed delay period, *Exp. Brain Res.,* 75, 183, 1989.
34. Evarts, E.V. and Tanji, J., Reflex and intended responses in motor cortex pyramidal tract neurons of monkey, *J. Neurophysiol.,* 39, 1069, 1976.
35. Kubota, K. and Funahashi, S., Direction-specific activities of dorsolateral prefrontal and motor cortex pyramidal tract neurons during visual tracking, *J. Neurophysiol.,* 47, 362, 1982.
36. Tanji, J. and Evarts, E.V., Anticipatory activity in motor cortex neurons in relation to direction of an intended movement, *J. Neurophysiol.,* 39, 1062, 1976.
37. Tanji, J. and Kurata, K., Contrasting neuronal activity in supplementary and precentral motor cortex of monkeys. I. Responses to instructions determining motor responses to forthcoming signals of different modalities, *J. Neurophysiol.,* 53, 129, 1985.
38. Tanji, J., Taniguchi, K., and Saga, T., Supplementary motor area: neuronal response to motor instructions, *J. Neurophysiol.,* 43, 60, 1980.

39. Niki, H., Prefrontal unit activity during delayed alternation in the monkey. I. Relation to direction of response, *Brain Res.,* 68, 185, 1974.
40. Niki, H., Prefrontal unit activity during delayed alternation in the monkey. II. Relation to absolute versus relative direction of response, *Brain Res.,* 68, 197, 1974.
41. Niki, H. and Watanabe, M., Prefrontal unit activity and delayed response: relation to cue location versus direction of response, *Brain Res.,* 105, 79, 1976.
42. Bruce, C.J. and Goldberg, M.E., Primate frontal eye fields. I. Single neurons discharging before saccades, *J. Neurophysiol.,* 53, 603, 1985.
43. Everling, S. and Muñoz, D.P., Neural correlates for preparatory set associated with prosaccades and anti-saccades in the primate frontal eye field, *J. Neurosci.,* 20, 387, 2000.
44. Crammond, D.J. and Kalaska, J.F., Neuronal activity in primate parietal cortex area 5 varies with intended movement direction during an instructed-delay period, *Exp. Brain Res.,* 76, 458, 1989.
45. Godschalk, M., Lemon, R.N., and Kuypers, H.G.J.M., Involvement of monkey inferior parietal lobule in preparation of visually guided arm movements, *Experientia,* 40, 1297, 1984.
46. Mackay, W.A. and Riehle, A., Correlates of preparation of arm reach parameters in parietal area 7a of the cerebral cortex, in *Tutorials in Motor Neuroscience*, Requin, J. and Stelmach, G.E., Eds., Kluwer Academic Publishers, Dordrecht, 1991, 347.
47. MacKay, W.A. and Riehle, A., Planning a reach: spatial analysis by area 7a neurons, in *Tutorials in Motor Behavior II*, Stelmach, G.E. and Requin, J., Eds., Elsevier, Amsterdam, 1992, 501.
48. Apicella, P., Scarnati, E., and Schultz, W., Tonically discharging neurons of monkey striatum respond to preparatory and rewarding stimuli, *Exp. Brain Res.,* 84, 672, 1991.
49. Jaeger, D., Gilman, S., and Aldridge, J.W., Primate basal ganglia activity in a precued reaching task: preparation for movement, *Exp. Brain Res.,* 95, 51, 1993.
50. Strick, P.L., The influence of motor preparation on the response of cerebellar neurons to limb displacements, *J. Neurosci.,* 3, 2007, 1983.
51. Basso, M.A. and Wurtz, R.H., Modulation of neuronal activity by target uncertainty, *Nature,* 389, 66, 1997.
52. Basso, M.A. and Wurtz, R.H., Modulation of neuronal activity in superior colliculus by changes in target probability, *J. Neurosci.,* 18, 7519, 1998.
53. Fuster, J.M., Unit activity in prefrontal cortex during delayed-response performance: neuronal correlates of transient memory, *J. Neurophysiol.,* 36, 61, 1973.
54. Fuster, J.M., Behavioral electrophysiology of the prefrontal cortex, *Trends Neurosci.,* 7, 408, 1984.
55. Funahashi, S., Bruce, C.J., and Goldman-Rakic, P.S., Mnemonic coding of visual space in the monkey's dorsolateral prefrontal cortex, *J. Neurophysiol.,* 61, 331, 1989.
56. Gnadt, J.W. and Andersen, R.A., Memory related motor planning activity in posterior parietal cortex of macaque, *Exp. Brain Res.,* 70, 216, 1988.
57. Mountcastle, V.B., Lynch, J.C., Georgopoulos, A., Sakata, H., and Acuña, C., Posterior parietal association cortex of the monkey: command functions for operations within extrapersonal space, *J. Neurophysiol.,* 38, 871, 1975.
58. Lynch, J.C., The functional organization of the posterior parietal association cortex, *Behav. Brain Sci.,* 3, 485, 1980.
59. Prochazka, A., Sensorimotor gain control: a basic strategy of motor systems, *Progr. Neurobiol.,* 33, 281, 1989.
60. Requin, J., Riehle, A., and Seal, J., Neuronal networks for movement preparation, in *Attention and Performance XIV*, Meyer, D.E. and Kornblum, S., Eds., MIT Press, Cambridge, MA, 1992, 745.

61. Riehle, A., Visually induced signal-locked neuronal activity changes in precentral motor areas of the monkey: hierarchical progression of signal processing, *Brain Res.*, 540, 131, 1991.
62. Georgopoulos, A.P., Schwartz, A.B., and Kettner, R.E., Neuronal population coding of movement direction, *Science,* 233, 1416, 1986.
63. Lee, C., Rohrer, W.H., and Sparks, D.L., Population coding of saccadic eye movements by neurons in the superior colliculus, *Nature,* 332, 357, 1988.
64. Georgopoulos, A.P., Caminiti, R., Kalaska, J.F., and Massey, J.T., Spatial coding of movement: a hypothesis concerning the coding of movement direction by motor cortical populations, *Exp. Brain Res. Suppl.,* 7, 327, 1983.
65. Georgopoulos, A.P., Crutcher, M.D., and Schwartz, A.B., Cognitive spatial-motor processes. 3. Motor cortical prediction of movement direction during an instructed delay period, *Exp. Brain Res.,* 75, 183, 1989.
66. Schöner, G., Kopecz, K., and Erlhagen, W., The dynamic neural field theory of motor programming: arm and eye movements, in *Self-Organization, Computational Maps, and Motor Control,* Morasso, P.G. and Sanguineti, V., Eds., Elsevier, Amsterdam Psychology Series, Vol. 119, 1997, 271.
67. Erlhagen, W. and Schöner, G., Dynamic field theory of movement preparation, *Psychol. Rev.,* 109, 545, 2002.
68. Bastian, A., Schöner, G., and Riehle, A., Preshaping and continuous evolution of motor cortical representations during movement preparation, *Eur. J. Neurosci.* 18, 2047, 2003.
69. Erlhagen, W., Bastian, A., Jancke, D., Riehle, A., and Schöner, G., The distribution of neuronal population activation as a tool to study interaction and integration in cortical representations, *J. Neurosci. Meth.,* 94, 53, 1999.
70. Miller, J., Riehle, A., and Requin, J., Effects of preliminary perceptual output on neuronal activity of the primary motor cortex, *J. Exp. Psychol. HPP,* 18, 1121, 1992.
71. Zhang, J., Riehle, A., and Requin, J., Analyzing neuronal processing locus in stimulus-response association tasks, *J. Math. Psychol.,* 41, 219, 1997.
72. Riehle, A., Kornblum, S., and Requin, J., Neuronal correlates of sensorimotor association in stimulus-response compatibility, *J. Exp. Psychol. HPP,* 23, 1708, 1997.
73. Zhang, J., Riehle, A., Kornblum, S., and Requin, J., Dynamics of single neuron activity in monkey primary motor cortex related to sensorimotor transformation, *J. Neurosci.,* 17, 2227, 1997.
74. Salinas, E. and Romo, R., Conversion of sensory signals into motor commands in primary motor cortex, *J. Neurosci.,* 18, 499, 1998.
75. von der Malsburg, C., *The Correlation Theory of Brain Function,* Internal Report 81-2: Abteilung Neurobiologie, MPI für Biophysikalische Chemie, Göttingen, 1981.
76. Abeles, M., *Local Cortical Circuits: An Electrophysiological Study,* Springer, Berlin, 1982.
77. Abeles, M., *Corticonics: Neural Circuits of the Cerebral Cortex,* Cambridge University Press, Cambridge, MA, 1991.
78. Gerstein, G.L., Bedenbaugh, P., and Aertsen, A.M.H.J., Neural assemblies, *IEEE Trans. Biomed. Eng.,* 36, 4, 1989.
79. Palm, G., Cell assemblies as a guideline for brain research, *Concepts Neurosci.,* 1, 133, 1990.
80. Singer, W., Neural synchrony: a versatile code for the definition of relations, *Neuron,* 24, 49, 1999.
81. Hebb, D.O., *The Organization of Behavior,* John Wiley & Sons, New York, 1949.

82. Abeles, M. Role of cortical neuron: integrator or coincidence detector? *Israel J. Med. Sci.*, 18, 83, 1982.
83. Softky, W.R. and Koch, C., The highly irregular firing of cortical cells is inconsistent with temporal integration of random EPSPs, *J. Neurosci.*, 13, 334, 1993.
84. Grün, S., Diesmann, M., Grammont, F., Riehle, A., and Aertsen, A., Detecting unitary events without discretization of time, *J. Neurosci. Meth.*, 94, 67, 1999.
85. Palm, G., Aertsen, A.M.H.J., and Gerstein, G.L., On the significance of correlations among neuronal spike trains, *Biol. Cybern.*, 59, 1, 1988.
86. Grün, S., Diesmann, M., and Aertsen, A., 'Unitary Events' in multiple single-neuron activity. I. Detection and significance, *Neural Comp.*, 14, 43, 2002.
87. Grün, S., Diesmann, M., and Aertsen, A., 'Unitary Events' in multiple single-neuron activity. II. Non-stationary data, *Neural Comp.*, 14, 81, 2002.
88. Grammont, F. and Riehle, A., Precise spike synchronization in monkey motor cortex involved in preparation for movement, *Exp. Brain Res.*, 128, 118, 1999.
89. Riehle, A., Grammont, F., Diesmann, M., and Grün, S., Dynamical changes and temporal precision of synchronized spiking activity in monkey motor cortex during movement preparation, *J. Physiol. Paris,* 94, 569, 2000.
90. Roux, S., *La dynamique de l'interaction neuronale: sa modulation en fonction du temps de réaction*, DEA de Neurosciences, Université de la Méditerranée, Marseille, France, 2001.
91. Melssen, W.J. and Epping, W.J.M., Detection and estimation of neural connectivity based on crosscorrelation analysis, *Biol. Cybern.*, 57, 403, 1987.
92. Gochin, P.M., Kaltenbach, J.A., and Gerstein, G.L., Coordinated activity of neuron pairs in anesthetized rat dorsal cochlear nucleus, *Brain Res.*, 497, 1, 1989.
93. Requin, J., Riehle, A., and Seal, J., Neuronal activity and information processing in motor control: from stages to continuous flow, *Biol. Psychol.*, 26, 179, 1988.
94. Shen, L. and Alexander, G.E., Neural correlates of a spatial sensory-to-motor transformation in primary motor cortex, *J. Neurophysiol.*, 77, 1171, 1997.
95. Grammont, F. and Riehle, A., Spike synchronization and firing rate in a population of motor cortical neurons in relation to movement direction and reaction time, *Biol. Cybern.*, 88, 360, 2003.

9 Is the Motor Cortex Only an Executive Area? Its Role in Motor Cognition

Marc Jeannerod

CONTENTS

9.1 INTRODUCTION

The classical view of the primary motor cortex (M1) holds that it is an area devoted to transferring to motor execution messages that have been elaborated upstream in the cerebral cortex. Anatomically, M1 is the site of the convergence of inputs from the premotor cortex and basal ganglia; it is also the main site of the origin of the pyramidal tract and of direct cortico-motoneuronal connections. Early functional studies using direct cortical stimulation had concluded that the role of the motor cortex is limited to selecting the proper muscular addresses and encoding muscular force for executing a movement. To quote Penfield's Ferrier Lecture,[1] "Movement produced by stimulation of the motor cortex takes place most often in those members of the body which are capable of dextrous and complicated voluntary activity, and yet the movements thus produced are never dextrous nor purposeful." And, in addition, "The conscious patient is never deceived into believing that he made the movement himself. He knows he did not plan it" (p. 344). This finding, which is

0-8493-1287-6/05/$0.00+$1.50

confirmed by the everyday practice of transcranial magnetic stimulation (TMS), contrasts with the effects of stimulation of other more rostral motor areas. Stimulation of lateral area 6 by Penfield and Boldrey,[2] although it elicited no overt movements, frequently elicited an intense "desire to move."

More recently, however, experimental data have pointed to the fact that the relation of motor cortex activity to the production of movements is not as simple as it was thought to be on the basis of early stimulation experiments. This revision of motor cortical function originated from two main lines of research, dealing first with the plasticity of the somatotopic organization of M1, and second with its involvement in cognitive functions.

9.2 RECENT FINDINGS REGARDING THE PRIMARY MOTOR CORTEX

The somatotopic organization of M1 is unstable. It can be radically altered in a number of situations, such as peripheral changes in neuromuscular connections or motor learning and training. It has been known for more than 10 years that motor cortical somatotopy in animals is subject to a vast amount of reorganization following amputation of a limb or peripheral nerve lesion.[3] In man, as in rats, the cortical territory controlling the amputated joints tends to shrink, whereas the territory controlling remnant adjacent joints tends to expand.[4] For example, following amputation of a hand, the territory of the fingers will be invaded by more proximal joints of the same limb (e.g., elbow and shoulder), or even by the face. It was suspected, but not proven until recently, that this plastic phenomenon is reversible. One case of hetero-transplantation of the two hands several years after bilateral amputation at the level of the mid-forearms was studied by Giraux et al.,[5] using fMRI for mapping the activation of the motor cortex. Before surgery, the areas corresponding to the two hands were mapped by asking the subject to "extend" or "flex" his (missing) fingers. Execution of these movements was controlled by palpating finger extensors and flexors at the level of the stump. Six months after surgery, the hand areas were mapped during movements of the grafted hands. The comparison of activation before and after surgery revealed that the hand areas on either side, which were initially reduced to the most lateral part of the normal hand area close to the face area, reexpanded medially to reoccupy its full territory. Similarly, the elbow area, which had invaded a large section of the hand area, was pushed back medially to its normal anatomical location.

Motor cortical reorganization following amputation is associated with subjective sensory phenomena, like phantom pain.[6,7] Partial restoration of the normal topography by training reduces phantom pain. Lotze et al.[8] showed that the extensive use of a myoelectric prosthetic device by the amputee, because it prevents cortical reorganization, has a positive effect on phantom pain. Giraux and Sirigu[9] also showed that reexpansion of the hand area by training resulted in a decrease in phantom pain. The visuomotor training method used by Giraux and Sirigu consisted in transferring, by way of mirrors, the image of the normally moving limb at the location of the paralyzed limb. After a few sessions, the patient imagined his paralyzed limb moving, and this resulted in reexpanding the atrophied corresponding primary motor cortex. The influence of imagined movements on M1 activity will be discussed again in another section.

Plastic modification of M1 somatotopy is not limited to peripheral changes such as amputation. It is also observed during motor training. Neuroimaging studies show that long-term training of finger movements produces not only an increase in the amplitude of the activation of the trained pixels, but also an enlargement of the finger cortical area.[10,11] Although this effect of training can be partly explained by peripheral factors (e.g., the increase in reafferent input from the moving limb during repetitive movements; see Reference 12), it is also clearly influenced by central factors. Indeed, as will be reviewed below, the same effect of training can be observed in the absence of overt movements from the trained limb. Thus, M1 is liable to long-term changes in its intrinsic arrangement and connectivity, an experimental fact that would not be expected from a system devoted to transmitting executive commands, and that opens new possibilities for reinterpreting the role of M1 in motor functions.

The second set of data that leads to a reconsideration of motor cortex function arises from experiments showing the role of the motor cortex in cognitive activities that are related to motor function, but where no movement occurs. Georgopoulos and his colleagues first demonstrated in the monkey the existence of orderly changes in activity of M1 neurons during a cognitive operation. In this experiment, a monkey was instructed to perform an arm movement directed to a virtual target different from the one shown to her. During this process of target selection preceding execution of the movement, the activity of the neuronal population coding for the direction of the movement (the population vector) progressively changed from the direction of the target shown to the monkey to the direction of the virtual target, suggesting that the animal was performing a mental rotation of the population vector until it matched the instructed direction.[13]

According to Georgopoulos,[14] the primary motor cortex could be considered a cognitive area — i.e., an area involved in cognitive motor processes — rather than simply as an area devoted to motor execution and control of the spinal level. The core of this chapter will be devoted to examining the role of the motor cortex in cognitive operations that remain covert — i.e., where the action itself is not executed. In this category are included a variety of mental states related to action which fit the broad concept of motor representations, such as motor imagery or action observation. The key question will be this: if M1 is an executive area and if no overt movement occurs in these situations, why should M1 be involved? The tentative answer that will be given to this question is that M1 activation during cognitive motor operations is part of a neural process of simulation that serves as the basis of action representations.

9.3 THE FUNCTIONAL ANATOMY OF MOTOR REPRESENTATIONS

Neuroimaging techniques have greatly contributed to the functional anatomy of those purely mental states devoted to the representation of action. Following the pioneer papers showing changes in regional cerebral blood flow during mental imagery,[15,16] the pattern of cortical activity during both motor imagery and action observation has been extensively investigated.

Many studies using functional brain imaging by magnetic resonance (fMRI) reported activation of the sensorimotor cortex during motor imagery.[17–22] Typically, M1 activation is not consistently found in every subject and, when present, is less intense than during motor execution of the same movement. The activated zone overlaps the zone activated during execution, with the same voxels involved in the two conditions.[21] The involvement of M1 during motor imagery can also be detected with the magnetoencephalographic (MEG) technique: in this case, the activity of the motor cortex is inferred from a specific change in cortical activity (suppression of the 20-Hz rebound induced by a peripheral nerve stimulation), which is observed in the precentral gyrus during manipulative finger movements, during motor imagery of the same movements,[23] and also during observation of an actor moving his fingers.[24] These MEG findings represent a direct demonstration of the existence of a cortical system for matching execution, imagination, and observation of the same movements. (See also Reference 25.)

Experiments reporting activation of M1 during action representation (like those reporting increased corticospinal excitability; see below) fall under the critique that the observed changes might simply reveal incomplete inhibition of action or incomplete relaxation of motor activity during the mental process of imagination. In fact, this critique can be ruled out by other findings showing primary motor cortical activation during imagined movements in subjects with an amputated or a paralyzed limb. Ersland et al.[26] found an increased metabolic activity in the contralateral primary motor cortex in a subject performing imaginary finger tapping with his amputated right hand. Lotze et al.[7] found in amputees with impressions of phantom limbs that imagination of moving the phantom hand produced an activation of the contralateral motor and somatosensory cortex that was higher than that produced by imagination of hand movements in control subjects. Brugger et al.[27] report the striking observation of a woman born without limbs, who presented strong sensations of phantom limbs. Although these sensations did not elicit activation of sensorimotor cortex (only premotor and parietal cortices were activated bilaterally), they were increased by stimulation of the sensorimotor cortex by TMS.

These observations raise a key question. Do motor images in amputees represent actions to the same extent that they do in normals, or are they real actions (i.e., if the muscles were still there or were still connected, would they contract)? This is an almost impossible question to answer. One could argue that the difference between the two situations is that motor images involve an inhibitory process, whereas blocked movements do not. Thus, in principle, one should be able to detect (by inspecting the activation pattern during a motor imagery task) whether the inhibitory process is present or not: if yes, this would mean that the subjects are generating a genuine motor image; if not, they are simply attempting to perform a movement in spite of the absence of the effector. In the latter case, however, one wonders why there should be an inhibitory process if there is no need to block the output.

The description of the (motor) brain activity during action representation strongly suggests that the same areas are involved during different types of representations.[28] Consciously evoking a motor image, making an evaluation of the feasibility of a movement, or observing an action performed by somebody else — to name only those factors — results in closely similar activation patterns. In addition, this same activation pattern can also be recorded during execution of the corresponding action. In other words, the neural correlates of representing an action are shared by different types of representations. A dramatic illustration of this concept of shared representation is offered by the finding of mirror neurons.[29] Mirror neurons were identified in the monkey premotor cortex. They are activated in two conditions: first, they fire when the animal is involved in a specific motor action, like picking a piece of food with a precision grip; second, they fire when the immobile animal watches the same action performed by an external agent (another monkey or an experimenter). In other words, mirror neurons represent one particular type of action, irrespective of the agent who performs it. At this point, it could be suspected that the signal produced by these neurons, and exploited by other elements downstream in the information processing flow, would be the same for an action performed by the self and by another agent: the two modalities of that action (executed and observed) would thus completely share the same neural representation. In fact, other premotor neurons (the canonical neurons), and presumably many other neuron populations as well, fire only when the monkey performs the action and not when it observes it from another agent. This is indeed another critical feature of the shared representations concept: they overlap only partially, and the part of a given representation that does not overlap can be the cue for attributing the action to the self or to another.

A mechanism similar to that of mirror neurons operates in humans. Brain activity during different conditions where subjects were self-representing actions (e.g., executing and imagining actions, inspecting tools, or observing actions performed by other people) was compared.[19,30–34] The outcome of these studies is twofold. First, there exists a cortical network common to all conditions. As shown in the preceding section, the motor cortex is part of this network, which also includes cortical areas located in the superior and inferior parietal lobules, the ventral premotor cortex, and the supplementary motor area (SMA). Second, motor representations for each individual condition are clearly specified by the activation of cortical zones, which do not overlap between conditions.[32,35,36]

9.4 CHANGES IN THE EXCITABILITY OF THE MOTOR PATHWAYS DURING MOTOR REPRESENTATIONS

Measuring the changes in the excitability of motor pathways during various forms of action representation can also provide further cues on the involved mechanisms. Indeed, it is a frequent finding that some degree of background electromyographic (EMG) activity persists in the muscular groups involved in the simulated action.[37–39] This finding suggests that during, for example, motor imagery, motor commands to muscles are only partially blocked, and that motoneurons are close to the firing threshold. Bonnet et al.[40] confirmed this point by measuring spinal reflexes during motor imagery tasks. They instructed subjects either to press isometrically on a pedal, or to mentally simulate the same action. Two levels of strength (weak and strong) were used. The H-reflexes in response to direct electrical stimulation of the popliteous nerve and the T-reflexes in response to a tap on the soleus tendon were measured. Both types of reflexes were increased during mental simulation and this increase correlated with the force of the simulated pressure. (See also Reference 39.)

The excitability of the corticospinal pathway was also extensively tested in several experiments using TMS. This method permits one to measure the amplitude of motor-evoked potentials (MEPs) produced in the muscles involved in mental simulation of an action, by the magnetically induced electrical stimulus applied to the corresponding area of the contralateral motor cortex. Authors consistently found a specific increase of MEPs in those muscles involved in an imagined task — e.g., in the flexor muscles during imagination of hand closure — whereas no such increase was found in the antagonist extensor muscles.[41–43] Not surprisingly, very similar results were obtained when MEPs were measured during the observation of actions. In this situation, where the subject is instructed simply to watch an actor, the MEP increase is also restricted to the muscle group involved in the observed action (e.g., the action of grasping an object[44,45]). Interestingly, the effects of observation are not limited to the visual domain. Fadiga et al.[46] were able to show that listening to specific phonemes increases the excitability of the motor pathway to the relevant tongue muscle. Finally, Baldissera et al.[47] found changes in upper limb H-reflexes during observation of finger flexion or extension.

These results add support to the view of an involvement of the motor system during different types of mental representation of actions. Indeed, in a recent study, Clark et al.[48] were able to compare MEP amplitude in the same subjects while explicitly imagining, observing, and physically executing the same hand gestures. They found that observation and imagery conditions led to a similar facilitation in MEP amplitude in the relevant hand muscle. In addition, during action observation, a condition of "active" observation (with the instruction to subsequently imitate) yielded larger MEPs than a purely passive observation. Although MEP facilitation was weaker during action representation than during physical execution of the same action, the finding clearly calls for a unitary mechanism based on action simulation. This point will be developed further in another section.

9.5 THE PROBLEM OF ACTION INHIBITION DURING THE REPRESENTATION OF ACTIONS

The comparison between the autonomic and the motor systems during action representation reveals that, whereas activation of the former leads to changes at the level of peripheral effectors, this is not the case for the latter, where the contraction of the involved muscles is inhibited. Considering the above body of data about the activity of the motor system during covert actions, there are two possible explanations for this absence of motor output. The first interpretation postulates that the transfer of the motor engrams elaborated within premotor or supramotor cortical areas (e.g., the dorsal and ventral premotor cortices, the parietal cortex) to M1 would be blocked by a central inhibitory mechanism. The prefrontal cortical areas, which are found to be active during motor imagery,[30] could represent a possible locus for this behavioral inhibition. An observation by Marshall et al.[49] of a patient with a hysterical paralysis of the left side of the body lends support to this possibility. Although in this patient a normal activation (mapped with PET) of the left sensorimotor cortex was observed during movements of the right "good" leg, no such activation was observed on the right side during unsuccessful attempts to move the left "bad" leg. Instead, the right anterior cingulate and orbitofrontal cortices were significantly activated. This result suggests that these prefrontal areas exerted a state-dependent inhibition on the motor system when the intention to move the left leg was formed. This point was specifically investigated by Brass et al.[50] in a neuroimaging experiment in normal subjects. Subjects were instructed to perform finger movements while they were observing another person executing either congruent or incongruent movements. When the observed movements were incongruent with respect to the instructed ones, the subjects had to inhibit their spontaneous tendency to imitate the movements of the other person. This task resulted in a strong activation of the dorsolateral and frontopolar areas of the prefrontal cortex. The hypothesis of a cortico-cortical "disconnection" is not compatible, however, with the simple fact that the motor cortex remains activated during action representation. A possible interpretation for the above data could be that the prefrontal cortex is involved, not in inhibiting the execution of represented actions, but rather in a process of selecting the appropriate representation. While executing an instructed action incompatible with an observed one, one has to select the endogenous representation and ignore the representation arising from the outside; in other words, one has to prevent oneself from being distracted by an external event.

In order to account for the empirical data showing the involvement of the motor cortex, we must conclude that the inhibitory mechanism is localized downstream of the motor cortex, possibly at the spinal cord or brainstem level. A tentative hypothesis would be that a dual mechanism operates at the spinal level. The subthreshold preparation to move, reflected by the increased corticospinal tract activity, would be paralleled by an inhibitory influence for suppressing the overt movement. The posterior cerebellum may play an important role in this inhibitory process.[22] Whereas during action execution, the activated cerebellar areas are located in anterior and lateral regions, those activated during imagery and action observation are located in the posterior cerebellum.[19] A similar explanation, but using a different site for inhibition, was put forward by Prut and Fetz[51] to explain motor inhibition in the monkey while the animal is waiting for execution of a learned action. They showed that, during the waiting period where the monkey is ready to move, spinal interneurons are activated, hence indicating that the spinal motor network is being primed by the descending cortico-motoneuronal input. Because the overt movement was suppressed during this period, Prut and Fetz hypothesized a superimposed global inhibition, possibly originating in the premotor cortex, and propagating to the spinal cord, parallel to the excitatory input. This hypothesis would account for both the increased motoneuron excitability and the block of muscular activity during action representation.

9.6 THE SIMULATION HYPOTHESIS AND ITS IMPLICATIONS

The simulation hypothesis is a rather broad framework accounting for the relationship of mental phenomena to the activity of the neural substrate. Its empirical basis accumulated from experiments in cognitive neuroscience in the past two decades. One of the most influential results showed that visual mental images rely on activation of the early stages of information processing of the visual system. The primary visual cortex (V1) is consistently involved in visual mental imagery,[52,53] with an additional selective involvement of the inferotemporal cortex during imagery of visual objects and of the occipitoparietal cortex in visual spatial imagery. The explanation put forward for an activation of low-level processing areas during a high-level cognitive activity is that activation of topographically organized areas, such as V1, is needed for replacing the image within a spatial frame of reference. Higher-order areas, because they lack topographical organization, would not be able, by themselves, to achieve this task. In other words, the processing of visual imagery would have to follow the same processing track as visual perception for giving an image its spatial layout, a process that requires the participation of V1.

This reasoning can be extended to the domain of action representations. The definition we gave at the beginning of this paper for represented actions is that they correspond to covert, quasi-executed actions, a definition that accounts for many of the properties of action representations that have been described here. Thus, by drawing a parallel with perceptual representations such as visual mental imagery, we come to the proposition that, if a represented action is a simulated action, then it should involve the mechanisms that normally participate in motor execution. In the above sections, we have seen a large amount of data that satisfy this proposition. Conversely, the content of motor images is explained by the involvement of neural structures such as M1, the premotor cortex, the basal ganglia, and the cerebellum, because this is where the aspects of action related to execution are normally processed. In other words, if the mental content of motor images is what it is, this is because the neural correlates include the structures required for execution. But this reasoning leads to another point, which can be set as a question: if motor images are not executed, why do they involve the activation of executive neural structures? The answer to this question can only be partial. The reason for this is that we do not know the precise function of all the activated neuron populations in these areas. Although some uncertainty remains, however, an answer can be given for those neurons that are connected to the spinal chord, the activation of which is responsible for the increased excitability at the motoneuron level that is observed during various types of action representation. During execution, these neurons are involved in selecting the proper muscular addresses for producing the action, for applying the required muscular force, for encoding the biomechanical limitations, for selecting the final endpoint postures, for optimizing the trajectory dynamics, and so on. Thus, by analogy with the activation of primary visual areas in visual mental imagery, which is thought to restore the topographical layout of the image, it could be proposed that M1 activation in motor representation is needed for providing the represented action with its motor format. In other words, in order for a represented action to be felt as a real action, it needs to be framed according to the constraints of a real action.

This proposition seems even easier to fulfill for action representations originating from observed actions. In that case, the simulation must provide a frame to the perception of the action: the observer understands the action he sees to the extent he can simulate it. In order for the perception to be veridical, the simulation must be complete and must involve all the aspects of the observed action. A mere observation, without activation of the motor system, would only provide a description of the visible aspects of the movements of the agent, but it would not give precise information about the intrinsic components of the observed action which are critical for understanding what the action is about, what its goal is, and how to reproduce it.

9.7 IMPLICATIONS OF THE SIMULATION THEORY

9.7.1 Mental Training

The strong relationships of motor imagery to the neural substrate lead to the logical expectation that the central changes produced in the motor system during imagery should affect subsequent motor performance. The sport psychology literature in the early 1960s offers a large number of studies reporting measurable effects of mental imagery on subsequent motor performance. (For review and meta-analysis, see References 54,55.) Mental training has been shown to affect several aspects of motor performance normally thought to be specific outcomes of physical training, such as the increase in strength of muscular contraction,[56] improvement in movement speed,[11] and reduction of variability and increase in temporal consistency of movements.[57]

Several explanations for these phenomena have been proposed in the literature. One possibility is that mental training could modify a new perceptual organization or a new insight of the action to be performed. Alternatively, minimal movements occurring during mental training would be sufficient to generate proprioceptive feedback and reactivation kinesthetic images stored from previous movements. Still another possibility would be that visual imagery is involved in providing an abstract representational system of spatial information. (For a review of these explanations, see Reference 12.) Finally, the most logical explanation would be that the motor activation observed during mental training rehearses the motor pathways and facilitates execution. This explanation is strongly supported by experimental data. Pascual-Leone et al.[11] used TMS to evaluate the changes in cortical excitability during mental training of finger movements. These authors found that the size of the excitable area devoted to the finger was increased as movements were repeated over training periods. The increase in the size of the excitable area produced by imaginal training was similar to that obtained during physical training. More recently, Lafleur et al.[58] showed that learning a motor task by using motor imagery induces a pattern of dynamic changes in cortical activation similar to that occurring during physical practice. In both conditions, a first phase is observed, with increase in activity in the premotor cortex and the cerebellum. Subsequently, this activation tends to disappear and to be replaced by activation in the basal ganglia and the prefrontal cortex (in the orbitofrontal and anterior cingular regions).

The simulation interpretation of the effects of mental training is confirmed by recent experimental evidence showing that subjects can learn to voluntarily increase the degree of activation of their motor cortex during an imagined manual action.[59] Subjects first need to receive an ongoing information about the level of activation during the training period. This information can be provided by monitoring a continuously updated fMRI signal taken from the cortical motor area. Subsequently, subjects become able to increase the level of activation in their motor cortex without recourse to the feedback signal. According to deCharms et al. this procedure yields a level of activity in the somatomotor cortex that is similar to, or higher than, the level of activity during actual manual action. Obviously, this type of result, showing the possibility of increasing activity in a specific brain area at will, opens a number of potential applications for designing new training techniques, not only in the domain of action, but also in the realm of the control of behavior in general. Rehabilitation procedures for motor impairments should greatly benefit from this possibility.

9.7.2 Coupling Motor Representations with Neuro-Prosthetic Devices

Evidence has been provided that represented actions are simulated real actions. More specifically, represented actions involve the orderly activation of the same neural structures that would be involved if the action were actually executed (the definition of action simulation). Following this line of thought, it seems logical that, if the neural in the motor areas of a subject imagining a movement could be properly monitored and connected to effectors, the imagined movement would become visible. This conjecture is the basis for building hybrid brain–machine interfaces that could be used to control artificial devices, with the ambition to restore motor function in patients with severe motor disabilities or paralysis.[60,61] Recent work has shown that a monkey can be trained to move a spot on a computer screen just by "thinking" the displacement of the spot. In the experiment of Serruya et al.,[62] the monkey was implanted with an electrode for recording the activity of a small neuronal population in area M1 controlling the animal's arm movement. The monkey first used a manipulandum for displacing a spot on the computer screen. Then the connection between the manipulandum and the computer was replaced by a connection between the output of the microelectrode and the spot: the monkey immediately used the neural-activity-based signal to carry out the task without any further training. During this time, the animal made intermittent arm movements, or no arm movements. The importance of this finding is twofold. Not only does it show that nonhuman primates, like humans, can generate motor representations that have properties similar to real actions; it is also of a high potential value for designing rehabilitation procedures. As already mentioned, the ability of human subjects to learn to increase their cortical activity at will indicates that it will become possible, for example, to train patients to learn to activate neuro-prosthetic devices and to recover in this way part of their motor autonomy.

9.7.3 Learning by Observation and Imitation

There is a wealth of data in the literature showing that observing a movement facilitates the execution of that movement by the observer. Brass et al.,[63] for example, noticed that the execution of instructed finger movements is influenced by the observation of another person performing compatible or incompatible finger movements. If the movements performed by the observer are compatible with those performed by the other person, they are clearly facilitated (e.g., their reaction time is shorter). Conversely, incompatibility of the movements of the observer with those of the other person yields to degraded performance. (See also Reference 64.) Craighero et al.[65] also found that observing pictures showing hand postures facilitates the execution of a grasping hand movement when the observed posture is congruent with the executed movement. These data stress the point that observing the performance of another person facilitates the formation of effector-specific motor representations. According to Sebanz et al.,[66] simply observing the action at the disposal of another person creates in the observer a representation of that action: this representation will facilitate the execution of the action by the observer. As for the action representation created during motor imagery, those that result from action observation seem to include information related to action execution. This is suggested by a series of experiments by Knoblich and his coworkers. Knoblich and Flach[67] presented subjects with videos of an action (throwing darts) that these subjects had previously performed and videos of the same action performed by other subjects. The subject's task consisted of predicting the accuracy of the observed actions. Prediction was more accurate when subjects observed their own actions than when they observed another person's actions. Thus, the observation of self-generated actions is more informative, because the mechanism that simulates the observed action is the same as that which produced it. These data represent a basis for the widespread phenomenon of learning by observation. Heyes and Foster,[68] for example, found that subjects simply watching an experimenter performing a sequence of movements can learn the sequence as efficiently as when practicing the task themselves.

These findings may have interesting implications in the domain of pathology. In patients with limb amputation or denervation, the activity of the corresponding part of the motor cortex tends to decrease over time. At the same time, the patients may experience subjective phenomena such as phantom limb or phantom pain. As already mentioned in the section on cortical plasticity, Giraux and Sirigu,[9] in two such patients with a unilateral brachial plexus avulsion, showed that a normal activity could be restored in their motor cortices by means of observational training. These patients were trained in a situation where the image of the valid hand was visually transposed (by way of mirrors; see Reference 69) at the location of the impaired hand. They were instructed to move the valid hand that they could see at the place of the impaired arm. Following a series of 24 daily visuomotor training sessions, the activity of the M1 area contralateral to the impaired (paralyzed) arm, monitored by fMRI, was greatly increased. This result shows that observation of a normally moving limb can activate the motor area controlling the homologous limb in the observer. A possible explanation for this phenomenon is that mirror neurons similar to those found in the monkey premotor cortex are activated by observation of the moving hand and trigger the activity of M1 neurons innervating the impaired hand. In these two patients, this visuomotor training procedure had a beneficial effect, not in improving their hand movements, which remained impaired due to the plexus brachial lesion, but in decreasing their phantom limb pain. This effect suggests that the motor cortex, in addition to its role in controlling movements, may also influence the processing of sensory input arising from the effector it controls.

The ability of human subjects to imitate each other derives from the possibility of forming action representations from the observation of other people. Neuroimaging studies of imitation in humans tend to favor what Iacoboni et al.[70] call the direct matching hypothesis. According to this hypothesis, brain areas that become active during execution of a movement should become more active when that movement is elicited by the observation of an identical movement made by another person. Areas in the left inferior frontal cortex and in the parietal cortex seem to fulfill this requirement. This result, showing that brain areas for execution and observation of a movement are matched, is compatible with the simulation theory. Furthermore, it raises again the point (already raised in the section about shared representations) of how two different representations of the same action are disentangled from one another. Other studies dealing with imitation[71] suggest that areas might be activated differently whether an action is internally produced or generated by another agent.

9.8 CONCLUSION: THE ROLE OF SIMULATION IN MOTOR COGNITION

The above pattern of results on the mechanisms of covert action corresponds to the central stages of action organization, uncontaminated by the effects of execution. As such, it represents a possible framework for motor cognition.

First, because all aspects of action appear to be involved during action representation, it seems a logical consequence of this rehearsal of the corresponding brain structures, and specifically the motor structures, that the subsequent execution will be facilitated. The presence of activity in the motor system during covert action puts the action representation in a true *motor* format, so that it can be regarded by the motor system as a real action. This facilitation explains various forms of training (e.g., mental training) and learning (e.g., observational learning) which occur as a consequence of self-representing an action. In addition, imitation seems to be based on directly matching the observed action onto an internal simulation of that action.

Second, activation of the motor cortex and of the descending motor pathway seems to fulfill several critical functions. First, this activation contributes to generating corollary signals that propagate upstream to the parietal and premotor cortices. This mechanism would allow for evaluating the potential consequences of the future action. (See Reference 72 for description of a control model that accounts for this function.) It could also provide the subject with information for consciously monitoring his S-states and realizing that he is the agent of this covert activity, in spite of the absence of overt behavior.

REFERENCES

1. Penfield, W., Ferrier Lecture. Some observations on the cerebral cortex of man, *Proc. R. Soc.*, B134, 329, 1947.
2. Penfield, W. and Boldrey, E., Somatic motor and sensory representation in the cerebral cortex of man as studied by electrical stimulation, *Brain*, 60, 389, 1937
3. Donoghue, J.P. and Sanes, J.N., Peripheral nerve injury in developing rats reorganizes representation pattern in motor cortex, *Proc. Nat. Acad. Sci. U.S.A.*, 84, 1123, 1987.
4. Pons, T.P., Garraghty, P.E., Ommaya, A.K., Kass, J.H., Taub, E., and Mishkin, M., Massive cortical reorganization after sensory deafferentation in adult macaques, *Science*, 252, 1857, 1991.
5. Giraux, P., Sirigu, A., Schneider, F., and Dubernard, J.M., Cortical reorganization in motor cortex after graft of both hands, *Nat. Neurosci.*, 4, 691, 2001
6. Flohr, H., Elbert, T., Knecht, S., Wienbruch, C., Pantev, C., Birbaumer, N., Larbig, W., and Taub, E., Phantom limb pain as a perceptual correlate of cortical reorganization following arm amputation, *Nature*, 375, 482, 1995.
7. Lotze, M., Flor, H., Grodd, W., Larbig, W., and Birbaumer, N., Phantom movement and pain. An fMRI study in upper limb amputees, *Brain*, 124, 2268, 2001.
8. Lotze, M., Grodd, W., Birbaumer, N., Erb, M., Huse, E., and Flor, H., Does use of a myoelectric prosthesis prevent cortical reorganization and phantom limb pain? *Nat. Neurosci.*, 2, 501, 1999.
9. Giraux, P. and Sirigu, A., Illusory movements of the paralyzed limb restore motor cortex activity, *NeuroImage*, 20, S107, 2003.
10. Karni, A., Myer, G., Jezzard, P., Adams, M.M., Turner, R., and Ungerleider, L.G., Functional MRI evidence for adult motor cortex plasticity during motor skill learning, *Nature*, 377, 155, 1995.
11. Pascual-Leone, A., Dang, N., Cohen, L.G., Brasil-Neto, J., Cammarota, A., and Hallett, M., Modulation of motor responses evoked by transcranial magnetic stimulation during the acquisition of new fine motor skills, *J. Neurophysiol.*, 74, 1037, 1995.
12. Johnson, P., The functional equivalence of imagery and movement, *Q. J. Exp. Psychol.*, 34A, 349, 1982.
13. Georgopoulos, A.P., Lurito, J.T., Petrides, M., Schwartz, A.B., and Massey, J.T., Mental rotation of the neuronal population vector, *Science*, 243, 234, 1989.
14. Georgopoulos, A.P., Neural aspects of cognitive motor control, *Curr. Opin. Neurobiol.*, 10, 238, 2000.
15. Ingvar, D. and Philipsson, L., Distribution of the cerebral blood flow in the dominant hemisphere during motor ideation and motor performance, *Annu. Neurol.*, 2, 230, 1977.
16. Roland, P.E., Skinhoj, E., Lassen, N.A., and Larsen, B., Different cortical areas in man in organization of voluntary movements in extrapersonal space, *J. Neurophysiol.*, 43, 137, 1980.
17. Leonardo, M., Fieldman, J., Sadato, N., Campbell, G., Ibanez, V., Cohen, L., Deiber, M-P., Jezzard, P., Pons, T., Turner, R., Le Bihan, D., and Hallett, M., A functional magnetic resonance imaging study of cortical regions associated with motor task execution and motor ideation in humans, *Hum. Brain Mapp.*, 3, 83, 1995.
18. Kim, S-G., Jennings, J.E., Strupp, J.P., Andersen, P., and Ugurbil, K., Functional MRI of human motor cortices during overt and imagined finger movements, *Int. J. Imaging Syst. Technol.*, 6, 271, 1995.

19. Grafton, S.T., Arbib, M.A., Fadiga, L., and Rizzolatti, G., Localization of grasp representations in humans by positron emission tomography. 2. Observation compared with imagination, *Exp. Brain Res.,* 112, 103, 1996.
20. Roth, M., Decety, J., Raybaudi, M., Massarelli, R., Delon-Martin, C., Segebarth, C., Morand, S., Gemignani, A., Décorps, M., and Jeannerod, M., Possible involvement of primary motor cortex in mentally simulated movement: a functional magnetic resonance imaging study, *NeuroReport,* 7, 1280, 1996.
21. Porro, C.A., Francescato, M.P., Cettolo, V., Diamond, M.E., Baraldi, P., Zuiani, C., Bazzochi, M., and di Prampero, P.E., Primary motor and sensory cortex activation during motor performance and motor imagery: a functional magnetic resonance study, *J. Neurosci.,* 16, 7688, 1996.
22. Lotze, M., Montoya, P., Erb, M., Hülsmann, E., Flor, H., Klose, U., Birbaumer, N., and Grodd, W., Activation of cortical and cerebellar motor areas during executed and imagined hand movements: an fMRI study, *J. Cogn. Neurosci.,* 11, 491, 1999.
23. Schnitzler, A., Selenius, S., Salmelin, R., Jousmäki, V., and Hari, R., Involvement of primary motor cortex in motor imagery: a neuromagnetic study, *NeuroImage,* 6, 201, 1997.
24. Hari, R., Forss, N., Avikainen, S., Kirveskari, E., Salenius, S., and Rizzolatti, G., Activation of human primary motor cortex during action observation: a neuromagnetic study, *Proc. Nat. Acad. Sci. U.S.A.,* 95, 15061, 1998.
25. Cochin, S., Barthelemy, C., Roux, S., and Martineau, J., Observation and execution of movement: Similarities demonstrated by quantified electroencephalography, *Eur. J. Neurosci.,* 11, 1839, 1999.
26. Ersland, L., Rosen, G., Lundervold, A., Smievoll, A.I., Tillung, T. Sundberg, H., and Hugdahl, K., Phantom limb imaginary fingertapping causes primary motor cortex activation: an fMRI study, *NeuroReport,* 8, 207, 1996.
27. Brugger, P., Kolias, S.S., Müri, R.M., Crelier, G., Hepp-Reymond, M.C., and Regard, M. Beyond re-membering: phantom sensations of congenitally absent limbs, *Proc. Nat. Acad. Sci. U.S.A.,* 97, 6167, 2000.
28. Jeannerod, M., The representing brain. Neural correlates of motor intention and imagery, *Behav. Brain Sci.,* 17, 187, 1994.
29. Rizzolatti, G., Fadiga, L., Gallese, V., and Fogassi, L., Premotor cortex and the recognition of motor actions, *Cogn. Brain Res.,* 3, 131, 1995.
30. Decety, J., Perani, D., Jeannerod, M., Bettinardi, V., Tadary, B., Woods, R., Mazziotta, J.C., and Fazio, F., Mapping motor representations with PET, *Nature,* 371, 600, 1994.
31. Rizzolatti, G., Fadiga, L., Matelli, M., Bettinardi, V., Paulesu, E., Perani, D., and Fazio, G. Localization of grasp representations in humans by PET. 1. Observation versus execution, *Exp. Brain Res.,* 111, 246, 1996.
32. Gérardin, E., Sirigu, A., Lehéricy, S., Poline, J.-B., Gaymard, B., Marsault, C., Agid, Y., and Le Bihan, D., Partially overlapping neural networks for real and imagined hand movements, *Cereb. Cortex,* 10, 1093, 2000.
33. Chao, L.L. and Martin, A., Representation of manipulable man-made objects in the dorsal stream, *NeuroImage,* 12, 478, 2000.
34. Buccino, G., Binkofski, F., Fink, G.R., Fadiga, L., Fogassi, L. Gallese, V., Seitz, R.J., Zilles, K., Rizzolatti, G., and Freund, H.J., Action observation activates premotor and parietal areas in a somatotopic manner: an fMRI study, *Eur. J. Neurosci.,* 13, 400, 2001.
35. Grèzes, J. and Decety, J., Functional anatomy of execution, mental simulation, observation and verg generation of actions: a meta-analysis, *Hum. Brain Mapp.,* 12, 1, 2001.
36. Jeannerod, M., Neural simulation of action: a unifying mechanism for motor cognition, *NeuroImage,* 14, S103, 2001.

37. Jacobson, E., Electrical measurements of neuro-muscular states during mental activities. III. Visual imagination and recollection, *Am. J. Physiol.,* 95, 694, 1930.
38. Wehner, T., Vogt, S., and Stadler, M., Task-specific EMG characteristics during mental training, *Psychol. Res.,* 46, 389, 1984.
39. Gandevia, S.C., Wilson, L.R., Inglis, J.T., and Burke, D., Mental rehearsal of motor tasks recruits alpha motoneurons but fails to recruit human fusimotor neurones selectively, *J. Physiol.,* 505, 259, 1997.
40. Bonnet, M., Decety, J., Requin, J., and Jeannerod, M., Mental simulation of an action modulates the excitability of spinal reflex pathways in man, *Cogn. Brain Res.,* 5, 221, 1997.
41. Hashimoto, R. and Rothwell, J.C., Dynamic changes in corticospinal excitability during motor imagery, *Exp. Brain Res.,* 125, 75, 1999.
42. Rossini, P.M., Rossi, S., Pasqualetti, P., and Tecchio, F., Corticospinal excitability modulation to hand muscles during movement imagery, *Cereb. Cortex,* 9, 161, 1999.
43. Fadiga, L., Buccino, G., Craighero, L., Fogassi, L., Gallese, V., and Pavesi, G., Corticospinal excitability is specifically modulated by motor imagery: a magnetic stimulation study, *Neuropsychologia,* 37, 147, 1999.
44. Fadiga, L., Fogassi, L., Pavesi, G., and Rizzolatti, G., Motor facilitation during action observation. A magnetic stimulation study, *J. Neurophysiol.,* 73, 2608, 1995.
45. Strafella, A.P. and Paus, T., Modulation of cortical excitability during action observation: a transcranial magnetic stimulation study, *NeuroReport,* 11, 2289, 2000.
46. Fadiga, L., Craighero, L., Buccino, G., and Rizzolatti, G., Speech listening specifically modulates the excitability of tongue muscles, *Eur. J. Neurosci.,* 15, 399, 2002.
47. Baldissera, F., Cavallari, P., Craighero, L., and Fadiga, L., Modulation of spinal excitability during observation of hand action in humans, *Eur. J. Neurosci.,* 13, 190, 2001.
48. Clark, S., Tremblay, F., and Ste-Marie, D., Differential modulation of corticospinal excitability during observation, mental imagery and imitation of hand actions, *Neuropsychologia,* 42, 105, 2004.
49. Marshall, J.C., Halligan, P.W., Fink, G.R., Wade, D.T., and Frackowiak, R.S.J., The functional anatomy of a hysterical paralysis, *Cognition,* 64, B1, 1997.
50. Brass, M., Zysset, S., and von Cramon, Y., The inhibition of imitative response tendencies, *NeuroImage,* 14, 1416, 2001.
51. Prut, Y. and Fetz, E.E., Primate spinal interneurons show premovement instructed delay activity, *Nature,* 401, 590, 1999.
52. Kosslyn, S.M., Alpert, N.M., Thompson, W.L., Maljkovic, V., Weise, S.B., Chabris, C.F., Hamilton, S.E., Rauch, S.L., and Buonanno, F.S., Visual mental imagery activates topographically organized visual cortex: PET investigations, *J. Cogn. Neurosci.,* 5, 263, 1993.
53. Le Bihan, D., Turner, R., Zeffiro, T.A., Cuénod, C.A., Jezzard, P., and Bonnerot, V., Activation of human primary visual cortex during visual recall: a magnetic resonance imaging study, *Proc. Nat. Acad. Sci. U.S.A.,* 90, 11802, 1993.
54. Feltz, D.L. and Landers, D.M., The effects of mental practice on motor skill learning and performance: a meta-analysis, *J. Sport Psychol.,* 5, 25, 1983.
55. Driskell, J.E., Cooper, C., and Moran, A., Does mental practice enhance performance? *J. Appl. Psychol.,* 79, 481, 1994.
56. Yue, G. and Cole, K.J., Strength increases from the motor program: comparison of training with maximal voluntary and imagined muscle contractions, *J. Neurophysiol.,* 67, 1114, 1992.
57. Vogt, S., On relations between perceiving, imagining and performing in the learning of cyclical movement sequences, *Brit. J. Psychol.,* 86, 191, 1995.

58. Lafleur, M.F., Jackson, P.L., Malouin, F., Richards, C.L., Evans, A.C., and Doyon, J., Motor learning produces parallel dynamic functional changes during the execution and imagination of sequential foot movements, *NeuroImage*, 16, 142, 2002.
59. DeCharms, R.C., Christoff, K., Glover, G.H., Pauly, J.M., Whitfield, S. and Gabrieli, J.D.E., Learned regulation of spatially localized brain activation during real-time fMRI, *NeuroImage,* 21, 436, 2004.
60. Nicolelis, M.A.L., Actions from thoughts, *Nature*, 409, 403, 2001.
61. Cincotti, F., Mattia, D., Babiloni, C., Carducci, F., Salinari, S., Bianchi, L., Marciani, M.G., and Babiloni, F., The use of EEG modifications due to motor imagery for brain–computer interfaces, *IEEE Trans. Neur. Syst. Rehab. Eng.*, 131, 2003.
62. Serruya, M.D., Hatsopoulos, N.G., Paninski, L., Fellows, M.R., and Donoghue, J.P., Instant neural control of a movement signal, *Nature*, 416, 141, 2002.
63. Brass, M., Bekkering, H., and Prinz, W., Movement observation affects movement execution in a simple response task, *Acta Psychol.*, 106, 3, 2001.
64. Kilner, J.M., Paulignan, Y., and Blakemore, S.J., An interference effect of observed biological movement on action, *Current Biol.,* 13, 522, 2003.
65. Craighero, L., Bello, A., Fadiga, L., and Rizzolatti, G., Hand action preparation influences the responses to hand pictures, *Neuropsychologia*, 40, 492, 2002.
66. Sebanz, N., Knoblich, G., and Prinz, W., Representing other's actions: just like one's own? *Cognition,* 88, B11, 2003.
67. Knoblich, G. and Fach, R., Predicting action effects: interactions between perception and action, *Psychol. Sci.*, 12, 467, 2001.
68. Heyes, C.M. and Foster, C.L., Motor learning by observation: evidence from a serial reaction time task, *Q. J. Exp. Psychol.*, 55A, 593, 2002.
69. Ramachandran, V.S. and Rogers-Ramachandran, D., Synaesthesia in phantom limbs induced with mirrors, *Proc. R. Soc. Lond.,* B 263, 377, 1996.
70. Iacoboni, M., Woods, R.P., Brass, M., Bekkering, H., Mazziotta, J.C., and Rizzolatti, G., Cortical mechanisms of human imitation, *Science,* 286, 2526, 1999.
71. Decety, J., Chaminade, T., Grèzes, J., and Meltzoff, A.N., A PET exploration of the neural mechanisms involved in reciprocal imitation, *NeuroImage*, 15, 265, 2002.
72. Wolpert, D.M., Ghahramani, Z., and Jordan, M.I., An internal model for sensorimotor integration, *Science,* 269, 1880, 1995.

Section III

Motor Learning and Performance

10 The Arbitrary Mapping of Sensory Inputs to Voluntary and Involuntary Movement: Learning-Dependent Activity in the Motor Cortex and Other Telencephalic Networks

Peter J. Brasted and Steven P. Wise

CONTENTS

0-8493-1287-6/05/$0.00+$1.50

ABSTRACT

Studies on the role of the motor cortex in voluntary movement usually focus on *standard sensorimotor mapping*, in which movements are directed toward sensory cues. Sensorimotor behavior can, however, show much greater flexibility. Some variants rely on an algorithmic transformation between a cue's location and that of a movement target. The well-known "antisaccade" task and its analogues in reaching serve as special cases of such *transformational mapping*, one form of *nonstandard mapping*. Other forms of nonstandard mapping differ from both of the above: they are arbitrary. In *arbitrary sensorimotor mapping*, the cue's location has no systematic spatial relationship with the response. Here we explore several types of arbitrary mapping, with emphasis on the neural basis of learning these behaviors.

10.1 INTRODUCTION

Many responses to sensory stimuli involve reaching toward or looking at them. Shifting one's gaze to a red traffic light and reaching for a car's brake pedal exemplify this kind of sensorimotor integration, sometimes termed *standard sensorimotor mapping*.[1] Other behaviors lack any spatial correspondence between a stimulus and a response, of which Pavlovian conditioned responses provide a particularly clear example. The salivation of Pavlov's dog follows a conditioned stimulus, the ringing of a bell, but there is no response directed toward the bell or, indeed, toward anything at all. Like braking at a red traffic light, Pavlovian learning depends on an arbitrary relationship between a response and the stimulus that triggers it. That is, it depends on *arbitrary sensorimotor mapping*.[1] Some forms of arbitrary mapping involve choosing among goals or actions on the basis of color or shape cues. The example of braking at a red light, but accelerating at a yellow one, serves as a prototypical (and sometimes dangerous) example of such behavior. In the laboratory, this kind of task goes by several names, including *conditional motor learning*, *conditional discrimination*, and *stimulus–response conditioning*. One stimulus provides the context (or "instruction") for a given response, whereas other stimuli establish the contexts for different responses.[2] Arbitrary mapping enables the association of any dimensions of any stimuli with any actions or goals.

The importance of arbitrary sensorimotor mapping is well recognized — a great quantity of animal psychology revolves around stimulus–response conditioning — but the diversity among its types is not so well appreciated. Take, once again, the example of braking at a red light. On the surface, this behavior seems to depend on a straightforward stimulus–response mechanism. The mechanism comprises an input, the red light, a black box that relates this input to a response, and the response, which consists of jamming on the brakes. This surface simplicity is, however, misleading. Beyond this account lies a multitude of alternative neural mechanisms. Using the mechanism described above, a person makes a braking response in the context of the red light regardless of the predicted outcome of that action[3] and without any consideration of alternatives.[2] Such behaviors are often called *habits*, but experts use this term with varying degrees of rigor. Experiments on rodents sometimes entail the assumption that all stimulus–response relationships are habits.[4,5] But other possibilities exist. Braking at a red light could reflect a voluntary decision, one based on an attended decision among alternative actions[2] and their predicted outcomes.[3] In addition, the same behavior might also reflect high-order cognition, such as a decision about whether to follow the rule that traffic signals must be obeyed.

Because the title of this book is *Motor Cortex in Voluntary Movements*, this chapter's topic might seem somewhat out of place. However, the motor cortex — construed broadly to include the premotor areas — plays a crucial role in arbitrary sensorimotor mapping, which Passingham has held to be the epitome of voluntary movement. In his seminal monograph, Passingham[2] defined a voluntary movement as one made in the context of choosing among alternative, learned actions based on attention to those actions and their consequences. We take up this kind of arbitrary mapping in Section 10.5, in which we discuss the premotor areas involved in this kind of learning. In addition, we summarize evidence concerning the contribution of other parts of the telencephalon — specifically the prefrontal cortex, the basal ganglia, and the hippocampal system — to this kind of behavior. Because of the explosion of data coming from neuroimaging methods, Section 10.5 also contains a discussion of that literature and its relation to neurophysiological and neuropsychological results. Before dealing with voluntary movement, however, we consider arbitrary sensorimotor mapping in three kinds of involuntary movements — conditioned reflexes (Section 10.2), internal models (Section 10.3), and habits (Section 10.4). Finally, we consider arbitrary mapping in relation to other aspects of response selection, specifically those involving response rules (Section 10.6). For a fuller consideration of arbitrary mapping, readers might consult Passingham's monograph[2] and previous reviews, which have focused on the changes in cortical activity that accompany the learning of arbitrary sensorimotor mappings,[6] the role of the hippocampal system[7,8] and the prefrontal cortex[9] in such mappings, and the relevance of arbitrary mapping to the life of monkeys.[10]

10.1.1 Types of Arbitrary Mapping

10.1.1.1 Mapping Stimuli to Movements

Stimulus–Reflex Mappings. Pavlovian conditioning is rarely discussed in the context of arbitrary sensorimotor mapping. Also known as classical conditioning, it requires the association of a stimulus, called the conditioned stimulus (CS), with a different stimulus, called the unconditioned stimulus (US), which is genetically programmed to trigger a reflex response, known as the unconditioned reflex (UR). Usually, pairing of the CS with the US in time causes the induction of a conditioned response (CR). For a CS consisting of a tone and an electric shock for the US, the animal responds to the tone with a protective response (the CR), which resembles the UR. The choice of CS is arbitrary; any neutral input will do (although not necessarily equally well). The two types of Pavlovian conditioning differ slightly. In one type, as described above, an initially neutral CS predicts a US, which triggers a reflex such as eye blink or limb flexion. This topic is taken up in Section 10.2.1. In another form of Pavlovian conditioning, some neural process stores a similarly predictive relationship between an initially neutral CS and the availability of substances like water or food that reduce an innate drive. Unlike the reflexes involved in the former variety of Pavlovian conditioning, the latter involves the triggering of consumatory behaviors such as eating and drinking. For example, animals lick a water spout after a sound that has been associated with the availability of fluid from that spout. This kind of behavior sometimes goes by the name Pavlovian-approach behavior (a topic taken up in Section 10.2.2). Both kinds of arbitrary sensorimotor mapping rely on the fact that one stimulus predicts another stimulus, one that triggers an innate, prepotent, or reflex response.

Stimulus–IM Mappings. Stimuli can also be arbitrarily mapped to motor programs. For example, Shadmehr and his colleagues (this volume[11]) discuss the evidence for internal models (IMs) of limb dynamics. These models involve predictions — computed by neural networks — about what motor commands will be needed to achieve a goal (and also about what feedback should occur). The IMs are not examples of arbitrary sensorimotor mapping per se. Arbitrary stimuli can, however, be mapped to IMs, a topic taken up in Section 10.3.

Stimulus–Response Mappings in Habits. When animals make responses in a given stimulus context, that response is more likely to be repeated if a reinforcer, such as water for a thirsty animal, follows the action. This fact lies at the basis of instrumental conditioning. According to Pearce,[12] many influential learning theories of the past 100 years or so[13–15] have held that after consistently making a response in a given stimulus context, the expected outcome of the action no longer influences an animal's performance. The instrumental conditioning has produced an involuntary movement, often known as a habit or simply as a stimulus–response (S–R) association.

Note, however, that many S–R associations are not habits. When used strictly, the term "habit" applies only to certain learned behaviors, those that are so "overlearned" that they have become involuntary in that they no longer depend on the predicted outcome of the response.[3] It is also important to note that the response in an S–R association is not a standard sensorimotor mapping. That is, it need not be directed toward either the reinforcers, their source (such as water spouts and feeding trays), or the conditioned stimuli. The response is spatially arbitrary. We take up this kind of arbitrary mapping in Section 10.4.

Stimulus–Response Mappings in Voluntary Movement. Section 10.5 takes up arbitrary stimulus–response associations that are not habits, at least as defined according to contemporary animal learning theory.[3]

10.1.1.2 Mapping Stimuli to Representations Other than Movements

Stimulus–Value Mappings. Although we focus here on arbitrary sensorimotor mappings, there are many other kinds of arbitrary mappings. Stimuli can be arbitrarily mapped to their biological value. For example, stimuli come to adopt either positive or negative affective valence, i.e., "goodness" or "badness," as a function of experience. This kind of arbitrary mapping is relevant to sensorimotor mapping because stimulus–value mappings can lead to a response,[16–19] as discussed in Section 10.2.2.2.

Stimulus–Rule Mappings. In addition to stimulus–response and stimulus–value mappings, stimuli can be arbitrarily mapping onto more general representations. For example, a stimulus could evoke a response rule, a topic explored in Section 10.4. Note that we focus here on the arbitrary mapping of stimuli to rules, not the representation of a rule per se, as reported previously in both the spatial[20–22] and nonspatial[22–24] domains.

Stimulus–Meaning Mappings. In Murray et al.,[10] we argued that evolution co-opted an existing arbitrary mapping ability for speech and language. Stimuli map to their abstract meaning in an arbitrary manner. For example, the phonemes and graphemes of language elicit meanings that usually have an arbitrary relationship with those auditory and visual stimuli. And this kind of arbitrary mapping leads to a type of response mapping not mentioned above. In speech production, the relationship between the meaning a speaker intends to express and the motor commands underlying vocal or manual gestures that convey that meaning reflects a similarly arbitrary mapping.

Given these several types of arbitrary mappings, what is known about the neural mechanisms that underlie their learning?

10.2 ARBITRARY MAPPING OF STIMULI TO REFLEXES

Cells in a variety of structures show learning-related activity for responses that depend upon Pavlovian conditioning, including the basal ganglia,[25–28] the amygdala,[29–31] the motor cortex,[32] the cerebellum,[33] and the hippocampus.[34] Why are there so many different structures involved? Partly, perhaps, because there are several types of Pavlovian conditioning. One type relies mainly on the cerebellum and its output mechanisms.[33] In response to potentially damaging stimuli, such as shocks, taps, and air jets, this type of conditioned response involves protective movements such as eye blinks and limb withdrawal. Another type, called Pavlovian approach behavior, depends on parts of both the basal ganglia and the amygdala, and involves consumatory behaviors such as eating and drinking. Although there are other types of Pavlovian conditioning, such as fear conditioning and conditioned avoidance responses, we will focus on these two.

10.2.1 Pavlovian Eye-Blink Conditioning

The many studies that describe learning-related activity in the cerebellar system during eye-blink conditioning and related Pavlovian procedures have been well summarized by Steinmetz.[33] The reader is referred to his review for that material. In addition, a number of studies have shown that cells in the striatum, the principal input structure of the basal ganglia, show learning-related activity during such learning. For example, a specific population of neurons within the striatum, known as tonically active neurons (TANs), have activity that is related in some way to Pavlovian eye-blink conditioning. At first glance, this result seems curious: Pavlovian conditioning of this type, which recruits protective reflexes, does not require the basal ganglia but instead depends on cerebellar mechanisms.[33] TANs, which are believed by many to correspond to the large cholinergic interneurons that constitute ~5% of the striatal cell population,[26,35,36] respond to stimuli that are conditioned by association with either aversive stimuli[37,38] or with primary rewards.[25–28] TANs also respond to rewarding stimuli.[37,39] However, studies that have recorded from TANs while monkeys performed instrumental tasks[40] tend to report less selectivity for reinforcers than in the Pavlovian conditioning tasks discussed above,[41,42] and it has been suggested that reward-related responses may reflect the temporal unpredictablility of rewards.[37] One current account of the function of TANs is that they serve to encode the probability that a given stimulus will elicit a behavioral response. Blazquez et al.[38] recorded from striatal neurons in monkeys during either appetitive or aversive Pavlovian conditioning tasks. In addition to finding that responses to aversive stimuli (air puffs) and reinforcers (water) can occur within individual TANs, they also noted that as monkeys learned each association (CS-air puff or CS-water), more TANs became responsive to the CS. Further analysis of the population responses of TANs revealed that they were correlated with the probability of occurrence of the conditioned response.

Given that eye-blink conditioning depends on the cerebellum rather than the striatum,[33] why would cells in the striatum reflect the probability of generating a protective reflex response? The most likely possibility, according to Steinmetz,[33] is that the basal ganglia uses information about the performance of these protective reflexes in order to incorporate them into ongoing sequences of behavior. Thus, recognizing the diversity of Pavlovian mechanisms can help us understand the learning-dependent changes in striatal activity. As is always the case with neurophysiological data, a cell's activity may be "related" to a behavior for many reasons, only one of which involves causing that behavior.

Does this imply that structures mentioned above, such as the amygdala and the basal ganglia, play no role in Pavlovian conditioning? Not at all. They participate instead in other types of Pavlovian conditioning, such as Pavlovian approach behavior.

10.2.2 Pavlovian Approach Conditioning

10.2.2.1 Learning-Related Activity Underlying Pavlovian Approach Conditioning

The properties of midbrain dopaminergic neurons are becoming reasonably well characterized,[43] as is their importance in reward mechanisms.[44,45] Dopaminergic neurons respond to unexpected rewards during the early stages of learning.[46,47] As learning progresses and rewards become more predictable, neuronal responses to reward decrease and neurons increasingly respond to conditioned stimuli associated with the upcoming reward.[46,48] Furthermore, the omission of expected rewards can phasically suppress the firing of these neurons.[49] In this context, Waelti et al.[50] predicted that dopaminergic neurons might reflect differences in reward expectancy and tested this prediction in what is known as a *blocking* paradigm. Understanding their experiment requires some background in the concepts underlying the *blocking effect*, also known as the Kamin effect.

As outlined above, the paired presentation of a US such as food with a CS such as a light or sound results in the development of an association between the representation of the US and the CS. However, the simple co-occurrence of a potential CS and the US does not suffice for the formation of such an association. Instead, effective conditioning also depends upon a neural prediction: specifically, whether the US is unexpected or surprising, and thus whether the CS can capture an animal's attention. Note that the concept of attention, in this sense, differs dramatically from the concept of *top-down* attention. Top-down attention leads to an enhancement in the neural signal of an object or place attended; it results from a stimulus (or aspect of a stimulus) being predicted and its neural signal enhanced. By contrast, the kind of "attention" studied in Pavlovian conditioning results from a stimulus, the US, *not* being predicted and its signal not cancelled by that prediction. Top-down attention, which corresponds to attention in common-sense usage, is volitional: it results from a decision and a choice among alternatives.[2] The other usage of the term refers to a process that is completely involuntary. A number of prominent theories of learning[51–53] stress this aspect of expectancy and surprise, as demonstrated in a classic study by Kamin.[54] In that study, one group of rats experienced, in a "pretraining" stage of the experiment, pairings of a noise (a CS) and a mild foot shock (the US), whereas a second group of rats received no such pretraining. Then, both groups subsequently received an equal number of trials in which a compound CS composed of a noise and a light was paired with shock. Finally, both groups were tested on trials in which only the light was presented. The presentation of the light stimulus alone elicited a conditioned response in rats that had received no pretraining, i.e., the group that had never experienced noise–shock pairing. Famously, rats that had received pretraining exhibited no such response. Kamin's *blocking effect* indicated that the animals' exposure to the noise–shock pairings had somehow prevented them from learning about the light–shock pairings.

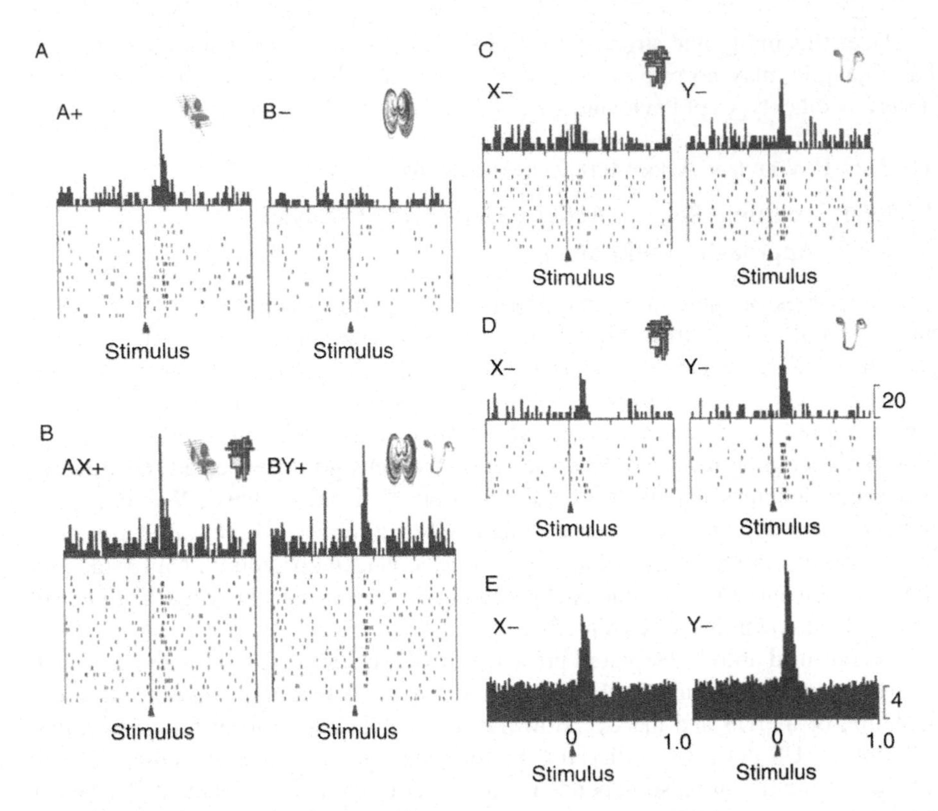

FIGURE 10.1 The response of dopaminergic neurons to conditioned stimuli in the Kamin blocking paradigm. (A) In the pretraining stage one stimulus was paired with reward (A+) and one stimulus (B-) was not. This dopaminergic neuron responded to the stimulus that predicted the reward, A+. The visual stimuli presented to the monkey appear above the histogram, which in turn appears above the activity raster for each presentation of that stimulus. (B) During compound-stimulus training, stimulus A+ was presented in conjunction with a novel stimulus (X), whereas stimulus B- was presented in conjunction with a second novel stimulus (Y). Both compound stimuli were paired with reward at this stage (AX+, BY+), and both stimulus pairs elicited firing. (C) The activity showed that compound-stimulus training had prevented the association between stimulus X and reward but not that between Y and reward. The association between stimulus X and reward was *blocked* because it was paired with a stimulus (A) that already predicted reward. Stimulus X was thus redundant throughout training. In contrast, the association between stimulus Y and reward was not blocked because it was paired with a stimulus (B) that did not predict reward. (D) Other dopamine neurons demonstrated a similar effect, but stimulus X elicited a weak increase in firing rate rather than no increase at all (as in C). (E) Average population histograms for 85 dopamine cells that were tested with stimuli X and Y in the format of C and D. (From Reference 50, with permission.)

The study of Waelti et al.[50] tested the hypothesis that the activity of dopaminergic midbrain neurons would reflect such blocking (Figure 10.1). As in the classic study of Kamin, the paradigm comprised three stages. During a "pretraining" stage, monkeys were presented with one of two stimuli on a given trial (Figure 10.1A), one of which was followed by a juice reward (designated A+, where + denotes reward) and one of which was not paired with reward (B–, where – denotes the lack of reward). Then, during compound stimulus conditioning (Figure 10.1B), stimulus X was presented in conjunction with the reward-predicting stimulus A+, whereas stimulus Y was presented in conjunction with stimulus B–. Both compound stimuli were now paired with reward (AX+, BY+), and trials of each type were interleaved. Because of the Kamin blocking effect, learning the association of X with reward was prevented because A already predicted reward, and thus rendered X redundant, whereas the association of Y with reward was learned because B did not predict reward (and therefore Y was not redundant). In the third stage, stimuli X and Y were presented in occasional unrewarded trials as a probe to test this prediction (Figures 10.1C and 10.1D). (There were other trial types as well.) An analysis of anticipatory licking was used as the measure of Pavlovian approach conditioning. This analysis demonstrated that in the third stage of testing (Figures 10.1C and 10.1D), the monkeys did not expect a reward when stimulus X was presented (i.e., learning had been blocked), but were expecting a reward when stimulus Y appeared, as predicted by the Kamin blocking effect.

Also as predicted, the Kamin blocking effect was faithfully reflected in the activity of dopaminergic neurons in the midbrain.[50] A total of 85 presumptive dopaminergic neurons were tested for the responses to probe-trial presentations of stimuli X and Y (Figures 10.1C and 10.1D). Nearly half of these (39 cells) responded to the nonredundant stimulus Y, but were not activated by the redundant stimulus X (as in Figure 10.1C). No neuron showed the opposite result. Some cells showed the same effects quantitatively (Figure 10.1D), rather than in an all-or-none manner (Figure 10.1C). As a population, therefore, dopamine neurons responded much more vigorously to stimulus Y than to X (Figure 10.1E). This finding demonstrated that the dopaminergic cells had acquired stronger responses to the nonredundant stimulus Y, compared to the redundant stimulus X, even though both stimuli had been equally paired with reward during the preceding compound stimulus training. These cells apparently predicted reward in the same way that the monkeys predicted reward.

10.2.2.2 Understanding Pavlovian Approach Behavior as a Type of Arbitrary Mapping

The data reported by Waelti et al.[50] are consistent with contemporary learning theories that posit a role for dopaminergic neurons in reward prediction.[55] This system shows a close similarity to those involved in other forms of Pavlovian conditioning, such as eye-blink conditioning. For eye-blink conditioning (and for other protective reflexes), cells in the inferior olivary nuclei compare predicted and received neuronal inputs, probably concerning predictions about the US.[33,56,57] The outcome of this prediction then becomes a "teaching" signal, transmitted by climbing-fiber inputs to the cerebellum, that induces the neural plasticity that underlies this form of learning. Why should there be two such similar systems? One answer is that the cerebellum subserves arbitrary stimulus–response mappings for protective responses, whereas the dopamine system plays a similar role for appetitive responses. The paradigmatic example of Pavlovian conditioning surely falls into the latter category:

the bell that triggered salivation in Pavlov's dog did so because of its arbitrary association with stimuli that triggered autonomic and other reflexes involved in feeding.

What is the neural basis for this Pavlovian approach behavior? This issue has been reviewed recently,[58,59] so we will only briefly consider this question here. The central nucleus of the amygdala, the nucleus accumbens of the ventral striatum, and the anterior cingulate cortex appear to be important components of the arbitrary mapping system that underlies certain (but not all) types of Pavlovian approach behavior in rats. Initially neutral objects, when mapped to a positive value, trigger ingestive reflexes, such as those involved in procurement of food or water (licking, chewing, salivation, etc.), and lesions of the central nucleus of the amygdala, the nucleus accumbens, or the anterior cingulate cortex block such learning.[60,61]

There are related mechanisms for arbitrary mapping of stimuli to biological value that involve other parts of the amygdala, the basal ganglia, and the cortex, at least in monkeys. As reviewed by Baxter and Murray,[59] these mechanisms involve different parts of the frontal cortex and amygdala than the typical Pavlovian approach behavior described above: the orbital prefrontal cortex (PF) instead of the anterior cingulate cortex and the basolateral nuclei of the amygdala instead of the central nucleus of the amygdala. These structures, very likely in conjunction with the parts of the basal ganglia with which they are interconnected, underlie the arbitrary mapping of stimuli to their value in a special and highly flexible way.

This flexibility is required when neutral stimuli map arbitrarily to food items and the value of those food items changes over a short period of time. Stimuli that map arbitrarily to specific food items can change their current value because of several factors, for example, when that food item has been consumed recently in quantity. Normal monkeys can use this information to choose stimuli that map to a higher current value. This mechanism appears to depend on the basolateral nucleus of the amygdala and the orbital PF: when these structures are removed or their interconnections severed, monkeys can no longer use the stimuli to obtain the temporarily more valued food item.[59] Separate analyses showed that the monkey remembered the mapping of the arbitrary cue to the food item, so the deficit involved mapping the stimulus to the food's current value. Furthermore, monkeys with those lesions remained perfectly capable of choosing the currently preferred food items. (Presumably, the preserved food preference is due to other mechanisms, probably hypothalamic ones, that are involved in foraging and food procurement.) Hence, the lesioned monkeys seemed to know which arbitrary stimulus mapped to which food item and they appeared to know which food they wanted. Their deficit — and therefore the contribution of the basolateral amygdala's interaction with orbital PF — involved the arbitrary mapping of otherwise neutral stimuli to their current biological value. The use of updated stimulus–value mappings allows animals to predict the current, biologically relevant outcome of an action produced in the context of that stimulus. This mechanism permits animals to make choices that lead to the best possible outcome when several possible choices with positive outcomes are available, and to choose appropriately in the face of changing values.

10.3 ARBITRARY MAPPING OF STIMULI TO INTERNAL MODELS

The previous section deals with arbitrary mappings in Pavlovian conditioning. In this section, we examine a different form of arbitrary mapping. Rao and Shadmehr[62] and Wada et al.[63] have recently shown that people can learn to map arbitrary spatial cues and colors onto the motor programs needed to anticipate the forces and feedback in voluntary reaching movements.

As summarized by Shadmehr and his colleagues in this volume,[11] in their experiments people move a robotic arm from a central location to a visual target. When, during the course of these movements, the robot imposes a complex pattern of forces on the limb, the movement deviates from a straight line to the target. With practice in countering a particular pattern of forces, the motor system learns to produce a reasonably straight trajectory. The system is said to have learned (or updated) an IM of the limb's dynamics. There is nothing arbitrary about such IMs; they reflect the physics of the limb and the forces imposed upon the limb.

People can, however, learn to map visual inputs arbitrarily onto such IMs. In the experiments that first demonstrated this fact, Rao and Shadmehr[62] presented participants with two different patterns of imposed force. They gave each person a cue indicating which of these force patterns would occur on any given trial. This cue could be either to the left or to the right of the target, and its location varied randomly from trial to trial, but in neither case did the cue serve as a target of movement or affect the trajectory of movement directly. Instead, the location of the cue was arbitrary with respect to the forces imposed by the robot. The participants in this experiment learned to use this arbitrary cue to call up the appropriate IM for the pattern of imposed forces associated with that cue location. That is, they could select the motor program needed to execute reasonably straight movements for either of two different patterns of perturbations, as long as an arbitrary visual cue indicated what the robot would do to the limb.

Having learned this mapping, the participants in these experiments could transfer this ability to color cues. For example, a red cue indicated that the same forces would occur as when the left cue appeared in the previous condition; a blue cue indicated that the other pattern of forces would occur. Interestingly, in the experiments of Shadmehr and his colleagues,[11,62] people could transfer the stimulus–IM mapping from the arbitrary spatial cue to the color (nonspatial) cue, but not the reverse. That is, if the color cues were presented first, participants were unable to learn how to counteract the forces imposed by the robot, even after 3 days of practice. Wada et al.[63] have recently shown, however, that color cues can be used to predict the pattern of forces encountered during a movement. It takes extensive practice, over days, not minutes, to learn this skill, and perhaps the people studied by Shadmehr and his colleagues would have learned, if given more time to do so. Although the two studies do not fully agree about color–IM mappings, both show that people can learn the mapping of arbitrary visual cues to internal models of limb dynamics.

10.4 ARBITRARY MAPPING OF STIMULI TO INVOLUNTARY RESPONSE HABITS

Like the arbitrary mapping of stimuli to IMs, other types of arbitrary sensorimotor mapping also involve involuntary aspects of movement. The finding that lesions of the striatum impair performance guided by certain types of involuntary stimulus–response associations[4] has encouraged neurophysiologists to examine learning-related activity in that structure.[64] Jog et al.[5] trained rats in a T-maze. To receive reinforcement, the rats were required to move through the left arm of the maze in the presence of one auditory tone and to turn right in the presence of a tone of a different frequency. Striatal cell activity was recorded using chronically implanted electrodes in the dorsolateral striatum. This study reported that the proportion of cells showing task relations increased over days as the rats gradually learned the task. This increase was primarily the effect of more cells showing a relationship with either the start of the trial or the end of the trial, when reward was gained. Interestingly, relatively few cells were reported to respond to the stimuli (the tones) per se, although the percentage of tone-related cells also increased with training. In contrast, the number of cells that were related to the response decreased with training. Jog et al.[5] also obtained activity data from a number of cells over multiple sessions, as training progressed. These individual neurons showed the same changes that had been noted in the population generally: the percentage of cells related to the task increased as performance improved. Although the authors noted that such neuronal changes could reflect the parameters of movement, a videotape analysis of performance was used in an attempt to rule out such an account.

The results of Jog et al.[5] for the learning of arbitrary stimulus–response mappings contrast somewhat with those of Carelli et al.,[65] who recorded neuronal activity from what seems to be the same dorsolateral part of the striatum as rats learned the instrumental response of pressing a lever in response to the onset of a tone. In contrast to the rats studied by Jog et al., those of Carelli et al. were not required to discriminate between different stimuli or make responses to receive a reward. Nevertheless, rats required hundreds of trials on the task to become proficient. Carelli et al.[65] reported the activity of 53 neurons that were both related to the lever-press and also showed activity related to contralateral forepaw movement outside of the task setting. However, the extent of the activity related to the conditioned lever-pressing (compared to a premovement baseline) decreased with learning, leading the authors to suggest that this population of cells in the dorsolateral striatum may be necessary for the acquisition, but not the performance, of learned motor responses.

How can these apparently contrasting results be reconciled? It is always difficult to compare studies performed in different laboratories with different behavioral methods, but the results seem to be at odds. In the task of Jog et al., cells in the dorsolateral striatum increased activity and task relationship during learning, whereas in the task of Carelli et al., cells in much the same area decreased activity and task relationship with learning. Much of the interpretation turns on the assumption that what was learned in the task used by Jog et al. was a "habit," as they assumed. However, Jog et al. provided no evidence that their arbitrary sensorimotor mapping task (two tones mapped arbitrarily to two responses, left and right) was learned as a habit and, as we have seen, there are many types of arbitrary stimulus–response relationships.

Any arbitrary sensorimotor mapping could be a habit, in the sense used in animal learning theory,[3] but many are not. A commonly cited view concerning the functional organization of the brain is that the basal ganglia, or more specifically the corticostriatal system, underlies the acquisition and performance of habits. This view remains popular, but there is considerable weakness and ambiguity in the evidence cited in support of it.[66,67] It remains an open question whether the basal ganglia plays the central role in the performance of habits; it seems more likely that it plays a role in the acquisition of such behaviors before they have become routine or relatively automatic.

Taken together, the data of Jog et al.[5] and Carelli et al.[65] seem to support this suggestion. It seems reasonable to presume that rats presented with only a few hundred trials of an arbitrary sensorimotor mapping task, as in the study of Jog et al.,[5] had not (yet) developed a stimulus–response habit, but rats presented with a much larger number of trials pressing a bar in response to a single tone had done so.[65] If one accepts this assumption, then their results can be interpreted jointly as evidence that neuronal activity in the dorsolateral striatum reflects the acquisition of learned instrumental behaviors and that this activity decreases once the learning reaches the habitual stage in the overlearned condition. Note that this conclusion is the reverse of the one most prominently asserted for this part of the striatum, namely that the dorsolateral striatum subserves habits.[4,68,69] It is, however, consistent with competing views of striatal function.[50,55,66,70] Thus, as with the other learning-related phenomena considered in this chapter, the recognition that arbitrary sensorimotor mappings come in many types provides interpretational benefits.

10.5 ARBITRARY MAPPING OF STIMULI TO VOLUNTARY MOVEMENT

Up to this point, we have mainly considered arbitrary mapping for involuntary movements and a limited amount of neurophysiological data on learning-related activity during the acquisition of such mappings. The title of this book, however, is *Motor Cortex in Voluntary Movements*, and consideration of that arbitrary mapping for voluntary movement will consume most of the remainder of this chapter.

10.5.1 Learning Rate in Relation to Implicit and Explicit Knowledge

Arbitrary sensorimotor mappings clearly meet Passingham's[2] definition of voluntary action — learned actions based on context, with consideration of alternatives based on expected outcome — but it is a definition that skirts the issue of consciousness. Of course, it remains controversial whether nonhuman animals possess a human-like consciousness,[71] and may always remain so. Regardless, the knowledge available to consciousness is often called declarative or explicit. For example, if one is aware of braking in response to a red traffic light, that would constitute explicit knowledge, but one could also stop at the same red light in an automatic way, using implicit knowledge. In a previous discussion of these issues, one of the authors presented the case for considering arbitrary sensorimotor mappings — as observed under certain circumstances — as explicit memories in monkeys.[7] We will not repeat that discussion here, but in very abbreviated form we outlined two basic ways to approach this problem: (1) identify the attributes of explicit learning that distinguish it from implicit learning; or (2) assume that, when damage to a given structure in the brain causes an inability to store new explicit memories in humans, damage to the homologous (and presumably analogous) structure in nonhuman brains does so as well. We termed these alternatives the *attribute approach* and the *ablation approach*, respectively.

The attribute approach is based partly on the speed of learning. Explicit knowledge is said to be acquired rapidly, implicit knowledge slowly, over many repetitions of the same input. But how rapid is rapid enough to earn the designation explicit? As pointed out recently by Reber,[72] some implicit knowledge can be acquired very rapidly indeed. What characterizes explicit knowledge in humans is the *potential* for the information to be acquired after a single presentation. The learning rates observed previously for arbitrary visuomotor mappings in rhesus monkeys (Figure 10.2B) are fast, but are they fast enough to warrant the term explicit?

Figure 10.2 compares the learning rates for two different forms of motor learning. Figure 10.2A shows some results for a traditional form of motor learning, described in Section 10.3, in which human participants adapt to forces imposed on their limbs during a movement. Figure 10.2B presents a learning curve for arbitrary sensorimotor learning in rhesus monkeys. In that experiment, monkeys had to learn to map three novel visual stimuli onto three spatially distinct movements of a joystick: left, right, and toward the monkey. The stimulus presented on any given trial was randomly selected from the set of three novel stimuli. Note that the learning rate τ was approximately 8 trials for both forms of learning. At first glance, this finding seems odd: most experts would hold that traditional forms of motor learning are slower than that shown in Figure 10.2A. In fact, under most experimental circumstances it takes dozens if not hundreds of trials for participants to adapt to the imposed forces. The curve shown in Figure 10.2A is unusual because it comes from a participant performing a single out-and-back movement on every trial, rather than varying the direction of movement among many targets, as is typically the case. When the participants make movements in several directions, the learning that takes place for a movement in one direction interferes to an extent with learning about movements in other directions.[11] This interference slows learning. As Figure 10.2A shows, traditional forms of motor learning need not be especially slow.

Figure 10.2B is also unusual, but in a different way than Figure 10.2A. Although the learning rate is virtually identical, Figure 10.2B illustrates the concurrent learning of *three* different sensorimotor mappings. In this task, each mapping can be considered a *problem* for the monkeys to solve. Thus, for a learning rate of ~8 trials, the learning rate for any given problem less than 3 trials. Hence, to make the traditional and arbitrary sensorimotor learning curves identical, the force adaptation problem has to be reduced to one reaching direction, back and forth, and the arbitrary mapping task has to be increased to three concurrently learned problems. This implies that

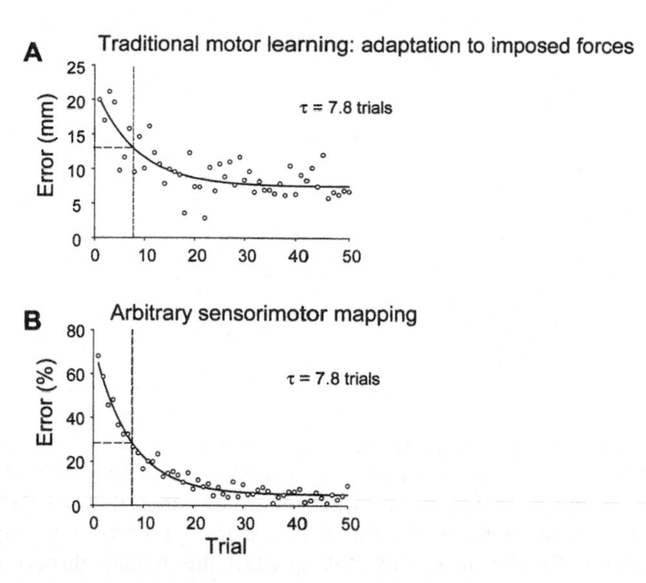

FIGURE 10.2 Learning curves for two forms of motor learning. (A) Adaptation to imposed forces in human participants, in experiments similar to those described in the legend for Figure 10.1. In these experiments, participants adapted a novel pattern of imposed forces, which perturbed their reaching movements. Motor learning is measured as a reduction in the error — i.e., less deviation from a straight hand path to the target. For the data presented here, participants moved back and forth to a single target trial after trial. (Data from O. Donchin and R. Shadmehr, personal communication.) (B) Concurrent learning of three arbitrary visuomotor mappings in rhesus monkeys. Three different, novel stimuli instructed rhesus monkeys to make three different movements of a joystick. The plot shows the average scores of four monkeys, each solving 40 sets of three arbitrary visuomotor mappings over the course of 50 trials. τ is a time constant that corresponds to learning rate, a reduction of error to e^{-1}. (Data from Reference 86.)

the learning of arbitrary visuomotor mappings in experienced rhesus monkeys is faster than the fastest motor learning of the traditional kind.

By one attribute, fast learning, a learning rate of less than 3 trials per problem conforms reasonably well with the notion that arbitrary sensorimotor mappings in monkeys, at least under certain circumstances, might be classed as explicit. But what about one-trial learning, a hallmark of explicit learning?[72] In Figure 10.3, the average error rate is plotted for four rhesus monkeys, each solving three-choice problems concurrently and doing so many times. For reasons described in a previous review,[7] we plot only trials in which the stimulus on one trial has changed from that on the previous trial. Then, we examine only responses to the stimulus (of the three) that appeared on the first trial. For obvious reasons, the monkeys performed at chance levels on the first trial of a 50-trial block. (Trial two is not illustrated because we exclude all trials that repeat the stimulus of the previous trial.) On trial three (the second presentation of the stimulus that had appeared on the first trial), one trial learning is significant[10] and is followed by a gradual improvement in performance.

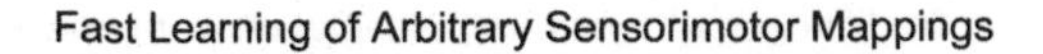

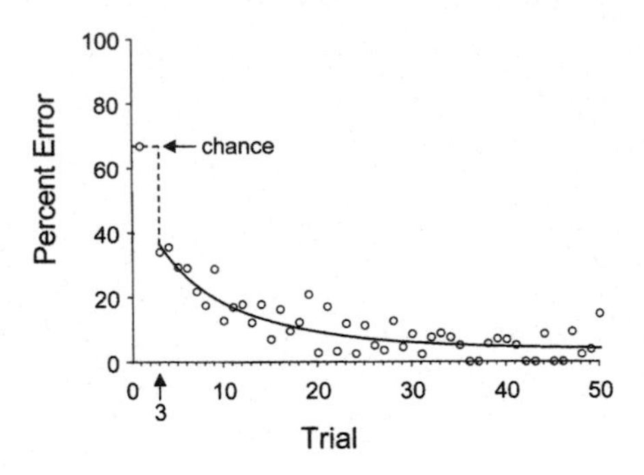

FIGURE 10.3 Fast learning of a single arbitrary visuomotor mapping in rhesus monkeys. The plot shows the average of four monkeys, each solving 40 sets of three arbitrary visuomotor mappings over the course of 50 trials. For whichever of the three stimuli in the set that was presented on trial one, the monkeys' percent error is shown for all subsequent presentations of the same stimulus. The plot shows only trials in which the stimulus changed from that on the previous trial. Therefore, no trial-two data are shown: the stimulus on trial two could not have both changed and been the same as that presented in trial one. (Data from Reference 86.)

What about the ablation approach? Data reviewed in detail elsewhere[7,8] show that ablations that include all of the hippocampus in both hemispheres abolish the fast learning illustrated in Figure 10.2B and Figure 10.3. Because it is thought that the hippocampal system subserves the recording of new explicit knowledge in humans,[73] these data also support the view that arbitrary sensorimotor mapping represents explicit knowledge and that remaining systems, possibly neocortical, remain intact to subserve the slower improvement.

Taking all of these data into account, one can argue that arbitrary sensorimotor mappings of the type learned quickly by experienced animals differs, in kind, from that learned slowly, and that this difference may correspond to the distinction between explicit and implicit knowledge in humans. This understanding informs the results obtained by lesion-, neurophysiological-, and brain-imaging methods for studying arbitrary sensorimotor mapping. The next sections address the structures, in addition to the hippocampal system, that support this kind arbitrary mapping.

10.5.2 Neuropsychology

Surgical lesions of a number of structures have produced deficits in arbitrary sensorimotor mapping, either in learning new arbitrary mappings (acquisition) or in performing according to preoperatively learned ones (retention).

10.5.2.1 Premotor Cortex

Severe deficits result from removal of the dorsal aspect of the premotor cortex (PM). For instance, Petrides[74] demonstrated that monkeys with aspiration lesions that primarily removed dorsal PM were unable to emit the appropriate response (choosing to open either a lit or an unlit box) when instructed to do so, and never reached criterion in this two-choice task, although they were given 1,020 trials. In contrast to this poor performance, control monkeys mastered the same task in approximately 300 trials. The lesioned monkeys were able to choose the responses normally, however, during sessions in which only one of the two responses was allowed, showing that the monkeys were able to detect the stimuli and were able to make the required movements.

Halsband and Passingham[75] produced a similarly profound deficit in monkeys that had undergone bilateral, combined removals of both the dorsal and ventral PM. Their lesioned monkeys could not relearn a preoperatively acquired arbitrary visuomotor mapping task in which a colored visual cue instructed whether to pull or turn a handle. Unoperated animals relearned this task within 100 trials; lesioned monkeys failed to reach criterion after 1,000 trials. However, lesioned monkeys were able to learn arbitrary mappings between different visual stimuli. This pattern of results confirms that the critical mapping function mediated by PM is that between a cue and a motor response, rather than arbitrary mappings generally. Putting the results of Petrides and Passingham together, the critical region for arbitrary sensorimotor mapping appears to be dorsal PM. Subsequently, Kurata and Hoffman[76] confirmed that injections of a GABAergic agonist, which transiently disrupts cortical information processing, impair the performance of arbitrary visuomotor performance for sites in the dorsal, but not the ventral, part of PM.

10.5.2.2 Prefrontal Cortex

There is also evidence indicating that the ventral and orbital aspects of the prefrontal cortex (PF) are crucial for arbitrary sensorimotor mapping.[77] Compared to their preoperative performance, monkeys were slower to learn arbitrary sensorimotor mappings after disrupting the interconnections between these parts of PF and the inferotemporal cortex (IT), either by the use of asymmetrical lesions[78,79] or by transecting the uncinate fascicle,[80] which connects the frontal and temporal lobes. These findings suggest that the deficits result from an inability to utilize visual information properly in the formation of arbitrary visuomotor mappings.

Both Bussey et al.[81] and Wang et al.[82] have directly tested the hypothesis that the ventral or orbital PF is integral to efficient arbitrary sensorimotor mapping. In the study of Bussey et al.[81] monkeys were preoperatively trained to solve mapping problems comprising either three or four novel visual stimuli, and then received lesions of both the ventral and orbital aspects of PF. The rationale for this approach was that both areas receive inputs from IT, which processes color and shape information. Postoperatively, the monkeys were severely impaired both at learning new mappings (Figure 10.4A) and at performing according to preoperatively learned ones. The same subjects were unimpaired on a visual discrimination task, which argues against the possibility that the deficit resulted from an inability to distinguish the stimuli from each other. A recent study by Rushworth and his colleagues[83] has demonstrated that the learning impairment seen in monkeys with ventral PF lesions reflects both the attentional demands inherent in the task and the acquisition of novel arbitrary mappings.

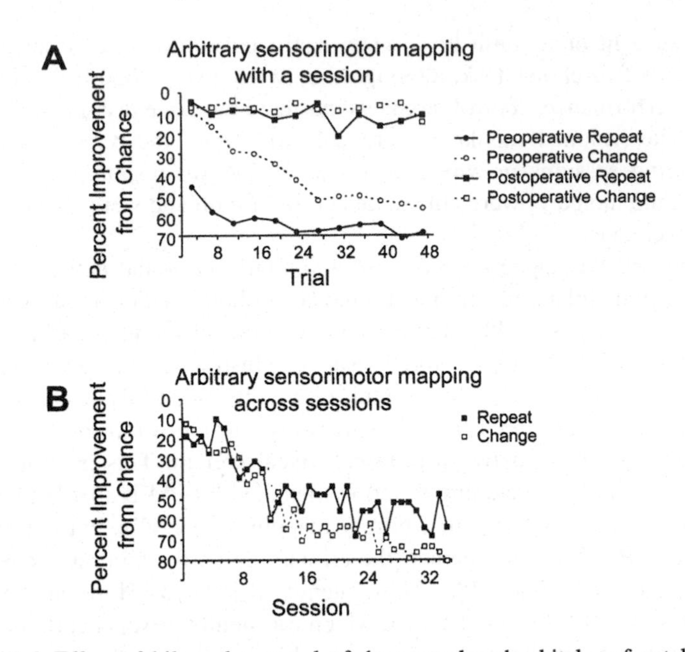

FIGURE 10.4 Effect of bilateral removal of the ventral and orbital prefrontal cortices on arbitrary visuomotor mapping and response strategies. (A) Preoperative performance is shown in the curves with circles for four rhesus monkeys. Note that over a small number of trials, the monkeys improve their performance, choosing the correct response more frequently. Note also that for repeat trials (filled circles, solid line), in which the stimulus was the same as the immediately preceding trial, the monkeys performed better than for change trials (unfilled circles, dashed line), in which the stimulus differed from that on the previous trial. The difference between these curves is a measure of the application of repeat-stay and change-shift strategies (see text); change-trial curve shows the learning rate. After removal of the orbital and ventral prefrontal cortex (postoperative), the animals remain at chance levels for the entire 48 trial session (curves with square symbols) and the strategies are eliminated. (B) Two of those four monkeys could, postoperatively, learn the same arbitrary sensorimotor mappings over the course of several days (sessions). (Data from Reference 81.)

Although this fast learning of arbitrary sensorimotor mappings was lost, and the monkeys performed at only chance levels for the first few dozen trials for given sets of stimuli, if given the same stimuli across days the monkeys slowly learned the mappings (Figure 10.4B). This slow, across-session visuomotor learning after bilateral lesions of ventral and orbital PF contrasts with the impairment that follows lesions of PM. Recall that those monkeys could not learn (or relearn) a two-choice task within 1,000 trials across several days.[74,75] This finding provides further evidence that different networks subserve fast and slow learning of these arbitrary stimulus–response mappings.[7] Whether this distinction between fast, within-session learning and slow, across-session learning corresponds to explicit and implicit learning, respectively, remains unknown.

In addition, Bussey et al. noted that lesioned monkeys lost the ability to employ certain cognitive strategies, termed the repeat-stay and change-shift strategies (Figure 10.4A). According to these strategies, if the stimulus changed from that on the previous trial, then the monkey shifted to a different response; if the stimulus was the same as on the previous trial, the monkey repeated its response. Application of these strategies doubled the reward rate, as measured in terms of the percentage of correct responses, prior to learning any of the sensorimotor mappings. Bilateral ablation of the orbital and ventral PF abolished those strategies (Figure 10.4A, squares). Could the deficit shown in Figure 10.4A be due entirely to a disruption of the monkeys' high-order strategies? This possibility is supported by evidence that strategies depend on PF function.[84,85] But the evidence presented by Bussey et al.[81] on familiar problem sets indicates otherwise. The repeat-stay and change-shift strategies are relatively unimportant for familiar mappings, but the monkeys' performance was also impaired for them. Further, there was evidence — from studies that disrupted the connection between ventral and orbital PF and IT in monkeys that did not employ the high-order strategies — that learning across sessions was impaired.[79]

Wang et al.[82] have also reported deficits in learning a two-choice arbitrary sensorimotor mapping task after local infusions of the GABAergic antagonist bicuculline into the ventral PF. The monkeys in that study, however, showed no impairment in performing the task with familiar stimuli, in contrast with the monkeys of Bussey et al.,[81] which had permanent ventral and orbital PF lesions. This difference could potentially reflect differences in task difficulty, in the temporary nature of the lesion made by Wang et al.,[82] or both.

10.5.2.3 Hippocampal System

In addition to lesions of the dorsal PM and the ventral and orbital PF, which substantially impair arbitrary sensorimotor mapping in terms of both acquisition and performance, disruption of the hippocampal system (HS) also impairs this behavior.[86–88] However, lesioned monkeys can perform mappings learned preoperatively. This finding supports the idea that HS functions to store mappings in the intermediate term, as opposed to the short term (seconds) or the long term (weeks or months). The general idea[89] is that repeated exposure to these associations results in consolidation of the mappings in neocortical networks.[9] Impairments in learning new arbitrary visuomotor mappings result from fornix transection, the main input and output pathway for the HS, even when both the stimuli and responses are nonspatially differentiated.[88] In contrast, monkeys with excitotoxic hippocampal lesions are not impaired in learning these "nonspatial" visuomotor mappings.[90] This finding implies that the impairment on this nonspatial task seen after fornix transection reflects either the disruption of cholinergic inputs to areas near the hippocampus, such as the entorhinal cortex, or dysfunction within those areas due to other causes.

10.5.2.4 Basal Ganglia

The ventral anterior nucleus of the thalamus (VA) receives input from a main output nucleus of the basal ganglia, the internal segment of the globus pallidus, and projects to PF and rostral PM. In an experiment reported by Canavan et al.,[91] radiofrequency lesions were centered in VA. Monkeys in this experiment first learned a single, two-choice arbitrary sensorimotor mapping problem to a learning criterion of 90% correct. The experiment involved a lesion group and a control group. After the "surgery," the control group retained the preoperatively learned mappings; they made only an average of ~20 errors to the learning criterion as they were retested on the task. After lesions centered on VA, monkeys averaged ~1,340 errors in attempting to relearn the mappings, and two of the three animals failed to reach criterion.

Nixon et al.[92] reported that disrupting the connections (within a hemisphere) between the dorsal part of PM and the globus pallidus had little effect on the acquisition of novel arbitrary mappings. This procedure, which involved lesions of dorsal PM in one hemisphere and of the globus pallidus in the other, led instead to a selective deficit in the retention and retrieval of familiar mappings. This finding provides further evidence for the hypothesis that premotor cortex and the parts of the basal ganglia with which it is connected play an important role in the storage and retrieval of well-learned, arbitrary mappings.

10.5.2.5 Unnecessary Structures

Nixon and Passingham[93] showed that monkeys with cerebellar lesions are not impaired on arbitrary sensorimotor mapping tasks. Similar observations have been made in patients with cerebellar lesions,[94] but this conclusion remains somewhat controversial. Lesions of the medial frontal cortex, including the cingulate motor areas and the supplementary and presupplementary motor areas, also fail to impair arbitrary sensorimotor mapping.[95,96] Similarly, lesions of the dorsolateral PF have been shown to have either mild impairments in arbitrary sensorimotor mapping,[97,98] or no effects.[99] Arbitrary sensorimotor mapping also does not require an intact posterior parietal cortex.[100] Along the same lines, a patient with a bilateral posterior parietal cortex lesion has been reported to have nearly normal timing for correcting reaching movements when these corrections were instructed by changes in the color of the targets.[101]

Two issues of connectivity arise from the lesion literature. First, it is interesting to note that the most severe deficits in arbitrary sensorimotor mapping are apparent after dorsal PM lesions and ventral PF lesions, and yet there is said to be little in the way of direct cortical connectivity between these two regions. Second is the issue of how and where the nonspatial information provided by a sensory stimulus is associated with distinct responses within the motor system. Perhaps the information underlying arbitrary sensorimotor mappings is transmitted via a third cortical region, for which the dorsal PF would appear to be a reasonable candidate. However, preliminary data indicated that lesions of dorsal PF do not cause the predicted deficit.[102] Similarly, the medial frontal cortex and the posterior parietal cortex would appear to be ruled out by the data presented in the preceding paragraph. It is possible that the basal ganglia play a pivotal role, as suggested by Passingham,[2] but the precise anatomical organization of inputs and outputs through the basal ganglia and cortex militates against this interpretation. The parts of basal ganglia targeted by IT and PF do not seem to overlap much with those that involve PM. Specific evidence that high-order visual areas project to the parts of basal ganglia that target PM — via the dorsal thalamus, of course — would contribute significantly to understanding the network underlying arbitrary visuomotor mapping. Unfortunately, clear evidence for this connectivity has not been reported.

10.5.2.6 Summary of the Neuropsychology

The hippocampal system, ventral and orbital PF, premotor cortex, and the associated part of the basal ganglia are involved in the acquisition, retention, and retrieval of arbitrary sensorimotor mappings.

10.5.3 Neurophysiology

10.5.3.1 Premotor Cortex

There is substantial evidence for premotor neurons showing learning-related changes in activity,[105–107] and these data have been reviewed previously.[6,8–10] Electrophysiological evidence for the role of dorsal PM in learning arbitrary mappings derived from a study by Mitz et al.[105] in which monkeys were required to learn which of four novel stimuli mapped to four possible joystick responses (left/right/up/no-go). In more than half of the cells tested, there was shown to be learning-dependent activity. Typically, but not exclusively, these learning-related changes were the result of increases in activity that correlated with an improvement in performance. Moreover, 46% of all learning-related changes were observed in cells that demonstrated directional selectivity, which would argue against such changes reflecting nonspecific factors such as reward expectancy. One finding of particular interest was that the evolution of neuronal activity during learning appeared to lag improved performance levels, at least slightly. This raised the possibility that the arbitrary mappings may be represented elsewhere in the brain prior to neurons in dorsal PM reflecting this sensorimotor learning. This idea is consistent with the findings, mentioned above, that HS damage disrupts the fastest learning of arbitrary sensorimotor (and other) associations, but slower learning remains possible. It is also consistent with models suggesting that the neocortex underlies slow learning and consolidation of associations formed more rapidly elsewhere. However, it should be noted that, as illustrated in Figure 10.4, PF damage also disrupts the fastest arbitrary visuomotor mapping, while allowing across-session learning to continue,[81] albeit at a slower rate.[79] Accordingly, fast mapping is not the exclusive province of the HS.

Chen and Wise[106] used a similar experimental approach to demonstrate learning-related changes in other parts of the premotor cortex, specifically the supplementary eye field (SEF) and the frontal eye field (FEF). In their experiment, some results of which are illustrated in Figure 10.5, the monkeys were required to fixate a novel visual stimulus, which was an instruction for an oculomotor response to one of four targets. The suggestion that subsequent changes in neuronal activity could reflect changes in motor responses could now be rejected with more confidence than in the earlier study of learning-dependent activity:[105] saccades do not vary substantially as a function of learning. Changes in activity during learning were common in SEF, but less so in FEF.

10.5.3.2 Prefrontal Cortex

Asaad et al.[108] recorded activity from the cortex adjacent and ventral to the principal sulcus (the ventral and dorsolateral PF) as monkeys learned to make saccades to

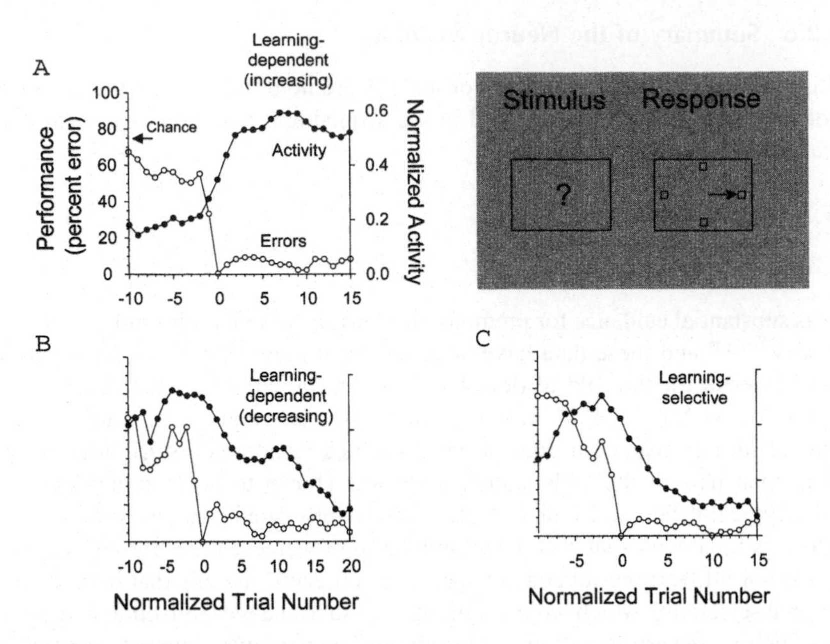

FIGURE 10.5 Three subpopulations of cells in the supplementary eye field (SEF), showing their change in activity modulation during learning (filled circles, right axis). Also shown is the monkeys' average learning rate over the same trials (unfilled circles, left axis). In the upper right part of the figure is a depiction of the display presented to the monkeys. The monkeys fixated the center of a video screen, and at that fixation point an initially novel stimulus (?) appeared. Later, four targets were presented, and the monkey had to learn — by trial and error — which of the four targets was to be fixated in order to obtain a reward on that trial in the context of that stimulus. The arrow illustrates a saccade to the right target. (A) The average activity (filled circles) of a population of neurons showing learning-dependent activity that increases with learning, normalized to the maximum for each neuron in the population. Learning-dependent activity was defined as significant modulation, relative to baseline activity, for responses to both novel and familiar stimuli. Unfilled circles show mean error rate (for a moving average of three trials), aligned on the first occurrence of three consecutive correct responses. Note the close correlation between the improvement in performance and increase in population activity. (B) Learning-dependent activity that decreases during learning. (C) Learning-selective activity, defined as neuronal modulation that was only significant for responses to novel stimuli. (Data from Reference 129.)

one of two targets in response to one of two novel stimuli. In addition to recording cells that demonstrated stimulus and/or response selectivity (80%), they observed many cells (44%) in which activity for a specific stimulus–response association was greater than the additive effects of stimulus and response selectivity. Such "nonlinear" cells could therefore represent the sensorimotor mapping per se, and the occurrence of such nonlinearity was essentially constant as a trial progressed: the percentage of cells showing this effect was 34% during the cue period, 35% during the delay period, and 33% during the presaccadic period. In contrast, cells showing cue selectivity decreased during the trial from 45% during the cue period to 32% and 21% during the delay and presaccadic periods, respectively; and cells showing response selectivity increased during the trial, from 14% during the cue period to 21% and 34% during the delay and presaccadic periods, respectively. Neuronal changes during learning were reported for these directional-selective cells in the delay period, with such selectivity becoming apparent at earlier time points within the trial as learning progressed. Also of note was the fact that the activity for novel stimuli typically exceeded that shown for familiar stimuli, even during the delay period.

10.5.3.3 Basal Ganglia

Tremblay et al.[109] studied the activity of cells in the anterior portions of the caudate nucleus, the putamen, and the ventral striatum while animals performed an arbitrary visuomotor mapping task using either familiar or novel stimuli. In this task, there were three trial types, signaled by one of three stimuli: a rewarded movement trial, in which a lever touch would result in reward; a rewarded nonmovement trial, in which the monkey maintained contact with a resting key (and thus did not move toward the lever) and consequently gained reward; or an unrewarded movement trial, in which a lever touch would result in the presentation of an auditory conditioned reinforcer (which also signaled that the next trial would be of the rewarded variety). Thus reward-related activity and movement-related activity could be compared across trials to demonstrate the specificity of the cell's activity modulations.

When the task was performed using familiar stimuli, 17% of neurons showed task-related activity.[110] When the activity between novel and familiar stimuli was compared,[109] 44% of neurons (90/205) exhibited significant decreases in task-related activation, while 46% of cells (95/205) demonstrated significant increases in task-related activity. These increases and decreases in activity were either transient in nature or were sustained for long after the association had been learned. This pattern of activity is reminiscent of changes observed in the neocortex,[105,106] and neurons that showed such task relations were distributed nonpreferentially over the caudate nucleus, the putamen, and the ventral striatum. Recent data from Brasted and Wise[111] not only confirm the presence of learning-related activity in striatal (putamen) neurons, but also showed that the time course of these changes in striatal neurons is similar to that seen in the dorsal premotor cortex, with changes in activity typically occurring in close correspondence with the learning curve.

Finally, there is evidence for learning-related changes in neuronal activity in cells in the globus pallidus.[112] Monkeys learned to perform a three-choice arbitrary visuomotor mapping task in which one of three cues presented on a monitor could instruct subjects to push, pull, or rotate a manipulator. Monkeys were required to maintain a center hold position with the manipulator until the cue appeared in the center of the screen. The cue was then replaced by a neutral stimulus for a variable delay period before the appearance of a trigger cue instructed monkeys to make their response. In a control condition, monkeys performed the task using three familiar stimuli that instructed well-learned associations. In a learning condition performed in a separate blocks of trials, one of the familiar stimuli was replaced by a novel stimulus, which required the same response as the replaced familiar stimuli. Inase et al. focused their efforts on delay-period activity and found about one-third of cells (49/157) to have delay-period activity, about half of which reflected a decrease in firing (inhibited neurons) and half an increase in firing (excited neurons) during the delay period. A difference between learning and control conditions was seen for 17/23 inhibited neurons, and for 10/26 excited neurons. The majority of the cells (21/27) that were sensitive to novel stimuli were located in the dorsal medial aspect of the internal segment of the globus pallidus, which projects indirectly to the dorsolateral PF according to Middleton and Strick.[113]

10.5.3.4 Hippocampal System

Cahusac et al.[114] reported changes in cell activity in the hippocampus and parahippocampal gyrus while monkeys learned arbitrary visuomotor tasks. Animals were presented with one of two visual stimuli on a monitor and were required either to tap the screen three times or to withhold such movement. Both responses were rewarded if performed appropriately. Cahusac et al.[114] reported similar types of learning as seen in the dorsal PM and in the SEF. Thus, 22% (19/87) of neurons demonstrated differential changes in activity for the two trial types as learning progressed, while 45% (39/87) of neurons showed only a transient difference between the two trial types, akin to the learning-selective neurons reported in SEF,[106] as illustrated in Figure 10.5C.

More recently, Wirth et al.[115] have also presented reports on arbitrary visual motor mapping. In their experiment, a complex scene filled a video monitor and four potential eye-movement targets were superimposed on that scene. Each scene instructed an eye movement to one and only one of the four targets. They describe their task as a scene–location association, but it does not differ from the tasks described above for studies of the premotor cortex. Wirth et al. report that 47% of hippocampal cells sampled showed activity related to the stimulus or delay period, and 36% of these cells showed learning-related changes in activity. On average, changes in neural activity were shown to lag behind changes in behavior. Nevertheless, Wirth et al. report that ~38% of their learning cases show changes in neural activity prior to behavioral changes. A more precise comparison of those data with the learning-related changes in activity observed in other parts of the brain, using similar analytical procedures, remains to be undertaken.

10.5.3.5 Summary of the Neurophysiology

There appears to be a close correspondence between the parts of the brain in which learning-related changes in activity are observed during the acquisition of arbitrary sensorimotor mappings and the areas necessary for those mappings. How, then, does this network correspond to that observed for comparable tasks in humans, as their brains are imaged with positron emission tomography (PET) or functional magnetic resonance imaging (fMRI)?

10.5.4 Neuroimaging

10.5.4.1 Methodological Considerations

A number of neuroimaging studies have sought to identify the neural network involved in arbitrary sensorimotor mapping in our species. These reports provide an interesting parallel to the neurophysiological and neuropsychological work summarized above, although any comparison requires assumptions about the homologies between cortical areas. In addition to potential species differences, there are substantial differences between examining single-cell activity and changes in blood flow rates or other local hemodynamic events. The relationships between the signals obtained in PET and fMRI studies to neural discharge rates are becoming better understood,[116,117] and we believe that a consensus is emerging that neuroimaging signals mainly reflect synaptic events rather than neural discharge rates. Further, difficult as it might be to resist doing so, negative results in neuroimaging cannot be interpreted in any meaningful way.[118] Take the example of learning-related activity changes during the acquisition of arbitrary sensorimotor mappings. As illustrated in Figure 10.5, some cells increase activity during learning, but others decrease, and they are so closely intermingled that the synaptic inputs causing these changes must contribute to a single fMRI or PET voxel. So it is likely that no change in PET and fMRI signals would be observed, despite the occurrence of important changes in information processing. In addition, of course, there are always issues about whether human and nonhuman participants approach learning in the same way.

Putting such general methodological issues aside for the sake of discussion, there remains the issue of how to identify the processes related to the formation of stimulus–response mappings, as opposed to its many covariants. Three approaches seem most popular. It is possible (1) to compare activation at different stages of arbitrary sensorimotor mapping;[119,120] (2) to compare arbitrary sensorimotor mapping with other kinds of learning, such as sequence learning;[121] or (3) to compare the learning of novel associations with performance controlled by established associations.[122] As for the first and third approaches, comparing activation in the early stages of learning with later stages or with established associations demonstrates that the experimental and control conditions are well matched for factors relating to stimulus processing, response preparation, selection, and execution. However, it is reasonable to assume that the early stages of learning may be associated with greater demands on attention, novelty detection, motivation levels, and task difficulty. As for the second approach, comparing arbitrary sensorimotor mapping to other kinds of learning with similar attentional demands and difficulty, the differences between the tasks create other interpretational difficulties.

Taking these interpretational problems into account, can we make sense of the brain-imaging data bearing on arbitrary sensorimotor mapping and its relation to the neurophysiological and neuropsychological data summarized here?

10.5.4.2 Established Mappings

One of the first studies to examine activity related to arbitrary sensorimotor mapping was a PET study of Deiber et al.[123] Participants were required to make one of four joystick movements, with each movement arbitrarily associated with a distinct tone. In a separate block of trials, the task was performed using the same auditory stimuli but with the contingencies reversed. Thus, both of these conditions involved arbitrary sensorimotor mappings. Participants received ~100 trials of training prior to the scanning in the former task and about 13 trials in the latter. Activity in these two conditions was compared with a condition in which subjects always moved the joystick in the same direction in response to a tone. The arbitrary mapping task with reversal showed some activity increases in the dorsal and dorsolateral PF, and both arbitrary mapping conditions resulted in significant increases in superior parietal areas, but neither showed activity increases in PM. There were no reported changes in other structures associated with arbitrary sensorimotor learning and performance, such as the basal ganglia and the hippocampal region.

Grafton et al.[124] examined arbitrary visuomotor associations by requiring subjects in a PET scanner to make one of two different grasping actions in response to the presentation of a red or green light. This condition was compared to an average of two control conditions in which subjects had to perform only one of the two grips, with the color light providing only response execution ("go") rather than response selection instructions. Thus, although the control task lacked the element of response selection that was required in the arbitrary mapping task, neither the stimuli nor the responses could be readily described as spatially differentiated. The arbitrary mapping task produced greater activation in the posterior parietal cortex and also resulted in significant increases in the rostral extent of dorsal PM contralateral to the arm used.

Toni et al.[125] conducted a PET study to compare activity in two tasks in which one of four objects instructed one of four movements. In one task, the stimuli instructed a spatially congruent grasping movement, while in the other task, stimuli cued an arbitrarily associated hand movement. The task that used arbitrary visuomotor associations was associated with significantly differential regional cerebral blood flow (rCBF) in the ventral PF, the dorsal PM, and the putamen and/or globus pallidus. It should be noted, however, that this dorsal PM activity was located in its medial aspect, which lesion studies suggest may not be necessary for effecting arbitrary mappings.[95,96] Such activity thus may reflect a more general role in monitoring visuomotor transformations.

Ramnani and Miall[126] found a selective increase in dorsal PM activity in an fMRI study of arbitrary visuomotor mapping. In their study, stimulus shape indicated which of four buttons to depress (and therefore which finger to move) and stimulus color indicated whether the participant, another person in the room, or a computer should perform the task. Their findings not only indicated that dorsal PM showed a significant hemodynamic response during performance of the arbitrary mapping task, but also that it was selective for a specific instruction (as opposed to a nonspecific warning stimulus) and for the scanned participant performing the task (as opposed to the other person in the room). Ramnani and Miall suggested that predictions about the actions of another person rely on a different brain system, one commonly activated when people attribute mental states to others. This brain system included part of Broca's area, among other regions of medial PF and the superior temporal cortex.

The above studies seem to provide supportive evidence for PF and basal ganglia involvement in arbitrary sensorimotor mappings, especially ventral PF and putamen.[125] There is, by contrast, a notable lack of consistent activation in PM in such tasks. In the studies outlined above, only the findings of Toni et al.[125] and Ramnani and Miall[126] provided support for the neuropsychological results, which have indicated that dorsal PM is necessary for performing arbitrary mappings. The PM activity reported in the study of Grafton et al.[124] (see also Sweeney et al.[127]) could simply have reflected greater response-selection demands. The potential reasons for false-negative PET or fMRI results for dorsal PM include the methodological issues mentioned above, as well as one important additional problem. PM mediates a diverse range of response-related functions in addition to arbitrary sensorimotor mappings. Thus, many control tasks may fail to yield contrasts because dorsal PM is involved in those tasks as well. For example, Deiber et al.[123] reported no PM activity at all during performance of arbitrary mappings, even when the mappings were reversed. But there is evidence from monkeys that PM might be involved in the control condition of their task: repetitive movement. Indeed, cells in dorsal PM show movement-related modulation in activity for arm movements in total darkness,[128] so a genuine control task might be difficult to devise.

10.5.4.3 Learning New Mappings

A number of imaging studies have also attempted to identify the neural substrates of arbitrary sensorimotor mapping during learning. Deiber et al.[119] measured rCBF as participants performed two kinds of arbitrary mapping tasks. One required them to perform a joystick movement that depended on arbitrary cues, and the other required them to report whether an arrow matched the arbitrary stimulus-to-place mapping for the same cues. Increases in rCBF during learning were reported for the putamen in the latter condition and in ventrocaudal PM in the former. Decreases in rCBF were more extensive: ventral PF, dorsal PF, and dorsal PM showed rCBF decreases for the reporting task.

Toni and his colleagues have undertaken a series of imaging studies designed to identify the neural network involved in arbitrary visuomotor mapping. Toni and Passingham[121] conducted a PET study in which they compared arbitrary sensorimotor learning with motor sequence learning. The responses were the same in the two tasks (one of four finger movements) and the stimuli also were of the same type, though not identical, for each task. Thus, although the two tasks were matched for sensory and motor components, the stimulus pattern only instructed the correct response in the arbitrary sensorimotor mapping task. (In a baseline task, subjects passively viewed four categorically similar stimuli.) The critical comparison between arbitrary visuomotor mapping and sequential motor learning revealed learning-related increases near the cingulate sulcus and in the body of the caudate nucleus in the left hemisphere and in orbital PF in the right hemisphere. The two tasks also showed differential patterns of activity in the left superior parietal cortex, with activity decreasing during arbitrary sensorimotor mapping and increasing during sequence learning. Increases in the parahippocampal gyrus and the putamen and globus pallidus, as well as decreases in the dorsolateral PF, were only seen when arbitrary sensorimotor mapping was compared to the passive, baseline condition. These data seem to confirm a role for the ventral PF and basal ganglia in arbitrary sensorimotor mapping, but the evidence for hippocampal involvement in this study remains weak because it was revealed only in comparison with a passive baseline condition. The lack of PM activity in the arbitrary sensorimotor learning condition, although seemingly inconsistent with the neuropsychological and neurophysiological literature on monkeys, may reflect the involvement of dorsal PM in the visuomotor transformations underlying both arbitrary visuomotor mapping and motor sequence learning.

In a subsequent fMRI study, Toni et al.[122] compared the learning of arbitrary sensorimotor mappings with performance using already established mappings. There were four stimuli in each condition, which mapped to the four finger movement responses mentioned above. Blocks of familiar and novel stimuli were mixed. Toni et al.[122] reported hemodynamic events in many regions in the frontal and temporal lobes, as well as in the HS and basal ganglia. Their results included signal increases in the ventral PF, the ventral PM, the dorsal PF, the orbital PF, the parahippocampal and hippocampal gyri, and the caudate nucleus, among other structures. Interestingly, whereas most time-related changes consisted of the hemodynamic signal for learning and control conditions converging over time, this was not true of increases in the caudate nucleus, in which the signal became greater in the learning than in the control condition as the task progressed. Subsequent structural equation modeling of the same data set led the authors to suggest that corticostriatal interactions strengthen when arbitrary sensorimotor mappings are learned.[130] An analysis of "effective connectivity" suggested that variation in the fMRI signal in the striatum showed a stronger correlation, as learning progressed, with changes seen in PM, the inferior frontal cortex, and the medial temporal lobe. On the basis of these analyses, the investigators inferred that the learning of arbitrary mappings was the result of an increase in activity in corticostriatal connections, although it should be noted that such inferences assume a level of corticostriatal convergence that remains to be shown neuroanatomically.

As in the previous section, the results summarized here also show a surprising lack of consistency regarding dorsal PM. Although Toni et al.[122] report learning-associated increases in ventral PM, Toni and Passingham[121] did not report any changes in PM activity when subjects learn arbitrary sensorimotor mappings that require skeletomotor responses (see also Paus et al.[131]). Similarly, while Deiber et al.[119] reported learning-associated increases in ventral PM during learning in one of their arbitrary mapping tasks, they also found no changes in PM activation in the other version of the task. Such negative results are at variance with neurophysiological (Section 10.5.3) and neuropsychological (Section 10.5.2) studies. As mentioned earlier, this lack of correspondence may reflect interpretational problems with neuroimaging techniques. As Toni and Passingham[121] suggest, "[I]t could be that the learning-related signal measured with PET is diluted by the contribution of a neuronal subpopulation that is related to the execution of movements" (p. 29). Alternatively, it may be that dorsal PM mediates both the learning of new mappings and the retrieval of established mappings, thus nullifying any contrast between the two conditions. In this regard, Toni et al.[122] state that "our results suggest that, at the system level, dorsal premotor regions were similarly involved in both the [retrieval] and [learning] tasks," and that "the discrepancy between these results and those obtained at the single-unit level ... calls for further investigation of the specific contribution of ventral and dorsal premotor cortex to visuomotor association." (p. 1055) That such caution is warranted is confirmed by the observation that in the PF,[108] SEF, and FEF[129] of monkeys, a number of cells discharge preferentially for either familiar or novel mappings. Also, to repeat the argument presented above, some cells increase activity, but others decrease activity during learning. Given that a myriad of synaptic signals that must have driven these changes, which also increased or decreased during learning, and given that these synaptic signals (rather than neuronal discharge rate) probably dominated the neuroimaging results,[116,117] one can argue that nothing specific can be predicted for either PET or fMRI learning-related activity. Another possibility is that human participants approach these learning tasks differently than do monkeys. Specifically, it seems possible that participants in neuroimaging studies may preferentially employ only fast-learning mechanisms, and for that reason fail to show PM activation.[79]

There is, however, a greater consensus in the neuroimaging literature that ventral PF and the basal ganglia mediate arbitrary visuomotor mapping. Toni et al.[122] and Toni and Passingham[121] (see also Paus et al.[131]) all reported learning-related changes in the ventral PF. Although Deiber et al.[119] did not find such changes in the visuomotor version of their learning task, they did see learning-related decreases in ventral PF in the version in which participants reported such relations. It has been suggested that the ventral PF plays a critical role in mediating arbitrary sensorimotor mapping, since it is in a position to represent knowledge about stimuli, responses, and outcomes,[77] a view supported by lesion studies.[81,82] Regarding the role of the basal ganglia, Toni et al.[122] reported increases in caudate as learning progresses, whereas Deiber et al.[119] reported learning-related increases in the putamen (see also Toni and Passingham[121] and Paus et al.[131]).

10.5.4.4 Summary

This section has focused on the neural mechanisms for learning arbitrary mappings in voluntary movement. Neurophysiological, neuropsychological, and neuroimaging findings appear to agree that ventral prefrontal cortex and parts of the basal ganglia play an important role in such learning. Neuroimaging findings are less consistently supportive of the neurophysiological and neuropsychological evidence that dorsal premotor cortex and the hippocampal system also play necessary roles in this kind of learning. Although others will surely disagree, we think that the limitations of neuroimaging methods make negative results of this kind uninterpretable. Accordingly, we conclude that the hippocampal system is necessary for the rapid learning of arbitrary mappings, but not for slow learning and not for the retention or retrieval of familiar mappings. Ventral and orbital PF are necessary for fast learning and the application of at least certain strategies, if not strategies in general. These parts of PF also contribute, in part through their interaction with IT, to the slow learning of arbitrary visuomotor mappings. The dorsal premotor cortex and the associated part of the basal ganglia are involved in the retention and retrieval of familiar mappings, not the learning of new ones. It seems likely that the role of the basal ganglia is diverse,[104] with some parts involved in fast learning, much like the hippocampal system and the prefrontal cortex, and other parts involved in slow learning and long-term retention, as postulated here for the premotor cortex.

10.6 ARBITRARY MAPPING OF STIMULI TO COGNITIVE REPRESENTATIONS

In addition to the many specific stimulus–response associations, it is also possible that a stimulus can map to a response rule. The arbitrary mapping of stimuli to rules is relevant to this chapter for two reasons: not only does the relationship between the stimulus and the rule represent an example of arbitrary mapping in its own right, but such rules also allow correct responses to be chosen even if the stimulus that will cue the response has never been encountered.

Wallis et al.[23] examined the arbitrary mapping of stimuli to response rules, and showed that abstract rules are encoded in PF (Figure 10.6). They trained two monkeys to switch flexibly between two rules: a matching-to-sample rule and a nonmatching-to-sample rule. Monkeys were presented with a sample stimulus and then, after a delay, were shown a test stimulus, and were required to judge whether the sample stimulus matched the test stimulus, and to respond (i.e., release or maintain bar press) accordingly. The rule applicable to each trial was indicated by a cue that was presented at the same time as the sample stimulus. In order to rule out the sensory properties of the cue as a confounding factor, each rule type could be signaled by one of two distinct cues from two different modalities (e.g., juice or low tone for the matching rule, no juice or high tone for the nonmatching rule). Thus, the specific event that occurred at the same time that the sample appeared served as an arbitrary cue that mapped onto one of the two rules. The authors reported the presence of rule-selective cells, such as those that were preferentially active during match trials, regardless of whether the match rule had been signaled a drop of juice or a low tone. Of 492 cells recorded from in dorsolateral or ventral PF, 200 cells (41%) showed such selectivity for either the match rule (101 cells) or the nonmatch rule (99 cells). There was no obvious segregation of rule cells in any one area of PF for either of the rule types, and rule specificity was recorded in both the stimulus and delay task periods, although there was a higher incidence of rule-selective neurons in the dorsolateral PF (29%) than in the ventrolateral PF (16%) and the orbitofrontal PF (18%). The same authors have subsequently shown such abstract rules to be encoded in premotor areas during the sample period.[134] These results are consistent with those of Hoshi et al.,[24] who also recorded from cells in the prefrontal cortex that were related to the abstract rules that govern responding.

While such studies demonstrate the ability of PF to *encode* abstract rules, Strange et al.[133] conducted an fMRI study during which subjects were required to *learn* rules.

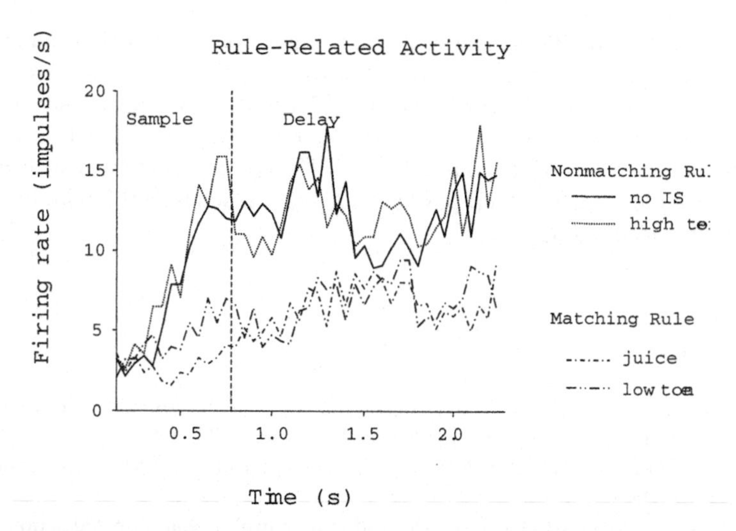

FIGURE 10.6 Neuronal activity in the prefrontal cortex, reflecting arbitrary stimulus–rule mappings. This cell's modulation was relatively high when a high-pitched tone or no additional stimulus appeared at the same time as a sample stimulus. The monkey later, after a delay period, was required to respond to a stimulus other than the sample in order to receive a reward. This response rule is called a nonmatching-to-sample rule. The cell showed much less modulation when the low-pitched tone or a reward occurred at the same time as the sample. In that eventuality, the monkey was required to respond to the sample when it reappeared later in the trial (called a matching-to-sample rule). Note that the cells firing rate was greater when the arbitrary cue signaled the nonmatching-to-sample rule, regardless of which cue mapped to that rule. (Data from Reference 23.)

Subjects would be presented with a string of four letters, and would have to respond according to whether the letters conformed or not with an unstated rule, such as "the second and fourth letters must match." While subjects were given examples of such rules prior to scanning, they were not informed of the exact rule. During scanning, the experimenters could then change either the exemplars, or the rule, or both the rule and the exemplars. A factorial analysis revealed that, after a rule change, increases in activity were observed in rostral aspects of PF, including dorsal, ventral, and polar PF. A slightly different pattern of activity was seen in the left hippocampal region: a decrease in activity was observed and it was greater when the rule changed than when only the exemplars changed.

Thus, although research on learning to map stimuli onto arbitrary response rules has only recently begun, it seems likely that it has much the same character and involves many of the same neural structures as the other kinds of arbitrary sensorimotor mapping outlined in this chapter.

10.7 CONCLUSION

Arbitrary sensorimotor mapping occurs in many types, which appear to be special cases of a more general arbitrary mapping capacity. Advanced brains can map stimuli arbitrarily to (1) reflex-like responses, (2) internal models of limb dynamics, and response choices that either (3) habitually follow a stimulus or (4) follow a stimulus based on a prediction about response outcome. They can also map stimuli arbitrarily to biological value, response rules, and abstract meaning, which, in turn, can be mapped to the four kinds of action listed above. The recurrent nature of such arbitrary mappings provides much of their power to enable the behavioral flexibility characteristic of advanced animals.

REFERENCES

1. Wise, S.P., di Pellegrino, G., and Boussaoud, D., The premotor cortex and nonstandard sensorimotor mapping, *Can. J. Physiol. Pharmacol.*, 74, 469, 1996.
2. Passingham, R.E., *The Frontal Lobes and Voluntary Action*, Oxford University Press, Oxford, 1993.
3. Dickinson, A. and Balleine, B., Motivational control of goal-directed action., *Animal Learn. Behav.*, 22, 1, 1994.
4. McDonald, R.J. and White, N.M., A triple dissociation of memory systems: hippocampus, amygdala, and dorsal striatum, *Behav. Neurosci.*, 107, 3, 1993.
5. Jog, M.S. et al., Building neural representations of habits, *Science*, 286, 1745, 1999.
6. Wise, S.P., Evolution of neuronal activity during conditional motor learning, in *Acquisition of Motor Behavior in Vertebrates*, Bloedel, J.R., Ebner, T.J., and Wise, S.P., Eds., MIT Press, Cambridge, MA, 1996, 261.
7. Wise, S.P. and Murray, E.A., Role of the hippocampal system in conditional motor learning: mapping antecedents to action, *Hippocampus*, 9, 101, 1999.
8. Wise, S.P. and Murray, E.A., Arbitrary associations between antecedents and actions, *Trends Neurosci.*, 23, 271, 2000.
9. Murray, E.A., Bussey, T.J., and Wise, S.P., Role of prefrontal cortex in a network for arbitrary visuomotor mapping, *Exp. Brain Res.*, 133, 114, 2000.
10. Murray, E.A., Brasted, P.J., and Wise, S.P., Arbitrary sensorimotor mapping and the life of primates, in *Neuropsychology of Memory*, Squire, L.R. and Schacter, D.L., Eds., Guilford, New York, 2002, 339.
11. Shadmehr, R. et al., Learning dynamics of reaching, in *Motor Cortex in Voluntary Movements*, Riehle, A. and Vaadia, E., Eds., CRC Press, Boca Raton, FL, 2005.
12. Pearce, J.M., *Animal Learning and Cognition: An Introduction*, Psychology Press, UK, 1997.
13. Thorndike, E.L., *Animal Intelligence*, Macmillan, New York, 1911.
14. Guthrie, E.R., *The Psychology of Learning*, Harper, New York, 1935.
15. Hull, C.L., *Principles of Behavior*, Appleton-Century-Crofts, New York, 1943.
16. Adams, C.D. and Dickinson, A., Actions and habits: variations in associative representations during instrumental learning, in *Information Processing in Animals: Memory Mechanisms*, Spear, N.E. and Miller, R.R., Eds., Lawrence Erlbaum Associates, Inc., Hillsdale, NJ, 1981, 143.
17. Colwill, R.M. and Rescorla, R.A., Instrumental responding remains sensitive to reinforcer devaluation after extensive training, *J. Exp. Psychol. Animal Behav. Proc.*, 11, 520, 1985.
18. Balleine, B.W. and Dickinson, A., Goal-directed instrumental action: contingency and incentive learning and their cortical substrates, *Neuropharmacology*, 37, 407, 1998.

19. Corbit, L.H. and Balleine, B.W., The role of the hippocampus in instrumental conditioning, *J. Neurosci.*, 20, 4233, 2000.
20. di Pellegrino, G. and Wise, S.P., Visuospatial vs. visuomotor activity in the premotor and prefrontal cortex of a primate, *J. Neurosci.*, 13, 1227, 1993.
21. Riehle, A., Kornblum, S., and Requin, J., Neuronal coding of stimulus-response association rules in the motor cortex, *NeuroReport*, 5, 2462, 1994.
22. White, I.M. and Wise, S.P., Rule-dependent neuronal activity in the prefrontal cortex, *Exp. Brain Res.*, 126, 315, 1999.
23. Wallis, J.D., Anderson, K.C., and Miller, E.K., Single neurons in prefrontal cortex encode abstract rules, *Nature*, 411, 953, 2001.
24. Hoshi, E., Shima, K., and Tanji, J., Neuronal activity in the primate prefrontal cortex in the process of motor selection based on two behavioral rules, *J. Neurophysiol.*, 83, 2355, 2000.
25. Kimura, M., The role of primate putamen neurons in the association of sensory stimuli with movement, *Neurosci. Res.*, 3, 436, 1986.
26. Aosaki, T. et al., Responses of tonically active neurons in the primate's striatum undergo systematic changes during behavioral sensorimotor conditioning, *J. Neurosci.*, 14, 3969, 1994.
27. Aosaki, T., Graybiel, A.M., and Kimura, M., Effect of the nigrostriatal dopamine system on acquired neural responses in the striatum of behaving monkeys, *Science*, 265, 412, 1994.
28. Raz, A. et al., Neuronal synchronization of tonically active neurons in the striatum of normal and parkinsonian primates, *J. Neurophysiol.*, 76, 1996.
29. Rosenkranz, J.A. and Grace, A.A., Dopamine-mediated modulation of odour-evoked amygdala potentials during Pavlovian conditioning, *Nature*, 417, 282, 2002.
30. Schafe, G.E. et al., Memory consolidation of Pavlovian fear conditioning: a cellular and molecular perspective, *Trends Neurosci.*, 24, 540, 2001.
31. Repa, J.C. et al., Two different lateral amygdala cell populations contribute to the initiation and storage of memory, *Nat. Neurosci.*, 4, 724, 2001.
32. Aou, A., Woody, C.D., and Birt, D., Increases in excitability of neurons of the motor cortex of cats after rapid acquisition of eye blink conditioning, *J. Neurosci.*, 12, 560, 1992.
33. Steinmetz, J.E., Brain substrates of classical eyeblink conditioning: a highly localized but also distributed system, *Behav. Brain Res.*, 110, 13, 2000.
34. McEchron, M.D. and Disterhoft, J.F., Hippocampal encoding of non-spatial trace conditioning, *Hippocampus*, 9, 385, 1999.
35. Kimura, M., Rajkowski, J., and Evarts, E., Tonically discharging putamen neurons exhibit set-dependent responses, *Proc. Nat. Acad. Sci. U.S.A.*, 81, 4998, 1984.
36. Aosaki, T., Kimura, M., and Graybiel, A.M., Temporal and spatial characteristics of tonically active neurons of the primate's striatum, *J. Neurophysiol.*, 73, 1234, 1995.
37. Ravel, S., Legallet, E., and Apicella, P., Tonically active neurons in the monkey striatum do not preferentially respond to appetitive stimuli, *Exp. Brain Res.*, 128, 531, 1999.
38. Blazquez, P.M. et al., A network representation of response probability in the striatum, *Neuron*, 33, 973, 2002.
39. Apicella, P., Legallet, E., and Trouche, E., Responses of tonically discharging neurons in the monkey striatum to primary rewards delivered during different behavioral states, *Exp. Brain Res.*, 116, 456–466.
40. Ravel, S. et al., Reward unpredictability inside and outside of a task context as a determinant of the responses of tonically active neurons in the monkey striatum, *J. Neurosci.*, 21, 5730, 2001.

41. Lebedev, M.A. and Nelson, R.J., Rhythmically firing neostriatal neurons in monkey: activity patterns during reaction-time hand movements, *J. Neurophysiol.*, 82, 1832, 1999.
42. Shimo, Y. and Hikosaka, O., Role of tonically active neurons in primate caudate in reward-oriented saccadic eye movement, *J. Neurosci.*, 21, 7804, 2001.
43. Schultz, W. and Dickinson, A., Neuronal coding of prediction errors, *Annu. Rev. Neurosci.*, 23, 473, 2000.
44. Robbins, T.W. and Everitt, B.J., Neurobehavioral mechanisms of reward and motivation, *Curr. Opin. Neurobiol.*, 6, 228, 1996.
45. Schultz, W., Dopamine neurons and their role in reward mechanisms, *Curr. Opin. Neurobiol.*, 7, 191, 1997.
46. Ljungberg, T., Apicella, P., and Schultz, W., Responses of monkey dopamine neurons during learning of behavior reactions, *J. Neurophysiol.*, 67, 145, 1992.
47. Hollerman, J.R. and Schultz, W., Dopamine neurons report an error in the temporal prediction of reward during learning, *Nat. Neurosci.*, 1, 304, 1998.
48. Schultz, W., Apicella, P., and Ljungberg, T., Responses of monkey dopamine neurons to reward and conditioned stimuli during successive steps of learning a delayed response task, *J. Neurosci.*, 13, 900, 1993.
49. Schultz, W., Predictive reward signal of dopamine neurons, *J. Neurophysiol.*, 80, 1, 1998.
50. Waelti, P., Dickinson, A., and Schultz, W., Dopamine responses comply with basic assumptions of formal learning theory, *Nature*, 412, 43, 2001.
51. Mackintosh, N.M., A theory of attention: variations in the associability of stimuli with reinforcement, *Psychol. Rev.*, 82, 276, 1975.
52. Pearce, J.M. and Hall, G., A model for Pavlovian learning: variations in the effectiveness of conditioned but not of unconditioned stimuli, *Psychol. Rev.*, 87, 532, 1980.
53. Rescorla, R.A. and Wagner, A.R., A theory of Pavlovian conditioning: variations in the effectiveness of reinforcement and non-reinforcement, in *Classical Conditioning: Current Research and Theory*, Black, A.H. and Prokasy, W.F., Eds., Appleton-Century-Crofts, New York, 1972, 64.
54. Kamin, L.J., Predictability, surprise, attention, and conditioning, in *Punishment and Aversive Behavior*, Campbell, B.A. and Church, R.M., Eds., Appleton-Century-Crofts, New York, 1969, 279.
55. Suri, R.E. and Schultz, W., Learning of sequential movements by neural network model with dopamine-like reinforcement signal, *Exp. Brain Res.*, 121, 350–354.
56. Kim, J.J., Krupa, D.J., and Thompson, R.F., Inhibitory cerebello-olivary projections and blocking effect in classical conditioning, *Science*, 279, 570, 1998.
57. Medina, J.F. et al., Parallels between cerebellum- and amygdala-dependent conditioning, *Nat. Rev. Neurosci.*, 3, 122, 2002.
58. Everitt, B.J., Dickinson, A., and Robbins, T.W., The neuropsychological basis of addictive behaviour, *Brain Res. Rev.*, 36, 129, 2001.
59. Baxter, M.G. and Murray, E.A., The amygdala and reward, *Nat. Rev. Neurosci.*, 3, 563, 2002.
60. Cardinal, R.N. et al., Effects of selective excitotoxic lesions of the nucleus accumbens core, anterior cingulate cortex, and central nucleus of the amygdala on autoshaping performance in rats, *Behav. Neurosci.*, 116, 553, 2002.
61. Parkinson, J.A. et al., Disconnection of the anterior cingulate cortex and nucleus accumbens core impairs Pavlovian approach behavior: further evidence for limbic cortical-ventral striatopallidal systems, *Behav. Neurosci.*, 114, 42, 2000.
62. Rao, A.K. and Shadmehr, R., Contextual cues facilitate learning of multiple models of arm dynamics, *Soc. Neurosci. Abstr.*, 26, 302, 2001.

63. Wada, Y. et al., Acquisition and contextual switching of multiple internal models for different viscous force fields, *Neurosci. Res.*, 46, 319, 2003.
64. Packard, M.G., Hirsh, R., and White, N.M., Differential effects of fornix and caudate nucleus lesions on two radial maze tasks: evidence for multiple memory systems, *J. Neurosci.*, 9, 1465, 1989.
65. Carelli, R.M., Wolske, M., and West, M.O., Loss of lever press-related firing of rat striatal forelimb neurons after repeated sessions in a lever pressing task, *J. Neurosci.*, 17, 1804, 1997.
66. Wise, S.P., The role of the basal ganglia in procedural memory, *Sem. Neurosci.*, 8, 39, 1996.
67. Gaffan, D., Memory, action and the corpus striatum: current developments in the memory-habit distinction, *Sem. Neurosci.*, 8, 33, 1996.
68. Mishkin, M. and Petri, H.L., Memories and habits: some implications for the analysis of learning and retention, in *Neuropsychology of Memory*, Squire, L.R. and Butters, N., Eds., Guilford Press, New York, 1984, 287.
69. Knowlton, B.J., Mangels, J.A., and Squire, L.R., A neostriatal habit learning system in humans, *Science*, 273, 1399, 1996.
70. Houk, J.C. and Wise, S.P., Distributed modular architectures linking basal ganglia, cerebellum, and cerebral cortex: their role in planning and controlling action, *Cereb. Cortex*, 5, 95, 1995.
71. Tulving, E., Episodic memory and common sense: how far apart? *Phil. Trans. R. Soc. Lond. B Biol. Sci.*, 356, 1505, 2001.
72. Reber, P.J., Attempting to model dissociations of memory, *Trends Cog. Sci.*, 6, 192, 2002.
73. Squire, L.R. and Zola, S.M., Structure and function of declarative and nondeclarative memory systems, *Proc. Nat. Acad. Sci. U.S.A.*, 93, 13515, 1996.
74. Petrides, M., Deficits in non-spatial conditional associative learning after periarcuate lesions in the monkey, *Behav. Brain Res.*, 16, 95, 1985.
75. Halsband, U. and Passingham, R.E., Premotor cortex and the conditions for a movement in monkeys, *Behav. Brain Res.*, 18, 269, 1985.
76. Kurata, K. and Hoffman, D.S., Differential effects of muscimol microinjection into dorsal and ventral aspects of the premotor cortex of monkeys, *J. Neurophysiol.*, 71, 1151, 1994.
77. Passingham, R.E., Toni, I., and Rushworth, M.F., Specialisation within the prefrontal cortex: the ventral prefrontal cortex and associative learning, *Exp. Brain Res.*, 133, 103, 2000.
78. Gaffan, D. and Harrison, S., Inferotemporal-frontal disconnection and fornix transection in visuomotor conditional learning by monkeys, *Behav. Brain Res.*, 31, 149, 1988.
79. Bussey, T.J., Wise, S.P., and Murray, E.A., Interaction of ventral and orbital prefrontal cortex with inferotemporal cortex in conditional visuomotor learning, *Behav. Neurosci.*, 116, 703, 2002.
80. Eacott, M.J. and Gaffan, D., Inferotemporal-frontal disconnection: the uncinate fascicle and visual associative learning in monkeys, *Eur. J. Neurosci.*, 4, 1320, 1992.
81. Bussey, T.J., Wise, S.P., and Murray, E.A., The role of ventral and orbital prefrontal cortex in conditional visuomotor learning and strategy use in rhesus monkeys (*Macacca mulatta*), *Behav. Neurosci.*, 115, 971, 2001.
82. Wang, M., Zhang, H., and Li, B.M., Deficit in conditional visuomotor learning by local infusion of bicuculline into the ventral prefrontal cortex in monkeys, *Eur. J. Neurosci.*, 12, 3787, 2000.
83. Rushworth, M.F. et al., Attentional selection and action selection in the ventral prefrontal cortex, *Soc. Neurosci. Abstr.*, 28, 722.11, 2003.

84. Gaffan, D., Easton, A., and Parker, A., Interaction of inferior temporal cortex with frontal cortex and basal forebrain: double dissociation in strategy implementation and associative learning, *J. Neurosci.*, 22, 7288, 2002.
85. Collins, P. et al., Perseveration and strategy in a novel spatial self-ordered sequencing task for nonhuman primates: effects of excitotoxic lesions and dopamine depletions of the prefrontal cortex, *J. Cog. Neurosci.*, 10, 332, 1998.
86. Murray, E.A. and Wise, S.P., Role of the hippocampus plus subjacent cortex but not amygdala in visuomotor conditional learning in rhesus monkeys, *Behav. Neurosci.*, 110, 1261, 1996.
87. Rupniak, N.M.J. and Gaffan, D., Monkey hippocampus and learning about spatially directed movements, *J. Neurosci.*, 7, 2331, 1987.
88. Brasted, P.J. et al., Fornix transection impairs conditional visuomotor learning in tasks involving nonspatially differentiated responses, *J. Neurophysiol.*, 87, 631, 2002.
89. McClelland, J.L., McNaughton, B., and O'Reilly, R., Why there are complementary learning systems in the hippocampus and neocortex: Insights from the successes and failures of connectionist models of learning and memory, *Psychol. Rev.*, 102, 419, 1995.
90. Brasted, P.J. et al., Bilateral excitotoxic hippocampal lesions do not impair nonspatial conditional visuomotor learning in a task sensitive to fornix transection, *Soc. Neurosci. Abstr.*, 28, 129.4, 2003.
91. Canavan, A.G.M., Nixon, P.D., and Passingham, R.E., Motor learning in monkeys (*Macaca fascicularis*) with lesions in motor thalamus, *Exp. Brain Res.*, 77, 113, 1989.
92. Nixon P.D. et al., Corticostriatal pathways in conditional visuomotor learning, *Soc. Neurosci. Abstr.*, 27, 282.1, 2002.
93. Nixon, P.D. and Passingham, R.E., The cerebellum and cognition: cerebellar lesions impair sequence learning but not conditional visuomotor learning in monkeys, *Neuropsychologia*, 38, 1054, 2000.
94. Tucker, J. et al., Associative learning in patients with cerebellar ataxia, *Behav. Neurosci.*, 110, 1229, 1996.
95. Thaler, D. et al., The functions of the medial premotor cortex (SMA) I. Simple learned movements, *Exp. Brain Res.*, 102, 445, 1995.
96. Chen, Y.-C. et al., The functions of the medial premotor cortex (SMA) II. The timing and selection of learned movements, *Exp. Brain Res.*, 102, 461, 1995.
97. Gaffan, D. and Harrison, S., A comparison of the effects of fornix transection and sulcus principalis ablation upon spatial learning by monkeys, *Behav. Brain Res.*, 31, 207, 1989.
98. Petrides, M., Motor conditional associative-learning after selective prefrontal lesions in the monkey, *Behav. Brain Res.*, 5, 407, 1982.
99. Petrides, M., Conditional learning and the primate frontal cortex, in *The Frontal Lobes Revisited*, Perecman, E., Ed., IRBN Press, New York, 1987, 91.
100. Rushworth, M.F.S., Nixon, P.D., and Passingham, R.E., Parietal cortex and movement. I. Movement selection and reaching, *Exp. Brain Res.*, 117, 292, 1997.
101. Pisella, L. et al., An 'automatic pilot' for the hand in human posterior parietal cortex: toward reinterpreting optic ataxia, *Nat. Neurosci.*, 3, 729, 2000.
102. Passingham, R.E., From where does the motor cortex get its instruction? in *Higher Brain Functions*, Wise, S.P., Ed., John Wiley & Sons, New York, 1987, 67.
103. Wise, S.P., Murray, E.A., and Gerfen, C.R., The frontal cortex–basal ganglia system in primates, *Crit. Rev. Neurobiol.*, 10, 317, 1996.
104. Hikosaka, O. et al., Parallel neural networks for learning sequential procedures, *Trends Neurosci.*, 22, 464, 1999.

105. Mitz, A.R., Godschalk, M., and Wise, S.P., Learning-dependent neuronal activity in the premotor cortex of rhesus monkeys, *J. Neurosci.*, 11, 1855, 1991.
106. Chen, L.L. and Wise, S.P., Neuronal activity in the supplementary eye field during acquisition of conditional oculomotor associations, *J. Neurophysiol.*, 73, 1101, 1995.
107. Germain, L. and Lamarre, Y., Neuronal activity in the motor and premotor cortices before and after learning the associations between auditory stimuli and motor responses, *Brain Res.*, 611, 175, 1993.
108. Asaad, W.F., Rainer, G., and Miller, E.K., Neural activity in the primate prefrontal cortex during associative learning, *Neuron*, 21, 1399, 1998.
109. Tremblay, L., Hollerman, J.R., and Schultz, W., Modifications of reward expectation-related neuronal activity during learning in primate striatum, *J. Neurophysiol.*, 80, 964, 1998.
110. Hollerman, J.R., Tremblay, L., and Schultz, W., Influence of reward expectation on behavior-related neuronal activity in primate striatum, *J. Neurophysiol.*, 80, 947, 1998.
111. Brasted, P.J. and Wise, S.P., Comparison of learning-related neuronal activity in the dorsal premotor cortex and striatum, *Eur. J. Neurosci.*, 19, 721, 2004.
112. Inase, M. et al., Pallidal activity is involved in visuomotor association learning in monkeys, *Eur. J. Neurosci.*, 14, 897, 2001.
113. Middleton, F.A. and Strick, P.L., Anatomical evidence for cerebellar and basal ganglia involvement in higher cognitive function, *Science*, 266, 458, 1994.
114. Cahusac, P.M. et al., Modification of the responses of hippocampal neurons in the monkey during the learning of a conditional spatial response task, *Hippocampus*, 3, 29, 1993.
115. Wirth, S. et al., Single neurons in the monkey hippocampus and learning of new associations, *Science*, 300, 1578, 2003.
116. Logothetis, N.K. et al., Neurophysiological investigation of the basis of the fMRI signal, *Nature*, 412, 150, 2001.
117. Logothetis, N.K., MR imaging in the non-human primate: studies of function and of dynamic connectivity, *Curr. Opin. Neurobiol.*, 13, 630, 2003.
118. Aguirre, G.K. and D'Esposito, M., Experimental design for brain fMRI, in *Functional MRI*, Moonen, C.T.W. and Bandettini, P.A., Eds., Springer-Verlag, Berlin, 2000, 369.
119. Deiber, M.P. et al., Frontal and parietal networks for conditional motor learning: a positron emission tomography study, *J. Neurophysiol.*, 78, 977, 1997.
120. Eliassen, J.C., Souza, T., and Sanes, J.N., Experience-dependent activation patterns in human brain during visual-motor associative learning, *J. Neurosci.*, 23, 10540, 2003.
121. Toni, I. and Passingham, R.E., Prefrontal-basal ganglia pathways are involved in the learning of arbitrary visuomotor associations: a PET study, *Exp. Brain Res.*, 127, 19, 1999.
122. Toni, I. et al., Learning arbitrary visuomotor associations: temporal dynamic of brain activity, *NeuroImage*, 14, 1048, 2001.
123. Deiber, M.P. et al., Cortical areas and the selection of movement: a study with positron emission tomography, *Exp. Brain Res.*, 84, 393, 1991.
124. Grafton, S.T., Fagg, A.H., and Arbib, M.A., Dorsal premotor cortex and conditional movement selection: a PET functional mapping study, *J. Neurophysiol.*, 79, 1092, 1998.
125. Toni, I., Rushworth, M.F., and Passingham, R.E., Neural correlates of visuomotor associations. Spatial rules compared with arbitrary rules, *Exp. Brain Res.*, 141, 359, 2001.
126. Ramnani, N. and Miall, R.C., A system in the human brain for predicting the action of others, *Nat. Neurosci.*, 7, 85, 2004.

127. Sweeney, J.A. et al., Positron emission tomography study of voluntary saccadic eye movements and spatial working memory, *J. Neurophysiol.*, 75, 454, 1996.
128. Wise, S.P., Weinrich, M., and Mauritz, K.-H., Movement-related activity in the premotor cortex of rhesus macaques, *Prog. Brain Res.*, 64, 117, 1986.
129. Chen, L.L. and Wise, S.P., Supplementary eye field contrasted with the frontal eye field during acquisition of conditional oculomotor associations, *J. Neurophysiol.*, 73, 1122, 1995.
130. Toni, I. et al., Changes of cortico-striatal effective connectivity during visuomotor learning, *Cereb. Cortex*, 12, 1040, 2002.
131. Paus, T. et al., Role of the human anterior cingulate cortex in the control of oculomotor, manual, and speech responses: a positron emission tomography study, *J. Neurophysiol.*, 70, 453, 1993.
132. Dolan, R.J. and Strange, B.A., Hippocampal novelty responses studied with functional neuroimaging, in *Neuropsychology of Memory*, Squire, L.R. and Schacter, D.L., Eds., Guilford, New York, 2002, 204.
133. Strange, B.A. et al., Anterior prefrontal cortex mediates rule learning in humans, *Cereb. Cortex*, 11, 1040, 2001.
134. Wallis, J.D. and Miller, E.K., From rule to response: neuronal processes in the premotor and prefrontal cortex, *J. Neurophysiol.*, 90, 1790, 2003.

11 Learning Dynamics of Reaching

Reza Shadmehr, Opher Donchin, Eun-Jung Hwang, Sarah E. Hemminger, and Ashwini K. Rao

CONTENTS

ABSTRACT

When one moves one's hand from one point to another, the brain guides the arm by relying on neural structures that estimate the physical dynamics of the task. For example, if one is about to lift a bottle of milk that appears full rather than empty, the brain takes into account the subtle changes in the dynamics of the task and this is reflected in the altered motor commands. The neural structures that compute the task's dynamics are "internal models" that transform the desired motion into motor commands. Internal models are learned with practice and are a fundamental part of voluntary motor control. What do internal models compute, and which neural structures perform that computation? We approach these problems by considering a task where physical dynamics of reaching movements are altered by force fields that act on the hand. Experiments by a number of laboratories on this paradigm suggest that internal models are sensorimotor transformations that map a desired sensory state

0-8493-1287-6/05/$0.00+$1.50

of the arm into an estimate of forces, i.e., a model of the inverse dynamics of the task. If this computation is represented as a population code via a flexible combination of basis functions, then one can infer activity fields of the bases from the patterns of generalization. We provide a mathematical technique that facilitates this inference by analyzing trial-to-trial changes in performance. Results suggest that internal models are computed with bases that are directionally tuned to limb motion in intrinsic coordinates of joints and muscles, and this tuning is modulated multiplicatively as a function of the static position of the limb. That is, limb position acts as a gain field on directional tuning. Some of these properties are consistent with activity fields of neurons in the motor cortex and the cerebellum. We suggest that activity fields of these cells are reflected in human behavior in the way that we learn and generalize patterns of dynamics in reaching movements.

11.1 INTRODUCTION

The arm has inertial dynamics that dictate a complex relationship between motion of the joints and torques. In order to reliably produce even the most simple movements — for example, flexion of the elbow — the brain must activate not only elbow flexors, but also shoulder flexors that counter the shoulder extension torque that is produced by the acceleration of the elbow. The importance of these interaction forces was quite apparent when engineers were trying to control motion of robots.[1] Yet the principle is the same for control of biological limbs, as has been confirmed in electromyographic (EMG) recordings from the human arm.[2] This has led to the idea that, contrary to earlier hypotheses,[3] passive properties of muscles are not enough to compensate for the complex physics of our limbs. Rather, the brain must *predict* the specific force requirements of the task in generating the motor commands that eventually reach the muscles.

To illustrate this idea, consider picking up an opaque carton of milk that appears full but has been drained empty. The brain overestimates the mass of the carton by only a couple of pounds (the weight of the missing milk) yet the error is sufficient so that the resulting motor commands produce a jerky motion of the hand. The visual appearance of the bottle apparently retrieves a motor memory in a neural system that predicts the forces that are necessary to move the bottle. Motor commands are constructed based on this prediction and the predicted forces must be accurate if we are to produce smooth movements.

The accuracy of force prediction is particularly important for control of our arm because our hands evolved in large part to support manipulation. For example, a trip to your local natural history museum will confirm that the hand of a chimpanzee has a much longer palm length as compared to a human hand. This means that while we can easily touch our index finger to our thumb and hold an object, say a string that is attached to a yo-yo, a chimpanzee's hand is poorly suited for this. Holding different objects can dramatically change the mechanical dynamics of our arm. The neural system that predicts force properties would have to be able to accommodate this variability and adapt. To study the properties of the neural system with which the brain learns to predict forces, we have used a paradigm (Figure 11.1) where arm

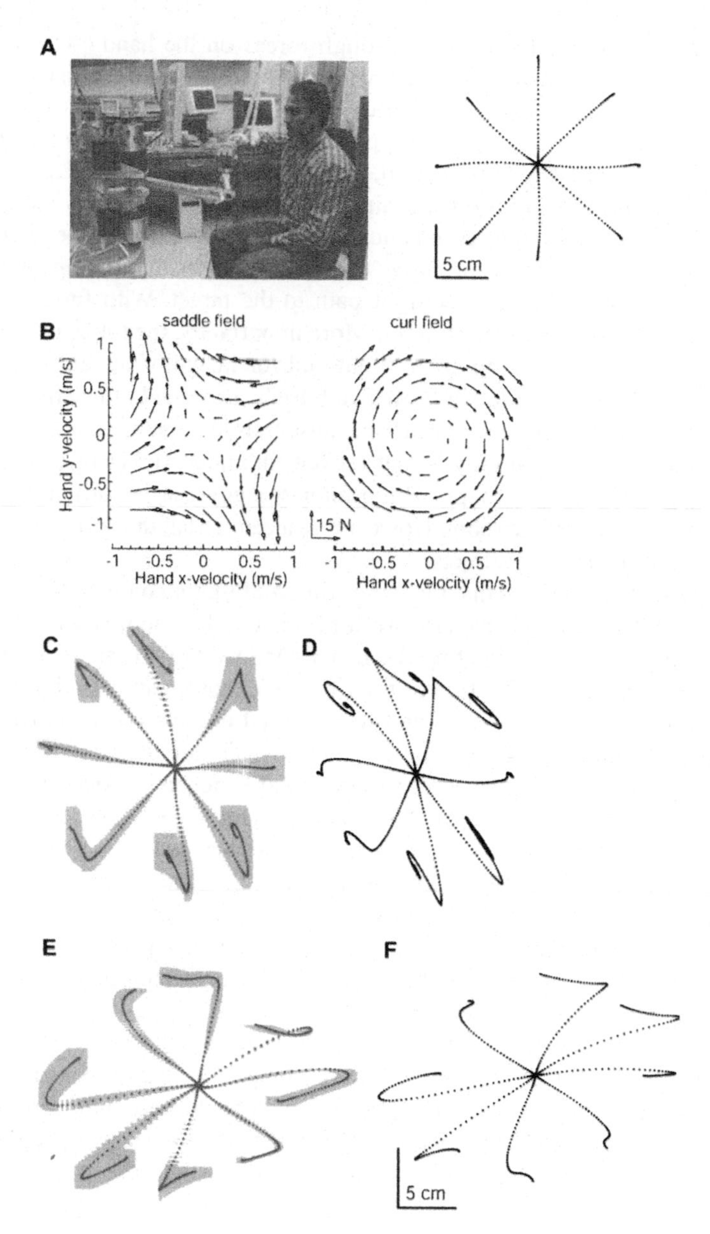

FIGURE 11.1 Experimental setup and typical data. (A) Subjects hold the handle of the robot and reach to a target. The plot shows hand trajectory (dots are 10 msec apart) for typical movements to eight targets. (B) Examples of two force fields produced by the robot. (C) Average hand trajectories (±SD) for movements during the initial trials in the saddle field. (D) Simulation results for movements in the saddle field. (E) Hand trajectories during catch trials. (F) Simulation results during catch trials. The controller in this simulation had fully adapted to the field and was expecting the field to be present in these movements. (Adapted from Reference 4, with permission.)

dynamics are systematically changed through forces on the hand.[4] The subject is provided with a target and asked to reach while holding the handle of a robot. When the robot's motors are off (null field condition), movements are straight (Figure 11.1A). The forces in the field typically depend on the velocity of the hand (Figure 11.1B). When the field is applied, movements are perturbed (Figure 11.1C). With practice, hand trajectories once again become smooth and nearly straight. The brain's ability to modify motor commands and predict the novel forces is revealed as a sudden removal of force in *catch trials*. Very early in training, the hand's trajectory in the catch trials is a straight path to the target. With further training, trajectories in field trials become straight. More importantly, the trajectories in catch trials (Figure 11.1E) become an approximate mirror image of the early field trials (Figure 11.1C). The trajectories in these catch trials are called after-effects.

Improvement in performance occurs because training results in a change in the motor commands. One possibility is that movements improve because subjects co-contract antagonist muscle groups. This motor strategy can be sufficient to resist perturbations imposed by the robot. However, in a catch trial, this kind of adaptation would not produce any after-effects.

An alternate hypothesis is that the composition of motor commands by the brain relies on a neural system that, for any given movement direction, predicts the forces that will be imposed on the hand by the robot. One way to do this is to imagine a tape that is played out as a function of time for each movement direction. This tape may be an average record of forces that were sensed in the previous movements in that direction. Mathematically, the inputs to this system are direction and time and the output is force. To test this idea, Conditt et al.[5] trained subjects to reach to a small number of targets in a force field and then suddenly asked them to draw a circle in the same field. They reasoned that if what was learned was like a tape recording of the forces encountered in reaching to each target, then the neural system that had been trained to predict forces in short, brief reaching movements should contribute little to longer, circular movements. However, they found that performance was quite good in circular movements when the field was on and, importantly, the subjects showed after-effects when the field was off.

This suggested that the neural system did not predict forces explicitly as a function of time. Rather, in performing the reaching movements, the neural system had learned to associate the sensory states of the limb — especially limb position and velocity — to forces. The particular order in which those states were visited and the trajectory at the time they were visited (e.g., in a straight line trajectory or in a curved movement) was immaterial. What was important was the region of the state space — the limb's velocity at a given position — that the reaching movements had visited. If the temporal order of the states were changed from the "training set" in which the system had experienced the forces, the neural system could still predict forces because the states themselves were part of the initial training set.

However, one could argue that the reason why the subjects learned to associate states to forces, rather than some other input that explicitly included time, was because the force field that was imposed on the hand was itself not explicitly time-dependent. Rather, it was dependent on hand velocity. Conditt and Mussa-Ivaldi[6] tested this by asking whether subjects could adapt to force fields that explicitly depended on time. Remarkably, the experimental results indicated that they could not. When a predictable, time-dependent pattern of force was imposed on reaching movements, generalization trials (circular movements) suggested that subjects still learned to associate states of the arm to forces. Therefore, the brain's ability to predict force did not explicitly depend on movement time. Rather, that prediction depended on an input that described the desired state of the arm.

These experiments suggested that with practice, participants learned a sensory to motor transformation where a velocity-like input signal was transformed into a force-like output signal. This is an *internal model* of the force field.

11.2 NEURAL CORRELATES OF LEARNING INTERNAL MODELS OF DYNAMICS

We have not specified how information is represented in this internal model, or how this information is acquired through experience. All we can say at this point is that at the start of training the internal model is "empty" (i.e., it predicts zero force for all input states) and, after a long period of training, it has adapted in the sense that it correctly predicts forces that are produced for typical states visited in reaching movements. However, there is sufficient information in this statement to allow us to test whether our formulation thus far is consistent with measurements.

If a simulation of the dynamics of the arm acquires an internal model of a force field, what will its trajectories of motion look like? The dynamics of the arm (in this case, a two-joint planar system) are derived from Newton's laws and are written as equations that describe how the limb's acceleration depends on forces. They describe how the mass of the limb responds to force input from the muscles. To represent the error feedback system of the muscles and the spinal reflexes, we add to the equations a simple low-gain spring-damper element that stabilizes the limb about the desired trajectory (the straight line). To produce a movement, we assume that the joint torques are commanded based on knowledge of the inverse dynamics of the limb, i.e., a map that transforms the desired sensory state of the limb into torques so that it compensates for the arm's inertial dynamics. This is an internal model of the arm's physical dynamics. These equations have been detailed in Shadmehr and Mussa-Ivaldi.[4]

Initially in training, the simulated internal model has no knowledge of the robot-imposed forces. Because of this, the simulated arm does not move straight to the target (Figure 11.1D). Rather, it moves along a trajectory that is similar to what we have recorded in our participants, that is, a peculiar hooking pattern.[4,7] Now we change the internal model so that it completely takes into account the added dynamics of the force field. That is, we assume that the internal model is fully trained. If we now simulate a catch trial, the resulting movement (Figure 11.1F) is an approximate mirror image of the field trials early in training. Therefore, the trajectories that we had recorded in the reaching movements of our subjects are consistent with learning an internal model that transformed desired sensory states into forces.

It is an easy next step to extend the mathematical formulation and predict not just the limb's trajectory before and after the internal model adapts to imposed forces, but also a correlate of that adaptation at the level of the neural commands to the muscles. The equations of motion that produced trajectories in Figure 11.1 included torque generators but not specific muscles. In another study[8] we added to these equations constraints regarding the function of muscles. The most important constraint was that opposing torques in two antagonistic muscles should be inversely proportional. That is, the more one muscle was activated, the less the antagonist was activated. This assumption allowed us to translate a pattern of expected forces on the hand onto changes in muscle activations. To visualize the changes, we plotted the average magnitude of activation for each simulated muscle as a function of movement direction in hand-centered coordinates and computed a preferred direction (PD) for each muscle.[9,10] The PD for a muscle was the direction of movement in Cartesian coordinates centered on the hand for which the modeled muscle was most active. For example, the simulations predicted that adaptation of the internal model to a clockwise curl field should accompany a clockwise rotation by ~27° for elbow muscles and ~18° for shoulder muscles. A curl field is a particular force field where the forces are always pushing the hand perpendicular to its current direction of motion. In the clockwise version of this, the force vectors are pointing in the clockwise direction perpendicular to the direction of hand motion. EMG in biceps, triceps, and anterior and posterior deltoids in a group of participants confirmed this prediction.[8] This confirmation of the model's prediction illustrated that one way to represent the change in motor commands due to adaptation of the internal model was as rotations in the PD of muscles.

The result is not surprising, because of course the brain would have to change the commands to the muscles if forces are to be produced to counter the effects of the robot-imposed field. But the results are useful for the following reasons. First, because the model predicts that for any given field, formation of an internal model should accompany a specific rotation in the PD of certain simulated muscles, it provides a compact way in which to quantitatively predict the experience-dependent change that should occur in the motor commands. Second, because in the monkey motor cortex, in certain conditions where the PD of muscle activation functions changed, the PD of some cells in the primary motor cortex (M1) changed as well,[11] one can suggest that the rotation in EMG that the model predicts is echoing a similar change in the PD of some motor cortical cells.

In an experiment where monkeys learned reaching movements in a clockwise curl field,[12] Bizzi and colleagues reported that task-related cells in M1 underwent a median clockwise shift in PD of 16°. However, whereas we had found that the EMG patterns returned to baseline conditions once the field was turned off (i.e., a washed out phase), many cells in the motor cortex kept the change in their PD. Bizzi and colleagues labeled these "memory" cells. Remarkably, whereas the memory cells kept their clockwise change in PD when the force field was turned off, another population of cells that had not changed their PD in the force field now changed their PD in a counterclockwise direction when the field was removed. Therefore, after completion of training and return to a null field condition, a distinct population of cells maintained the effect of the field in terms of rotations in their PD. As demonstrated many times previously, the cells in M1 are not simply upper motorneurons.

Rather, in this case they are likely involved in representing the memory of the internal model.

What might be a testable behavioral consequence of the hypothesis that the motor cortex is involved in representing the internal model? One of the consistent properties of task-related cells in M1 is that if a cell is active for reaching movement to a group of targets at one arm configuration, it is also likely to be active when the configuration of the arm is changed and the targets are moved to the new workspace. However, the change in the workspace often results in a change in the PD of the cells. The PD of a typical M1 cell will rotate approximately with the shoulder angle.[13,14] The reason for this rotation possibly lies in the observation that many M1 cells are sensitive to the force requirements of the task. Sergio and Kalaska[11] trained monkeys to generate isometric force ramps in 8 spatially constant directions on a horizontal plane while holding the arm in 9 different locations in a 16-cm diameter workspace. Typically, M1 activity was directionally tuned for the direction of isometric force in any given arm location in the workspace. However, many cells showed small but systematic shifts of directional tuning at different workspace locations, even though the output force was in a constant spatial direction. On average, there was a significant clockwise rotation of cell PDs from the central hand location to locations to the right, and a significant counterclockwise rotation of cell PDs for hand position to the left. These rotations were consistent with the rotation of PDs in the shoulder and elbow muscles of the arm in the same task.

Because the memory cells that Bizzi and colleagues[12] found were sensitive to changes in force properties of the task, i.e., their PDs rotated as the task was changed from a null field to a curl field, we can hypothesize that these cells might be "muscle-like." By this, we mean that their PDs will change as the workspace in which the reaching movements are performed changes. We would expect that their PDs will rotate with the shoulder in a way similar to the rotation of PDs in arm muscles performing the same task. Imagine that the force field–related changes in the PD and the posture-related changes in PD are cumulative. Then training in one workspace should result in the rotation of PDs by a certain amount, and translation of the arm to a new workspace should result in an additional rotation by an amount approximately equal to the rotation in the shoulder joint. At the new workspace, despite the fact that no prior training had taken place there, an effect of the training elsewhere should be observed, i.e., we should observe generalization. This is a behavioral prediction of the model.

However, it is certainly not the case that all M1 cells are "muscle-like" in their tuning properties. In many instances, experiments have demonstrated that a significant portion of cells in M1 code for parameters of reaching movements in extrinsic coordinates.[15,16] Indeed, in their force field learning experiment, Bizzi and colleagues[12] found that 34% of the M1 cells they recorded had tuning properties that remained invariant despite the changes in force properties of the task. (These were labeled as "kinematic" cells.) One would predict that these kinematic cells would not change their PD with the configuration of the arm. Therefore, our hypothesis assumes that M1 cells that have more muscle-like properties — i.e., cells that change their discharge patterns in a way that correlates with changes in muscle activations — would be predicted to be the ones that contribute most to the representation of internal models for the dynamics of reaching movements.

11.3 GENERALIZATION AS A FUNCTION OF ARM POSITION

Because M1 cells that have muscle-like properties in their tuning tend to rotate their PD with the shoulder angle, learning an internal model with these cells should result in a specific pattern of generalization. For example, consider adaptation to a force field described by $\mathbf{f} = B_1\dot{\mathbf{x}}$, where $\mathbf{f}$ is a force vector acting on the hand, $\dot{\mathbf{x}}$ is a hand velocity vector, and $B_1 = [-11, -11; -11, 11]$ N·sec/m. This is a "saddle field" (Figure 11.1B) where movements toward 120° and 300° encounter a resistive force and movements toward 210° and 30° encounter an assistive force. If the right arm is near the horizontal plane and the shoulder is flexed so that the hand is at a "left" workspace (meaning that reaching movements are performed in a flexed posture for the shoulder), the PD of the triceps is about 90°. When a subject trains in the field, one observes a 30° clockwise rotation in the PD of the triceps. Now imagine that there are cells in the motor cortex that also rotate their PD by an amount similar to this. If we now take the subject's arm and extend the shoulder so that the hand is at a right workspace, we would expect that M1 cells that were directionally tuned with the arm in the left workspace to be also directionally tuned when the hand is at the right workspace. Furthermore, we would expect that on average, the 90° clockwise rotation in the shoulder joint should cause the PD of these cells to rotate by an average of 90°. So for a motor cortical cell that was "muscle-like" and had a PD of, say, 180° at the left workspace, adaptation to the field at that workspace should cause the PD to change to 150° (i.e., 180° – 30°), and movement of the hand to the right workspace should bring the PD to 60° (i.e., 180° – 30° – 90°). If the subject had not practiced movements in the field, this cell would have a PD of 90°. Therefore, the effect of training at the left workspace should be observable in terms of the behavior of the hand at the right workspace if the "memory cells" that rotated their PD at the left workspace maintain their relative rotation at the right workspace. In terms of forces, this corresponds to a field where the relative rotation of the muscle PDs is maintained as a function of the shoulder angle.

One can approximate such a force field by transforming forces on the hand at the "left" workspace to joint torques, and then transforming the torques back to hand forces at the "right" workspace.[17] For our saddle field, this procedure produces the surprising result that a 90°-rotation in the shoulder results in a 180°-rotation of the matrix B_1. This theoretical result means that the force field described by B_1 should be generalized to $-B_1$ at the right workspace. We were intrigued by this prediction because we had observed earlier that if one adapts to field B and then is given field $-B$ in the same workspace, performance in $-B$ is absolutely terrible. In fact, performance in $-B$ for these subjects is far worse than performance of naïve subjects in the same field.[18] The model now predicted that if after training in B we simply moved the subject's arm to a new location, we would see that performance in $-B$ is quite good. Experiments confirmed this prediction.[17] The results suggested the intriguing theory that not only might the motor cortex take part in representing the memory of the internal model, but also the properties of activity fields (or tuning) of cells in M1 might be related to the behavioral patterns of generalization in force fields. The property of activity fields that is relevant in this case is the change in PD as a function of shoulder angle.

11.4 COMPUTING AN INTERNAL MODEL WITH A POPULATION CODE

How does one test the idea that activity fields of certain cells influence patterns of generalization during the learning of reaching movements? Alternatively, how does one infer the shape of the activity fields from the patterns of behavioral generalization? We need to advance beyond a description of the input–output variables that are encoded by internal models (sensory state of the arm and force, respectively) and consider how the transformation from input to output might take place. That is, we must first consider how the central nervous system might compute internal models.

One of the most widely used models of neural computation is population coding. While the idea of using populations of neurons to code variables of interest is old,[19] it has become a compelling tool since it was combined with a simple decoding strategy called a population vector to reconstruct the direction of reaching movements from cells in M1.[20] To motivate our approach, let us put aside for now the problem of predicting force as a function of velocity and consider the simpler problem of representing direction of movement of the hand. Georgopoulos et al.[20] recorded from a collection of cells in M1 and asked whether one could estimate the direction of a reaching movement from the discharge of cells. Each cell had a PD of movement, a vector of unit length $\mathbf{w}_i$. The movements were on a plane. Therefore, $\mathbf{w}$ is a two-dimensional vector that might point in any direction about a unit circle. In a given trial, imagine that the movement direction is θ, and each cell i discharges by amount r_i. This discharge can be decomposed into two terms. The first term is an average response $g_i(\theta)$ which represents the cell's tuning curve as computed over many movements to various directions. The second term is noise n_i that we might encounter at any given trial i:

$$r_i = g_i(\theta) + n_i \tag{11.1}$$

In this equation, the first term is the tuning curve of the cell and the second term is noise. Experiments show that the tuning curve is typically a cosine-like function of movement direction and has a half-width at half-height value of approximately 56°.[21] The second term is noise that cannot be accounted for by the "input" (target direction). Experiments suggest that this noise term (for neurons in the visual cortex) is often normally distributed with a variance that is proportional to the mean value of the tuning function.[22] If cells did not have this noise and we could record from a large number of cells at the same time, we would note that cell j happened to fire most during some movement and estimate the movement direction $\hat{\theta}$ to be the PD of that cell:

$$\hat{\theta} = \mathbf{w}_j \tag{11.2}$$

This is a winner-take-all coding. However, because cells are noisy, our estimate would have a large variance from trial to trial, even though the actual direction of movement did not change. A better approach is a population code.[20] In this approach, each cell's discharge is weighted by its PD vector. The sum of these vectors produces the estimate of movement direction:

$$\hat{\theta} = \sum_i \mathbf{w}_i r_i = \sum_i \mathbf{w}_i g_i(\theta) + \mathbf{w}_i n_i \tag{11.3}$$

This approach is better in the sense that it produces a smaller variance in its estimate from trial to trial (when the movement direction is fixed) than in the winner-take-all approach. In fact, if the tuning curves were exactly cosine functions, the estimate would be optimal in the sense that its variance would be as small as possible.[23] Therefore, the success of population coding depends on computing with neurons that broadly encode the input variable. Where this condition has been approximately met, experiments have generally demonstrated that a population code could successfully be used to estimate the input variable from noisy neuronal discharge.[20,24]

The example of population coding above is an instance of the neural computation of an *identity map*, i.e., a map where the output is an estimate of the input variable. In general, a population code could also be used to map an input variable x into any other variable y.[25,26] In this case, the tuning curves of the neurons that participate in this computation become the basis functions with which the output is approximated. Basis functions are a set of functions such that when they are linearly combined, they can approximate almost any linear or nonlinear function. For example, Pouget and Sejnowski[27] suggested that neurons in the parietal cortex might serve as basis functions with which the brain could compute the position of a visual target with respect to the head. Cells in this region of the brain typically have a discharge r that is modulated by both the position of the eye x_e in the orbit and the position of the target on the retina x_r. These cells have a preferred position on the retina where discharge is at a maximum, and this discharge is modulated approximately linearly with the position of the eye.[28] The tuning function of a cell i can be labeled as $g_i(x_e, x_r)$. Using a weighted sum of these functions, one could estimate position of the target with respect to the head:

$$\hat{y} = \sum_i \mathbf{w}_i g_i(x_e, x_r) + \mathbf{w}_i n_i \tag{11.4}$$

The appropriate weighting $\mathbf{w}_i$ would have to be learned to form this map. However, Pouget and Sejnowski[27] point out that because the tuning functions are the bases by which the map is constructed, the same bases can be used to form any other representation — for example, a shoulder-centered representation of the target. This idea is important because it demonstrates that population coding, a method that can be used to form neural computations of identity maps, is equally suited to the more general problem of computing nonlinear maps. Another point is that whereas in the population code described for decoding of movement direction the weights were vectors that were static and pointed in the PD of a cell, here, if the bases are to be used for learning arbitrary maps, then the weight vectors will change and will have no specific relationship with the tuning function.

Let us now return to our problem of how the brain might compute an internal model. One can think of an internal model as a map that transforms sensory input regarding the desired state of the arm $\mathbf{x}$ (i.e., the position and velocity of the arm) into force $\mathbf{f}$. Let us assume this neural computation is performed via a population code. Each neuron that participates in this computation has a tuning curve g_i that describes the average discharge of that cell as a function of hand position and velocity. Each cell also has a *preferred force vector* $\mathbf{w}_i$. The population vector response of the network is as follows:

$$\hat{\mathbf{f}} = \sum_i \mathbf{w}_i g_i(\mathbf{x}) + \mathbf{w}_i n_i \tag{11.5}$$

We now have a framework for relating tuning properties with behavioral generalization. Consider the following experiment: participants are initially trained with force field $\mathbf{f}_1$ for movement along arm state $\mathbf{x}_1$. The error that they experience in a movement changes the preferred force vector $\mathbf{w}$. Assuming Hebbian learning rules, the weight change will be maximum for those neurons that happened to be most active at about state $\mathbf{x}_1$. The subject is then asked to make a movement with the arm along state $\mathbf{x}_2$, an arm position (or velocity) for which the subject has not been trained. If performance is different from naïve, then the function g_i for which the weights adapted for movements along $\mathbf{x}_1$ must have been broad enough not only to be active for $\mathbf{x}_1$, but also for $\mathbf{x}_2$. Therefore, if the internal model is represented via a population code, then generalization is described by the shape of the tuning curves of the elements of computation.

From the experiment in which we observed generalization in one arm from one workspace to another workspace,[17] we can now conclude that the tuning functions could not have had "preferred positions" of the hand, in the sense that this implies sharply tuned activity functions around that position. If they did, then experiencing force at one hand position could not be generalized to another hand position far away.

11.5 INFERRING CODING OF LIMB POSITION AND VELOCITY FROM PATTERNS OF GENERALIZATION

The idea is that the tuning properties of "muscle-like" cells in M1 may be the function g in this population coding (Equation 11.5). To describe mathematically how discharge varies with arm position and velocity, we note that cell activity in

M1 is modulated globally and often linearly as a function of limb position,[29] and cells have PDs of movement that often change as a function of the shoulder angle.[13] To capture these observations, Hwang et al.[30] hypothesized that cells that are involved in computing the internal model have tuning functions that are described as follows:

$$
\begin{aligned}
g(\underline{q},\dot{\underline{q}}) &= g_{position,i}(\underline{q}) \cdot g_{velocity,i}(\dot{\underline{q}}) \\
g_{position,i}(\underline{q}) &= k^T \cdot \underline{q} + b \\
g_{velocity,i}(\dot{\underline{q}}) &= \exp\left(\left\| \dot{\underline{q}} - \dot{\underline{q}}_i \right\|^2 / 2\sigma^2 \right)
\end{aligned}
\tag{11.6}
$$

The above function represents output of a basis function. The position-dependent term is a linear function that encodes joint angles, $\underline{q} = (\theta_{shoulder}, \theta_{elbow})$, while the velocity-dependent term encodes joint velocities. Figure 11.2C plots this function for reaching movements to various directions at various starting positions. The basis is sensitive to both the static position of the limb and its velocity. It combines the two via a gain field, i.e., directional tuning is modulated multiplicatively as a function of limb position. As a result, both the PD of the tuning and the depth of modulation vary with the starting position of the reach. The gradient vector *k* reflects sensitivities for the shoulder and elbow displacement, and *b* is a constant. The velocity-dependent term is a Gaussian that encodes joint velocity $\dot{\underline{q}}$ centered on the preferred velocity $\dot{\underline{q}}_i$.

The multiplicative nature of this encoding is one of the requirements of basis functions.[27] However, we should note that the properties that we assigned to the tuning function are not unique to cells in M1. For example, in addition to M1, linear modulation of discharge with respect to limb position has been observed in the spinocerebellar tract[31] and the somatosensory cortex.[32] Tuning functions that have PDs or velocities of movement have been reported in the cerebellar cortex.[33] Indeed, it appears that a linear modulation of discharge with respect to limb position and a Gaussian tuning with respect to arm velocity may be a fundamental property of many cells in the motor system.

Consider a situation in which the internal model is constructed as a linear combination of these nonlinear bases. How would their activity fields be reflected in behavioral generalization? With training at a given arm configuration, the preferred force vector of some of these bases will change. The change will occur in those bases that happen to be most active at this arm configuration. The way that these active bases change their static discharge with arm position dictates how far in position space the learning will generalize.

Next, note that because the bases encode joint velocity and not hand velocity, the PD of movement (which is represented in Cartesian coordinates) will rotate for some of the cells as the shoulder joint changes position. The way that the elements change their PDs with arm position dictates the coordinate system of the generalization.

Let us first examine how adaptation with these elements results in generalization in terms of the spatial displacement of the hand. Suppose one trains subjects in a

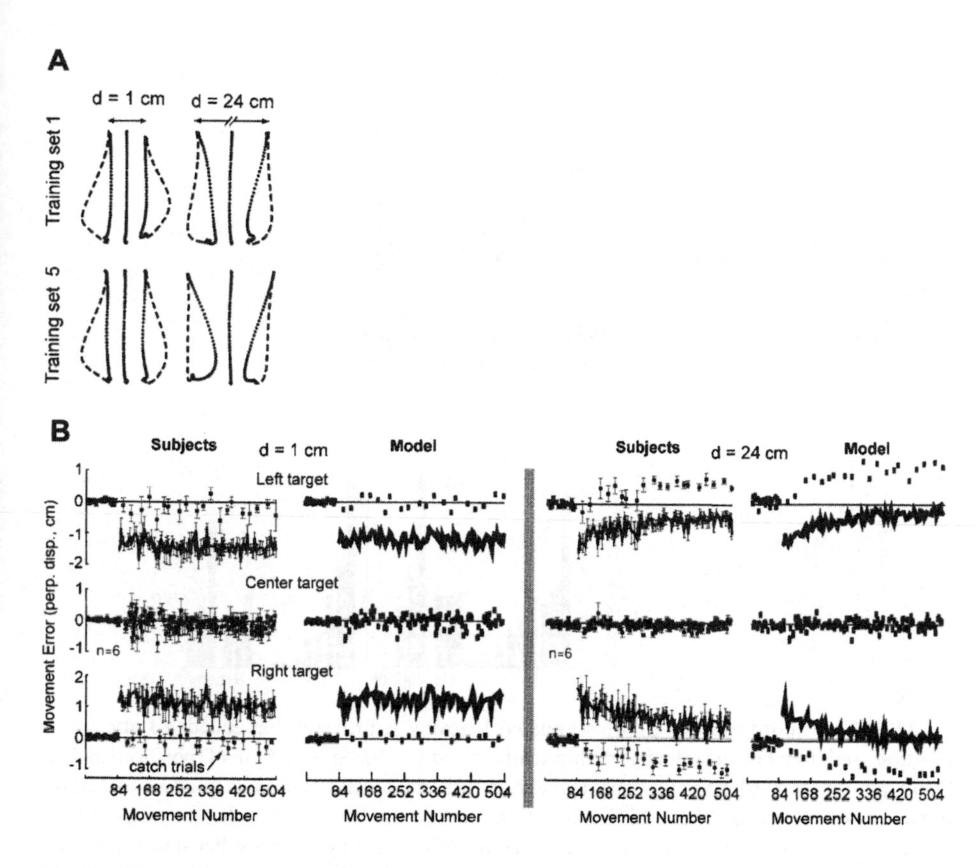

FIGURE 11.2 A gain field coding of limb position and velocity in the internal model of arm dynamics. (A) Subjects made parallel, 10-cm movements at left, center, and right toward target at –90°. At the left and right targets a curl field was present (clockwise at left, counterclockwise at right). The field at the center target was always null. Subjects were divided into four groups based on the distance between the left and right targets: $d = 1$, $d = 6$, $d = 14$, and $d = 24$ cm. Typical movements for single subjects in the $d = 1$ and $d = 24$ cm groups are shown (field trials are dashed, catch trials are dotted). The 24-cm group could learn the task: they had clear after-effects and movements in the field improved. The 1-cm group did not show after-effects and movements did not improve. (B) Group data (mean ± SD) for the 1- and 24-cm groups ($n = 6$ in each group). Field trials are connected with a solid line; catch trials are dots. Note the increased interference in the center target for the 1 cm group as field trials start.

small workspace (of reaching) with the arm at one configuration, and then tests them at another workspace. The gain k dictates how close two workspaces have to be (in position space) before learning of conflicting fields becomes impossible. When the gain is high, output of the bases changes greatly as a function of hand position. This results in poor generalization between neighboring positions of the hand, making it possible to learn two different patterns of force at two different hand positions. When

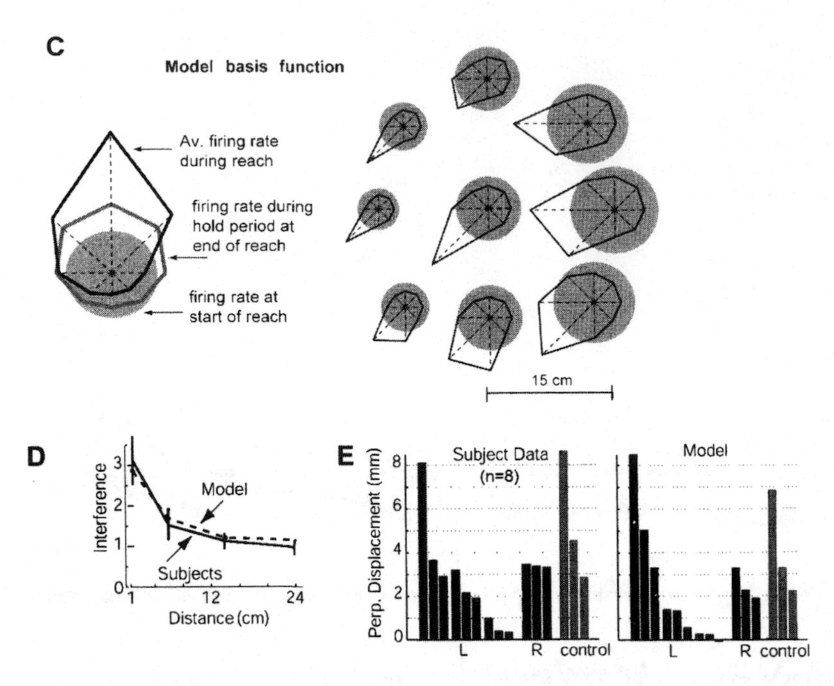

FIGURE 11.2 (continued) (C) We simulated learning with bases that encoded static hand position and movement velocity multiplicatively via a linear function of position and a Gaussian function of velocity. The firing rate of one basis function is plotted for eight directions of movement during movement time, and at the hold time at the start and end of the movement. Firing rate is also plotted for another basis for movements that start from nine different start positions. Note that the preferred direction of the basis rotates with start position. Firing at start of reach varies with start of reach. Because of multiplicative interaction of position sensitivity with directional tuning, depth of modulation varies with start position. (D) An interference measure that quantifies how the left and right movements affected the movements at the center. This is the ratio of standard deviations (SD) for the center movements 420–504 versus 1–84. As distance between the targets decreases, interference increases. The bases appear to account for the generalization pattern of our subjects over this small displacement of the hand. (E) We asked whether the bases could also account for the generalization pattern that we had observed in Shadmehr and Moussavi.[17] In that report, we observed that after subjects were trained in a small "left" workspace, performance was significantly better than naïve when they were tested at a "right" workspace (80 cm away) in a rotated version of the same field. The bases produced similar patterns of generalization as our subjects over this large displacement. (Adapted from Reference 30, with permission.)

the gain is low, output of the bases changes slowly as a function of hand position. At its limit, output changes not at all as a function of hand position and there is effectively no coding of hand position. In this situation, forces generalize globally in hand position space and it is not possible to learn two different forces for the same spatial direction of movement in two different spatial locations.

To quantify how people generalize forces as a function of hand position, Hwang et al.[30] performed an experiment in which participants made reaching movements from different starting locations to targets in the same direction (i.e., the direction of movement was the same in joint velocity space). The arm was covered by a semitransparent screen, upon which an overhead projector painted targets. The handle of the robot, held by the subject, housed an LED that was visible on the screen. Therefore, subjects had visual and proprioceptive cues regarding hand position. In these parallel movements, opposite curl fields acted on the hand in the *left* and *right* targets (Figure 11.2A). A null field was always present in the *middle* target. The robot brought the hand to a random start position after completion of each movement. When the left and right targets were 24 cm apart, subjects could learn these movements, and few after-effects were present in the middle target (Figure 11.2B). As targets were brought closer, learning became more difficult and interference became apparent in the middle target.

Figure 11.2D plots the dependence of interference on the spatial distance. The shape of this interference pattern constraints the gain k, i.e., the slope of the gain field. We simulated learning with Equations 11.5 and 11.6 and kept k the same for all bases. We found the gain k that produced an interference pattern similar to what we had seen in our subjects. Next, we took these same bases and asked whether they could also explain the global generalization pattern that we had seen earlier: that is, when subjects learned a field at the left workspace and generalized to 80 cm away at the right workspace.[17] Remarkably, we found that the same k also explained the amount of generalization that we had seen in that experiment (Figure 11.2E). Furthermore, the forces were generalized in terms of torques on the joints, rather than forces on the hand. Therefore, bases that are linearly modulated by arm position (Figure 11.2C) and encode joint velocity rather than hand velocity appear to explain the pattern of interference in Figure 11.2D, the intrinsic coordinate system of generalization, along with its large spatial generalization in Figure 11.2E.

The reason why we assumed that the bases linearly coded hand position space was because discharge of cells in the spinocerebellar tract,[31] the somatosensory cortex,[32] and M1[29] is modulated linearly with hand position. The reason for assuming that the bases encoded joint velocity (rather than hand velocity) was because such encoding can account for the observation that the PD of many task-related cells rotates with a rotation in the shoulder angle.[34,35] Therefore, generalization as a function of static position of the arm seems consistent with bases that encode limb velocity and position in intrinsic, joint-like coordinates. The bases appear to be tuned to direction of movement and that tuning is multiplicatively modulated as a linear function of limb position.

11.6 GENERALIZATION FROM ONE ARM TO THE OTHER

We can take this argument a step further and predict generalization patterns from one arm to another based on the tuning properties of cells in M1. One of the remarkable properties of many of these cells is that if their discharge is modulated as a function of movement direction for the contralateral arm, that discharge is also modulated when the reaches are performed with the ipsilateral arm. In fact, one of the authors recently observed that many cells in M1 maintained their PD of motion (calculated with the contralateral arm) even when the ipsilateral arm was performing the reaching movements and the contralateral arm was at rest.[36] Kalaska and colleagues also observed similar properties of tuning functions for cells in the premotor cortex.[37] An important technical point in both of these experiments is that both arms performed reaching movements directly in front of the animal in the same workspace.

Let us now imagine that this invariance of PD with respect to the right and left arms is also a property of the cells that we hypothesized were involved in computing an internal model. Consider a subject who trained with her right hand in a curl force field, resulting in a rotation of PDs in some cells by approximately 30°. We now ask the subject to use her left hand and make reaching movements. Because the neurons in the left hemisphere changed their PDs due to training with the right hand, and because these same cells are also tuned for movements with the left hand, they could potentially influence movements with the left hand. In fact, the model predicts that there should be generalization from the right arm to the left arm. Furthermore, it makes the surprising prediction that the generalization from one arm to another should be in extrinsic, Cartesian-like coordinates!

The prediction is surprising because we noted before that PDs rotate with the shoulder of the trained arm, causing the training to generalize in intrinsic coordinates within the same arm. The theory now predicts that because the PDs are invariant to the arms, if we looked for generalization between arms, we would see transfer in extrinsic coordinates.

Criscimagna-Hemminger et al.[38] tested this using the standard reaching movement paradigm with curl fields. Hand position was directly in front of the subject centered on the midline. We considered two coordinate systems for generalization: intrinsic (joint) coordinates and extrinsic (Cartesian) coordinates. In the *intrinsic coordinate* system, if a movement to a given direction required increased activity in the biceps (for example), then the same movement direction with the other arm should also require increased activity in biceps. Mathematically, this results in a mirror transformation of the force field to the other hand. In the *extrinsic coordinate* representation, the system would expect the same forces to act on the other hand in terms of direction of movement in Cartesian space. Was there generalization from one arm to the other, and if so, was the transfer in extrinsic coordinates?

We first quantified generalization in right-handed individuals from right to left. In comparing performance of the extrinsic and intrinsic groups to a control group (Figure 11.3), we found significant interlimb generalization in extrinsic coordinates only. In the transfer trials, the extrinsic group's performance with the left hand was significantly better than controls (Figure 11.3C), whereas in the intrinsic group performance with the left hand was significantly worse than controls (Figure 11.3C).

We next quantified generalization in right-handed individuals from left to right. Subjects trained in a procedure similar to that of Figure 11.3A, except that they trained with the left arm in a curl field and were then tested with the right arm on either the extrinsic or the intrinsic representation of the same field. We found that performance during the test of generalization was not significantly different from controls on either the extrinsic or the intrinsic representation (data not shown). This suggested that in humans, generalization of arm dynamics in right-handed individuals occurred only from the dominant right to the left arm, and its coordinate system was extrinsic in the workspace that we tested.

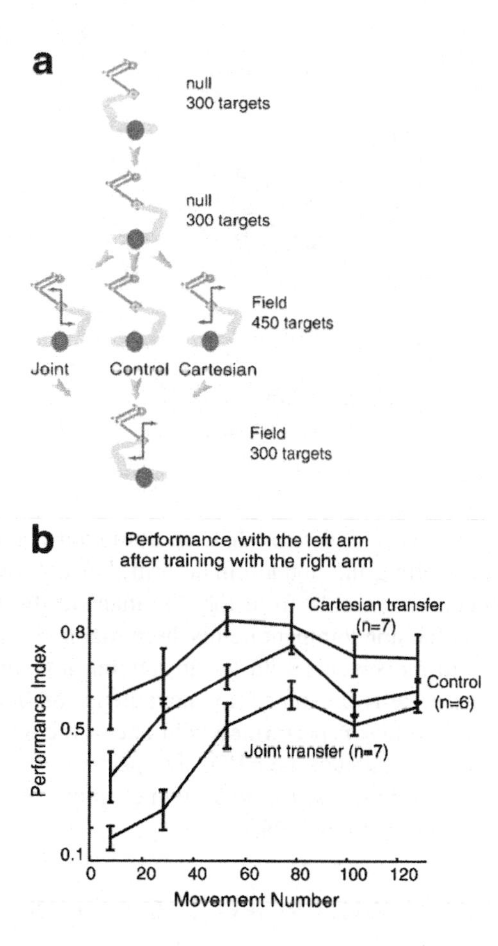

FIGURE 11.3 Interlimb generalization. (a) Right-handed subjects (n = 20) trained with the left and then the right arm in a null field. They were then assigned to one of three groups: training with the right arm in a clockwise curl or in a counterclockwise curl, or a control group that received further training in the null field. All of these groups were then tested with their left arms on a clockwise curl field. For subjects who trained with their right arms on a clockwise curl field, this was a test of interlimb transfer in extrinsic (Cartesian) coordinates. For subjects who trained with their right arms in a counterclockwise curl field, this was a test of interlimb transfer in intrinsic (joint) coordinates. (b) Performance index in the test trials (mean ± SEM). Cartesian coordinate group shows transfer. Joint coordinate group shows interference. (c) Performance of subject J.W. (a split-brain patient) on the Cartesian and joint transfer of the field from right arm to left. This subject also showed generalization from right arm to left. (Adapted from Reference 38, with permission.)

Interestingly, the same pattern of generalization was observed in a right-handed callosotomy patient. In callosotomy patients, when visual information is restricted to one hemisphere, that hemisphere can produce a reaching movement with the ipsilateral arm.[39] This is because a small but significant number of corticospinal projections to the proximal muscles of the arm are from the ipsilateral hemisphere.[40]

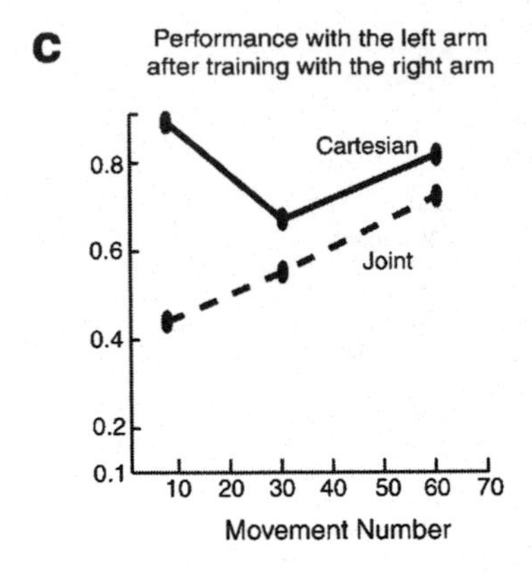

FIGURE 11.3 (continued)

However, converging evidence[41] indicates that the dominant hemisphere may have a significant role in controlling the nondominant arm, but not vice versa. The fact that we observed generalization only from the dominant to the nondominant arm suggests that the cells in the nondominant hemisphere that participate in learning in this task are not tuned to movements with the ipsilateral arm. In contrast, cells in the dominant hemisphere (left) are tuned to movements of both arms and they maintain their PDs when the task is performed with one or the other arm. Therefore, the fact that some M1 cells maintain their PDs irrespective of the arm that is used for reaching is consistent with the coordinate system of interlimb generalization that we observed in the learning of force fields.

11.7 ACTIVITY FIELDS WITH RESPECT TO COLOR OF THE TARGET

The main claim of the hypothesis is that tuning properties of cells in the motor cortex can strongly influence behavior. In particular, the tuning properties predict how we learn dynamics of reaching movements. In most of the examples that we have considered thus far, forces that were imposed on reaching movements explicitly depended on the proprioceptive state of the arm. This made sense because cells in the motor cortex are sensitive to these states, and we wished to infer how this sensitivity influences learning. Let us now consider a task where the forces do not depend on state of the arm. For example, imagine a reaching movement where position or velocity (or any other kinematic variable of the arm) does not uniquely describe the forces in the task. A very simple case is one where a target is presented at a given direction, but the forces that will be presented during that movement depend on the color of the target.

If the cells that take part in learning this task are strongly tuned with respect to position or velocity of the arm and not to the color of the target, then this apparently simple task should be, in fact, extremely difficult to learn. Gandolfo et al.[42] asked

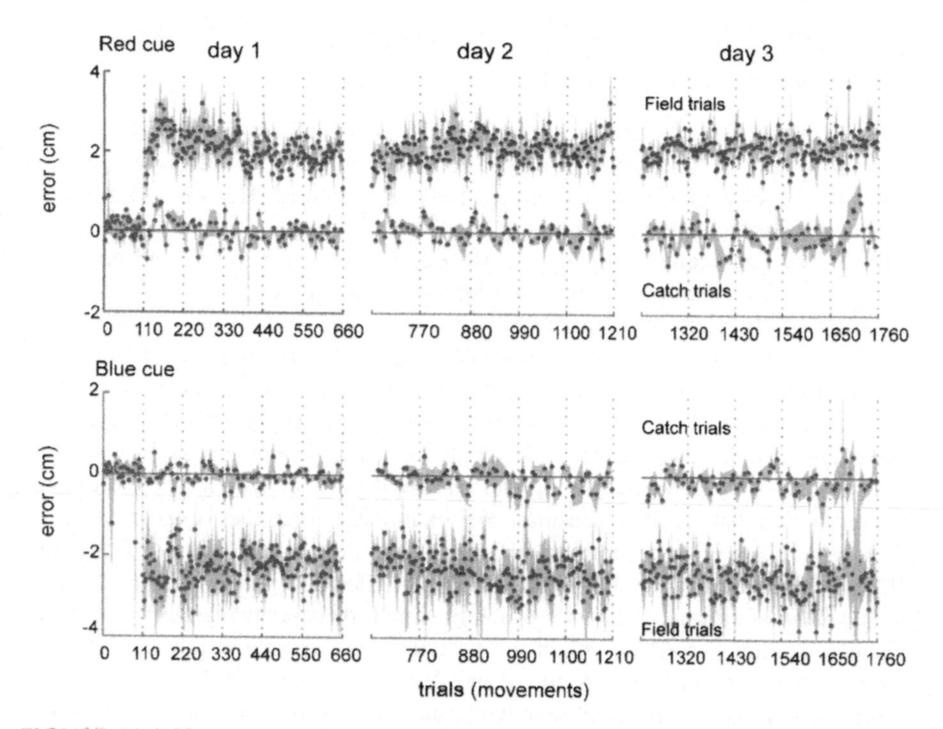

FIGURE 11.4 Naïve subjects ($n = 3$) were trained to associate color cues (blue or red colored squares appearing in the direction of the primary target) to force field B (a clockwise curl field) or $-B$ (a counter-clockwise curl field). Error in each trial was measured as perpendicular displacement from a straight-line trajectory. Data are means ± SD. Performance remained poor and no field specific after-effects developed in catch trials despite 3 days of training.

subjects to make movements to various directions. For the initial 48 movements, a velocity-dependent field, labeled B_1, was present. For the next block of 48 movements, field $-B_1$ was present, and so on. During presentation of each field, the room was flooded with a specific color of light. Despite hundreds of movements, subjects never learned to use the color as a cue to predict the pattern of forces.

We recently simplified this experiment by limiting movements to only one direction.[43] On any given trial, the color of the target was randomly chosen as either red or green. Because the movement was always in the same direction, the pattern of forces on that movement depended exclusively on this cue. We trained subjects ($n = 3$) extensively on this task, providing them with more than 3000 trials, spread over 3 days. Remarkably, in catch trials we consistently found no evidence of after-effects (Figure 11.4) and performance showed no suggestion of adaptation. However, with longer training[44] or with explicit instruction about the nature of the forces,[45] it is possible to associate color with force fields. The remarkable difficulty in learning this apparently simple task leads to the prediction that the activity fields of the bases are typically only weakly modulated by the color of the target.

11.8 INFERRING SHAPE OF THE TUNING CURVES FROM PATTERNS OF GENERALIZATION

The assumption about the formation of an internal model via a population code means that when one measures generalization, one might be able to infer activity fields of the bases with which the internal model is computed. However, it is useful to outline the problems that are inherent in this approach.

To measure generalization, subjects are trained with an input $\mathbf{x}_1$ and are then tested with a new input $\mathbf{x}_2$. The first problem with this approach is that it requires an experimentally naïve set of participants to be trained in each pairing of $\mathbf{x}_1$ and $\mathbf{x}_2$. As a result, behavioral experiments are often limited to training and testing with one or two pairs of inputs, and conclusions are in terms of qualitative statements regarding the shape of the bases (wide or narrow). The second problem is that in motor control, we have to consider coordinate systems. Generalization depends not only on the distance between training and test locations, but also on the coordinate system in which that space is measured. For example, a force that is experienced at a given location may be generalized in terms of torques on the joints or forces on the hand. These two coordinates predict different patterns of generalization in terms of the position of the hand. The third problem is that the bases that are inferred from one generalization experiment might not be consistent with those that are inferred in another. In other words, adaptation to one force field might result in a pattern of generalization that is inconsistent with the pattern observed in adaptation to another field. It would indeed be remarkable if behavioral data from a wide variety of force adaptation experiments suggested a consistent shape to the bases. If this were the case, then one could argue that one has estimated the basic motor primitives with which internal models are computed. Finally, even if we are lucky enough to solve all of these problems, we would still have the problem of interpretation: we would hope that the bases that are inferred by this abstract model not only explain behavior, but also are interpretable in terms of the neurophysiology of the motor system. In this section, we show that all four questions can be approached.

To approach the first two questions — being limited to naïve subjects and needing to consider coordinate systems — we have been developing a new mathematical method to estimate the shape of the bases from the trial-to-trial variations in performance.[46,47] We leverage the fact that the shape of the bases determines how error experienced during a movement will generalize to the subsequent movement. That is, the preferred force vector associated with a basis is likely to change most for those bases that are most active. This means that if error on one movement effects behavior on the next movement in a different direction, then some of the bases must be reasonably active during both of these movements. That is because they must be active in the first movement to be influenced by the error and active in the second movement in order to have an influence on behavior. Thus, generalization of error from one movement to the next can tell us whether the bases are wide enough to encompass both the movements, or, alternatively, whether they are so narrow that they cannot span the gap. The generalization function depends on how the tuning curves encode movements, and one can acquire critical information regarding generalization from the trial-to-trial variations in behavior.

The importance of this idea is that it suggests that it is not necessary to train in one set and then test in another in order to estimate generalization. Rather, all possible inputs should be presented in a random sequence. From the movement-to-movement changes in performance, one can estimate how error in one movement affected the subsequent movement as a function of the distance of the two movements in the state space (for example, angular distance in the directions of the two movements). The result is a generalization function.

Begin with the assumption that the internal model is composed of a linear combination of an unknown set of bases (Equation 11.5). These bases encode the state of the arm (velocity only in the current case, because we limit movement to a small spatial workspace). Assume that the purpose of learning is to minimize the difference between the expected force in a movement and the actual force, and that adaptation is through gradient descent that results in modification of the preferred force vector associated with each basis. How does the shape of the bases affect the pattern of trial-to-trial errors?

Donchin et al.[48] demonstrated a method for quantifying generalization from the trial-to-trial measures of behavior. The idea was to represent adaptation with basis functions as a hidden-state dynamical system. Once that system was expressed mathematically, fitting it to the data would provide an estimate of the generalization function.

As an example, consider a task where subjects make movements to eight directions in a random order. We are interested in estimating how force experienced in a given direction is generalized to all other directions and would like to understand how that generalization depends on the shape of the basis functions g. To simplify matters, let us ignore the noise in Equation 11.5, assume that the bases are only a function of velocity because our reaching movements will all be performed in a small spatial workspace, and rewrite that equation in terms of vector quantities:

$$\begin{aligned}
\hat{\mathbf{f}} &= W\mathbf{g}(\dot{\mathbf{x}}) \\
W &= \begin{bmatrix} w_{11} & \cdots & w_{1m} \\ w_{21} & \cdots & w_{2m} \end{bmatrix} \\
\mathbf{g}(\dot{\mathbf{x}}) &= \begin{bmatrix} g_1(\dot{\mathbf{x}}) & \cdots & g_m(\dot{\mathbf{x}}) \end{bmatrix}^T
\end{aligned} \tag{11.7}$$

Here $\hat{\mathbf{f}}$ is a 2×1 vector. It is an estimate of the actual force $\mathbf{f}$. The error in our estimate is as follows:

$$\tilde{\mathbf{f}} = \mathbf{f} - \hat{\mathbf{f}} = \begin{bmatrix} \tilde{f}_1 \\ \tilde{f}_2 \end{bmatrix} \tag{11.8}$$

Our objective is to change W so that we minimize the "squared" error e:

$$e = \frac{1}{2}\tilde{\mathbf{f}}^T\tilde{\mathbf{f}} \tag{11.9}$$

To do so, we need the gradient of e with respect to W. After some algebra, we find:

$$\frac{de}{dw_{ij}} = -\tilde{f}_i g_j \tag{11.10}$$

After performing trial n, the error in that trial $\tilde{f}^{(n)}$ will be used to change the weights $W^{(n)}$ of the internal model. That change will be in a direction opposite that of the gradient, and will be weighted by a small constant α:

$$w_{ij}^{(n+1)} = w_{ij}^{(n)} + \alpha \tilde{f}_i^{(n)} g_j(\dot{\mathbf{x}}^{(n)}) \tag{11.11}$$

Writing this in vector form, we have the following:

$$W^{(n+1)} = W^{(n)} + \alpha \tilde{\mathbf{f}}^{(n)} \mathbf{g}(\dot{\mathbf{x}}^{(n)})^T \tag{11.12}$$

If we multiply both sides of this equation by $\mathbf{g}(\dot{x}^{(n+1)})$, the result is as follows:

$$W^{(n+1)} \mathbf{g}(\dot{\mathbf{x}}^{(n+1)}) = W^{(n)} \mathbf{g}(\dot{\mathbf{x}}^{(n+1)}) + \alpha \tilde{\mathbf{f}}^{(n)} \mathbf{g}(\dot{\mathbf{x}}^{(n)})^T \mathbf{g}(\dot{\mathbf{x}}^{(n+1)}) \tag{11.13}$$

This is equivalent to the following:

$$\hat{\mathbf{f}}^{(n+1)}(\dot{\mathbf{x}}^{(n+1)}) = \hat{\mathbf{f}}^{(n)}(\dot{\mathbf{x}}^{(n+1)}) + \underbrace{\alpha \mathbf{g}(\dot{\mathbf{x}}^{(n)})^T \mathbf{g}(\dot{\mathbf{x}}^{(n+1)})}_{\text{generalization function}} \hat{\mathbf{f}}^{(n)} \tag{11.14}$$

Equation 11.14 says that the change in the internal model from trial n to $n + 1$ is completely described by the error in trial n times a generalization function. That generalization function is the projection of the bases in trial n upon the bases evaluated at trial $n + 1$. Intuitively, we see that the projection will be largest when the two consecutive trials are along the same velocity. The shape of the tuning functions will determine the generalization as the distance between the two movement changes in the state space.

In an artificial system, a "trial" would be an example data point. That is, the internal model would make a guess about the force at a particular velocity, a "teacher" would provide the actual force, and the error would be used to modify the weights. The shape of the bases would then dictate the generalization to the next trial. However, movements are not a single point in velocity space, but a trajectory. Because of delays in sensorimotor feedback, we could reasonably assume that feedback about the actual forces might not be available to update the internal model until the movement is completed. After completion of movement, the internal model would be updated along the entire trajectory of the desired velocity. Because this desired trajectory is along a straight line to the target, we will represent it simply as a direction. For example, if we have 8 directions of movement, the "generalization function" is a matrix of size 8×8. Element (i, j) of this matrix describes the fraction of error in movement direction i that is generalized to direction j. To simplify things, we can assume that what matters is not the specific directions of the two consecutive movements, but rather their angular distance. In this way, the generalization function becomes a vector of size 8×1, where each element indicates generalization between two consecutive movements that are separated by an angular distance of 0°, 45°, …, 270°. Let us call this generalization function b. With this approximation, a trial becomes a single reaching movement.

Now the important thing to notice is that despite the fact that the error in movement n potentially affects the internal model for all possible directions of movement, we can observe that effect for only one direction, the actual direction for which movement was made in trial $n + 1$. Therefore, while the effects of the generalization to the other seven possible directions are hidden from us, they nevertheless exist and we must account for them in order to represent accurately the trial-to-trial changes in the internal model. To do so, let k be an integer variable that can take a value from 1 to 8. This variable represents the possible directions of movement that could occur in trial $n + 1$. After movement n is completed, the internal model is updated in all these directions, and we have:

$$\hat{\mathbf{f}}_k^{(n+1)} = \hat{\mathbf{f}}_k^{(n)} + b(k - \dot{x}^{(n)})\tilde{\mathbf{f}}^{(n)} \qquad k = 1, \cdots, 8 \tag{11.15}$$

We see that if we could estimate the generalization function b from trial-to-trial changes in performance, we could have a reasonable idea of the kind of bases that are being used for computation of the internal model. However, our problem is that we can only record people's reaching movements, not $\hat{\mathbf{f}}$. Movements will be straight when the internal model has a correct estimate of force. There will be an error in the hand's trajectory when this estimate is incorrect. Let us assume that this error in the hand's trajectory is computed simply as a vector that describes where the hand is at peak velocity with respect to where it "should be" if the internal model were perfect. Let us call that position error vector $\mathbf{y}$. Let us further assume that it will be related to the force error $\tilde{\mathbf{f}}$ in the estimate of the internal model via a compliance matrix D. This matrix relates how force error produces a displacement from the intended trajectory. We now have the following:

$$\begin{aligned} &\mathbf{y}^{(n)} = D\tilde{\mathbf{f}}^{(n)} \\ &\hat{\mathbf{f}}_k^{(n+1)} = \hat{\mathbf{f}}_k^{(n)} + b(k - \dot{x}^{(n)})\tilde{\mathbf{f}}^{(n)} \qquad k = 1, \cdots, 8 \end{aligned} \tag{11.16}$$

Now let us introduce a new variable z, and define it as follows:

$$\mathbf{z}_k^{(n)} \equiv D\hat{\mathbf{f}}_k^{(n)} \tag{11.17}$$

With substitution of the above equation into Equation 11.16, we arrive at a coupled dynamical system:

$$\mathbf{y}^{(n)} = D\mathbf{f}^{(n)} - \mathbf{z}^{(n)}_{k(n)}$$
$$\mathbf{z}^{(n+1)}_{k} = \mathbf{z}^{(n)}_{k} + b(k - \dot{x}^{(n)})\mathbf{y}^{(n)} \qquad k = 1, \cdots, 8 \tag{11.18}$$

$\mathbf{y}^{(n)}$ is the error vector on n^{th} movement, made in direction k; $\mathbf{f}^{(n)}$ is the force experienced in that movement, and is scaled by a compliance-like matrix D. Compliance is the inverse of stiffness. Whereas stiffness describes force produced when a body is displaced, compliance describes displacement produced when a body experiences force. When an error occurs in a movement, the internal model is updated (as reflected in the eight equations). b is the generalization function that characterizes the effect of error that was experienced in a given state on all other states. We measure a sequence of movement errors $\mathbf{y}^{(n)}$ and fit them to the system in Equation (11.18) in order to find the best fit for matrix D and vector b. There are 12 unknown parameters in these two variables. The procedure for fitting these equations to a sequence of movements is provided in Donchin et al.[48] If the model is correct, it should describe all the trial-to-trial changes in performance that take place during adaptation and provide us with an estimate of the generalization function.

We begin by considering the fit of these equations to human data (Figure 11.5A). Data from a large group of subjects (n = 75) was collected as they learned to make movements in a curl force field. The target pattern was out-and-back in a half-pinwheel pattern. That is, movements began at center; a target was presented at 0°, 40°, 90°, or 135°. Upon completion of that movement the center target was lit, and the pattern was repeated. In this way, the movements were in eight directions, but all outward movements were followed by a movement back to the center. We found that (1) the equations fit the trial-to-trial variations in performance remarkably well (Figure 11.5A) and (2) the generalization function B and compliance matrix D remained consistent across repeated measures (Figure 11.5B). Interestingly, the generalization function was wide and bimodal. That is, generalization dropped off as angular distance of movements increased and reached a minimum at a distance of 90°, but then rose to approximately 50% of its peak value at 180°.

It was possible that this bimodality was an artifact of our out-and-back target sequence. We tested a new group of subjects (n = 8) in a random target sequence where the robot brought the hand to the start position of each movement (the second row in Figure 11.5). B and D maintained their shape (Figure 11.5B).

We next tested another group of subjects (n = 11) in a target sequence where not only the directions of movement, but also the force fields at each direction, were random (the third row in Figure 11.5). In this condition, at any given trial the field was either null, clockwise curl, or counterclockwise curl. As the field was random, we did not expect any adaptation. Remarkably, analysis of the trial-to-trial changes in performance produced a generalization function similar to that which we had estimated from trials in which subjects learned a "constant" field. In all cases, the generalization function was bimodal, consistent with bases that encode direction of movement with a bimodal activation pattern. The shape of the basis function that is consistent with our behavioral data is shown in Figure 11.5C.

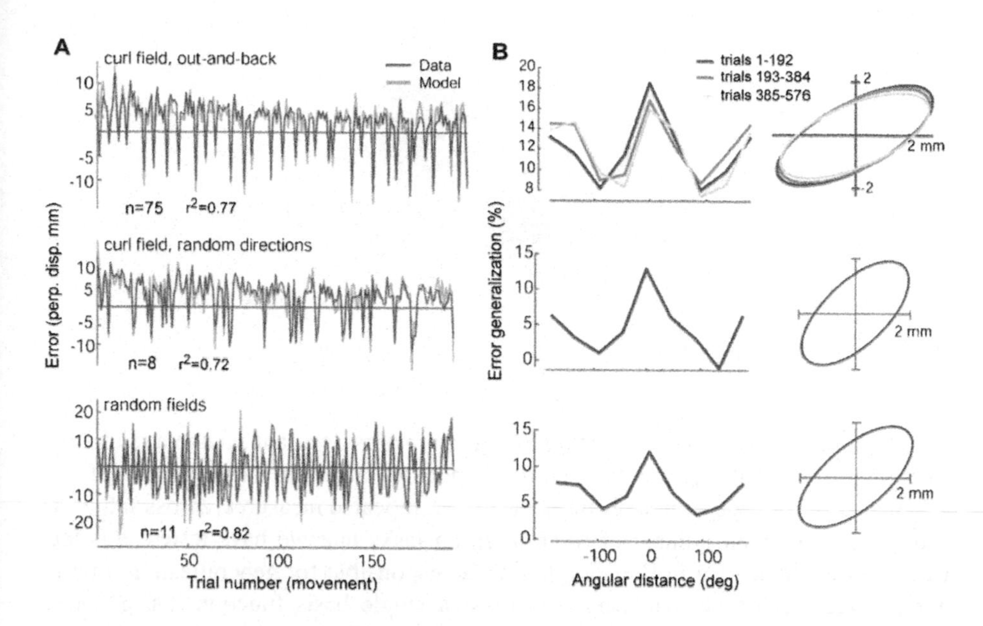

FIGURE 11.5 Generalization as a function of direction. Top row: (A) Black lines are movement errors during 192 movements (out-and-back pattern) in a standard curl field paradigm to 8 directions of targets (n = 72 subjects). Sharp negative spikes are catch trials. Gray lines are $y^{(n)}$ as fit to Equation 11.18, where the influence of error in any given movement on subsequent movements is estimated. Subjects performed 3 × 192 movements (3 target sets), but data for only one set is shown. (B) The estimated generalization function (b in Equation 11.18) and estimated compliance matrix D for each target set. The generalization function implies that ~18% of the error that was recorded for a movement toward any given direction updated the internal model for that same direction. About 12% of the error was generalized to neighboring directions at 135° and 180°. The same subjects were again tested on the same field a second and a third time (second and third target sets, each set 192 movements). The generalization functions for all three sets of targets are shown in (B). Little change is seen in these repeated measures. The matrix D is plotted as a transformation of a circle. The estimate changes little in repeated measures and its orientation and shape are consistent with previous estimates of the stiffness of the arm (Mussa-Ivaldi et al., 1985). Second row: In this experiment, a group of subjects (n = 8) practiced in a target set that was not out-and-back, but in random directions. The shape of the generalization function and compliance are similar to that obtained in the first row. Third row: In this experiment, a group of subjects (n = 11) trained in a force field that randomly changed from movement to movement. Despite no obvious learning of this field, the generalization function is similar to other "learnable" tasks. (C) The shape of the basis function implied by the generalization functions. This particular basis has a preferred velocity at [0.21, 0.21] m/s, corresponding to the peak velocity for a 10 cm movement toward 45°. Dark regions indicate higher activation. Velocity dependent component parameter values in Equation 11.19: σ = 0.15, s = 2. (Adapted from Reference 48, with permission.)

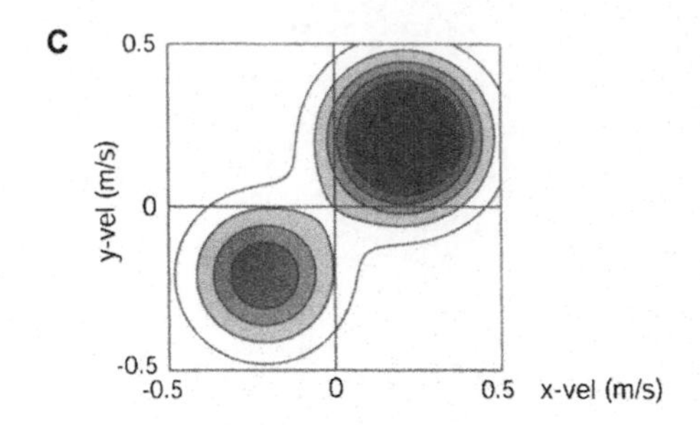

FIGURE 11.5 (continued)

Our finding that the generalization function remains invariant even in a randomly changing force field suggests that the fundamental computational properties of the internal model are approximately the same across repeated measures, across subjects, and across the small number of force learning tasks that we have tested thus far. Because the shape of g in Equation (11.5) is responsible for generalization, this is our strongest evidence that there may exist a single basis function that encodes movement kinematics and explains learning in all of these tasks. Our best guess today is that this function encodes hand position of the contralateral arm linearly and hand velocity with a bimodal activation function:

$$g_{velocity,i}(\dot{\underline{q}}) = \exp\left(-\frac{\left\|\dot{\underline{q}} - \dot{\underline{q}}_i\right\|^2}{2\sigma^2}\right) + \frac{1}{s}\exp\left(-\frac{\left\|\dot{\underline{q}} + \dot{\underline{q}}_i)\right\|^2}{2\sigma^2}\right) \tag{11.19}$$

11.9 RELATING THE INFERRED ACTIVITY FIELDS TO THE NEUROPHYSIOLOGY OF THE MOTOR SYSTEM

Are the bases that we inferred with this abstract model interpretable in terms of the neurophysiology of the motor system? From the patterns of generalization, we inferred the following:

1. The bases encode hand velocity with a function that has a PD and is modulated broadly but is bimodal.
2. The bases encode the position of the arm linearly in the horizontal plane. This position coding multiplicatively modulates tuning with respect to direction.
3. The PD of the bases rotates with the shoulder angle.
4. The bases are tuned to movements of the ipsilateral arm such that the PD remains arm invariant if the workspace is near the midline.
5. The bases are only weakly modulated by the color of the target.

All of these properties except the bimodality can be found among task-related cells in the primary motor cortex, the basal ganglia, and the cerebellum.[14,14,29,49,50] The invariance of the PD with respect to movements of the contralateral and ipsilateral arms was recently observed in the cells of the motor cortex,[36] the premotor cortex,[37] and the cerebellum (Bradley Greger and Tom Thach, personal communication). However, to our knowledge bimodality has only been observed in the cerebellum during reaching movements: Purkinje cell discharge during reaching movements shows a weak but consistent bimodal activation pattern as a function of hand velocity,[33] whereas no such bimodality is reported in the same task in M1.[50]

In reaching movements, a muscle that provides the agonist burst to reach in a particular direction (say, 0°) also provides the antagonist burst for a movement in the opposite direction (180°), but is generally not modulated very much when a movement is made to 90°. The antagonist burst is generally significantly smaller than the agonist burst. Therefore, bimodality is a fundamental characteristic of muscle activation functions, and generalization patterns in terms of direction of movement suggest that the bases are likely to have muscle-like tuning functions. We saw earlier that generalization patterns in terms of spatial configuration of the arm suggested the same thing. Taken together, this suggests that the neural computation of the internal model contains elements that have muscle-like tuning properties with respect to the contralateral arm during reaching movements.

The one aspect of the model that is not muscle-like is the particular encoding of velocity. In Equation 11.19, we have chosen to represent velocity with Gaussians. This means that each basis has a preferred velocity of movement. Purkinje cells in the cerebellar cortex appear to encode movement velocity in this way, whereas cells in M1 generally increase their discharge with increased movement speed.[15,50] Only one study has considered how the internal model generalizes in terms of speed of movement.[51] In that study, force adaptation at a given average velocity generalized less than linearly to neighboring velocities. However, the precise shape of the generalization function is not known. If it generalizes globally, then that representation would be muscle-like and consistent with the tuning of cells in M1. If it generalizes locally, then that representation implies a coding of velocity that peaks at a particular value and then declines — that is, a preferred velocity. This later generalization would be consistent with the tuning of task-related cells in the cerebellum.

11.10 CONSOLIDATION

Thus far we have been describing the learning of internal models using a mathematical framework where acquisition of information is one and the same as memory. In this framework, preferred force vectors associated with the bases change to minimize error in the task. Once the task is over, presumably these changes are maintained, and this forms the basis of long-term memory.

The scope of our naiveté was plainly demonstrated when we found that acquisition of memory of an internal model is merely the first step in a sequence of events that eventually results in a long-term representation of motor memory. Our behavioral measurements suggested that the internal model changed not only during the training session, but also in the hours that followed completion of training.[7] The motor memory appeared to gradually change from an initially fragile state to a state more resistant to change during a period of ~5 hours. Some of these results have recently been extended: Ghez and colleagues reported that in a task where subjects learned internal models of an inertial object, motor memory of inertial object one could be disrupted if practice was immediately followed by movements with inertial object two.[52] Using transcranial magnetic stimulation (TMS), Hallett and colleagues reported that stimulation of M1 immediately after practicing a thumb flexion task resulted in marked retention deficits, whereas stimulation of M1 at 5 hours post-practice did not affect retention.[53] Using functional imaging, we have observed that, at comparable levels of motor performance, the map of activation patterns in the brain differed significantly near the end of training on day 1, as compared with patterns at 6 hours[54,55] and at 2 or 4 weeks after initial practice.[56]

Therefore, passage of time changes the neural representation of the internal model. We currently have no theory to account for this. One hope is that we eventually might be able to track changes in neural representation by measuring their influence on patterns of generalization.

11.11 MAJOR SHORTCOMINGS OF THE THEORY

Both our measures of performance and the construction of the internal model focused on the early component of the reach (typically up to 250 msec), a period when one expects little influence from feedback. Therefore, even if the theory is successful, it only addresses adaptation associated with the motor commands that initiate the movement. However, our recent work[57,58] and that of our colleagues[59,60] has found that with training, the brain also learns to respond to feedback during a movement by producing appropriate motor responses. We currently have no model to account for trial-by-trial adaptation or generalization of this form of adaptation.

The alert reader will also note that while we started our story with the problem of using the visual appearance of objects to estimate their dynamics, the theory that we developed made little mention of these cues but rather focused on proprioceptive measures of limb state. How do nonproprioceptive cues like color, spatial cues about the pattern of forces, sequential cues regarding movement order, or cognitive cues affect the computation of internal models? In our daily interaction with the environment, it is these cues that must dominate the selection and adaptation of internal models. This important question remains poorly understood.

11.12 SUMMARY

The specific coding of movement parameters in the neurons that compose internal models has a significant, measurable influence on behavior. That influence can be observed in how our brain learns to predict forces in control of reaching movements. Training to make reaching movements in a force field results in a specific, highly reproducible pattern of force generalization to other movements. If we assume that the neural computation of an internal model is via a population code, then the tuning curves of the neurons that participate in this computation are the bases with which the force field is approximated. From the patterns of generalization one can infer the shape of these bases: (1) the bases are modulated as a function of hand velocity with a broad function that has a PD but is bimodal; (2) the bases are modulated linearly with arm position in the horizontal plane and this position coding multiplicatively modulates directional tuning, resulting in a gain field; (3) the PD of the bases rotates with the shoulder angle; (4) the bases are tuned to movements of the ipsilateral arm such that the PD remains arm invariant if the workspace is near the midline; or (5) the bases are weakly modulated by the color of the target. These are also some of the properties of cells in M1 and the cerebellum.

REFERENCES

1. Hollerbach, J.M. and Flash, T., Dynamic interactions between limb segments during planar arm movement, *Biol. Cybern.*, 44, 67, 1982.
2. Gribble, P.L. and Ostry, D.J., Compensation for interaction torques during single- and multijoint limb movement, *J. Neurophysiol.*, 82, 2310, 1999.
3. Flash, T., The control of hand equilibrium trajectories in multi-joint arm movements, *Biol. Cybern.*, 57, 257, 1987.
4. Shadmehr, R. and Mussa-Ivaldi, F.A., Adaptive representation of dynamics during learning of a motor task, *J. Neurosci.*, 14, 3208, 1994.
5. Conditt, M.A., Gandolfo, F., and Mussa-Ivaldi, F.A., The motor system does not learn the dynamics of the arm by rote memorization of past experience, *J. Neurophysiol.*, 78, 554, 1997.
6. Conditt, M.A. and Mussa-Ivaldi, F.A., Central representation of time during motor learning, *Proc. Nat. Acad. Sci. U.S.A.*, 96, 11625, 1999.
7. Shadmehr, R. and Brashers-Krug, T., Functional stages in the formation of human long-term motor memory, *J. Neurosci.*, 17, 409, 1997.
8. Thoroughman, K.A. and Shadmehr, R., Electromyographic correlates of learning internal models of reaching movements, *J. Neurosci.*, 19, 8573, 1999.
9. Flanders, M. and Soechting, J.F., Arm muscle activation for static forces in three-dimensional space, *J. Neurophysiol.*, 64, 1818, 1990.
10. Sergio, L.E. and Kalaska, J.F., Changes in the temporal pattern of primary motor cortex activity in a directional isometric force versus limb movement task, *J. Neurophysiol.*, 80, 1577, 1998.
11. Sergio, L.E. and Kalaska, J.F., Systematic changes in directional tuning of motor cortex cell activity with hand location in the workspace during generation of static isometric forces in constant spatial directions, *J. Neurophysiol.*, 78, 1170, 1997.
12. Li, C.S.R., Padoa-Schioppa, C., and Bizzi, E., Neuronal correlates of motor performance and motor learning in the primary motor cortex of monkeys adapting to an external force field, *Neuron*, 30, 593, 2001.
13. Caminiti, R., Johnson, P.B., and Urbano, A., Making arm movements within different parts of space: dynamic aspects in the primate motor cortex, *J. Neurosci.*, 10, 2039, 1990.
14. Caminiti, R., Johnson, P.B., Galli, C., Ferraina, S., and Burnod, Y., Making arm movements within different parts of space: the premotor and motor cortical representation of a coordinate system for reaching to visual targets, *J. Neurosci.*, 11, 1182, 1991.

15. Moran, D.W. and Schwartz, A.B., Motor cortical representation of speed and direction during reaching, *J. Neurophysiol.*, 82, 2676, 1999.
16. Kakei, S., Hoffman, D.S., and Strick, P.L., Muscle and movement representations in the primary motor cortex, *Science*, 285, 2136, 1999.
17. Shadmehr, R. and Moussavi, Z.M.K., Spatial generalization from learning dynamics of reaching movements, *J. Neurosci.*, 20, 7807, 2000.
18. Brashers-Krug, T., Shadmehr, R., and Bizzi, E., Consolidation in human motor memory, *Nature*, 382, 252, 1996.
19. Humphrey, D.R., Schmidt, E.M., and Thompson, W.D., Predicting measures of motor performance from multiple cortical spike trains, *Science*, 170, 758, 1970.
20. Georgopoulos, A.P., Schwartz, A.B., and Kettner, R.E., Neural population coding of movement direction, *Science*, 233, 1416, 1986.
21. Amirikian, B. and Georgopoulos, A.P., Directional tuning profiles of motor cortical cells, *Neurosci. Res.*, 36, 73, 2000.
22. Tolhurst, D.J. and Thompson, I.D., Organization of neurones preferring similar spatial frequencies in cat striate cortex, *Exp. Brain Res.*, 48, 217, 1982.
23. Seung, H.S. and Sompolinsky, H., Simple models for reading neuronal population codes, *Proc. Nat. Acad. Sci. U.S.A.*, 90, 10749, 1993.
24. Schwartz, A.B., Direct cortical representation of drawing, *Science*, 265, 540, 1994.
25. Poggio, T., A theory of how the brain might work, *Cold Spring Harbor Symp. Quant. Biol.*, 55, 899, 1990.
26. Pouget, A., Dayan, P., and Zemel, R., Information processing with population codes, *Nat. Rev. Neurosci.*, 1, 125, 2000.
27. Pouget, A. and Sejnowski, T.J., Spatial transformations in the parietal cortex using basis functions, *J. Cog. Neurosci.*, 9, 222, 1997.
28. Andersen, R.A., Essick, G.K., and Siegel, R.M., Encoding of spatial location by posterior parietal neurons, *Science*, 230, 456, 1985.
29. Georgopoulos, A.P., Caminiti, R., and Kalaska, J.F., Static spatial effects in motor cortex and area 5: quantitative relations in a two-dimensional space, *Exp. Brain Res.*, 54, 446, 1984.
30. Hwang, E.J., Donchin, O., Smith, M.A., and Shadmehr, R., A gain-field encoding of limb position and velocity in the internal model of arm dynamics, *PLoS. Biol.*, 1, 209, 2003.
31. Bosco, G., Rankin, A., and Poppele, R.E., Representation of passive hindlimb postures in cat spinocerebellar activity, *J. Neurophysiol.*, 76, 715, 1996.
32. Helms Tillery, S.I., Soechting, J.F., and Ebner, T.J., Somatosensory cortical activity in relation to arm posture: nonuniform spatial tuning, *J. Neurophysiol.*, 76, 2423, 1996.
33. Coltz, J.D., Johnson, M.T.V., and Ebner, T.J., Cerebellar Purkinje cell simple spike discharge encodes movement velocity in primates during visuomotor arm tracking, *J. Neurosci.*, 19, 1782, 1999.
34. Scott, S.H. and Kalaska, J.F., Reaching movements with similar hand paths but different arm orientation. I. Activity of individual cells in motor cortex, *J. Neurophysiol.*, 77, 826, 1997.
35. Ajemian, R., Bullock, D., and Grossberg, S., A model of movement coordinates in the motor cortex: posture-dependent changes in the gain and direction of single cell tuning curves, *Cereb. Cortex*, 11, 1124, 2001.
36. Steinberg, O., Donchin, O., Gribova, A., Cardosa, D.O., Bergman, H., and Vaadia, E., Neuronal populations in primary motor cortex encode bimanual arm movements, *Eur. J. Neurosci.*, 15, 1371, 2002.

37. Kalaska, J.F., Cisek, P., and Crammond, D.J., Effector-independent activity in primate dorsal premotor cortex (PMd) during instructed delay tasks, *Soc. Neurosci. Abs.*, 26, 359.2, 2000.
38. Criscimagna-Hemminger, S.E., Donchin, O., Gazzaniga, M.S., and Shadmehr, R., Learned dynamics of reaching movements generalize from dominant to nondominant arm, *J. Neurophysiol.*, 89, 168, 2003.
39. Gazzaniga, M.S., Cerebral specialization and interhemispheric communication: does the corpus callosum enable the human condition, *Brain*, 123, 1293, 2000.
40. Galea, M.P. and Darian-Smith, I., Corticospinal projection patterns following unilateral section of the cervical spinal cord in the newborn and juvenile macaque monkey, *J. Comp Neurol.*, 381, 282, 1997.
41. Haaland, K.Y. and Harrington, D.L., Hemispheric asymmetry of movement, *Curr. Opin. Neurobiol.*, 6, 796, 1996.
42. Gandolfo, F., Mussa-Ivaldi, F.A., and Bizzi, E., Motor learning by field approximation, *Proc. Nat. Acad. Sci. U.S.A.*, 93, 3843, 1996.
43. Rao, A.K. and Shadmehr, R., Contextual cues facilitate learning of multiple models of arm dynamics, *Soc. Neurosci. Abs.*, 27, 302.4, 2001.
44. Krouchev, N.I. and Kalaska, J.F., Context-dependent anticipation of different task dynamics: rapid recall of appropriate motor skills using visual cues, *J. Neurophysiol.*, 89, 1165, 2003.
45. Wada, Y., Kawabata, Y., Kotosaka, S., Yamamoto, K., Kitazawa, S., and Kawato, M., Acquisition and contextual switching of multiple internal models for different viscous force fields, *Neurosci. Res.*, 46, 319, 2003.
46. Thoroughman, K.A. and Shadmehr, R., Learning of action through adaptive combination of motor primitives, *Nature*, 407, 742, 2000.
47. Donchin, O. and Shadmehr, R., Linking motor learning to function approximation: learning in an unlearnable force field, *Adv. Neural Inf. Process. Syst.*, 14, 197, 2002.
48. Donchin, O., Francis, J.T., and Shadmehr, R., Quantifying generalization from trial-by-trial behavior of adaptive systems that learn with basis functions: theory and experiments in human motor control, *J. Neurosci.*, 23, 9032, 2003.
49. Turner, R.S. and Anderson, M.E., Pallidal discharge related to the kinematics of reaching movements in two dimensions, *J. Neurophysiol.*, 77, 1051, 1997.
50. Johnson, M.T.V. and Ebner, T.J., Processing of multiple kinematic signals in the cerebellum and motor cortices, *Brain Res. Rev.*, 33, 155, 2000.
51. Goodbody, S.J. and Wolpert, D.M., Temporal and amplitude generalization in motor learning, *J. Neurophysiol.*, 79, 1825, 1998.
52. Krakauer, J.W., Pine, Z.M., and Ghez, C., Visuomotor transformations for reaching are learned in extrinsic coordinates, *Soc. Neurosci. Abs.*, 25, 2177, 1999.
53. Muellbacher, W., Ziemann, U., Wissel, J., Dang, N., Kofler, M., Facchini, S., Boroojerdi, B., Poewe, W., and Hallett, M., Early consolidation in human primary motor cortex, *Nature*, 415, 640, 2002.
54. Shadmehr, R. and Holcomb, H.H., Neural correlates of motor memory consolidation, *Science*, 277, 821, 1997.
55. Shadmehr, R. and Holcomb, H.H., Inhibitory control of motor memories: a PET study, *Exp. Brain Res.*, 126, 235, 1999.
56. Nezafat, R., Shadmehr, R., and Holcomb, H.H., Long-term adaptation to dynamics of reaching movements: a PET study, *Exp. Brain Res.*, 140, 66, 2001.
57. Bhushan, N. and Shadmehr, R., Computational architecture of human adaptive control during learning of reaching movements in force fields, *Biol. Cybern.*, 81, 39, 1999.

58. Wang, T., Dordevic, G.S., and Shadmehr, R., Learning the dynamics of reaching movements results in the modification of arm impedance and long-latency perturbation responses, *Biol. Cybern.*, 85, 437, 2001.
59. Burdet, E., Osu, R., Franklin, D.W., Milner, T.E., and Kawato, M., The central nervous system stabilizes unstable dynamics by learning optimal impedance, *Nature*, 414, 446, 2001.
60. Franklin, D.W., Osu, R., Burdet, E., Kawato, M., and Milner, T.E., Adaptation to stable and unstable dynamics achieved by combined impedance control and inverse dynamics model, *J. Neurophysiol.*, 90, 3270, 2003.

12 Cortical Control of Motor Learning

Camillo Padoa-Schioppa, Emilio Bizzi, and Ferdinando A. Mussa-Ivaldi

CONTENTS

ABSTRACT

The execution of the simplest gestures requires the accurate coordination of several muscles. In robotic systems, engineers coordinate the action of multiple motors by writing computer code that specifies how the motors must be activated for achieving the desired robot motion and for compensating for unexpected disturbance. Humans and animals follow another path. Something akin to programming is achieved in nature by the biological mechanisms of synaptic plasticity; that is, by the variation in efficacy of neural transmission brought about by past history of pre- and post-synaptic signals. However, robots and animals differ in another important way. Robots (at least those of the current generations) have fixed mechanical structure and dimensions. In contrast, the mechanics of muscles, bones, and ligaments change over time: the length of our limbs varies as we grow into adulthood; some part of our body may lose its functionality following a lesion or a degenerative process; muscle mechanics may vary over just a few minutes of intense activity. Because of these changes, the central nervous system must continuously adapt motor commands to the mechanics of the body. Adaptation — the ability to carry previously learned

0-8493-1287-6/05/$0.00+$1.50

motor skills into new mechanical contexts — is a form of motor learning. In this chapter, we present a view of motor learning that starts from the analysis of the computational problems associated with the execution of the simplest gestures. We discuss the theoretical idea of internal models and present some evidence and theoretical considerations suggesting that internal models of limb dynamics may be obtained by the combination of simple modules or "motor primitives." Then, we review some experimental results on the activity of neurons in the cortex during a learning task. These findings suggest that the motor cortical areas include neurons that process well-acquired movements as well as neurons that change their behavior during and after being exposed to a new task.

12.1 DYNAMICS

According to the laws of Newtonian mechanics, in order to impress a motion upon an object one must apply a force directly proportional to the desired acceleration. This is Newton's equation **f** = m **a**.

A desired motion of an object is a sequence of positions $x(t)$ that one wishes the limb to occupy at subsequent instants of time t. Such a sequence is called a trajectory and is mathematically represented as a function, $x = x(t)$. To use Newton's equation for deriving the needed time-sequence of forces, one must calculate the first temporal derivative of the trajectory, the velocity, and then the second temporal derivative, the acceleration. Finally, one obtains the desired force from this acceleration. This is an example of inverse dynamic computation. The problem of direct dynamics is to compute the trajectory resulting from the application of a force.

One of the central questions in motor control is how the central nervous system solves the inverse dynamics problem and generates the motor commands that guide our limbs.[1] A system of second-order nonlinear differential equations is generally considered to be an adequate representation for the passive dynamics of a limb. A compact expression for such a system is as follows:

$$D(q, \dot{q}, \ddot{q}) = \tau(t) \tag{12.1}$$

where q, $\dot{q}$ and $\ddot{q}$ represent the limb configuration vector — for example the vector of joint angles — and its first and second time derivatives. The term $\tau(t)$ is a vector of joint torques at time t — it plays the role of **f** in Newton's equation. In practice, the expression for D — which corresponds to m **a** — may have a few terms for a two-joint planar arm (Figure 12.1) or it may take several pages for more realistic models of the arm's multiple-joint geometry. The inverse dynamics approach to the control of multiple-joint limbs consists in solving explicitly for a torque trajectory $\tau(t)$ given a desired trajectory of the limb $q_D(t)$. This is done by replacing $q_D(t)$ for the variable q on the left side of Equation 12.1:

$$\tau(t) = D(q_D(t), \dot{q}_D(t), \ddot{q}_D(t)) \tag{12.2}$$

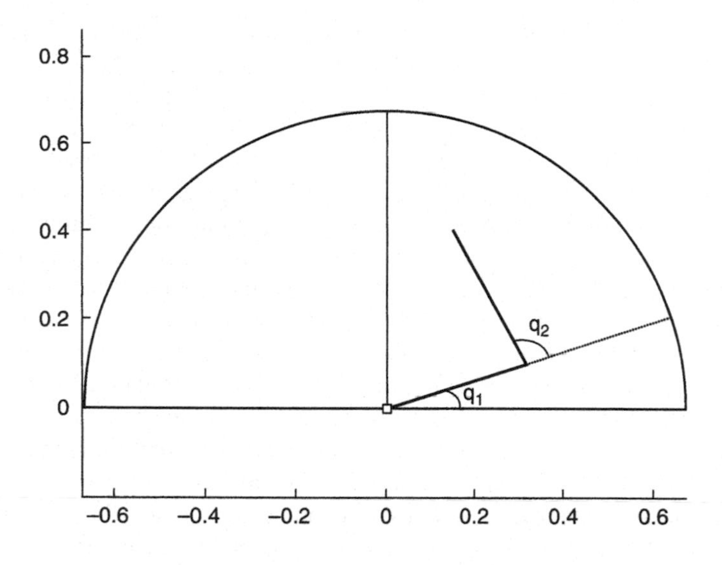

$$
\begin{cases}
D_1 = \left(I_1 + I_2 + m_2 l_1 l_2 \cos(q_2) + \dfrac{m_1 l_1^2 + m_2 l_2^2}{4} + m_2 l_1^2\right)\ddot{q}_1 + \\
\quad + \left(I_2 + \dfrac{m_2 l_1 l_2}{2}\cos(q_2) + \dfrac{m_2 l_2^2}{4}\right)\ddot{q}_2 - \dfrac{m_2 l_1 l_2}{2}\sin(q_2)\dot{q}_2^{\,2} + \\
\quad - m_2 l_1 l_2 \sin(q_2)\dot{q}_1\dot{q}_2 + \beta_1(q_1, q_2, \dot{q}_1, \dot{q}_2) \\
D_2 = \left(I_2 + \dfrac{m_2 l_1 l_2}{2}\cos(q_2) + \dfrac{m_2 l_2^2}{4}\right)\ddot{q}_1 + \left(I_2 + \dfrac{m_2 l_2^2}{4}\right)\ddot{q}_2 + \\
\quad - \dfrac{m_2 l_1 l_2}{2}\sin(\dot{q}_2)\dot{q}_1^{\,2} + \beta_2(q_1, q_2, \dot{q}_1, \dot{q}_2)
\end{cases}
$$

FIGURE 12.1 Simplified model of planar limb dynamics. The mechanics of the arm are approximated by a two-joint mechanism with angles q_1 (with respect to the torso) and q_2 (with respect to the forearm, respectively (*top*). The dynamics are described by two nonlinear equations (*bottom*) that relate the joint torques at the shoulder (D_1) and at the elbow (D_2) to the angular position velocity and acceleration of both joints. The parameters that appear in these expressions are the lengths of the two segments (l_1 and l_2); their masses (m_1 and m_2); their moments of inertia (I_1 and I_2). The numerical values used in the simulations are the same as those listed in Table 1 of Shadmehr and Mussa-Ivaldi[7] and correspond to values estimated from an experimental subject. The terms β_1 and β_2 describe the viscoelastic behavior of the resting arm. They are simulated here by linear stiffness and viscosity matrices. (From Reference 4, with permission.)

12.2 INTERNAL MODELS

Early models of motor control[2,3] suggested that the nervous system may store specific solutions of Equation 12.2 corresponding to the desired motions of the body. However, simple considerations about the geometrical space of meaningful behaviors are sufficient to establish that this approach would be inadequate.[4] An alternative approach postulates that the motor system solves the problems of dynamics by constructing internal representations of the way in which limbs respond to applied forces. These representations would allow us to generate new behaviors and to handle situations that we have not yet encountered. A vivid illustration of how explicit representations of dynamics, also called internal models, may facilitate motor learning is offered by the work of Schaal and Atkeson,[5] who studied the task of balancing an inverted pendulum with a robotic arm. They found that robots can be trained to carry out this task successfully when they can build an internal model of the dynamics associated with the balancing act. Such a model may be constructed using data derived from the observation of humans engaging competently in the same task.

The term *internal model* refers either to (1) the transformation from a motor command to the consequent behavior, or to (2) the transformation from a desired behavior to the corresponding motor command.[6] A model of the first kind is called a forward model. Forward models provide the control system with the means to predict the outcome of a command, and to estimate the current state in the presence of feedback delay. A representation of the mapping from planned actions to motor commands is called an inverse model. Strong experimental evidence for the existence of internal models has been offered by studies of the adaptation of arm movements to perturbing force fields.[7–12]

12.3 EVIDENCE FOR INVERSE INTERNAL MODELS

One way to test for the existence of inverse internal models is by changing the dynamics that the central nervous system must control in order to execute a desired movement. This approach was adopted by Shadmehr and Mussa-Ivaldi.[7] They asked subjects to make reaching movements in the presence of externally imposed forces. These forces were produced by a robot whose free endpoint was held as a pointer by the subjects (Figure 12.2A). The subjects were asked to execute reaching movements toward a number of visual targets. Since the force field produced by the robot (Figure 12.2B) significantly changed the dynamics of the reaching movements, the subjects' movements initially were grossly distorted (Figure 12.2D) when compared to the undisturbed movements (Figure 12.2C). However, with practice, the subjects' hand trajectories in the force field converged to a path similar to that produced in absence of any perturbing force (Figure 12.3).

Subjects' recovery of performance is due to learning. After the training had been established, the force field was unexpectedly removed for the duration of a single hand movement. The resulting trajectories (Figure 12.4), named after-effects, were approximate mirror images of those that the same subjects produced when they had initially been exposed to the force field. The emergence of after-effects indicates that the central nervous system had composed an internal model of the external field. The internal model was generating patterns of force that effectively anticipated the disturbing forces that the moving hand was encountering. The fact that these learned forces compensated for the disturbances applied by the robotic arm during the subjects' reaching movements indicates that the central nervous system programs these forces in advance. The after-effects demonstrate that these forces are not the products of some reflex compensation of the disturbing field.

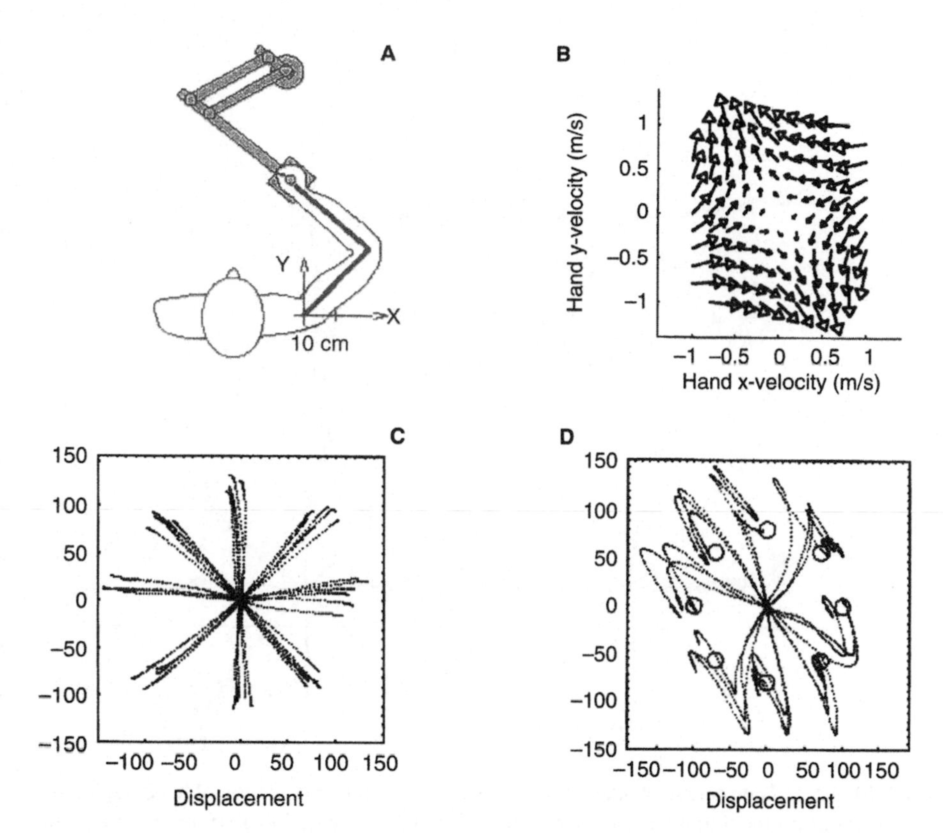

FIGURE 12.2 Adaptation to external force fields. (A) Experimental apparatus. Subjects executed planar arm movements while holding the handle of the manipulandum. A monitor (not shown) placed in front of the subjects and above the manipulandum displayed the location of the handle as well as targets of reaching movements. (B) Velocity-dependent force field generated by the manipulandum corresponding to the expression $F = B \cdot v$ with

$$B = \begin{bmatrix} -10.1 & -11.2 \\ -11.2 & 11.1 \end{bmatrix} \text{ Newton sec/m.}$$

The force **F** was linearly related to the velocity of the hand, $\mathbf{v} = [v_x, v_y]$. (C) Unperturbed reaching trajectories executed by a subject when the manipulandum was not producing disturbing forces. (D) Initial responses observed when the force field shown in (B) was applied unexpectedly. The circles indicate the target locations. (Modified from Reference 7.)

It is of interest to ask what the properties of the internal model are, and in particular whether the model could generalize to regions of the state space where the disturbing forces were not experienced. Recent experiments by Gandolfo and coworkers[13] were designed to test the generalization of motor adaptation to regions where training had not occurred. In these experiments, subjects were asked to execute point-to-point planar movements between targets placed in one section of the work-space. Their hand grasped the handle of the robot, which was used to record and perturb their trajectories. Again, as in the experiments of Shadmehr and Mussa-Ivaldi,

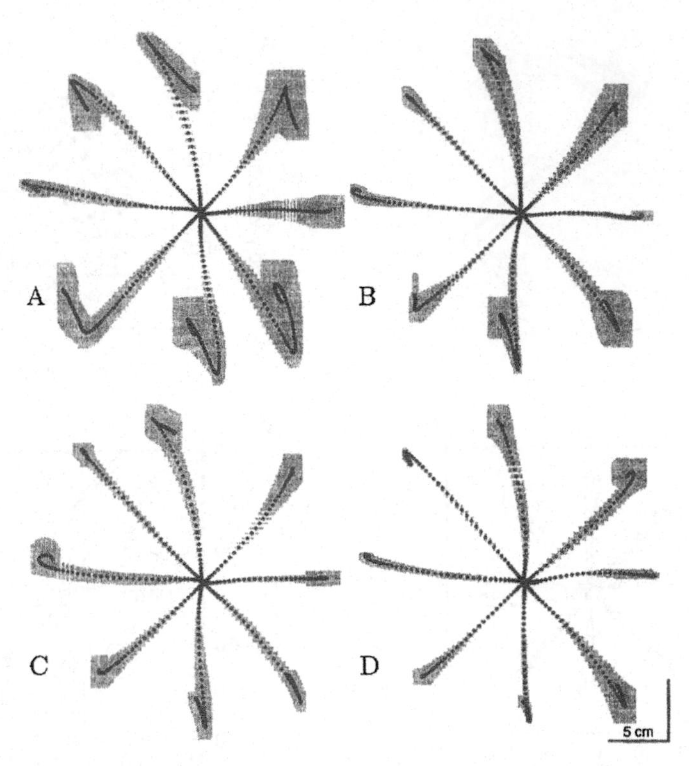

FIGURE 12.3 Time course of adaptation. Average and standard deviation of hand trajectories executed during the training period in the force field of Figure 12.7B. Performance is plotted during the first (A), second (B), third (C), and final (D) set of 250 movements. (From Reference 7, with permission.)

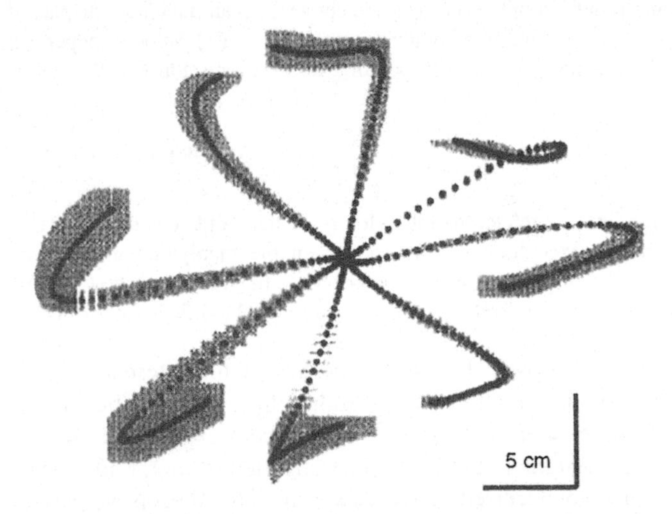

FIGURE 12.4 After-effects of adaptation. Average and standard deviations of hand trajectories executed at the end of training in the field when the field was unexpectedly removed on random trials. (From Reference 7, with permission.)

adaptation was quantified by the amount of the after-effects observed when the perturbing forces were discontinued.

As a way of establishing the generalization of motor learning, Gandolfo and coworkers[13] perturbed only the trajectories made to a subset of the targets and searched for after-effects in movements that had not been exposed to perturbations. The amount of the after-effect made it possible to quantify the force field that the subjects expected to encounter during their movements in the trained as well as in the novel directions. The same investigators found that the after-effects were present, as expected, along the trained directions, but the magnitude of the after-effects decayed smoothly with increasing distance from the trained directions. This finding indicates that the subjects were also able to compensate for forces experienced at neighboring workspace locations. Corroborating evidence was recorded by Thoroughman and Shadmehr.[14,15]

The experiments described above have shown that subjects adapt to a new environment by forming a representation of the external force field that they encounter when making reaching movements. Does this representation form an imprint in long-term memory? Brashers-Krug and coworkers[9] investigated this question by exposing their subjects to perturbing force fields that interfered with the execution of reaching movements (Figure 12.5). After practicing reaching movements, these subjects were able to compensate for the imposed forces (task A) and were able to guide the cursor accurately to the targets despite the disturbing forces. This group of subjects, which was tested 24 hours later with the same disturbing forces, demonstrated not only retention of the acquired motor skill, but also additional learning. Surprisingly, they performed at a significantly higher level on day two than they had on day one. A second group of subjects was trained on day one with a different pattern of forces (task B), immediately after performing task A. In task B the manipulandum produced forces opposite in direction to those applied during task A. When this second group of subjects was tested for retention of task A on day two, the investigators found that the subjects did not retain any of the skills that had been learned earlier. This phenomenon is known as retrograde interference. Next, Brashers-Krug and colleagues[9] investigated whether the susceptibility to retrograde interference decreased with time. They found that retrograde interference decreased monotonically with time as the interval between tasks A and B increased. When 4 hours passed before task B was learned, the skill learned in task A was completely retained — the initial learning had consolidated. What is remarkable in these results is that motor memory is transformed with the passage of time and in absence of further practice, from an initial fragile state to a more solid state.

In summary, the main findings of these studies are as follows: (1) when exposed to a deterministic field of velocity-dependent forces, arm movements are first distorted and, after repeated practice, the initial kinematics are recovered; (2) if, following adaptation, a field is suddenly removed, after-effects are clearly visible as mirror images of the initial perturbations; (3) adaptation is achieved by the motor system through the formation of a local map that associates the states (positions and velocities) visited during the training period with the corresponding forces; and (4) after adaptation this map — that is, the internal model of the field — undergoes a process of consolidation.

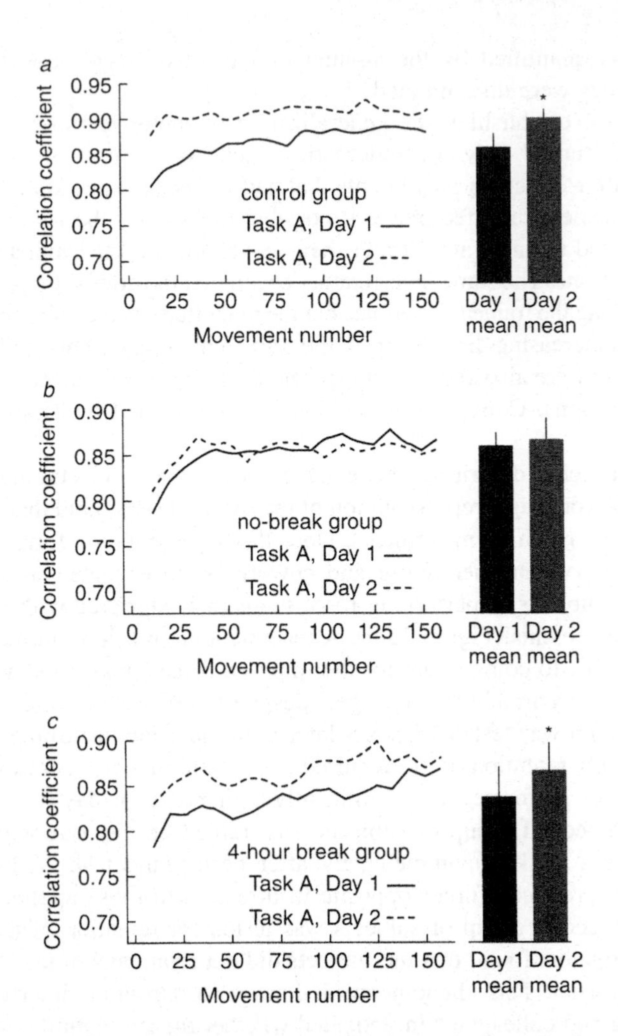

FIGURE 12.5 Motor memory consolidation. The left panels show the learning curves for three groups of subjects. Learning in a perturbing force field was quantified by a correlation coefficient between the trajectories in the field and the average trajectory before any perturbation had been applied. On the right are the mean performances in experiment days 1 and 2. Subjects in the control group (A) practiced reaching movements against a force field (task A) in the first day and then were tested again in the same field during the second day. Subjects in the no-break group (B) during the first day practiced movements in the field of task A. Then they immediately practiced movements in a different field (task B). In the second day they practiced again in the field of task A. Finally, subjects of the 4-hour break group (C) during the first day were exposed to the fields of tasks A and B but with a breaking interval of 4 hours between the two. Their performance was measured on task A on day 2. Learning curves and mean performance were significantly higher on day 2 both for the control group and for the 4-hour break group. In contrast, subjects in the no-break group did not display any difference in performance from day 1 to day 2. (From Reference 9, with permission.)

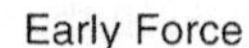

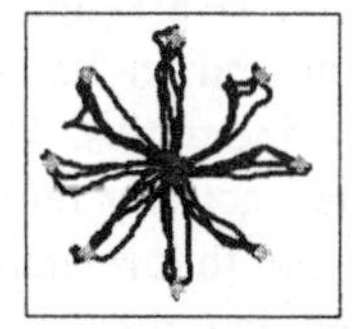

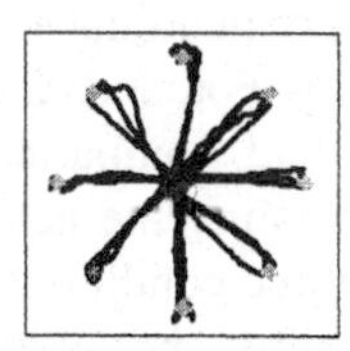

FIGURE 12.6 (see color figure) Psychophysics of motor learning. Data are shown from a representative experimental session. (A) Trajectories in real space. The trajectories are roughly straight when the movements are not perturbed (Baseline). When a counterclockwise (CCW) force field is turned on, trajectories are deviated at first (Early Force). After the perturbing force is turned off, the first movements show an after-effect, inasmuch as they are deviated in the clockwise direction (Early Washout). Within a few trials, however, the monkey readapts to the unperturbed condition, and trajectories become straight again (Late Washout). (From Reference 41, with permission.)

In order to investigate the neural signals that occur during adaptation, we recently recorded the activity of single neurons from the motor cortical areas of monkeys.

12.4 NEURONAL ACTIVITY RECORDED DURING ADAPTATION: LEARNING A NEW DYNAMICS

In a series of recent studies, we investigated how the activity of neurons in the cortical motor areas of the frontal lobe is modified when monkeys learn a new dynamic. To this end, we used the experimental setup of Shadmehr and Mussa-Ivaldi (scaled down). During the experiments, the monkeys sat on a chair and executed reaching movements instructed by targets appearing on a computer monitor, while holding the handle of a robotic arm. Two motors at the base of the robot could exert perturbing forces upon the hand of the monkey. In each session, the monkeys performed center-out reaching movements in three subsequent conditions: Baseline (160 trials, no force); Force (160 trials); and Washout (160 trials, no force). In the Force condition, the monkeys were exposed to either a clockwise (CW) or to a counterclockwise (CCW) viscous force field $\mathbf{F} = \mathbf{BV}$ with $\mathbf{B} = [0\ -b;\ b\ 0]$ and $\mathbf{V}$ equal to the hand velocity.

The psychophysics of the task, illustrated in Color Figure 12.6*, are essentially the same as those in humans as described by Shadmehr and Mussa-Ivaldi.[7] The hand trajectories are essentially straight in the Baseline and initially become deviated

* See color insert following page 170.

when the external force is introduced. As the monkey adapts to the perturbation, however, the hand kinematics gradually converge to those observed in the Baseline. In other words, the hand trajectories become straight again and the speed profile returns to its original bell shape. In the Washout, when the force is removed, the monkey displays a few after-effects as the hand trajectories are deviated in a way that mirrors the initial deviation observed in the Force condition. After a short time, however, the hand kinematics return to those observed in the Baseline.

In the analysis of neuronal activity, we essentially disregarded the first adaptation phase in the Early Force condition and in Early Washout, and we focused on movements that had comparable kinematics. Hence, this experimental design allowed for dissociating the neuronal activity related to the movement kinematics (the same in the three conditions) from that related to the movement dynamics (the same in the Baseline and Washout, but different in the Force condition). Most importantly, the experimental design allowed us to dissociate the neuronal correlates of motor performance from plastic changes associated with motor learning. For this dissociation, we compared the activity of neurons recorded in the Washout with that recorded in the Baseline. Indeed, the performance of the monkey (kinematics and dynamics) was essentially identical in the two conditions. The only difference was that in the Washout the monkeys had previously adapted and learned a new dynamic. Hence, changes in the activity in the Washout compared to the Baseline were associated with that learning experience.

Our first study focused on the primary motor cortex (M1). In particular, we recorded and analyzed the activity of 162 individual neurons in a movement-related time window (from 200 msec before the movement onset to the end of movement). As first described by Georgopoulos and colleagues,[14] we found that a large proportion of neurons in M1 were directionally tuned in the Baseline; their activity differed for movements in different directions. Surprisingly, however, we found that some of the neurons that were initially not tuned in the Baseline acquired a new directional tuning in the Force condition following adaptation to the force field. In some cases, these "tune-in" cells maintained their newly acquired directional tuning in the Washout following readaptation to the unperturbed conditions. Conversely, other neurons that were initially tuned lost their directional tuning following adaptation ("tune-out" neurons). The presence of these two groups of cells is an indication of what seems to be an intrinsic property of cells in M1: to be shaped by experience and to undergo plastic changes in a relatively short period of time.[15]

The tune-in and tune-out groups accounted for 37% of the cells recorded in M1. A further analysis, however, revealed another variety of plastic changes associated with motor learning. Specifically, neurons that were directionally tuned throughout the three conditions (Baseline, Force, and Washout) generally changed their preferred direction (PD) as the monkey adapted to the perturbation and readapted to the unperturbed conditions in the washout. Interestingly, in some cases, the final PD in the Washout was different from that originally recorded in the Baseline.[16] These memory cells accounted for a total of 40% of the population of neurons directionally tuned throughout the three conditions.

In conclusion, these data strongly suggest that M1 plays a prime role in motor learning.

12.5 MOTOR PERFORMANCE AND MOTOR LEARNING IN THE PRIMARY MOTOR CORTEX

Although our M1 results are quite intriguing, they are also somewhat puzzling. They show a surprisingly high degree of plasticity in M1, an area that seems crucial for motor control (for instance, lesions to M1 dramatically disrupt movement generation). Moreover, they show that plastic changes can be induced by a relatively brief exposure to new forces. But how can the same population of neurons effectively support motor performance (after all, movements in the Washout are as good as in the Baseline) and at the same time be flexible enough to support motor learning? A closer inspection of the changes of PD recorded for individual neurons and for the entire population offers a glimpse into this fascinating question.

One of the advantages of our experimental design is that curl force fields (i.e., forces in a direction that is orthogonal to the instantaneous hand velocity) impose strong constraints onto the changes of electromyographic (EMG) muscle activity across conditions. Specifically, when monkeys adapt to a curl force field, the PD of muscles shifts in the direction of the external force (CW or CCW, depending on the force field). The reason for this shift is that the internal forces exerted by muscles sum with the external force field in the Force condition. As a result, the monkey maximally activates any given muscle in the Force condition to execute movements in a direction (the new PD) different from the direction that elicited maximal muscle activation in the Baseline (the old PD). Most importantly, the PD shifts for all the muscles in the same direction, namely the direction of the external force field, independently of the original PD. We verified these predictions empirically by recording in our monkeys the EMG of five muscles of the upper arm (pectorals, deltoid, triceps, biceps, and brachioradialis). We found that the PD of all muscles shifted in the direction of the external force, on average by 19.2° ($p < 0.005$, t test). In the Washout, the PD of muscles shifted back by –15.4° ($p < 0.05$, t test) so that there was no net shift of PD in the Washout compared to the Baseline (mean shift 4.4°, $p = 0.06$, t test).

These changes of PD observed for the muscle EMG offer a framework for interpreting the activity of neurons. For each neuron in M1 directionally tuned in both conditions, we computed the shift of PD in the Force as compared to the Baseline. Shifts in the direction of the external force were defined as positive. Considering the entire population, we found that the PD of M1 neurons shifted on average by 16.2° in the Force condition compared to the Baseline ($p < 10^{-5}$, t test). In the Washout, the PD of M1 neurons shifted back by 14.2° ($p < 0.001$, t test), so that no net shift was present when comparing the Washout and the Baseline ($p = 0.9$, t test). In other words, the changes across conditions recorded for neurons in M1 *as a population* matched the changes observed for muscles.

When individual neurons are taken into consideration, an interesting variety of behaviors appears. For one group of neurons, the PD did not change at all across conditions. This group of "kinematic" cells accounted for 34% of the neurons that were directionally tuned throughout the three conditions. For another group of cells, the PD shifted in the Force condition (typically in the direction of the external force field) and shifted in the opposite direction in the Washout, back to the original PD.

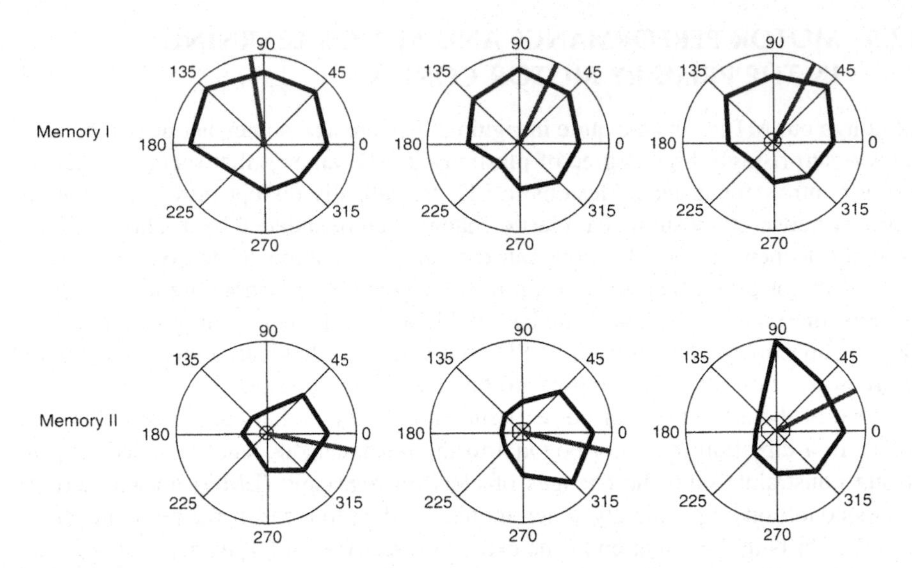

FIGURE 12.7 (see color figure) The tuning curves are plotted in polar coordinates. For each cell, the three plots represent the movement-related activity in the Baseline (left), in the Force epoch (center), and in the Washout (right). In each plot, the circle in the dashed line represents the average activity during the center hold time window, when the monkey holds the manipulandum inside the center square and waits for instructions. Examples of memory I and memory II cells, in terms of the modulation of the PD. All cells were recorded with a clockwise force field. (From Reference 16, with permission.)

In other words, this group of "dynamic" cells (22%) behaved very much like muscles. For the most interesting group of cells, named "memory" cells, the PD in the Washout was significantly different from that in the Baseline. More precisely, we found two groups of memory cells. For "memory I" cells, the PD shifted in the Force condition, typically in the direction of the external force field, and remained in the Washout oriented in the newly acquired direction. In contrast, for "memory II" cells, the PD did not change in the Force compared to the Baseline, and shifted in the Washout, typically in the direction opposite to the previously experienced force field. In total, the two classes of memory I and memory II cells accounted for 19% and 22% of the population, respectively. Thus, a large proportion of individual neurons in M1 maintained a trace of the learning experience outlasting exposure to the perturbation (Color Figure 12.7).

In our interpretation, the coexistence of memory I and memory II cells conforms well with the notion that the population of M1 supports both functions of motor performance and motor learning, and offers a glimpse into how it may do so. On the one hand, the PD of memory I cells shifted in the direction of the external force in the Force condition and remained shifted in the Washout. On the other hand, the PD of memory II cells did not shift in the Force condition but shifted in the opposite direction in the Washout. On average, the shifts of PD of memory I and memory II cells cancelled each other in the Washout. (Notably, the percentages of the two classes were similar.)

In order to subserve motor performance, M1 must provide a similar output in the Baseline and in the Washout. And indeed, in a statistical macroscopic sense the activity of M1 is the same in the Washout as in the Baseline, because the changes recorded for the entire population average to zero in the Washout. But in order to subserve motor learning, M1 must maintain after readaptation a trace of the previous learning experience. And indeed, at the microscopic level of individual neurons M1 was very different in the Washout and in the Baseline, because for 40% of neurons the Washout PD was significantly different in the two conditions. Thus, M1 as a population may subserve both functions of motor performance and motor learning by letting individual neurons change their activity when monkeys learn a new dynamic (motor learning), while reorganizing itself at any time to meet behavioral needs (motor performance).[16]

12.6 NEURONAL PLASTICITY IN OTHER MOTOR AREAS

Recent anatomical studies have identified some 10 or 12 motor areas in the primate frontal lobe.[17–19] According to the traditional view, several "premotor" areas host "high" sensorimotor processes and project to M1, which in turn controls movements through its cortico-spinal projections. More recent anatomical work, however, has found that direct projections to the spinal cord originate from a number of motor areas, including the dorsal premotor area (PMd), the ventral premotor area (PMv), the supplementary motor area (SMA), three or four cingulate motor areas, and M1. In a series of studies, we extended to SMA, PMd, and PMv the experiments first conducted on M1.[20–22] During the experiments, we imposed a randomly variable delay period between the instruction (*cue*) and the *go* signal. In total, we recorded and analyzed the activity of 798 neurons from the 4 areas during a delay time (DT) window (500 msec before the *cue*) and during the movement-related time (MT) window (from 200 msec before the movement onset to the movement end). Our results can be summarized as follows.

Considering neurons as populations, dynamics-related activity (i.e., significant shifts of PD) are observed during movement planning (DT time window) in PMd and SMA, but not in M1 and PMv. (In fact, very limited directional tuning is observed in M1 and PMv during the delay.) In contrast, during movement execution (MT time window), dynamics-related activity is significantly present in all four areas. Likewise, evidence of neuronal plasticity associated with the learning of a new dynamic is found in all four areas.

12.7 REVIEW OF STUDIES ON CORTICAL PLASTICITY

Vast evidence accumulated in the past two decades shows that sensory and motor areas of the cerebral cortex are plastic. Numerous studies have found extensive cortical reorganization associated with perceptual and motor learning. For instance, in the visual domain, Sakai and Miyashita[23] described neurons in the anterior temporal cortex that increased their activation in the delay following presentation of a (nonpreferred) visual stimulus arbitrarily associated with their preferred stimulus.

More recently Erickson and colleagues[24] were able to induce similar response preference in neurons of the perirhinal cortex after one day of exposure to complex visual stimuli, suggesting that clusters of neurons with similar stimulus preferences are shaped through experience. In the acoustic domain, the cortical representation of the frequency range that monkeys were trained to discriminate was found to be increased in the primary auditory cortex.[25] In the somatosensory domain, extensive reorganization of the somatosensory cortex was observed after removal of sensory afferent,[26] and after training.[27,28] Evidence of short-term neuronal plasticity was also found in the dorsolateral prefrontal cortex of monkeys learning a new conditional association. Asaad and coworkers[29] found that the latency of neuronal response (directional selectivity) of neurons progressively decreased over the course of learning.

Several studies also found evidence of neuronal plasticity in various areas when monkeys learned a new conditional motor association. In the task of Wise and coworkers, a novel visual stimulus instructed one of four movements, arbitrarily selected, and the monkeys learned the correct association by trial and error. The authors found extensive learning-related plasticity in PMd[30] and in the supplementary eye fields[31] for conditional associations that instructed limb and eye movements, respectively. Hikosaka and coworkers recently obtained similar results in presupplementary motor area (preSMA). In a first set of experiments in both humans functional magnetic resonance imaging (fMRI) and monkeys (single-cell recordings and reversible lesions; reviewed in Reference 32), the authors contrasted the activity recorded during execution of new versus learned sequences of arm movements instructed by targets appearing on a computer screen. In particular, they found that neurons in preSMA were preferably activated during the execution of new sequences.[33] Similar results were obtained by Germain and Lamarre[34] in the rostral PMd. Finally, plastic changes were also found in the motor cortex of rats learning new sensorimotor associations.[35]

With respect to motor learning, several studies found evidence of long-lasting changes (long-term plasticity) in M1 following skill acquisition. In humans, it was found that the digit representation of the left hand in the M1 of string players was significantly enlarged.[36] Similar effects were also found comparing the activation recorded during execution of a motor sequence practiced over a few weeks versus an unpracticed sequence.[37] Similar findings were obtained with transcranial magnetic stimulation (TMS).[38] In monkeys, Nudo and colleagues[39] mapped with microstimulation the cortical representation of digits and wrist/forearm in M1 before and after training in one of two tasks. They found that the digit representation was enlarged following training in a small-object retrieval task, when the digits were actively used. Conversely, the wrist/forearm representation was enlarged following training in a key-turning task. Work by Donoghue and coworkers has found evidence of long-term potentiation (LTP) and strengthening of horizontal connection in the motor cortex of rats after learning a new motor skill.[40–42]

In other studies investigators have described the changes in neuronal activity that intervene shortly after acquisition of a new motor skill (short-term plasticity). For instance, it was found that the training of one finger movements for 10 minutes changed the direction of movements evoked by focal TMS.[43] Wise and colleagues[44] recorded from M1, PMd, and SMA of one monkey adapting to new visuomotor mappings. In this task — sometimes referred to as acquisition of a new internal model for the kinematics[45,46] — the experimenters manipulated the association between the visual stimulus and the instructed movement. Wise and colleagues found evidence of learning-related plastic changes in the activity of all three areas.

12.8 CONCLUSIONS AND FUTURE DIRECTIONS ON MOTOR LEARNING

A priori motor learning could be achieved in at least two ways. One possible scenario could be that one or more areas (e.g., M1), "in charge of the usual business," process movements in already-learned conditions, and supports well-acquired motor skills. According to this hypothesis, other areas hierarchically higher or parallel (e.g., "premotor" areas) would activate when the normal system fails and would play a more direct role in motor learning. An alternative scenario is that learning-related activity is embedded in the motor system, and that the same areas and the same neurons that process well-acquired movements also accommodate the new conditions when necessary.

At least in part, the results of our studies seem more consistent with this second view, for two reasons. First, dynamics-related activity was present in multiple areas, and plastic changes associated with motor learning were similarly found in all of them. Second, plastic changes were often observed among cells that were already active and committed to the task prior to learning. Furthermore, we did not observe a sharp distinction between the classes of cells (kinematic, dynamic, and memory) in any dimension except for the changes of PD across epochs.

Clearly, the emerging view of "embedded memory" is in syntony with the neural networks model of associative memory, where the same variables that represent any given process modify themselves to execute new computations. Two important remarks should however be made in regard to this issue. First, in all areas we also found neurons that only became committed to the task when the monkeys learned the new dynamics (tune-in cells). Second, in our experiments monkeys were learning a new dynamic. The embedded-memory view may well fail for other instances of motor learning, for example when human subjects or monkeys learn more elaborate motor skills.

We conclude this chapter by indicating one important issue that remains open for future research. From a psychophysical standpoint, our task involves both short-term learning (the monkeys adapt to the force field within one session) and long-term learning (adaptation becomes better and faster across sessions). Furthermore, studies in humans have shown that in the hours immediately following training, the newly learned dynamic undergoes consolidation. Both imaging and TMS studies have suggested that M1 plays a somewhat specific role in the early phase of learning and consolidation. Our experiments essentially fail to address the important issue of whether and how the learning-related plasticity observed here plays a functional role in long-term learning. The techniques currently available allow recording from any one neuron reliably only for a limited time (a couple of hours). Thus, we cannot ascertain at this point whether the plastic changes recorded here are long-lasting and persist through consolidation. Advances in the recording techniques will hopefully help to address these questions.

REFERENCES

1. Hollerbach, M.J. and Flash, T., Dynamic interactions between limb segments during planar arm movement, *Biol. Cybern.,* 44, 67, 1982.
2. Albus, J., A theory of cerebellar function, *Math. Biosci.,* 10, 25, 1971.
3. Marr, D., A theory of cerebellar cortex, *J. Physiol. (Lond.),* 202, 437, 1969.
4. Mussa-Ivaldi, F.A. and Bizzi, E., Motor learning through the combination of primitives, *Phil. Trans. R. Soc. Lond. B Biol. Sci.,* 355, 1755, 2000.
5. Schaal, S. and Atkeson, C.G., Constructive incremental learning from only local information, *Neural. Comput.,* 10, 2047, 1998.
6. Kawato, M. and Wolpert, D.M., Internal models for motor control, in *Novartis Found. Symp.*, 1998, 291.
7. Shadmehr, R. and Mussa-Ivaldi, F.A., Adaptive representation of dynamics during learning of a motor task, *J. Neurosci.,* 14 (5 Pt. 2), 3208, 1994.
8. Flash, T. and Gurevich, I., Arm stiffness and movement adaptation to external loads, in *IEEE Eng. Med. Biol. Soc.*, 1992, 885.
9. Brashers-Krug, T., Shadmehr, R., and Bizzi, E., Consolidation in human motor memory, *Nature,* 382, 252, 1996.
10. Gottlieb, G.L., On the voluntary movement of compliant (inertial-viscoelastic) loads by parcellated control mechanisms, *J. Neurophysiol.,* 76, 3207, 1996.
11. Flanagan, J.R. and Wing, A.M., Effects of surface texture and grip force on the discrimination of hand-held loads, *Percept. Psychophys.,* 59, 111, 1997.
12. Sabes, P.N., Jordan, M.I., and Wolpert, D.M., The role of inertial sensitivity in motor planning, *J. Neurosci.,* 18, 5948, 1998.
13. Gandolfo, F., Mussa-Ivaldi, F.A., and Bizzi, E., Motor learning by field approximation, *Proc. Nat. Acad. Sci. U.S.A.,* 93, 3843, 1996.
14. Georgopoulos, A.P., Kalaska, J.F., Caminiti, R., and Massey, J.T., On the relations between the direction of two-dimensional arm movements and cell discharge in primate motor cortex, *J. Neurosci.,* 2, 1527, 1982.
15. Gandolfo, F., Li, C., Benda, B.J., Padoa-Schioppa, C., and Bizzi, E., Cortical correlates of learning in monkeys adapting to a new dynamical environment, *Proc. Nat. Acad. Sci. U.S.A.,* 97, 2259, 2000.
16. Li, C.S., Padoa-Schioppa, C., and Bizzi, E., Neuronal correlates of motor performance and motor learning in the primary motor cortex of monkeys adapting to an external force field, *Neuron,* 30, 593, 2001.
17. Geyer, S., Matelli, M., Luppino, G., and Zilles, K., Functional neuroanatomy of the primate isocortical motor system, *Anat. Embryol. (Berl.),* 202, 443, 2000.
18. He, S.Q., Dum, R.P., and Strick, P.L., Topographic organization of corticospinal projections from the frontal lobe: motor areas on the lateral surface of the hemisphere, *J. Neurosci.,* 13, 952, 1993.
19. He, S.Q., Dum, R.P., and Strick, P.L., Topographic organization of corticospinal projections from the frontal lobe: motor areas on the medial surface of the hemisphere, *J. Neurosci.,* 15 (5 Pt. 1), 3284, 1995.
20. Padoa-Schioppa, C., Li, C.S., and Bizzi, E., Neuronal correlates of kinematics-to-dynamics transformation in the supplementary motor area, *Neuron,* 36, 751, 2002.
21. Padoa-Schioppa, C., Li, C.S., and Bizzi, E., Neuronal activity in the supplementary motor area of monkeys adapting to a new dynamic environment, *J. Neurophysiol.,* 91, 449, 2004.

22. Xiao, J., Padoa-Schioppa, C., and Bizzi, E., Neuronal activity in the dorsal and ventral premotor areas of monkeys adapting to a new dynamic environment, 2004 (under review).
23. Sakai, K. and Miyashita, Y., Neural organization for the long-term memory of paired associates, *Nature,* 354, 152, 1991.
24. Erickson, C.A., Jagadeesh, B., and Desimone, R., Clustering of perirhinal neurons with similar properties following visual experience in adult monkeys, *Nat. Neurosci.,* 3, 1143, 2000.
25. Recanzone, G.H., Schreiner, C.E., and Merzenich, M.M., Plasticity in the frequency representation of primary auditory cortex following discrimination training in adult owl monkeys, *J. Neurosci.,* 13, 87, 1993.
26. Merzenich, M.M., Kaas, J.H., Wall, J., Nelson, R.J., Sur, M., and Felleman, D., Topographic reorganization of somatosensory cortical areas 3b and 1 in adult monkeys following restricted deafferentation, *Neuroscience,* 8, 33, 1983.
27. Recanzone, G.H., Merzenich, M.M., and Jenkins, W.M., Frequency discrimination training engaging a restricted skin surface results in an emergence of a cutaneous response zone in cortical area 3a, *J. Neurophysiol.,* 67, 1057, 1992.
28. Recanzone, G.H., Merzenich, M.M., Jenkins, W.M., Grajski, K.A., and Dinse, H.R., Topographic reorganization of the hand representation in cortical area 3b owl monkeys trained in a frequency-discrimination task, *J. Neurophysiol.,* 67, 1031, 1992.
29. Asaad, W.F., Rainer, G., and Miller, E.K., Neural activity in the primate prefrontal cortex during associative learning, *Neuron,* 21, 1399, 1998.
30. Mitz, A.R., Godschalk, M., and Wise, S.P., Learning-dependent neuronal activity in the premotor cortex: activity during the acquisition of conditional motor associations, *J. Neurosci.,* 11, 1855, 1991.
31. Chen, L.L. and Wise, S.P., Neuronal activity in the supplementary eye field during acquisition of conditional oculomotor associations, *J. Neurophysiol.,* 73, 1101, 1995.
32. Hikosaka, O., Sakai, K., Nakahara, H., Lu, S., Miyachi, S., Nakamura, K., and Rand, M.K., Neural mechanisms for learning of sequential procedures, in *The New Cognitive Neuroscience*, Gazzaniga, M.S., Ed., MIT Press, Cambridge, MA, 2000, 553.
33. Nakamura, K., Sakai, K., and Hikosaka, O., Neuronal activity in medial frontal cortex during learning of sequential procedures, *J. Neurophysiol.,* 80, 2671, 1998.
34. Germain, L. and Lamarre, Y., Neuronal activity in the motor and premotor cortices before and after learning the associations between auditory stimuli and motor responses, *Brain Res.,* 611, 175, 1993.
35. Laubach, M., Wessberg, J., and Nicolelis, M.A., Cortical ensemble activity increasingly predicts behaviour outcomes during learning of a motor task, *Nature,* 405, 567, 2000.
36. Elbert, T., Pantev, C., Wienbruch, C., Rockstroh, B., and Taub, E., Increased cortical representation of the fingers of the left hand in string players, *Science,* 270, 305, 1995.
37. Karni, A., Meyer, G., Jezzard, P., Adams, M.M., Turner, R., and Ungerleider, L.G., Functional MRI evidence for adult motor cortex plasticity during motor skill learning, *Nature,* 377, 155, 1995.
38. Pascual-Leone, A., Nguyet, D., Cohen, L.G., Brasil-Neto, J.P., Cammarota, A., and Hallett, M., Modulation of muscle responses evoked by transcranial magnetic stimulation during the acquisition of new fine motor skills, *J. Neurophysiol.,* 74, 1037, 1995.
39. Nudo, R.J., Milliken, G.W., Jenkins, W.M., and Merzenich, M.M., Use-dependent alterations of movement representations in primary motor cortex of adult squirrel monkeys, *J. Neurosci.,* 16, 785, 1996.

40. Rioult-Pedotti, M.S., Friedman, D., and Donoghue, J.P., Learning-induced LTP in neocortex, *Science,* 290, 533, 2000.
41. Rioult-Pedotti, M.S., Friedman, D., Hess, G., and Donoghue, J.P., Strengthening of horizontal cortical connections following skill learning, *Nat. Neurosci.,* 1, 230, 1998.
42. Sanes, J.N. and Donoghue, J.P., Plasticity and primary motor cortex, *Annu. Rev. Neurosci.,* 23, 393, 2000.
43. Classen, J., Liepert, J., Wise, S.P., Hallett, M., and Cohen, L.G., Rapid plasticity of human cortical movement representation induced by practice, *J. Neurophysiol.,* 79, 1117, 1998.
44. Wise, S.P., Moody, S.L., Blomstrom, K.J., and Mitz, A.R., Changes in motor cortical activity during visuomotor adaptation, *Exp. Brain Res.,* 121, 285, 1998.
45. Flanagan, J.R., Nakano, E., Imamizu, H., Osu, R., Yoshioka, T., and Kawato, M., Composition and decomposition of internal models in motor learning under altered kinematic and dynamic environments, *J. Neurosci.,* 19, RC34, 1999.
46. Krakauer, J.W., Ghilardi, M.F., and Ghez, C., Independent learning of internal models for kinematic and dynamic control of reaching, *Nat. Neurosci.,* 2, 1026, 1999.

Section IV

Reconstruction of Movements Using Brain Activity

13 Advances in Brain–Machine Interfaces

Jose M. Carmena and Miguel A.L. Nicolelis

CONTENTS

13.1 INTRODUCTION

Throughout history, the introduction of new technologies has significantly impacted human life in many different ways. Until now, however, each new artificial device or tool designed to enhance human motor, sensory, or cognitive capabilities has relied on explicit human motor behaviors (e.g., hand, finger, or foot movements), often augmented by automation, in order to translate the subject's intent into concrete goals or final products. The increasing use of computers in our daily lives provides a clear example of such a trend. Yet, the realization of the full potential of the "digital revolution" has been hindered by its reliance on low bandwidth and relatively slow user–machine interfaces (e.g., keyboard, mouse). Because these user–machine interfaces are far removed from how the brain normally interacts with the surrounding environment, the potential of such a tool is limited by its inherent inability to be assimilated by the brain's multiple internal representations as a continuous extension of our body appendices or sensory organs.

0-8493-1287-6/05/$0.00+$1.50

Two decades ago, an alternative method was proposed for restoring motor behaviors in severely paralyzed patients.[1] This method proposed a bypass of the spinal cord and started a new paradigm, namely the interface between brains and machines, or brain–machine interfaces (BMIs). This approach contends that paralyzed patients could enact their voluntary motor intentions through a direct interface between their brains and artificial actuators in virtually the same way that we see, walk, or grab an object. The studies in macaque monkeys conducted by Fetz and collaborators[2–5] were the first experimental support for a cortical driven BMI. In fact, recent BMI research in animals and humans has supported the contention that we are at the brink of a technological revolution, where artificial devices may be "integrated" in the multiple sensory, motor, and cognitive representations that exist in the primate brain. These studies have demonstrated that animals can learn to utilize their brain activity to control two-dimensional displacements of computer cursors,[6,7] one-dimensional to three-dimensional movements of simple and elaborate robot arms[8,9] and, more recently, reaching and grasping movements of a robot arm.[10] In addition to the current research performed in rodents and primates, there are also preliminary studies using human subjects.[11–13]

The ultimate goal of this emerging field of BMIs is to allow human subjects to interact seamlessly with a variety of actuators and sensory devices through the expression of their voluntary brain activity, either for augmenting or restoring sensory, motor, and cognitive function — e.g., after a traumatic lesion of the central nervous system. Moreover, by providing ways to deliver sensory (visual, tactile, auditory, etc.) feedback from these devices to the brain, one could establish a reciprocal (and more biologically plausible) interaction between large neural circuits and machines, hence fulfilling the requirements for artificial actuators to be recognized as simple extensions of our bodies.

In addition to the potential clinical application, BMIs also serve as a unique tool for systems neuroscience research. The combination of multiple-site, multiple-electrode recordings with the BMI paradigm provides the experimenter with a new way to quantify neurophysiological modifications occurring in cortical networks, as animals learn motor tasks of various complexities.[14]

13.1.1 Invasive and Noninvasive BMIs

The noninvasive approach in BMIs utilizes features of brain activity, such as event-related responses or continuous electroencephalogram (EEG) rhythms, to control a computer-based device. These devices, commonly known as brain–computer interfaces (BCIs), record brain activity from surface electrodes positioned on the scalp. They typically consist of a computer screen on which a subject, after training, can control the selection of characters by moving a cursor up, down, left, and right to operate simple word processing programs or indicate a particular action to a caregiver.[15] The main issue with this approach is the small communication bandwidth currently available. Still, with the aid of artificial intelligence and robotics, more complex tasks can be achieved using this approach. Such is the case with autonomous wheelchair navigation. Millán and colleagues have devised an EEG-based system that discriminates among the recorded signals that are generated in different mental activities, such as adding numbers, thinking of a family member, imagining geometric shapes, etc. Once identified, these signals are translated into high-level commands that actually control the navigation of the wheelchair. Lower-level actions, such as path planning and obstacle avoidance, are performed by the system through the reading of the sensors attached to the wheelchair.[16] This system exploits the diversity of brain waves during different mental exercises, which has no real correlation with behavioral outcome — i.e., no decoding of motor actions is performed by the BCI. Nevertheless, the noninvasive approach lacks the spatial resolution and bandwidth necessary for extracting the kind of time-varying motor signals that would be necessary to control accurate three-dimensional arm movements in real time, as would be needed for prosthetic devices.[17,18]

The invasive approach typically uses extracellular recordings of individual neurons through chronically implanted microelectrodes in the cortex.[6–10] Other researchers[19] use local field potentials, which offer more resolution than the EEG recordings on the scalp, but still contain less information than extracellular single-neuron recordings. In this chapter, however, we will focus on invasive BMIs that use arrays of microelectrodes chronically implanted in the cortex of macaque monkeys.

13.2 BMI DESIGN

In this section, the state of the art in BMI design will be reviewed, together with some discussion of specific issues. Figure 13.1 illustrates the different parts that form a closed-loop cortical BMI.

13.2.1 Chronic, Multisite, Multielectrode Recordings

The capability of recording the activity of many single cortical neurons for long periods of time in awake, behaving macaque monkeys or rodents is a powerful tool that permits neurophysiological investigation of learning, perception, and sensorimotor integration. Moreover, BMIs in humans will require electrodes to be implanted chronically for long periods of time, raising issues on the quality and stability of the recordings, and on the biocompatibility of the materials.

Recent attempts to obtain long-lasting, single-neuron recordings from macaques have employed the 100-electrode "Utah array" or arrays of individual sharp microwires.[6,7,20–23] However, these studies have provided relatively modest neuronal yields of uncertain longevity, and, in most cases, they have thus far been limited to just one or two cortical areas per animal.

Nevertheless, progress in the development of high-density microwire arrays during the past years has resulted in the standardization of this technique in rodents[24] and primates.[25] This technique permits high quality single-unit recordings for long periods of time in macaque monkeys.[25] A multineuron acquisition processor (MAP) (Plexon, Inc., Dallas, TX) cluster, formed by three 128-channel MAPs synchronized by a common clock signal, was specially built for simultaneous recordings from hundreds of neurons in real time as reported in this study. This 384-channel recording system has a theoretical capacity of recording up to 1536 single neurons simultaneously (e.g., 4 neurons per channel), at a 25-µs precision. Among other results, the

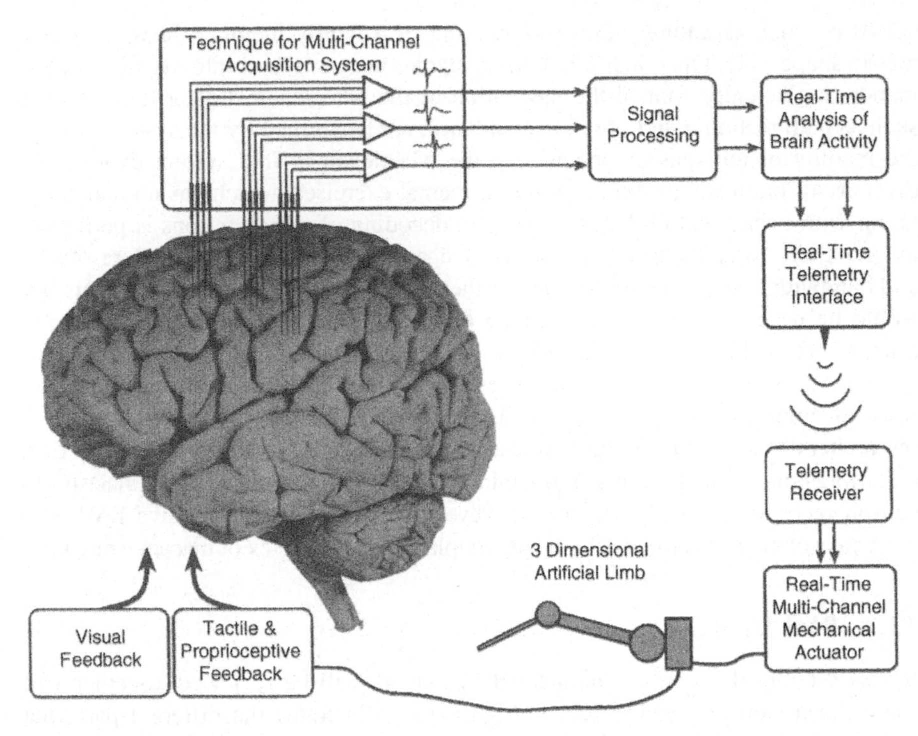

FIGURE 13.1 General architecture of a closed-loop control brain–machine interface.

study demonstrates the simultaneous recording of extracellular activity of 247 single cortical neurons from 384 microwires implanted in multiple cortical areas of the brain of a macaque monkey 30 days after the implantation surgery. In a different monkey, recordings were obtained from up to 58 isolated neurons 18 months after surgery. The success of this technique has been crucial for our BMI work.[9,10]

13.2.2 Data Acquisition and Telemetry

The next step toward a final BMI product will be to perform unsupervised (i.e., automatic) spike detection in real time from the extracellular recordings. This detection should be reliable so that spikes can be separated from the background noise before they are sorted and transmitted via a transcutaneous wireless telemetry device. Spike information will then be sent to a processing device implanted or carried elsewhere on the subject's body.

Since the critical issues in the components of a BMI system are small size and low power consumption, the computational bandwidth of the system will be limited. Obeid and colleagues are working on a wearable multichannel neural telemetry system that would suit the needs of a BMI. The current version of their system allows sampling, digitizing, processing, and transmission of 32 channels.[26,27] Further versions of this system currently under development will increase the number of channels substantially. Obeid et al. have also investigated different classes of real-time spike detection algorithms. A computationally cheap method, such as the estimation of the absolute value of the neural signal before applying a threshold, was found to be as efficient for detecting spikes as more complex energy-based detectors (e.g., matched filters).[28]

One of the key issues that will permit an increase of channels in a wireless telemetry system is the optimization of the transmission bandwidth. In order to address this issue, Bossetti and colleagues[29] are investigating how variability in the firing rate of neural ensembles, due to intrinsic neuronal properties such as bursting, or due to the behavioral task performed by the subject, causes wide fluctuations in the instantaneous data rate that is being transmitted. Bossetti et al. found that an efficient reduction of the transmission bandwidth (i.e., avoiding significant transmission delays) requires a transmission rate three to five times higher than the mean firing rate of the ensemble.

13.2.3 Bidirectional BMIs: Decoding and Encoding

After the neural spikes are sampled, digitized, and transmitted to the signal processing unit of the BMI, decoding algorithms are used to predict the intention of movement. The decoding algorithm used in our BMI work is a simple linear model, namely a multidimensional linear regression of times series, or Wiener filter.[30] We have successfully used it to estimate the hand position of owl monkeys reaching for food[9] and the hand position, velocity, and grasping force of rhesus macaques performing a reaching and grasping task.[10]

The basic form of this filter is

$$\mathbf{y}(t) = \mathbf{b} + \sum_{u=-m}^{n} \mathbf{a}(u)\mathbf{x}(t-u) + \varepsilon(t) \tag{13.1}$$

in which $\mathbf{x}(t-u)$ is an input vector of neuronal firing rates at time t and time-lag u; $\mathbf{y}(t)$ is a vector of kinematic and dynamic variables (e.g., position, velocity, and gripping force) at time t; $\mathbf{a}(u)$ is a vector of weights at time-lag u; $\mathbf{b}$ is a vector of y-intercepts; and $\varepsilon(t)$ are the residual errors. There is one weight for each neuron in $\mathbf{x}(t)$ and dimension in $\mathbf{y}(t)$. In general, the lags in the summation can be negative (in the past) or positive (in the future) with respect to the present time t. For on-line applications, as in the case of BMIs, only lags into the past can be considered.

Several others decoding algorithms can be found in the BMI literature. These range from linear — Kalman filter,[31] Least Mean Squares[32] — to nonlinear models, such as Recurrent Artificial Neural Networks[32] and Echo State Networks,[33] among many others. Although it is beyond the scope of this chapter to go into the details of such models, it is important to mention that none of these algorithms significantly outperforms the Wiener filter. This has been confirmed in previous works.[9,10,32,34] Recently, Kim and colleagues[35] performed an extensive comparison study of the accuracy of several linear and nonlinear models based on three quantitative performance measures on data from owl and macaque monkeys performing different tasks.[9,10] These measures are the correlation coefficient (CC) between the estimated and observed behavioral output; the signal-to-error ratio (SER), defined as the ratio of the powers of the actual behavioral output and the error; and the cumulative error metric (CEM), which estimates the cumulative distribution function of the error radius. Results of the comparison showed no significant differences in performance between the Wiener filter and the rest of the models tested.

Another of the reasons for using a linear model is that it permits us to understand the contributions of the individual neurons of the ensemble to the derived control signals, in contrast to the "black box" nature of the nonlinear algorithms.[36] While this linear model approach is very practical at the research and development stage, there are several problems inherent in its current form that prevent it from being implemented in a low-power, portable hardware BMI. The main one is the overfitting introduced by the explosion in the number of free parameters as the number of sampled neurons increases — e.g., for 100 neurons, 3-dimensional movements, and 10 time-lags, there are 3000 free parameters. One way of minimizing this problem is by optimizing the size of the ensemble, selecting the most contributing neurons in a particular moment of time. For this purpose, we have developed methods for ascertaining the importance of neurons using single neuron correlation analysis, sensitivity analysis through a vector linear model, and directional tuning analysis.[37] While these methods allow us to rank an ensemble of neurons in a simple and practical way, this ranking needs to be continuously updated because of neuronal variability along time. Evidence of this variability is shown in Section 13.3.1. Another way of minimizing the number of free parameters is to reduce the number of time-lags by selecting them depending on the cortical area to which a neuron belongs; e.g., the time-lag associated with primary motor cortex (M1) neurons will be smaller than the one for dorsal premotor area (PMd) neurons.

Also, since the purpose of a BMI is to work on-line, the hardware implementation of the models will need to be iterative; i.e., they will generate predictions as they are being trained. For this purpose, models based on the Wiener filter solution, such as the least-mean squares (LMS) adaptive filter, are ideal candidates.

As indicated in Figure 13.1, the predictions from these decoding algorithms are used as motor commands to control artificial actuators (e.g., cursor, robot), and the information from these actuators is sent back to the brain via visual feedback, and by direct stimulation in somatosensory pathways. The latter is an important, yet largely unexplored avenue of research within the BMI field. The most common example of an "encoding BMI" is the widely known cochlear implant,[38] in which an implanted device converts the frequency of sound waves into electrical impulses that stimulate the auditory nerve. Another example of encoding BMI is the visual neuroprosthesis, both at the retinal[39] and cortical[40] levels. However, the state of the art in these neuroprostheses is not as advanced in restoring sensory functionality as in the cochlear implant.

One remarkable example of an encoding BMI is the work of Talwar and colleagues.[41] They demonstrated that rats can be controlled remotely as mobile robots or even as biosensors, by stimulating cue and reward signals directly in their brains. Stimulating electrodes were implanted in the somatosensory cortex (S1) and the medial forebrain bundle (MFB), and stimulation was delivered by a remote-controlled microstimulator mounted on a backpack. Rats were guided through mazes and other environments by a combination of *left* and *right* stimulation cues in the

S1 whisker area of the right and left hemispheres, respectively, and with a reward signal in the MFB that enacted forward movements.

These results are interesting as a proof of concept of an encoding BMI. However, for a realistic somatosensory neuroprosthesis, a larger set of "encoding commands" will be needed. For example, a motor task will require the encoding of sensory information from the artificial limb, including parameters such as limb position, velocity, and gripping force, among others. In order to be able to encode these parameters directly into the brain, a much deeper understanding about how sensory information is encoded in the brain is needed. In a BMI context we could think of a "library" of spatiotemporal stimulation patterns that would be applied to evoke particular sensory information in the brain. In this direction, Xu and colleagues are working on stimulation patterns in the rat thalamus that, when applied, will evoke selective and "natural" somatic perceptions.[42] The rationale for this approach comes from good statistical correlation between touch stimuli in the rat forepaw/forelimb and S1 neural responses. This finding suggests using these cortical responses as the target criteria for optimizing the thalamic stimuli.

An alternative to direct brain stimulation is vibrotactile stimulation. This form of somatosensory feedback allows the encoding of spatiotemporal patterns of vibration in the skin. Sandler and colleagues[43] are currently looking at the electrophysiological changes that occur during conditional motor learning in owl monkeys using this kind of feedback. Vibrotactile stimulation could be integrated in a BMI and applied in regions of the subject's body not affected by trauma, such as the neck or the face. After training, the subject could learn to use this source of feedback as a source of information that is supplementary to visual feedback. Availability of this "somatosensory" feedback in a BMI could be very advantageous in real life situations where a clear visual perception of the artificial limb is absent.

13.3 REACHING AND GRASPING WITH A BMI

There are electrophysiological and engineering issues regarding the operation of BMIs that are a matter of considerable debate. These include the type of brain signals[17–19,44] (single unit, multiple unit, or field potentials) that would provide the optimal input for a such a device, and the number of single units (small [8–30][6,7] or substantially larger [hundreds to thousands][9,10]) that may be necessary to operate a BMI efficiently for many years. These and other questions were investigated in our recent study in which we showed how macaque monkeys learned to use a BMI to reach and grasp virtual objects with a robot even in the absence of overt arm movement signals.[10] Some of the findings of this study that are relevant to these questions are summarized below.

Monkeys were implanted with multiple arrays (96 in monkey 1, and 320 in monkey 2) in several frontal and parietal cortical areas (PMd, M1, supplementary motor area [SMA], S1, and posterior parietal [PP]). In this study we used multiple linear models, similar to the one described in Section 1.2.3, to simultaneously extract a variety of motor parameters (hand position [HPx, HPy, HPz], velocity [HVx, HVy, HVz], and gripping force [GF]) and multiple muscle electromyographs (EMGs) from the activity of cortical neural ensembles while macaque monkeys performed several motor tasks (see color Figure 13.2*). Although all these parameters were extracted in real time in each session, only some of them were used to control the BMI, depending on each of the three tasks the monkeys had to solve in a given day. In each recording session, an initial 30-minute period was used for training of these models. During this period, monkeys used a handheld pole either to move a cursor on the screen or to change the cursor size by application of GF to the pole. This period is referred to as the "pole control" mode. As the models converged to an optimal performance, their coefficients were fixed and the control of the cursor position (tasks 1 and 3) and/or size (tasks 2 and 3) was obtained directly from the output of the linear models. This period is referred to as the "brain control" mode. During the brain control mode, animals initially produced arm movements, but they soon realized that these were not necessary and ceased to produce them for periods of time. This is shown in Color Figure 13.2D, in which EMG modulations were absent during brain control.

Accurate performance was possible because large populations of neurons from multiple cortical areas were sampled, showing that large ensembles are preferable for efficient operation of a BMI. This conclusion is consistent with the notion that motor programming and execution is represented in a highly distributed fashion across frontal and parietal areas, and that each of these areas contains neurons that represent multiple motor parameters. We suggest that, in principle, any of these areas could be used to operate a BMI, provided that a large enough neuronal sample was obtained. This is supported by the analysis of neuron dropping curves[9,10] shown in Color Figures 13.3A–C, which indicate the number of neurons that are required to achieve a particular level of model prediction for each cortical area. Although all cortical areas surveyed contained information about any given motor parameter, for each area different numbers of neurons were required to achieve the same level of prediction. Although a significant sample of M1 neurons consistently provides the best predictions of all motor parameters analyzed, neurons in areas such as SMA, S1, PMd, and PP contribute to BMI performance as well.

Another important finding of this study is that accurate real-time prediction of all motor parameters as well as a high level of BMI control can be obtained from multiple-unit signals. This observation is essential because it eliminates the need to develop elaborated real-time spike-sorting algorithms, a major technological challenge, in the design of a future cortical neuroprosthesis for clinical applications. Color Figures 13.3D–F show that the linear predictions of hand position, velocity, and GF were somewhat better when single units were used (by 17, 20, and 17%, respectively). That difference could be compensated for by increasing the number of channels. For example, as seen in Color Figure 13.3D, around 30 additional multiple units compensate for the difference in prediction of hand position provided by adding 20 single units. That difference was, however, not critical, as the animals could still maintain high levels of BMI performance in all tasks using multiple-unit activity only.

Our experiments demonstrated, for the first time, that monkeys can learn to control a BMI to produce a combination of reaching and grasping movements in

* See color insert following page 170.

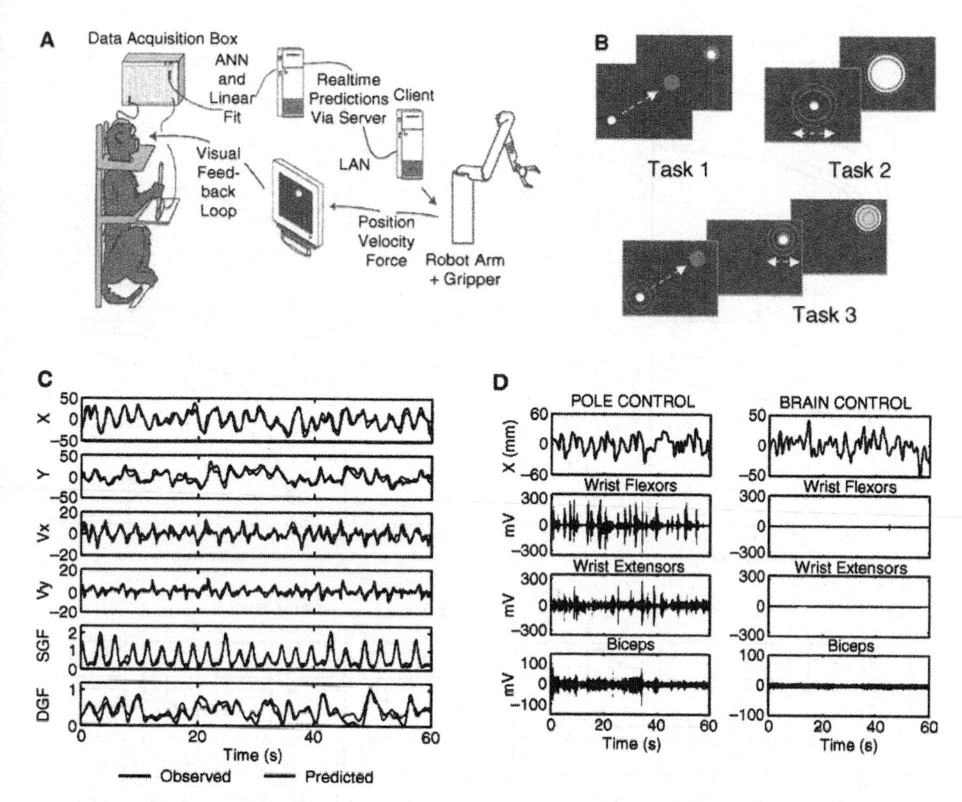

FIGURE 13.2 (see color figure) (A) Experimental setup and control loops, consisting of a data acquisition system, a computer running multiple linear models in real time, a robot arm equipped with a gripper, and a monkey visual display. The pole was equipped with a gripping force transducer. Robot position was translated into cursor position on the screen, and feedback of the gripping force was provided by changing the cursor size. (B) Schematics of three behavioral tasks. In task 1 the monkey's goal was to move the cursor to a visual target (green) that appeared at random locations on the screen. In task 2 the pole was stationary, and the monkey had to grasp a virtual object by developing a particular gripping force instructed by 2 red circles displayed on the screen. Task 3 was a combination of tasks 1 and 2. The monkey had to move the cursor to the target and then develop a gripping force necessary to grasp a virtual object. (C) Motor parameters (blue) and their prediction using linear models (red). From top to bottom, hand position (HPX, HPY) and velocity (HVx, HVy) during execution of task 1, and gripping force (GF) during execution of tasks 2 and 1. (D) Surface EMGs of arm muscles recorded in task 1 for pole control (left) and brain control without arm movements (right). Top plots show X-coordinate of the cursor; plots below display EMGs of wrist flexors, wrist extensors, and biceps. EMG modulations were absent in brain control. (Extracted from Reference 10.)

order to locate and grasp virtual objects. The major challenge in task 3 was to predict hand position and gripping force simultaneously using the activity recorded from the same neuronal ensemble. This problem could not be reduced to predicting only hand position as in task 1 or gripping force in task 2, because the animal had to reach for and grasp the target sequentially. The monkeys' performance in brain

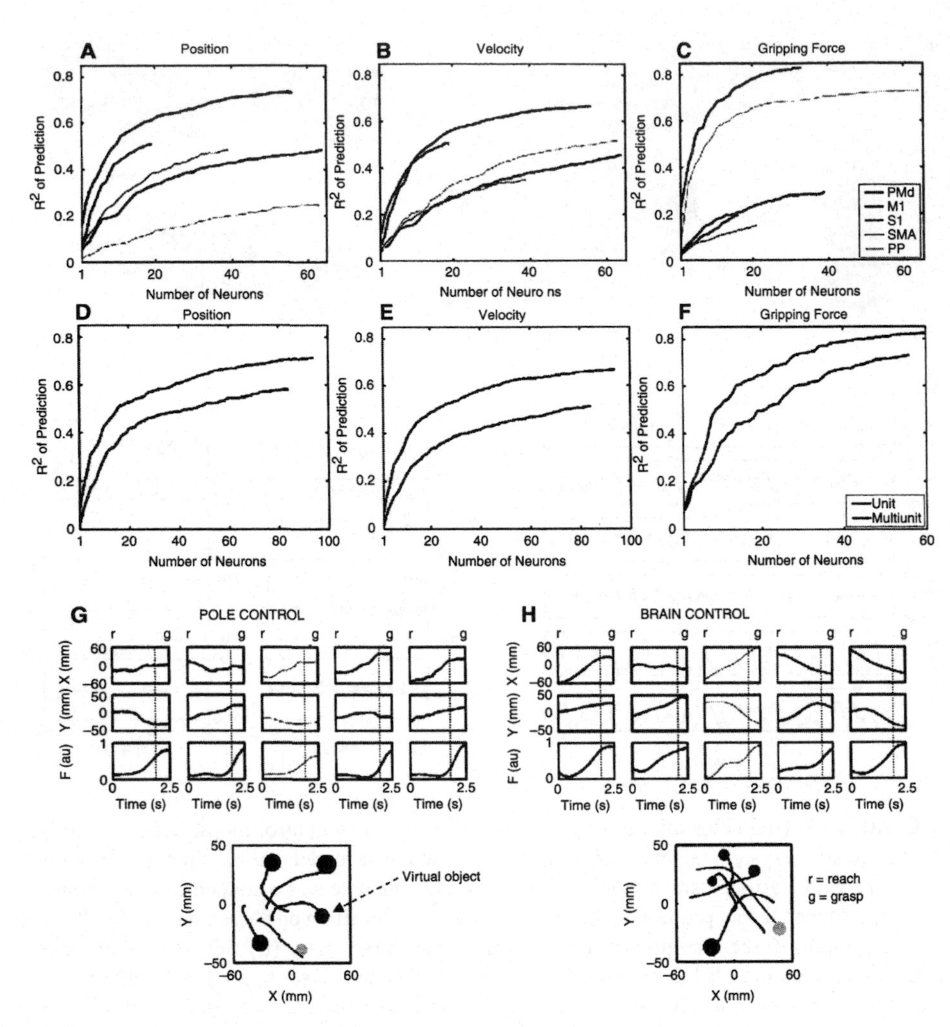

FIGURE 13.3 (see color figure) (A–F) Contribution of different cortical areas to model predictions of hand position, velocity (task 1), and force (task 2). For each area, neuronal dropping curves represent average prediction accuracy (R^2) as a function of the number of neurons needed to attain it. Contributions of each cortical area vary for different parameters. Typically more than 30 randomly sampled neurons were required for an acceptable level of prediction. (G–I) Comparison of the contribution of single units (blue) and multiple units (red) to predictions of HP, HV, and GF. Single units and multiple units were taken from all cortical areas. Single units' contribution exceeded that of multiple units by ~20%. (G, H) Representative robot trajectories and gripping force profiles in an advanced stage of training in task 3 during both pole and brain control. The bottom graphs show trajectories and the amount of the gripping force developed during grasping of each virtual object. The dotted vertical lines in the panels indicate the end of reach. Note that during both modes of BMI operation, the patterns of reaching and grasping movements (displacement followed by force increase) were preserved. (Extracted from Reference 10.)

control in task 3 approximated that during pole control, with characteristic robot displacement (reach) followed by force increase (grasp). Color Figures 13.3G and 13.3H show several representative examples of reaching and grasping during pole and brain control in task 3 by monkey 1. Hand position (X,Y) and gripping force (F) records are shown. In the display of hand trajectories, the size of the disc at the end of each hand movement shows the gripping force produced by the monkey (Color Figure 13.3G) or by the BMI (Color Figure 13.3H) to grasp a virtual object. The reach (r) and grasp (g) phases are clearly separated, demonstrating that the monkeys could gracefully use the same sample of neurons to produce distinct motor outputs at different moments in time. Thus, during the reaching phase, X and Y changed while F remained relatively stable. However, as the monkey got closer to the virtual object, F started to increase while X and Y stabilized to maintain the cursor over the virtual object.

Our study also demonstrated that the initial introduction of a mechanical device, such as the robot arm, in the control loop of a BMI significantly affects learning and task performance. After the robot was introduced in the control loop, the monkey had to adjust to the dynamics of this artificial actuator. As a result, there was an immediate drop in performance (data not shown). With further training, however, the animals were able to overcome the difficulties. Thus, in order to test the limitations and challenges involved in operating a clinically relevant BMI, we must include the incorporation in the system of the mechanical actuator designed to enact the subject's motor intentions, as well as training the subject to operate it.

13.3.1 Neuronal Variability

In our aforementioned studies[9,10] the motivation for sampling from multiple areas came primarily from the notion that motor programming and execution is represented in a highly distributed fashion across frontal and parietal cortical and subcortical areas of the primate brain, and that each of these areas contains neurons that represent multiple motor parameters. Moreover, the nonstationary nature of the neural code, expressed in the form of neuronal variability over time, suggests the use of large neuronal ensembles to guarantee task performance.

Evidence of this neuronal variability is shown in Color Figure 13.4, in which analysis of neuron dropping curves during learning of the reach and grasp task revealed a gradual increase in the contribution of different cortical areas to the prediction of HP, HV, and GF. The most dramatic changes occurred during the first five sessions. As shown in the figure, more significant changes were observed for the prediction of HP and HV than GF. From the onset of training, the M1 sample gave the highest predictions for all motor parameters. M1 contributions to the prediction of HP and HV increased significantly more than its contribution to GF, but the latter was already very high from the first day of training. Even though their contribution was lower than that of M1 cells, PMd neurons also provided a very stable source for predicting GF. In addition, PMd contribution to HP and HV prediction increased significantly during the first 5 days. Similar changes in contribution to HP and HV were observed for S1 neurons. Yet this latter area had a more complex behavior in relation to GF. First, it showed a gradual reduction in prediction

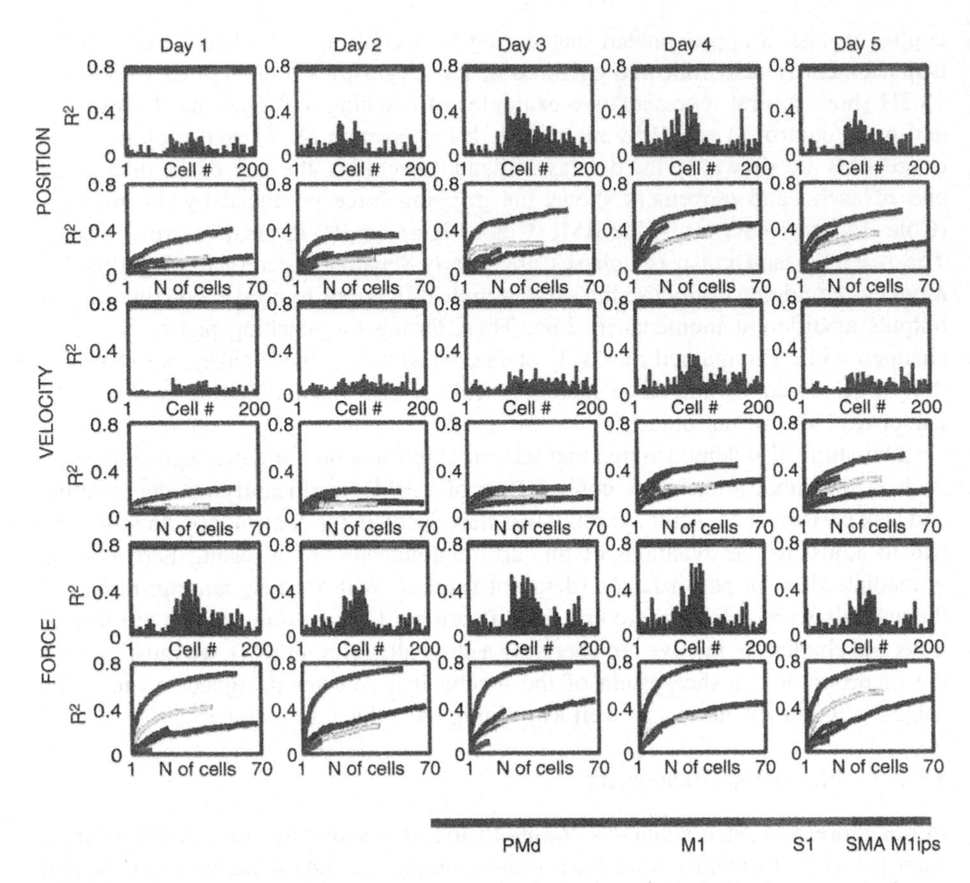

FIGURE 13.4 (see color figure) Variability in contributions of individual neurons and cortical areas to the representation of multiple motor parameters (from top to bottom: hand position, hand velocity, and gripping force). Note the clear increase of accuracy in predictions for individual neurons and cortical areas during the 5-day period. During the same period, a high degree of variability in both neuronal and real contributions was observed. The color bar indicates the sample size for each cortical area.

over the first 3 days, followed by a slight increase in contribution. Both the SMA sample (20 neurons) and a relatively small sample of ipsilateral M1 (5 neurons) showed a remarkable increase in their contribution to all motor parameters during the first 4 days of training. While SMA neurons maintained their level of performance on the fifth day, the contribution of ipsilateral M1 neurons during that day decreased. Thus, considerable variations in the contribution of distinct cortical areas were observed even after the animals mastered the task. Color Figure 13.4 also shows significant variability in the contribution of individual units. Variations were seen in the contribution of individual channels to real-time predictions of HP, HV, and GF. The most important conclusion derived from this analysis was that neurons that provided the highest contribution to a given parameter on one day were not necessarily the best contributors for predictions of the same parameter on other days. This effect was observed for all cortical areas, matching the results obtained with single neurons in the reaching task. Given this variability, a high level of long-term BMI performance could only be attained because large samples of neurons, from multiple cortical areas, were simultaneously recorded. These results suggest the importance of recording from large neuronal ensembles for achieving reliable BMI performance.[45]

13.4 FUTURE DIRECTIONS

This final section of the chapter presents continuous shared control as a new paradigm in BMI research. This new approach incorporates an artificial intelligence module for improving the control of artificial devices (robots, prosthetic limbs, etc.). This will be followed by a discussion of translating BMI experiments to the clinic.

13.4.1 Continuous Shared-Control BMIs

In our study[10] we showed how an artificial device (cursor/robot) could be controlled by the brain using multiple linear models on-line. Such decoding algorithms perform a linear mapping between the neuronal activation in several areas of the primate's brain and its behavioral output. However, in order to restore control of upper limb movements, a neuroprosthetic device will need to incorporate general physiological principles of how motor signals underlying these movements are encoded in the brain.[18] In other words, the decoding algorithms will need to incorporate physiological *knowledge*. Still, this may not be enough to reach the performance level an injured patient would desire. In fact, dexterous manipulation in humans is one of the most impressive examples of motor control, and requires a significant amount of skill. The relatively low bandwidth in current BMI works (~10 Hz) and the lack of sensory feedback makes the task of restoring hand dexterity using an artificial limb (or robotic actuator) extremely challenging, and perhaps not feasible with current technology.

On the other hand, there is the availability of the robotic domain, namely exploiting the fields of control theory and artificial intelligence, among others, and creating a hybrid BMI that will incorporate both real (neuronal) and artificial signals in a way that would allow a patient to accomplish tasks more accurately than when using neuronal signals only. For example, we could think of a BMI that will decode the intention of movement directly from neuronal signals, and leave the path-planning execution, obstacle avoidance, and final refinement on grasping to a control module incorporated into the robot. This control module would have inputs from neuronal signals as well as readings from sensors embedded in the robot. At this point, the following question arises: what ratio of neuronal versus artificial signal is needed for optimal control of a BMI? In response, Kim and colleagues[46] introduced the concept of continuous shared control (CSC). As the authors indicate, "the control is *continuous* because the interaction is immediate and does not have the 'wait and see' characteristics of a planner-based approach or the switching characteristic of a traded-control. The control is *shared* because it always reflects input of both brain and sensor, as distinguished from traded control where control switches discreetly from direct operator control to the autonomy of the robot depending on task and situation."[46] This is illustrated in Figure 13.5.

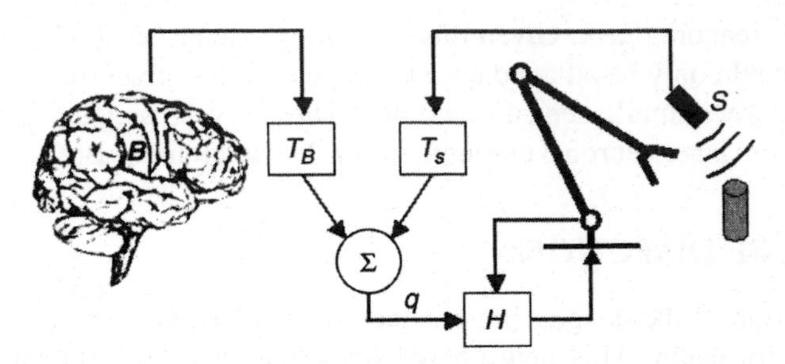

FIGURE 13.5 The general problem of shared control in brain–machine interfaces. The goal is to share control between commands arising from the user's brain (B) and sensors (S) on the slave device in a way that improves task performance. Both types of signals must be transformed (e.g., by *TB*, *TS*) into a "normal" command (e.g., a set of joint angles *q*) on which a control system (H) may operate.

Kim et al. tested this idea on real data from our previous study[10] in which macaque monkeys were reaching and grasping virtual objects using only brain-derived signals. However, to reach and squeeze a real object with the required force at the right location is an extremely difficult task. A 3 degrees of freedom (DOF) robot with a pneumatic gripper that incorporated optical sensors was used to replicate the task using CSC. This gripper produced reflex-like reactions to augment the brain-controlled trajectories, providing obstacle avoidance and stabilized grasping. Different levels of sensor-based reflex effort were tested, and the ratio of 70% brain command and 30% sensor command was the optimal level, resulting in a sevenfold increase in task performance. This significant improvement in performance suggests the use of CSC to be critical in the future development of BMI systems.

13.4.2 Neuroprostheses: Translating BMIs to the Clinic

The ultimate goal of the neuroprosthetic experiments in monkeys is the translation of the technique to the clinic; i.e., the chronic implant of such devices in humans with motor disabilities. As discussed in this chapter, further primate studies have yet to address the fundamental question of whether current BMI technology and approaches may be applied successfully to human subjects.[17,18,44] It is also important to identify which regions of the brain may provide an effective motor control signal for task prediction, and ultimately for neuroprosthetic control. In a recent report[13] we showed how neuronal acute recordings from subcortical motor regions of the human brain, such as motor thalamus (ventralis oralis posterior [VOP]) and subthalamic nucleus (STN), serve as predictors of motor function. During deep brain stimulation surgery, 11 Parkinsonian patients performed a motor task in the form of squeezing a sensorized ball to control the one-dimensional movement of a vertical bar in a computer screen while acute ensemble recordings in STN and VOP were obtained. Offline analysis of the data revealed that ensembles of 3 to 55 simultaneously recorded neurons were informative enough to predict gripping force during 30-second test periods with accuracy of up to $R^2 = 0.68$. These results suggest that a larger scale, multiple-electrode human BMI might ultimately function as a human brain–machine interface (HBMI) in patients with neurological injury.

There is evidence in the literature that motor areas in the brain remain functional in human subjects for several years following paralysis caused by spinal cord injuries.[47] However, HBMIs face the major problem of lacking a behavioral signal to map with their neuronal correlates — i.e., the paralyzed condition of the subjects will not permit a direct measure of their intended movements, as in the monkey paradigm. One method for circumventing this unsupervised learning problem could be using a hybrid paradigm in which subjects will imagine making task movements (e.g., moving a cursor on a screen to a target) while a trajectory planner moves the cursor to the target. This trajectory will be the one used in conjunction with the simultaneously recorded neural data to train the models online. The outputs of the model (predicted movement) and the trajectory planner (imposed movement) can be combined in various proportions. At first, when the model is completely untrained, the task will be totally under the control of the trajectory planner. But as the models are trained and their predictive power improves, that ratio will be gradually shifted until the task is completely under the control of the subject. At this point, the training phase will be considered finished, and the subject will be able to control the device with the brain.

In conclusion, recent advances in BMIs allow us to feel optimistic about the dream of restoring basic motor functions in patients with neuromuscular disorders. One of the key elements is the contention that many of the BMI principles derived from work with nonhuman primates are highly relevant to the human intraoperative setting, as we have shown.[13] Nevertheless, further advances need to be made in both the engineering and neurophysiology domains. In particular, one of the components that needs dramatic development is the artificial actuators field. Current off-the-shelf actuators are not designed for neuroprosthetic applications and lack a way of delivering broadband sensory feedback information to the subject's brain. A new generation of neuroprosthetic devices providing muscle-like actuation properties would be very beneficial for the advancement of the BMI field.

ACKNOWLEDGMENTS

This work was supported by the Christopher Reeve Paralysis Foundation (contract number CA2-0308-2) (Jose Carmena), and the Defense Advanced Research Projects Agency (contract number N66001-02-C-8022), NIH, and James S. McDonnell foundation (Miguel Nicolelis). The authors would like to thank Misha Lebedev, Joey O'Doherty, Huyn Kim, James Biggs, Erin Phelps, and Susan Halkiotis for their contributions to this work.

REFERENCES

1. Schmidt, E.M., Single neuron recording from motor cortex as a possible source of signals for control of external devices, *Ann. Biomed. Eng.*, 8, 339–349, 1980.
2. Fetz, E.E., Operant conditioning of cortical unit activity, *Science,* 163, 955–957, 1969.

3. Fetz, E.E. and Finocchio, D.V., Operant conditioning of specific patterns of neural and muscular activity, *Science*, 174, 431–435, 1971.
4. Fetz, E.E. and Baker, M.A., Operantly conditioned patterns of precentral unit activity and correlated responses in adjacent cells and contralateral muscles, *J. Neurophysiol.*, 36, 179–204, 1973.
5. Fetz, E.E. and Finocchio, D.V., Correlations between activity of motor cortex cells and arm muscles during operantly conditioned response patterns, *Exp. Brain Res.*, 23, 217–240, 1975.
6. Serruya, M.D., Hatsopoulos, N.G., Paninski, L., Fellows, M.R., and Donoghue, J.P., Instant neural control of a movement signal, *Nature*, 416, 141–142, 2002.
7. Taylor, D.M., Tillery, S.I., and Schwartz, A.B., Direct cortical control of 3D neuroprosthetic devices, *Science*, 296, 1829–1832, 2002.
8. Chapin, J.K., Moxon, K.A., Markowitz, R.S., and Nicolelis, M.A.L., Real-time control of a robot arm using simultaneously recorded neurons in the motor cortex, *Nature Neurosci.*, 2, 664–670, 1999.
9. Wessberg, J. et al., Real-time prediction of hand trajectory by ensembles of cortical neurons in primates, *Nature*, 408, 361–365, 2000.
10. Carmena, J.M. et al., Learning to control a brain–machine interface for reaching and grasping by primates, *PLoS Biol.*, 2, 1–16, 2003.
11. Birbaumer, N. et al., A spelling device for the paralysed, *Nature*, 398, 297–298, 1999.
12. Kennedy, P.R. and King, B., Dynamic interplay of neural signals during the emergence of cursor related cortex in a human implanted with the neurotrophic electrode, in *Neural Prostheses for Restoration of Sensory and Motor Functions*, Chapin, J. and Moxon, K., Eds., CRC Press, Boca Raton, FL, 2000, 221–233.
13. Patil, P.G., Carmena, J.M., Nicolelis, M.A.L., and Turner, D.A., Ensembles of human subcortical neurons as a source of motor control signals for a brain–machine interface, *Neurosurgery*, 55, 1–10, 2004.
14. Nicolelis, M.A.L. and Ribeiro, S., Multi-electrode recordings: the next steps, *Curr. Opin. Neurobiol.*, 12, 602–606, 2002.
15. Wolpaw, J.R., Birbaumer, N., McFarland, D.J., Pfurtscheller, G., and Vaughan, T.M., Brain–computer interfaces for communication and control, *Clin. Neurophysiol.*, 113, 767–791, 2002.
16. Millán, J.R. and Mouriño, J., Asynchronous BCI and local neural classifiers: an overview of the Adaptive Brain Interface project, *IEEE Trans. Neural Syst. Rehab. Eng.*, 11, 159–161, 2003.
17. Nicolelis, M.A.L., Actions from thoughts, *Nature*, 409, 403–407, 2001.
18. Nicolelis, M.A.L., Brain–machine interfaces to restore motor function and probe neural circuits, *Nature Rev. Neurosci.*, 4, 417–422, 2003.
19. Pesaran, B., Pezaris, J.S., Sahani, M., Mitra, P.P., and Andersen, R.A., Temporal structure in neuronal activity during working memory in macaque parietal cortex, *Nature Neurosci.*, 5, 805–811, 2002.
20. Hatsopoulos, N.G., Ojakangas, C.L., Paninski, L., and Donoghue, J.P., Information about movement direction obtained from synchronous activity of motor cortical neurons, *Proc. Nat. Acad. Sci. U.S.A.*, 95, 15706–15711, 1998.
21. Maynard, E.M., Hatsopoulos, N.G., Ojakangas, C.L., Acuna, B.D., Sanes, J.N., Normann, R.A., and Donoghue, J.P., Neuronal interactions improve cortical population coding of movement direction, *J. Neurosci.*, 19, 8083–8093, 1999.
22. Isaacs, R.E., Weber, D.J., and Schwartz, A.B., Work toward real-time control of a cortical neural prothesis, *IEEE Trans. Rehab. Eng.*, 8, 196–198, 2000.

23. Hoffman, K.L. and McNaughton, B.L., Coordinate reactivation of distributed memory traces in primate neocortex, *Science*, 297, 2070–2073, 2002.
24. Nicolelis, M.A.L., Ghanzanfar, A.A., Faggin, B.M., Votaw, S., and Oliveira, L.M.O., Reconstructing the engram: simultaneous, multisite, many single neuron recordings, *Neuron*, 18, 529–537, 1997.
25. Nicolelis, M.A.L. et al., Chronic, multi-site, multi-electrode recordings in macaque monkeys, *Proc. Nat. Acad. Sci. U.S.A.*, 100, 11041–11046, 2003.
26. Obeid, I., Nicolelis, M., and Wolf, P.D., A low power multichannel analog front end for portable neural signal recordings, *J. Neurosci. Meth.*, 133, 27–32, 2004.
27. Obeid, I., Nicolelis, M.A.L., and Wolf, P.D., A multichannel telemetry system for single unit neural recordings, *J. Neurosci. Meth.*, 133, 33–38, 2004.
28. Obeid, I. and Wolf, P.D., Evaluation of spike detection algorithms for a brain–machine interface application, *IEEE Trans. Biomed. Eng.*, 51, 905–911, 2004.
29. Bosseti, C.A., Carmena, J.M., Nicolelis, M.A.L., and Wolf, P.D., Transmission latencies in a telemetry linked brain-machine interface, *IEEE Trans. Biomed. Eng.*, 51, 919–924, 2004.
30. Haykin, S., *Adaptive Filter Theory,* Prentice Hall, Upper Saddle River, N.J., 2002.
31. Wu, W., Black, M.J., Gao, Y., Bienenstock, E., Serruya, M., Shaikhouni, A., and Donoghue, J.P., Neural decoding of cursor motion using a Kalman filter, *NIPS*, 15, 133–140, 2003.
32. Sanchez, J.C., Kim, S.P., Erdogmus, D., Rao, Y.N., Principe, J.C., Wessberg, J., and Nicolelis, M.A.L., Input-output mapping performance of linear and nonlinear models for estimating hand trajectories from cortical neuronal firing patterns, *Proceedings of NNSP2002*, Martigny, Switzerland, 139–148, 2002.
33. Rao, Y.N., Kim, S-P., Sanchez, J.S., Rao, Y.N., Erdogmus, D., Principe, J.C., Carmena, J.M., Lebedev, M.A., and Nicolelis, M.A.L., Learning mappings in brain–machine interfaces with echo state networks, *IEEE International Conference on Acoustics, Speech, and Signal Processing*, Philadelphia, PA, 2004.
34. Sanchez, J.C., Erdogmus, D., Rao, Y.N., Principe, J.C., Nicolelis, M.A.L., and Wessberg, J., Learning the contributions of the motor, premotor, and posterior parietal cortices for hand trajectory reconstruction in a brain–machine interface, *Proceedings of IEEE EMBS Neural Engineering Conference*, 59–62, Capri Island, Italy, March 2003.
35. Kim, S-P., Sanchez, J.C., Rao, Y.N., Erdogmus, D., Principe, J.C., Carmena, J.M., Lebedev, M.A., and Nicolelis, M.A.L., A comparison of optimal MIMO linear and nonlinear models for brain–machine interfaces, submitted to *Neural Computation*.
36. Wessberg, J. and Nicolelis, M.A.L., Optimizing a linear algorithm for real-time robotic control using chronic cortical ensemble in recording monkeys, *J. Cogn. Neurosci.*, 16, 1022–1035, 2004.
37. Sanchez, J.S., Carmena, J.M., Lebedev, M.A., Nicolelis, M.A.L., Harris, J.G., and Principe, J.C., Ascertaining the importance of neurons to develop better brain–machine interfaces, *IEEE Trans. Biomed. Eng.*, 51, 943–953, 2004.
38. Advanced Bionics Corporation, Sylmar, CA, <http://www.advancedbionics.com>.
39. Optobionics Corporation, Naperville, IL, <http://www.optobionics.com>.
40. The Dobelle Institute Lda., Lisboa, Portugal, <http://www.artificialvision.com>.
41. Talwar, S.K., Xu, S., Hawley, E.S., Weiss, S.A., Moxon, K.A., and Chapin, J.K., Rat navigation guided by remote control, *Nature*, 417, 37–38, 2002.
42. Xu, S., Li, L., Francis, J.T., Talwar, S.K., and Chapin, J.K., Spatiotemporal encoding in rat somatosensory cortex of electrical stimulation in sensory thalamus, *Soc. Neurosci. Abstr.*, 29, 60.14, 2003.

43. Sandler, A.J., Kralik, J.D., Shanklin, K.A., Phelps, E.E., and Nicolelis, M.A.L., Neuronal correlates of primate somatosensorimotor learning, *Soc. Neurosci. Abstr.*, 29, 279.24, 2003.
44. Donoghue, J.P., Connecting cortex to machines: recent advances in brain interfaces, *Nature Neurosci.*, Supp. 5, 1085–1088, 2002.
45. Carmena, J.M., Lebedev, M.A., and Nicolelis, M.A.L., Variability in correlation between neuronal firing and motor parameters in the fronto-parietal cortex of Macaque monkeys, in preparation.
46. Kim, H.K., Biggs, S.J., Schloerb, D.W., Carmena, J.M., Lebedev, M.A., Nicolelis, M.A.L., and Srinivasan, M.A., Continuous shared control stabilizes reaching and grasping with brain–machine interfaces, *IEEE Trans. Biomed. Eng.*, under revision.
47. Shoham, S., Halgren, E., Maynard, E.M., and Normann, R.A., Motor-cortical activity in tetraplegics, *Nature*, 413, 793, 2001.

14 Human Brain–Computer Interface

Gert Pfurtscheller, Christa Neuper, and Niels Birbaumer

CONTENTS

ABSTRACT

A brain–computer interface (BCI) transforms signals originating from the human brain into commands that can control devices or applications. In this way, a BCI provides a new nonmuscular communication channel, which can be used to assist patients who have highly compromised motor functions, as is the case with patients

0-8493-1287-6/05/$0.00+$1.50

suffering from neurological diseases such as amyotrophic lateral sclerosis (ALS) or brainstem stroke. The immediate goal of current research is to provide these users with an opportunity to communicate with their environment. Present-day BCI systems use different electrophysiological signals such as slow cortical potentials, evoked potentials, and oscillatory activity recorded from scalp or subdural electrodes, and cortical neuronal activity recorded from implanted electrodes. Due to advances in methods of signal processing, it is possible that specific features automatically extracted from the electroencephalogram (EEG) and electrocorticogram (ECoG) are used to operate computer-controlled devices. The interaction between the BCI system and the user, in terms of adaptation and learning, is a challenging aspect of any BCI development and application. This chapter outlines and explains current approaches and methods used in BCI research.

14.1 INTRODUCTION

A technical system that permits direct communication between brain and computer is known as a BCI.[1,2] In this case, the normal communication channels, such as speech and movement, are not used, but instead the brain activity is directly recorded and transformed into a control signal. Therefore, a BCI provides a new communication channel that can be used to convey messages and commands directly from the brain to the external world. The use of a BCI depends on the interaction of two adaptive controllers, the user's brain and a computer, which has to produce an action that accomplishes the user's intention.

One general base for a BCI is that music or visual or motor imagery modifies neuronal activity and can result in measurable changes in firing patterns of cortical neurons and in the ongoing EEG and ECoG.[3,4] Furthermore, focused or selective attention can enhance different brain signals or components of evoked potentials and in turn can be used in a BCI system.[5] Beside electrical potentials, the blood oxygen level dependent (BOLD) response measured by real-time functional magnetic resonance imaging (fMRI) can be used as an input signal for a BCI.[6]

The current and most important applications of a BCI are the restoration of a communication channel for patients with a locked-in syndrome and the control of neuroprosthesis in patients with spinal cord injury. Aside from these, there are the important and established field of neurofeedback therapy and the upcoming field of multimedia and virtual reality applications. In this context, a BCI could be used for diverse tasks such as playing games or providing multidimensional feedback in virtual reality. Concerning the last point, BCI adds a new dimension in man–machine interaction and may be of great importance in multimedia applications.

A BCI system is, in general, composed of the following components: signal acquisition, preprocessing, feature extraction, classification (detection), and application interface (Figure 14.1). The signal acquisition component is responsible for recording the electrophysiological signals providing the input to the BCI. Depending on the type of analyzed brain signals and the processing mode, sensory stimulation may be necessary. The task of preprocessing is to enhance the signal-to-noise ratio.

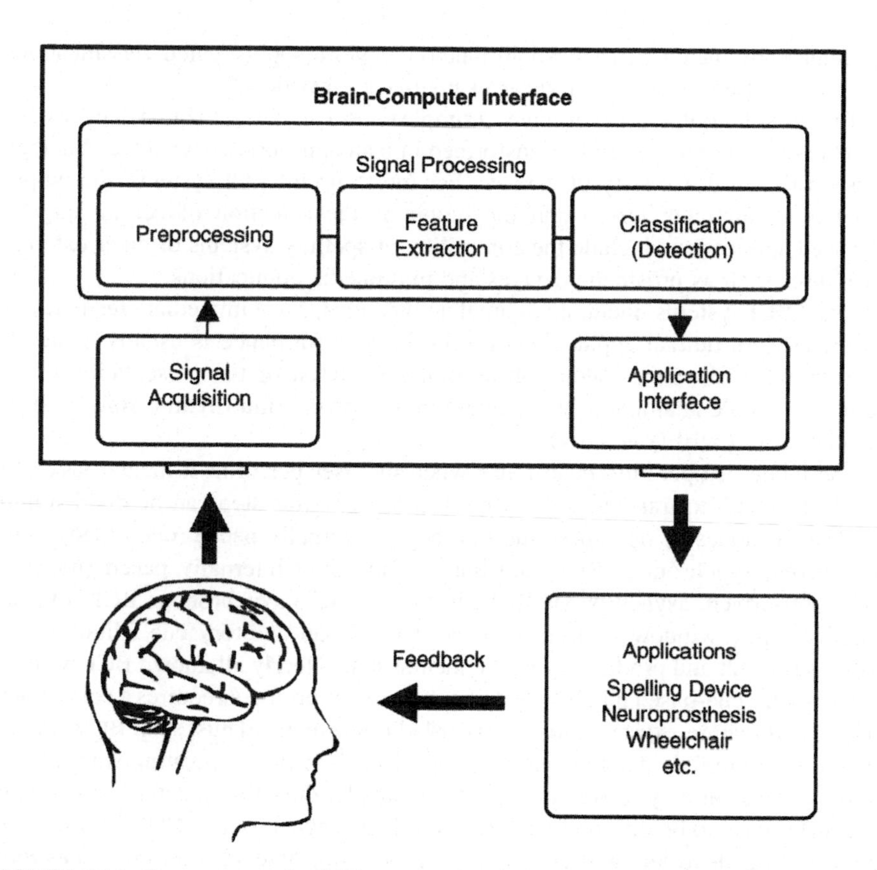

FIGURE 14.1 Components of a BCI system: signals from the user's brain are acquired and processed to extract specific features used for classification. The classifier output is transformed into a device command, which, at the same time, provides feedback to the user.

This can include artifact reduction methods and the application of advanced signal processing methods. After preprocessing, the signal is subjected to the feature extraction algorithm. The goal of this component is to find a suitable representation (signal features) of the electrophysiological data that simplifies the subsequent classification or detection of brain patterns. That is, the signal features should encode the commands sent by the user, but should not contain noise and other patterns that can impede the classification process. There are a variety of feature extraction methods used in current BCI systems. A list (not exhaustive) of these methods includes amplitude measures, band power, Hjorth parameters, autoregressive models, and wavelets.

The task of the classifier component is to use the signal features provided by the feature extractor to assign the recorded samples of the signal to a category of brain patterns. In the simplest form, detection of a single brain pattern is sufficient, for instance, by means of a threshold method.[7,8] More sophisticated classifications of different patterns depend on linear or nonlinear classifiers.[9]

The classifier output, which can be a simple on–off signal or a signal that encodes a number of different classes, is transformed into an appropriate signal that can then be used to control a variety of devices. For most current BCI systems, the output device is a computer screen and the output is the selection of certain targets. Advanced applications include the controlling of spelling systems or other external apparatuses such as prosthetic devices and multimedia applications.

Many BCI systems, including animal applications,[10] use immediate feedback of performance. In human applications feedback of performance is usually given by visualization of the brain signal on a computer screen or the presentation of an auditory or tactile analogue of the actual brain response (mu rhythm, slow cortical potential, or other EEG activity).

The mode of operation determines when the user performs a mental task and intends therewith to transmit a message. In principle, this step can be divided into two distinct modes of operation, the first being externally paced (cue-based, computer-driven, synchronous BCI) and the second being internally paced (noncue-based, user-driven, asynchronous BCI). In the case of a synchronous BCI, a fixed, predefined time window is used. After a visual or an auditory cue stimulus, the subject has to act and produce a specific mental state. Nearly all known BCI systems work in such a cue-based mode.[9,11,12] An asynchronous protocol requires a continuous analysis and feature extraction of the recorded brain signal. Thus, such BCIs are, in general, even more demanding and more complex than BCIs operating with a fixed timing scheme. In a synchronous BCI, for example, only two mental tasks or two brain states have to be differentiated, whereas in an asynchronous BCI, a third brain state has to be identified, which is the resting or idling state, also referred to as zero class. To date, only a few BCI research groups are working on an asynchronous BCI.[8,13–18]

14.2 BRAIN SIGNALS

In principle, there are invasive and noninvasive methods used to record brain signals. The EEG, a noninvasive method, records electrical potential changes and reflects the common activity of several millions of neurons extending over some square centimeters of the cortical tissue. Invasive methods are exemplified by the ECoG as well as by intracortical recordings. In contrast to the EEG, the ECoG represents integrated bioelectrical activity over a much smaller cortical area, but still constitutes the common activity of many thousands of neurons. The multichannel intracortical recordings reflect extracellular activity generated by small neuronal populations in the order of about 100 cells or fewer.[19,20]

In the EEG, as well as in the EcoG, two types of phenomena need to be differentiated:

1. Event-related potentials (ERPs), including evoked potentials and slow cortical potential (SCP) shifts
2. Event-related changes in ongoing EEG/ECoG activity in specific frequency bands

Event-related desynchronization (ERD) defines an amplitude (power) decrease of a rhythmic component, whereas event-related synchronization (ERS) characterizes an amplitude (power) increase.[21]

In a simplified form it may be assumed that ERPs represent the summative responses of cortical neurons due to changes of the afferent activity, while ERD and ERS reflect changes in the activity of local interactions between main neurons and interneurons that control the frequency components of ongoing EEG/ECoG components.

Slow cortical potentials are slow shifts of the EEG with a duration from 300 msec to several seconds. The amplitude on the scalp may vary between 1 and 100 μV in the case of pre-epileptic or epileptic activity. The family of SCPs includes the so-called contingent negative variation (CNV), premovement potentials, the Bereitschaftspotential,[22] and expectancy waves.[23]

The family of SCPs originates in the upper layers (layers I and II) of the cortex. However, slow potential changes, probably of the same origin, can be found in all parts of the nervous system. Whether the mechanisms of physiological generation are the same is still a matter of debate. The apical dendrites receive most of their input from intracortical fibers, callosal input from the other hemisphere, and input from the medial and reticular thalamus, the so-called nonspecific ascending activation system. Cholinergic inflow to the apical dendrites arrives primarily from structures in the basal forebrain and the basal ganglia.[24] The amplitudes of slow cortical potentials are highly sensitive to the manipulation of cholinergic, monoaminergic, and glutamatergic transmission. SCPs tending toward the negative reflect slow EPSP and glial potentials and, therefore, indicate longer-lasting depolarization of the dendritic network. SCPs tending toward the positive are more difficult to analyze because they may result from a reduction of inflow to the apical dendrites, or, probably in rare occasions, from direct inhibitory activity at the level of layer I or II. Finally, SCPs tending toward the positive may indicate the inversion of the cortical dipole between the upper input layers and the lower output layers of the cortex.[25] Therefore, the interpretation of slow potentials tending toward the positive requires a description of the experimental and physiological context of recording. Note also that in areas where the convexity of the cortex results in an inversion of the usual cortical dipole directions, such as in the orbitofrontal cortex, the inferior temporal cortex, part of area 17 of the occipital cortex, and the interhemispheric sulcus, inversion of the polarity of slow cortical potentials may be found. Therefore, polarity changes recorded at electrodes from the scalp alone can only be interpreted with great caution. Birbaumer et al.[24] have described a frontal corticothalamic and basal ganglia network responsible for the attentional regulation of all types of preparatory SCPs. They have shown that slow cortical potentials are part of an excitatory and inhibitory threshold regulation mechanism involving an extended frontothalamo basal ganglia cortex loop, whose anatomy was described by Braitenberg and Schüz.[26] In sum, negative SCPs indicate a preparatory state of the underlying cortical network, but they reflect not only motor preparation but also preparation for cognitive and emotional activity.

Changes of the oscillatory brain activity, referred to as ERD/ERS, may reflect the general characteristic of cortical populations of neurons either to work in synchrony or not. Coherent or synchronous activity results in large potential amplitudes in EEG, whereas incoherent behavior leads to desynchronized patterns. One important prerequisite for the occurrence of oscillations in the EEG might be the interconnectivity between networks and the formation of feedback loops. In general, it can be stated that small synchronized populations of neurons with short feedback loops (e.g., intracortical) may be responsible for low-amplitude and high-frequency oscillations in the EEG (e.g., gamma activity), whereas large synchronized populations with relatively long feedback loops (e.g., thalamocortical) lead to high-amplitude oscillations (e.g., alpha rhythm). Therefore, various components in the EEG (alpha, beta, and gamma) show different reactivity patterns. For example, in contrast to the mu desynchronization (mu ERD), which lasts some seconds, the induced gamma oscillations (gamma ERS) are short lasting and clearly embedded in the mu ERD (see Color Figure 14.2*). It seems that the desynchronization of large population of neurons in widespread networks (alpha or mu ERD) might be the prerequisite for a synchronization of small populations in localized networks.

The gamma ERS occurring during movement execution in the EEG is small-banded and composed of frequency components around 40 Hz (Color Figure 14.2A; see also Pfurtscheller and Neuper[27]), whereas the gamma ERS in the ECoG is more broad-banded, including frequency components from 60–90 Hz (Color Figure 14.2C; see also Crone et al.[28] and Pfurtscheller et al.[29]). Short-lasting bursts of beta oscillations in the range of approximately 14–30 Hz (beta ERS) can be found as a rebound phenomenon after active movement and somatosensory stimulation,[30] but also after motor imagery.[31] An example for such a beta ERS is displayed in Color Figure 14.2B.

It is important to notice that the detection of an amplitude decrease within a single EEG (ECoG) trial is not always a simple task. In contrary, the detection of an amplitude increase in the form of bursts or induced oscillations (ERS pattern) within predefined frequency bands can be performed much more easily. This speaks strongly in favor of focusing a BCI on the detection of ERS patterns and ERPs. Successful detection of mentally induced beta oscillations has been already reported in BCI research. A tetraplegic patient, exemplarily, was able to induce 17-Hz oscillations by foot motor imagery, after some months of training. Consequently, he was able to control the opening and closing of a hand orthosis[32] and to perform grasping functions by means of functional electrical stimulation[33] (for details, see Section 14.9). Another example of a high-classification accuracy obtained in a specific BCI system is the induction of 10-Hz oscillations over the hand representation area during imagination of foot movements.[34] In this case a weak ERD over the foot representation area is accompanied by a phenomenon resembling a "lateral inhibition"

* See color insert following page 170.

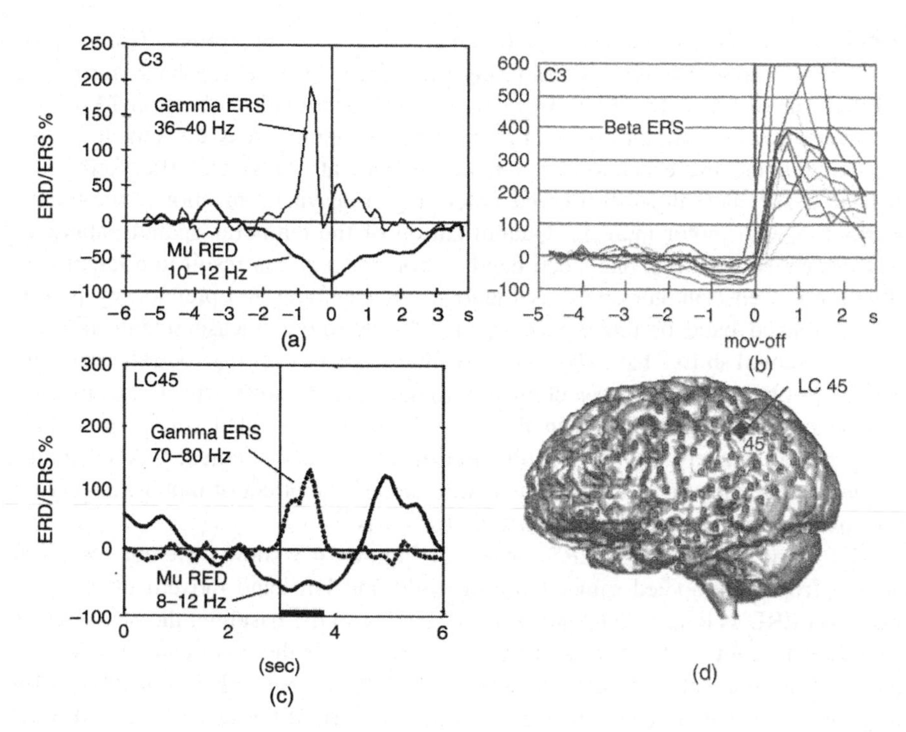

FIGURE 14.2 (see color figure) (a) Simultaneous ERD/ERS in the mu (10–12 Hz) and gamma bands (36–40 Hz) recorded from the left sensorimotor area and processed synchronous to the offset of voluntary right finger movements. (b) Superimposed ERD/ERS time courses from individual subjects (thin lines) and grand average curve (thick line) calculated for subject-specific beta frequency bands. Data were recorded on electrode C3 and processed synchronous to the offset of wrist movements. (c) Simultaneous ERD/ERS in the mu (8–12 Hz) and gamma (70–80 Hz) band in ECoG recordings during voluntary finger movement. (d) ECoG electrode locations.

mechanism in the form of laterally induced 10-Hz oscillations. This phenomenon has been referred to as "focal ERD/surround ERS" by Lopes da Silva and linked to a thalamic gating mechanism.[35]

14.3 MOTOR IMAGERY AS MENTAL STRATEGY FOR A BCI

Research work in recent years has shown that imagination of movement activates similar cortical areas and shares similar temporal characteristics as the execution of the same movement.[36] A number of fMRI studies reported similarities between the preparation of self-paced movement and motor imagery.[37–39] However, distinct processes may be involved: besides the activation of various motor areas during the imagination of movement, the importance of a perceptual or visuo-spatial component should be stressed, since motor imagery probably involves visual and kinesthetic internal representations.[40] In this respect it is of interest that even observation of a movement can activate premotor areas.[41,42] The latter finding underlines the importance of a visuospatial component in motor cortex function and provides evidence for the existence of an action observation/execution matching system. This point is of particular importance when considering the effects of visual feedback during the operation of a BCI, and will be discussed in more detail below (Section 14.6).

Several EEG studies indicate that primary sensorimotor areas are activated when subjects imagine the execution of a hand movement. Klass and Bickford[43] and Chatrian et al.[44] have already observed a blocking or desynchronization of the central mu rhythm with motor imagery. Quantification of the temporal–spatial pattern of ERD clearly showed that one-sided hand motor imagery can result in a lateralized activation of sensorimotor areas, as found in the planning and preparatory phases of a self-paced hand or finger movement.[31] Furthermore, measurements of slow cortical potential shifts[45] have shown that similar changes over the contralateral hand area can be observed during execution and imagination of movement. The evaluation of movement-related potentials in Parkinson's disease revealed similar potentials during motor imagery and during preparation of a real movement.[46] Also, multichannel neuromagnetic measurements demonstrated the effect of motor imagery on brain oscillations generated in primary motor areas.[47]

The grand average ERD/ERS curves in Figure 14.3 show different reactivity patterns from able-bodied subjects during right and left hand movement imagery. The alpha ERD is long-lasting and recovers slowly to the baseline; the beta ERD is of shorter duration and is followed by a beta ERS over the contralateral side. The fact that the imagination of a movement can result in a beta ERS is of interest for the general interpretation of this phenomenon, because of the lack of afferent input during motor imagery. The processing of somatosensory afferent input, though, has been assumed to play an important role for the beta ERS.[48]

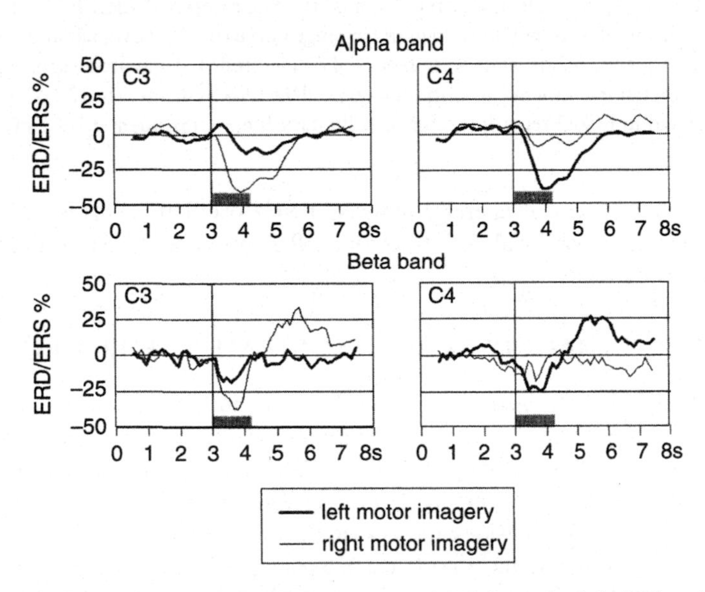

FIGURE 14.3 Grand average ERD/ERS curves recorded from the left (C3) and right sensorimotor cortex (C4) and calculated in the alpha/mu (upper panels, n = 16) and beta range (lower panel, n = 9). A gray bar indicates the time period of cue presentation. (From Reference 31, with permission.)

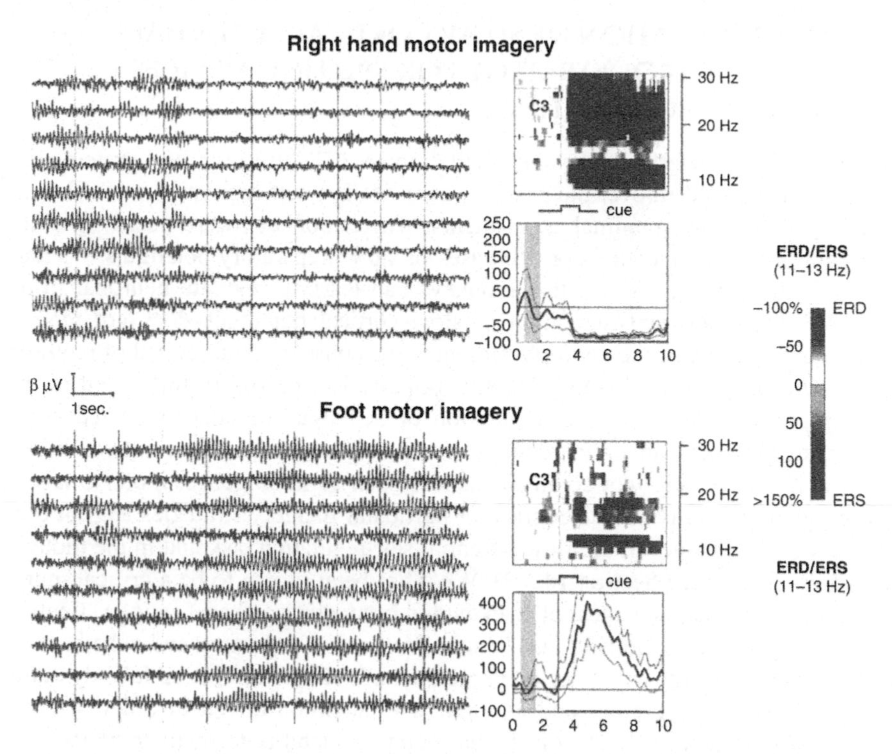

FIGURE 14.4 (see color figure) Left side: Examples of single EEG trials recorded from electrode position C3 during right hand (upper panel) and foot (lower panel) motor imagery. Right side: ERD/ERS time frequency maps and time curves of the frequency band 11–13 Hz recorded from electrode position C3 during right hand (upper panels) versus foot (lower panels) motor imagery. Onset of cue presentation at second 3.

Color Figure 14.4 illustrates how various types of motor imagery can result in different reactivity patterns on one electrode location over the sensorimotor area. Whereas, for instance, right hand motor imagery can block mu and beta rhythms over the left sensorimotor cortex (Color Figure 14.4, upper panel), foot motor imagery can have an opposite effect — namely, it can induce mu oscillations (Color Figure 14.4, lower panel). Parallel to the mu ERS, the beta activity is, in this case, moderately desynchronized. This example demonstrates that frequency-specific reactivity patterns in the ongoing EEG make it possible to distinguish between different types of motor imagery.

In summary, it can be stated that motor imagery can modify sensorimotor rhythms in a manner similar to that observed in the preparatory phase of a movement that is actually executed. Since motor imagery results in somatotopically organized activation patterns, mental imaginations of different movements (e.g., left versus right hand, hand versus foot) can be an efficient strategy for operating a BCI based on oscillatory EEG activity. The challenge is to detect the imagery-related changes in ongoing, not-averaged EEG recordings.

14.4 SELF-REGULATION OF SLOW CORTICAL POTENTIALS AS A CONTROL STRATEGY FOR SPELLING DEVICES AND NEUROPROSTHESES

The mechanisms underlying the self-regulation of SCPs used for BCI systems has been intensively studied over the past 25 years.[24,49] The self-regulation of SCPs was originally derived from animal experiments, where monkeys were positively reinforced for producing increases or decreases of slow preparatory potentials.[50,51] Earlier, Fetz[52] had already shown that monkeys can learn to regulate neuronal firing rates. As in animals, self-regulation of slow cortical potentials does not require continuous feedback of the neurophysiological response, but the reward of required amplitude changes in positive or negative polarity is a necessary ingredient of the learning process. Therefore, self-regulation of SCPs can be conceptualized as an implicit learning mechanism involving elements of classical and operant conditioning. It is not surprising that cognitive factors such as intelligence, motor imagery ability, age, personality characteristics, or particular imagery strategies are not critical for the performance in the slow potential self-regulation task and in the thought translation device (see Section 14.5.3). What has been shown to be a critical ingredient is the functional intactness of an extended frontocortical, basal ganglia attention regulation system and motivational factors such as reward value and schedules of reinforcement. Patients with a lesioned prefrontal lobe or with frontal lobe dysfunctions, such as schizophrenics and children with attention deficit disorder or other frontoregulatory deficits, show substantial delay or inability to regulate SCPs.

Color Figure 14.5 shows fMRI of ten highly successful subjects during the regulation of positive- and negative-tending SCPs (unpublished results; see also Hinterberger et al.[53]). As can be seen and as predicted from threshold regulation theory,[24] negative SCPs correlate with an increase of the BOLD response in several cortical areas of the frontal motor and nonmotor areas, while positive potentials covary with a decrease of the BOLD response in most cortical areas. Successful self-regulation of SCPs at the cortex can be predicted with high accuracy ($r > 0.9$) from the activation of the most likely inhibitory structures in the basal ganglia (putamen/pallidum) and deactivation of premotor areas. Subjects and patients therefore use an attentional threshold regulation mechanism to manipulate the excitation threshold of the cortex, as reflected in negative- and positive-tending SCPs.

In a study of motor imagery with healthy subjects[54] it was shown that subjects explicitly instructed to use activating motor imagery for negative SCPs and passive motor imagery for positive SCPs did not profit in the speed of acquisition of the learned cortical response. The groups with or without imagery achieved the same level of performance. However, in intractable epilepsy it was shown that patients with extremely high levels of negative polarized SCPs do not profit from self-regulation training.[55] In addition, operant learning of SCP control was shown to be a prerequisite for correct psychophysical scaling of the subjects' own cortical potentials.[55] Only subjects with a high level of performance were able to estimate their own cortical polarization subsequent to achieving high performance. Correct self-estimation and scaling of the ongoing brain state and verbal descriptions of imagery

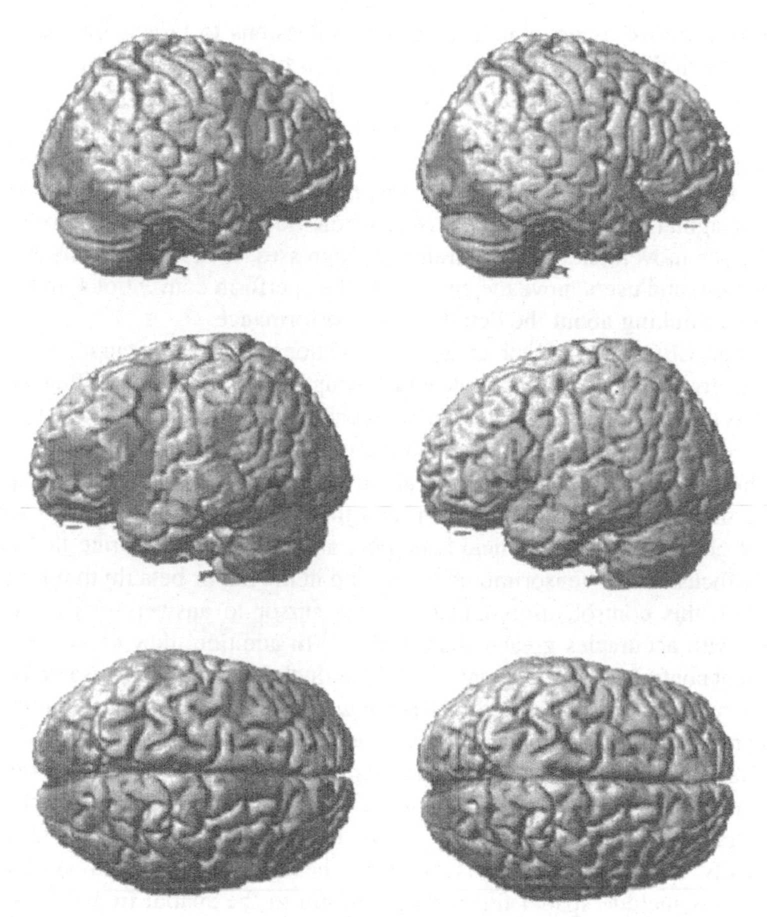

FIGURE 14.5 (see color figure) Averaged functional magnetic resonance images (fMRI) of ten subjects during self-regulation of cortical negativity (left) and positivity (right). Red symbolizes increase and green symbolizes decrease of BOLD response.

used for self-regulation of SCPs are thus a consequence and not a cause of successful learning.

14.5 A SHORT OVERVIEW OF EEG-BASED BCI SYSTEMS

EEG-based BCIs can be realized in an externally paced or an internally paced mode. On the one hand, changes of oscillatory EEG activity as ERD and ERS[9,56] are analyzed and classified; on the other, various types of ERPs are detected by BCI systems. To the latter belong the SCP,[7] the visual evoked potential,[57] the P300 component,[58] and the steady-state visual evoked potential.[59]

14.5.1 The Wadsworth BCI

With the BCI system of Wolpaw, McFarland, and their colleagues,[56,60,61] people with or without motor disabilities learn to control mu or beta rhythm amplitude and use that control to move a cursor in one or two dimensions to targets on a computer screen. For each dimension of cursor movement, a linear equation translates mu or beta rhythm amplitude at one or several scalp locations into cursor movement ten times per second. Users learn in a series of 40-minute sessions to control the cursor. They complete 2 to 3 sessions per week. Most (about 80%) gain significant control in two to three weeks. In their first sessions, most use some kind of motor imagery (such as imagination of hand movements, whole body activities, or relaxation) to control cursor movement. As their training progresses, imagery typically becomes less important, and users move the cursor like they perform conventional motor acts, i.e., without thinking about the details of the performance.

Although EEG from only 1 or 2 scalp locations determines cursor movement online, data from 64 locations over the entire scalp are stored for later offline analysis that defines the complete topography of EEG changes associated with target position and helps to design improvements in online operation. This analysis relies mainly on r^2, which is the proportion of the total variance in mu or beta rhythm amplitude that is accounted for by target position, and therefore indicates the user's level of EEG control. The r^2 topographical and spectral analyses demonstrate that control is sharply focused over sensorimotor cortex and in the mu or beta rhythm frequency bands. With this control, users can move the cursor to answer spoken "yes/no" questions with accuracies greater than 95%.[62,63] In addition, they can also achieve independent control of two different mu or beta rhythm channels and use that control to move a cursor in two dimensions.[12] Recent work has focused on further improving one-dimensional control, and on applying it to selecting among up to eight different targets. Users have reached information transfer rates up to 20–25 bits per minute.[64]

Research with the Wadsworth BCI has concentrated on defining the topographical, spectral, and temporal features of mu and beta rhythm control and on improving the mutually adaptive interactions between the user and the BCI system. The improvements include spatial filters that conform to the spatial frequencies of the user's mu or beta rhythms; autoregressive frequency analysis, which provides higher resolution for short time segments and therefore allows more rapid device control; and better selection of the constants in the equations that translate EEG control into cursor control.[61,65,66] Current studies are also exploring incorporation of other EEG features into this BCI. For example, in trained users, errors in target selection are accompanied by a positive potential centered at the vertex.[67] This error potential might be useful for recognizing and canceling mistakes. While work up until now has used cursor movement as a prototype BCI application and has focused on improving it, effort is also being committed to specific applications such as verbal communication.[62,63,68]

14.5.2 The Graz BCI

The Graz BCI is an EEG-based system using motor imagery as the appropriate mental control strategy. Therefore, it uses electrode positions located over sensorimotor areas. Another characteristic feature of the Graz BCI is the usage of a classifier to discriminate between brain states associated with different types of motor imagery. Such a classifier has to be set up by a great number of examples of imagery-induced

EEG patterns obtained in sessions without feedback, following the timing of the cue-based standard paradigm (as shown in Figure 14.6). During the feedback sessions, the EEG is analyzed and classified in predefined time windows and the feedback is then given in real time. In the course of the training, the computer (classifier) is repeatedly adapted to the user's brain signals (Figure 14.7).

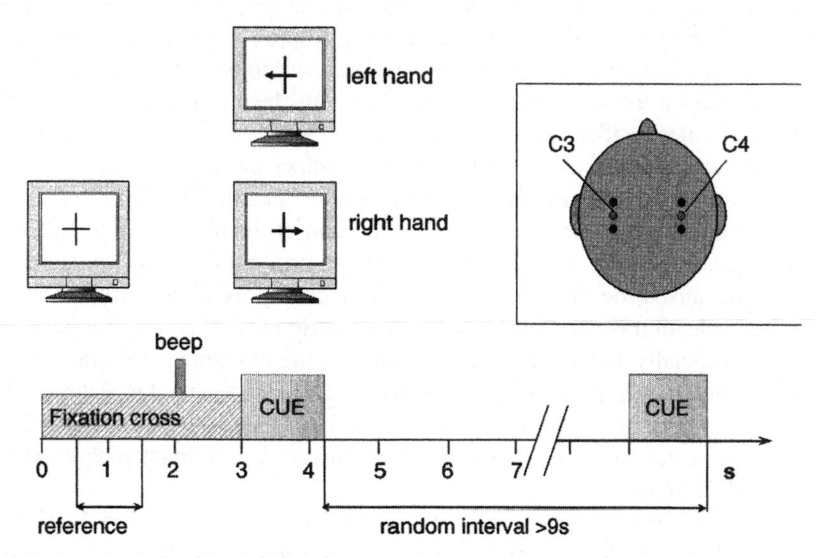

FIGURE 14.6 Imagination paradigm. The cue stimulus in the form of an arrow indicates the side of imagination. Insert: Bipolar channels close to electrode positions C3 and C4 as used in BCI experiments.

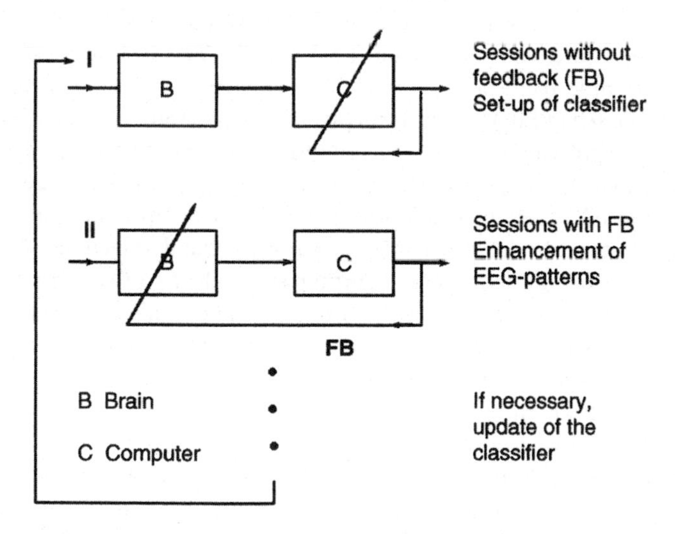

FIGURE 14.7 Types of brain–computer interactions. In a first stage (I), sessions without feedback are performed to collect data to set up a classifier. The computer learns to recognize different brain states. As soon as a classifier is available, feedback can be provided to the user (II), which modifies brain activity.

One of the first EEG-based BCI systems developed by Graz was able to discriminate between three brain states.[69] The subject was instructed to execute a specific movement (right hand, left hand, right foot) in the first three sessions and to imagine the same movement in the next two sessions. Band power (5–35 Hz) was estimated in intervals of 250 msec from three bipolar EEG channels. Four power estimates of each of the three channels were concatenated to form 12-dimensional vectors that were presented to a neural network-based classifier. The nonlinear classifier was trained with examples from the first sessions without feedback. The on-line results from the motor imagery sessions with feedback varied around 50% (the worst case of classification accuracy was 33.3%).

The next approach of the Graz BCI was to select the input features for each subject by using distinction-sensitive learning vector quantization (DSLVQ).[70] This method is a valuable tool for the automatic selection of the most relevant frequency components and the most relevant electrode positions. Subject-specific frequency band selection raised the on-line classification accuracy by almost 10%.[71] No frequency band selection is necessary when a parametric method is used for parameter estimation. Especially suitable for the modeling of the dynamics of brain signals is the adaptive autoregressive (AAR) parameter estimation using the Kalman algorithm.[72–74] Time courses demonstrating the separability between two motor imagery tasks by adaptive autoregressive parameters and linear discriminant analysis (LDA) are displayed in Figure 14.8.

Besides neural networks and linear classifiers, hidden Markov models (HMMs) are also suitable in BCI research.[75,76] The HMM itself could be seen as a finite automata containing discrete steps and emitting a feature vector that depends on the current state of every time point, whereby each feature vector is modeled by Gaussian mixtures. The transition probabilities from one state to the other state are described using a transistor matrix. A one-state HMM was used to implement an EEG-based spelling device, also called the "virtual keyboard."[76] The feature vector was composed of logarithmic band power values estimated in two bipolar EEG channels, recorded over the sensorimotor area during two types of motor imagery. The subject's task was to copy presented words (copy spelling) by selecting single letters out of a predefined alphabet. The structure of the virtual keyboard contains 5 decision levels for the selection of 32 letters and 2 further levels for confirmation and correction: in 5 successive steps the letter set was split into 2 equally sized subsets, until the subject selected a certain letter, and 2 further steps allowed the subject to confirm or correct this selection. In a first experiment, 3 able-bodied subjects achieved a spelling rate of 0.67 to 1.02 letters per minute.[76] Current studies focus on improving the spelling rate — for instance, by considering the probability of each letter and by shortening the trial length.

For the preprocessing of multichannel EEG recordings, the method of common spatial patterns (CSPs) was successfully used.[77] This is a supervised method and requires data sets divided into two groups from which each belongs to one of two brain states (i.e., motor imagery tasks).[78]

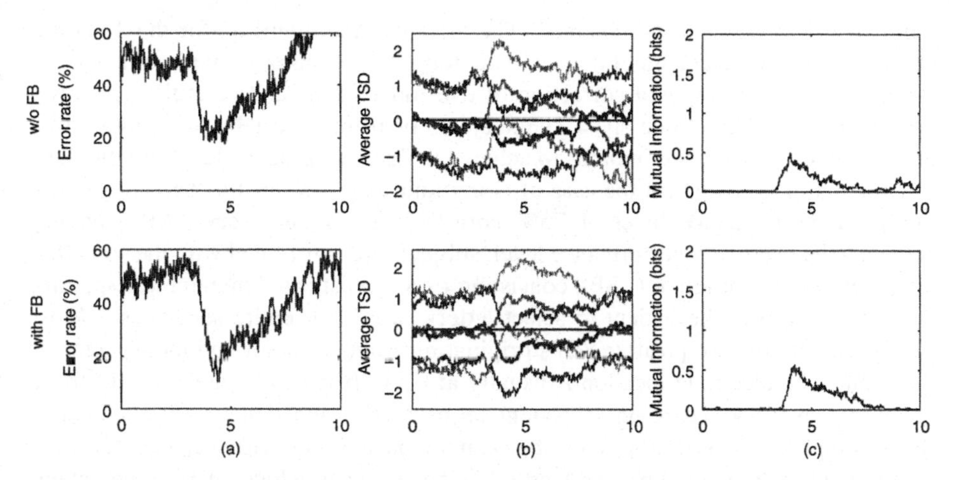

FIGURE 14.8 Time courses of one subject displaying the separability between two classes. Adaptive autoregressive (AAR) parameters (order $p = 6$) from bipolar channels close to C3 and C4, estimated with Kalman filtering, from the first session (without feedback; time interval 4.008–4.125 sec) were used to classify the EEG signals in the second session in order to provide online feedback. (a) Time courses of the error rate. ERR(t) gives the classification error with LDA at time t, calculated with a jackknife method (leaving one trial out for testing). (b) Averaged time-varying signal distance (TSD), calculated as linear combination of AAR(6) parameters, for the left (dark) and right (light) trials. The average TSD curves (thick lines) clearly show a different behavior during imagined left and right hand movement. The thin lines represent the within-class standard deviation (SD) of the TSD and indicate the inter-trial variability of the EEG patterns. (c) Mutual information between the TSD and the class relationship. The entropy difference of the TSD with and without class information was calculated for every time step. This gives a time course of the mutual information in bits per trial.

14.5.3 The Thought Translation Device

The thought translation device (TTD) is a brain–computer interface based on self-regulation of SCPs. The system was first described by Birbaumer et al.[79] and Elbert et al.[80] In the clinical context it was first applied in intractable and drug-resistant epilepsies[23,81] and was later used as a communication device for completely paralyzed, locked-in patients.[82] Other BCI systems using SCPs, primarily movement related potentials, are described by Müller et al.[83] These BCI systems do not include a training phase, but aim instead to classify the ongoing SCPs on-line with the use of motor imagery similar to the method employed by Pfurtscheller (Graz BCI; see Section 14.5.2) and Wolpaw (Wadsworth BCI; see Section 14.5.1). In the TTD, subjects receive visual, auditory, or tactile feedback from their SCPs on-line over a period from 1 to 10 sec, depending on the clinical or experimental purpose of the experiment.

Figure 14.9 shows the basic principle in the case of visual feedback. The illumination of the upper goal on a screen indicates that a negative cortical potential should be produced and the illumination of the lower part of the screen signals that a positive SCP needs to be obtained. The technical requirements for the TTD are DC amplifiers or EEG amplifiers with a long time constant (minimum 5 sec), simultaneous recording and correction of eye movements, and careful preparation of electrodes and electrode sites. The subjects or patients usually are trained over periods of several hours to months to achieve a high performance level of voluntary self-regulation of SCPs (in the case of communication devices for the paralyzed, a minimum performance level of 75% correct trials is necessary). After having achieved the necessary performance level, subjects are confronted with the so-called language support program (LSP), consisting of a sophisticated hierarchical spelling system that allows the patient to select letters, words, or entire sentences.[84] Most paralyzed patients use positive SCPs originating from premotor and motor areas of the brain. The electrode location is usually at Cz or between Fz and Cz. Electrode location may vary according to patient requirements and clinical characteristics. Since completely locked-in patients often can no longer follow visual signals, the TTD is equipped with an auditory and tactile feedback system which enables the patient to listen to his or her own brain activity, transformed into high- or low-pitched tones and harmonic tone sequences that function as reward.

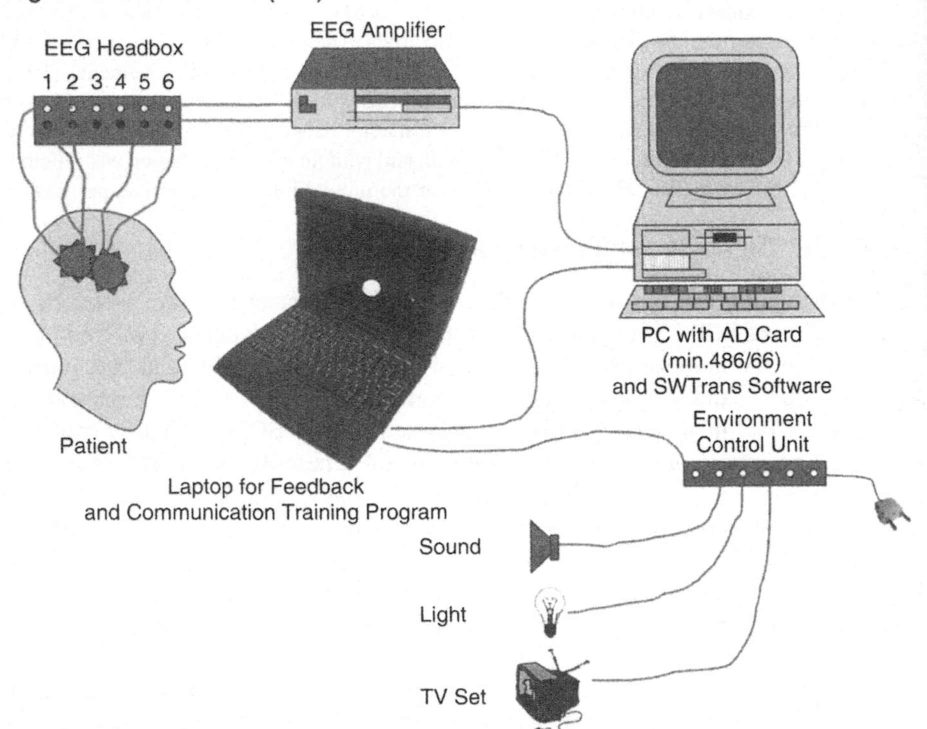

FIGURE 14.9 The thought translation device (TTD).

14.5.4 Other BCI Systems

An interesting approach to a communication system was introduced by Farwell and his colleagues.[5] It is based on the fact that infrequent or particularly significant auditory, visual, or somatosensory stimuli, when interspersed with frequent standard stimuli, typically evoke in the EEG over the parietal cortex a positive peak at about 300 msec termed the P300 component or "oddball" response.[5] The user is confronted with a 6 × 6 letter matrix. In short intervals one of the rows or one of the columns of the matrix is flashed. The user makes a selection by focusing the desired letter and counting how many times the respective row or column flashes. Only in the responses elicited by the desired choice a prominent P300 is visible, and the BCI uses this effect to determine the user's selection. With this system a communication rate of about seven items per minute can be obtained.[58] Another approach is to evaluate the amplitude of steady-state visual evoked potentials (SSVEPs). When, for example, two items on the monitor are flickering with slightly different frequencies in the alpha/beta range, the SSVEP amplitude is enhanced, when the participants shift their gaze to the corresponding light source.[59,85] The system of Cheng and co-workers[85] uses 13 buttons arranged as a virtual telephone keypad, all flickering at different frequencies. The subject has to pay special attention to one of the flashing buttons in order to enhance the amplitude of the corresponding flicker frequency (SSVEP).

The implantation of a special electrode into the outer layer of the neocortex allows us to record multiple unit activity in patients. This activity in form of spikes can be transformed into a pulse train and, in turn, can be used for cursor control and selection of letters or other items on a monitor. Spelling rates of about three letters per minute are reported with patients.[86]

14.6 IMPORTANCE AND IMPACT OF VISUAL FEEDBACK

A central issue in BCI development is the impact of the setup and presentation of feedback. The type of feedback used in early experiments with the Graz BCI paradigm provided information of a correct versus an incorrect response, as determined by the classifier, at the end of each trial.[71] In general, when using such a discrete feedback presented with delay, no increase of classification accuracy was found over the course of additional sessions. Further advances in feature extraction methods, such as the implementation of AAR algorithms,[73] allow one to calculate the parameters concurrent to the data acquisition. In other words, it was possible to provide instantaneous, classifier-based feedback during defined time periods. This type of feedback indicates in real time how well two EEG patterns (representing, for example, two classes) can be discerned. Using such a continuous feedback system, a general enhancement of performance could be achieved.[74]

In order to investigate the impact of a continuously present visual feedback on sensorimotor EEG rhythms during motor imagery, the ERD/ERS time courses of BCI sessions with high performance were analyzed. Figure 14.10 displays ERD curves of BCI feedback sessions with left versus right motor imagery of two subjects,

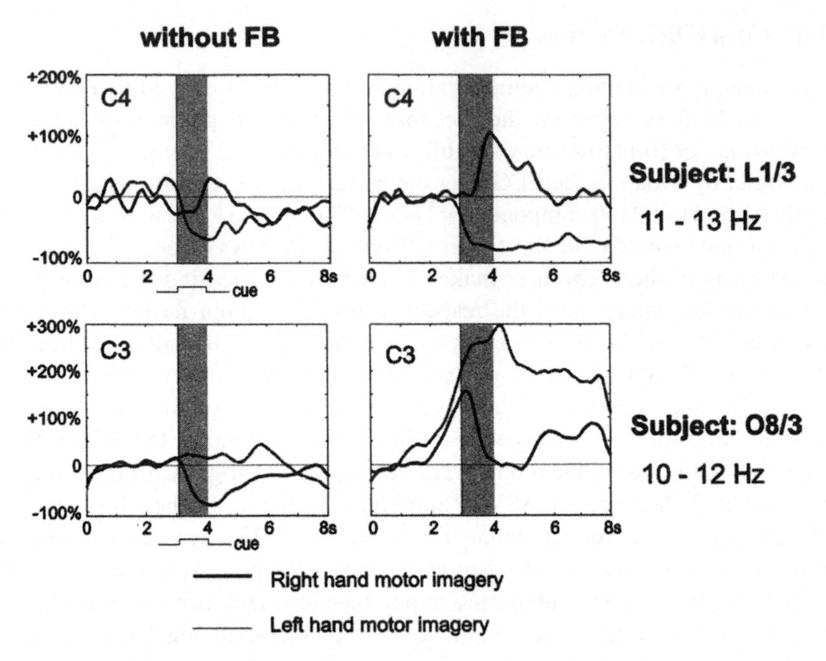

FIGURE 14.10 ERD/ERS curves of two subjects (L 1/3, 11–13 Hz; O 8/3, 10–12 Hz) for sessions without feedback (left panels) and with continuously present feedback (right panels). Data were recorded from the sensorimotor cortex (C3, C4) during imagined movements of the left (thin line) versus right (thick line) hand. The time period of cue presentation is indicated by a gray vertical bar.

and comparable data of control sessions without feedback. The initial recordings without feedback (left panels) reveal a transient desynchronization (ERD) of the mu rhythm over the contralateral hemisphere starting with the cue presentation. Continuously present feedback results in dramatic changes of the ERD/ERS patterns: both subjects show a clearly more pronounced difference between left and right motor imagery during feedback regulation. Instantaneous information may help the subject to test various strategies to manipulate the feedback bar and related EEG activity. Hence, these findings are in line with the general notion that feedback should be as fast as possible.[87]

On the other hand, it is known that visual feedback may have positive as well as negative effects, and that these effects vary across subjects.[88] Correct responses are a source of positive reinforcement, which may lead to an enhancement of the target EEG pattern. Feedback of false responses, however, may elicit a negative emotional reaction, which in turn is likely to alter the EEG patterns in an unpredictable way. Further, the processing of the (moving) feedback stimulus may interfere with the mental imagery task, and may therefore, in some cases, impair the development of EEG control.

Hence, we should take into consideration the possible impact of the visual display itself (cursor, letter, etc.), which may cause further variations of the EEG patterns. There is evidence that sensorimotor rhythms do not become desynchronized by visual stimulation.[89,90] In contrast, even a localized augmentation of alpha band activity over sensorimotor areas was reported during visual processing.[91] In a case where a subject watches a video film showing human movements, however, sensorimotor rhythms can display a significant decrease in power.[92]

In animal studies, it was demonstrated that the observation of movement actions results in a strong activation in premotor areas.[93] So-called mirror neurons in area F5 displayed similar firing patterns, both when a monkey observed the experimenter's grasping movements and when the same motor action was performed by the monkey itself. The above-mentioned studies give strong evidence that a visual input, either used as target or as feedback stimulus in a BCI experiment, can modify the neuronal activity and sensorimotor rhythms in motor or related areas. Since most BCI systems are based on the processing of visual information (target presentation, cursor movement, selection of letters, etc.), the impact of the visual display on the sensorimotor activity is of particular interest, especially when motor imagery is used as a control strategy.

In a movement imagination experiment, the time course of EEG reactivity (ERD/ERS) in sensorimotor areas was studied based on multiple-channel EEG recordings. By analyzing and classifying single-trial EEG data by the method of common spatial patterns,[78] researchers found that a correct discrimination between right- and left-hand movement imagination was already possible 250 msec after the onset of the visual cue signalling the side of motor imagery.[94] This fast increase of classification accuracy after cue presentation can be interpreted as the result of a stimulus-specific and lateralized modification of sensorimotor rhythms. This finding, that two different visual targets (e.g., an arrow pointing to the left or to the right) can result in two distinguishable spatiotemporal EEG patterns as early as 250 msec after stimulus onset, is of special importance for BCI research and, possibly, opens up new ways of installing a fast communication channel between brain and computer.

14.7 THOUGHT TRANSLATION DEVICE: A CASE REPORT OF A COMPLETELY LOCKED-IN PATIENT

Patient E.M., a wealthy businessman from Lima, Peru, suffers from end-stage motor nerve disease. After diagnosis of the disorder at the age of 48, E.M. deteriorated rapidly and has been artificially respirated and fed for 4 years. Six months before initiation of training with the TTD, E.M. was diagnosed as completely locked-in and communication using assisted communication devices with eye movement control, heart-rate control, and EMG control had to be discarded. No voluntary eye movement was possible. The eyes of the patient were closed. Opening of the eyes was only possible by means of adhesive tape. However, testing of vision was impossible, and therefore a combined auditory-visual feedback and spelling system was used.

Figure 14.11 demonstrates that even the external anal sphincter could not be voluntarily controlled by E.M. An anal sphincter electrode was used and the patient was instructed to systematically contract (C) and relax (R) anal sphincter muscles. No single voluntary correct response could be recorded. In addition, several EMG locations were tested and the patient was asked to voluntarily increase and decrease

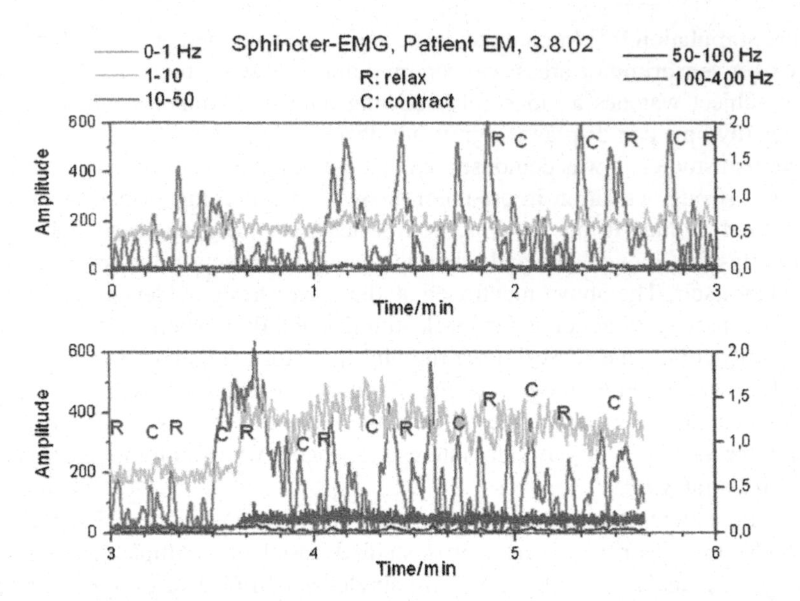

FIGURE 14.11 Recording of EMG from external anal sphincter of locked-in patient E.M. *R* indicates instruction to relax, *C* to contract.

EMG activity. Again, no successful voluntary motor control was possible. In order to test remaining cognitive abilities of locked-in patients a battery of cognitive tests using cognitive ERPs (as described by Kotchoubey et al.[55]) were used in the case of E.M. The results of highly complex cognitive stimuli presented to E.M. and the recorded cognitive ERPs demonstrated intact or nearly intact cognitive potentials to complex verbal and nonverbal material. Therefore the diagnosis of a completely locked-in state could be confirmed on the psychophysiological as well as the neurological level. (Two independent neurologists diagnosed E.M. as completely locked-in.)

E.M. was first trained with the Tübingen version of the TTD by local technicians over a period of 1 month with daily training sessions of 1 to 2 hours' duration. His performance varied between 30% correct and 70% correct with an average of 52% — that is, chance level. No communication is possible if patients are unable to achieve significant control of positive or negative SCPs. Therefore, the Tübingen team, consisting of one physicist (T.H.) and two psychologists/neuroscientists (N.B. and H.F.) continued the training at Lima over a period of 22 days. A thorough behavioral analysis and an analysis of the motivational context of the training were conducted during that period, suggesting that most of the training difficulties of E.M. were motivational in nature. The presence of several family members with unclear motivation negatively affected the training progress. Other family members and trainers, interested in communicating with the patient, had a positive effect on training success. In addition, the expectation of rewarding activities such as changing locations strongly affected successful training. This and other results of the behavioral analysis and the motivational system underscore the necessity of a careful clinical

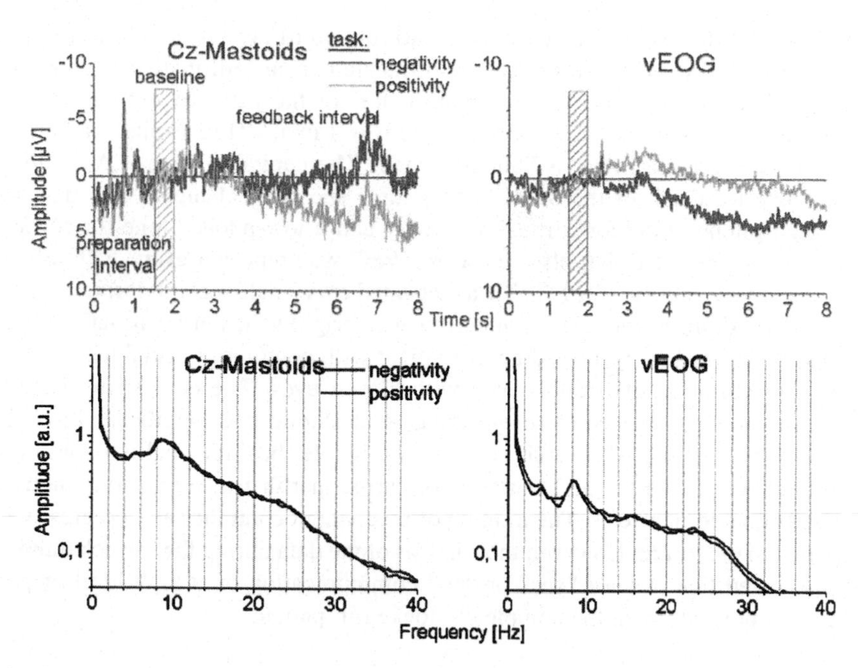

FIGURE 14.12 Averaged slow cortical potentials (SCPs) after 60 training sessions with locked-in patient E.M. (left, Cz), vertical electrooculogram (EOG) (right), EEG power spectrum (left below), and vertical electrooculogram (vEOG) power spectrum.

psychological evaluation of the patient and the patient's environment before BCI systems can be used in a clinical context.

In order to improve learning, a shaping procedure was introduced for training sessions before copy spelling or questioning of the patient was installed. Shaping consisted of rewarding the patient in a step-by-step manner for small improvements at the beginning and making the thresholds for achieving rewards more and more difficult. At the initial training session, the patient was rewarded for differentiation between positive and negative SCPs of only 1 µV, close to the noise level. After achieving a 70% correct response over 40 successive trials, the threshold was increased by 1 to 2 µV. Again the patient was trained up to 70% correct, and the next level of difficulty in SCP differentiation was introduced. If the patient did not succeed over a period of 40 trials, the threshold of reward was again reduced to the baseline. Figure 14.12 shows the average performance of the patient after 30 training sessions encompassing 200 to 400 trials per session. Average training success was significantly different from chance, but the average did not increase over 60% within the first 20 days of training by our team. Therefore, because of time constraints, copy spelling and questioning of the patient was introduced at a low but significant performance level of 60% correct differentiation between negative- and positive-tending cortical potentials.

Following the logic of lie detection and the assessment of criminal sceneries, a set of 40 questions was constructed (a BCI questioning system) and formulated in

Spanish so that the decisive and meaning-carrying word always arrived at the end of the question and each question was formulated in an affirmative or negative grammatical form. Questions of vital importance for the patient were mixed with neutral questions such as "The capital of Peru is Lima," "The capital of Peru is Bogota," "The capital of Peru is Santiago," and "The capital of Peru is Asunciòn." The auditory feedback of the SCPs lasting for 4 sec were changed such that not only a high-pitched tone for cortical negativity and a low-pitched tone for cortical positivity was presented, but also the word "yes" was repeated every half second during negativity and the word "no" was repeated every half second during cortical positivity. In addition, the television screen was brightly lit during negativity and darkened during positivity, providing the patient with continuous visual feedback of his ongoing cortical polarization even with closed eyes. The data show that the patient was able to answer these questions at a stabile and significant level. In addition, copy spelling of several words was possible, but free spelling could not be achieved because of the high variation of correct performance related to motivational and behavioral effects. The training of this patient continues and progress will be reported in the future. However, the analysis of the data during the 3-week training session showed that at least "yes" or "no" communication on a high level of performance is possible with the completely locked-in patient.

14.8 EEG-BASED COPY SPELLING: A CASE REPORT

The following case history represents an attempt to employ the Graz BCI to establish an alternative communication channel for a severely paralyzed patient. It was demonstrated that this is not an impossible task, when patients suffering from advanced ALS acquired the ability to operate a spelling device by regulating their SCPs[82,95] (see also Section 14.7). In some cases, however, even after several months of practice, no voluntary control of SCPs could be attained. Therefore, the question arose whether it was possible for such a patient to learn to control specific frequency components of the sensorimotor EEG by using an imagery strategy.[96]

The male patient, 32 years old and diagnosed with cerebral palsy, suffered from a severe spastic form of tetraparesis and had lost the ability to speak. Regular training sessions for the patient were carried out at a clinic for assisted communications over a period of 22 weeks, supervised from the technical laboratory with the help of a "telemonitoring system."[97] This system provided a direct access to the patient's computer (a BCI system), enabling, e.g., visual control of the EEG data on-line, as well as a video conference connection that transmitted direct instructions to the patient and the caregiver as well as providing visual control of their behavior.

The EEG signal (5–30 Hz) used for classification or feedback was recorded from one bipolar channel over the left sensorimotor cortex and sampled at 128 Hz. To generate the feedback based on oscillatory components of the ongoing EEG, two approaches were used: (1) direct band power feedback (20–30 Hz), and (2) feedback calculated by a linear discriminant classifier, which was developed to discriminate between two brain states.[71]

The training period was divided into several steps involving both learning of the patient as well as adaptation of the system (Figure 14.13).

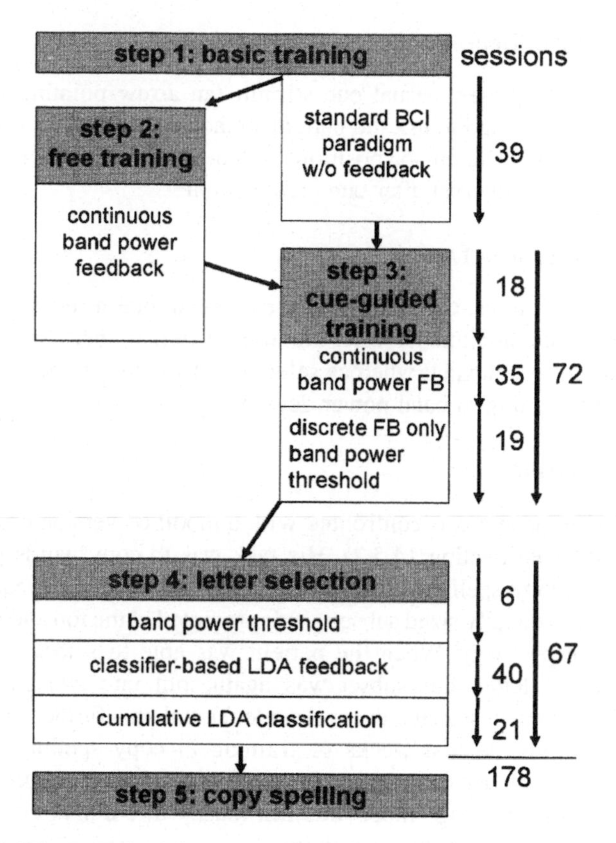

FIGURE 14.13 Diagram of training steps as described in the text, and corresponding number of performed sessions (right side). (From Reference 96, with permission.)

Step 1: Basic Training

In the first sessions the patient was asked to perform various mental imagination tasks in response to a cue stimulus (the standard BCI paradigm, as described in Section 14.5.2). This denotes a search for the most efficient imagination strategy. No feedback was provided at this stage. The discriminating feature was a prominent, long-lasting desynchronization (ERD) of higher beta band components during imagination of right-hand movement, which was not visible during other imagination tasks. This led us to utilize upper beta band (20–30 Hz) activity for BCI control.

Step 2: Free Training

In order to enhance the selected EEG components, a so-called "free training" was performed, where the band power (20–30 Hz) was continuously averaged over 4 sec and displayed on the screen as a vertically moving feedback dot ("cursor"). The patient was advised that imagination of right-hand movement moved the cursor downward (band power decrease, ERD). Relaxation, in contrast, either moved the cursor upward or caused it to remain in the center of the screen.

Step 3: Cue-Guided Training

The next step was to present visual cue stimuli (an arrow pointing up or down; standard BCI paradigm) and to ask the patient to move the feedback dot (cursor) in the indicated direction. The cursor position, based on the actual band power, was shown for a 4-sec time interval after cue presentation.

Step 4: Letter Selection Task

Instead of the cue stimulus, two letters were presented, one near the top, the other near the bottom of the monitor. To select the upper letter, an increase in band power had to be produced by relaxing, whereas selection of the lower letter was achieved by motor imagery leading to band power decrease.

Step 5: Copy Spelling

In the final step the patient was confronted with a modified version of the so-called virtual keyboard[76] (see Section 14.5.2). His task was to copy words presented by the experimenter (copy spelling). Instead of single characters, a predefined set of letters, split into two equally sized subsets, was presented at the top and at the bottom of the monitor, respectively. When the patient was able to select the subset that contained the target letter, this subset was again split into two parts. This was continued until the patient selected the desired letter and, in a further step, confirmed this selection. During the first weeks of training in copy spelling, only correct selections were accepted by the system; false selections were measured for off-line analyses. This "error ignoring" mode was introduced in order to avoid the consequences of a wrong selection during training.

The on-line performance of letter selection, quantified as percentages of correct responses according to the classifier-based discrimination, indicated a significant learning progress from the first ten sessions (61.6%, SD = 5.3) to the last ten sessions (68.9%, SD = 5.4) of the training period. At the end of the reported training procedure, this patient was able to produce voluntarily two distinct EEG patterns, associated with motor imagery versus intended relaxing, and to use this imagery strategy for BCI control. With the achieved level of 70% accuracy in letter selection training, verbal communication was possible by means of a spelling device. This allowed the patient to write with a rate of approximately one letter per minute.

14.9 EEG-BASED CONTROL OF FUNCTIONAL ELECTRICAL STIMULATION IN A TETRAPLEGIC PATIENT: A CASE REPORT

The tetraplegic patient who participated in this study is a 28-year-old man who has been suffering from a traumatic spinal cord injury since 1998. He participated in BCI training with different types of motor imagery in order to check whether he was able to operate an orthosis or functional electrical stimulation (FES). After a number of training sessions with variations of the motor imagery strategy over a time period of several months, imaginations of foot movement versus imagination of right hand movement achieved a classification accuracy of close to 100%.[32]

Inspecting the EEG signals, it was found that foot motor imagery induced long trains of 17-Hz beta oscillations focused to the electrode position on the vertex (Figure 14.14). After thousands of foot movement imaginations, the midcentrally localized neural networks in the foot representation area and/or supplementary motor areas may be modified by motor imagery and may become able to generate oscillating activity in the beta band. This underlines both the importance of repeated BCI sessions with feedback and the plasticity of the brain.

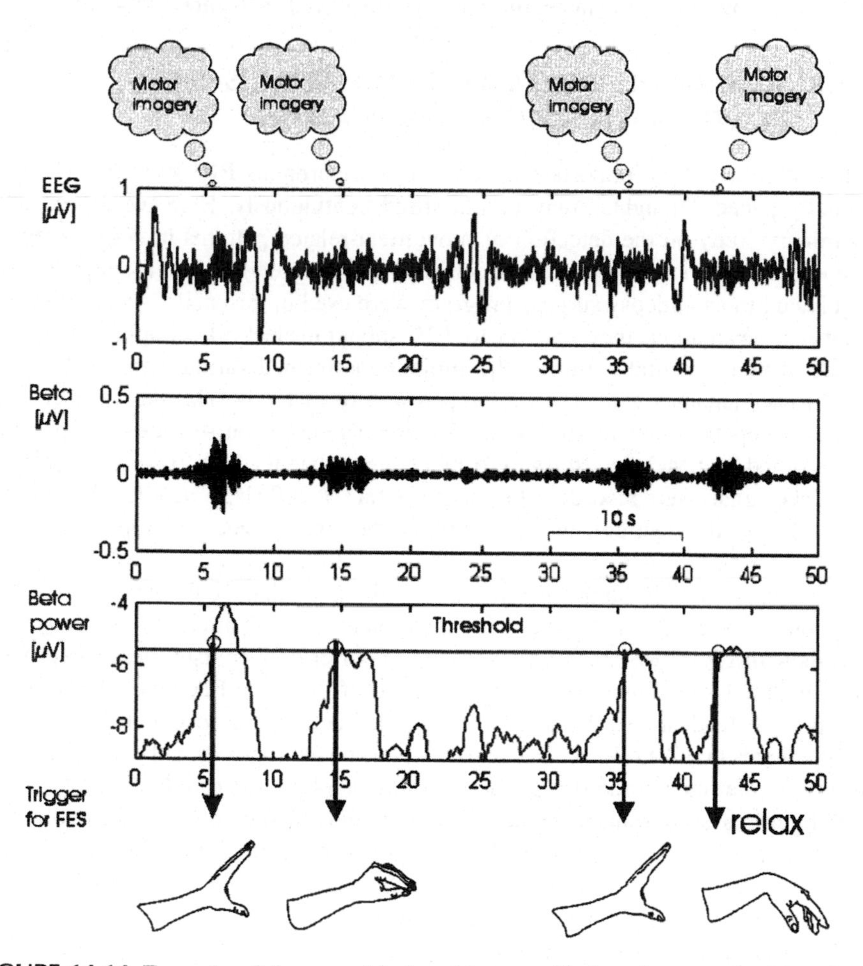

FIGURE 14.14 Example of the use of induced beta oscillations for control of functional electrical stimulation (FES) in a tetraplegic patient. The patient was able to induce bursts of beta oscillations by imagination of foot movement: Bipolar EEG recording from the vertex (overlying the foot representation area; upper trace), band pass filtered (15–19 Hz) EEG signal (middle trace), and band power time course (lower trace, arbitrary units) over a time interval of 50 sec. Threshold and trigger pulse generation according to FES and grasp phases are indicated.

With the mentally induced 17-Hz oscillations, a simple brain switch was constructed and used to control the functional electrical stimulation with three stimulation channels.[33] Whenever a 17 Hz beta burst was induced by motor imagery and the beta band power exceeded a predefined threshold, a trigger was generated. This trigger was used to switch the functional electrical stimulation. Surface electrodes placed on the forearm were used for stimulation of the paralyzed muscles. To realize a hand grasp, four basic muscle groups have to be activated: the finger and thumb extension for hand opening, the finger flexions for hand closing, the thumb flexion for grasping, and the wrist extension for stabilization of the hand. The individual grasp phases by induced beta oscillations are displayed in Figure 14.14.

14.10 ECOG-BASED BCI: PRELIMINARY RESULTS ON CONTINUOUS ECOG CLASSIFICATION

In the future it will be important to develop asynchronous BCI systems, whereby internally paced "thoughts" may be classified continuously. Here we report on a preliminary study on the detection of movement-related patterns in single-channel ECoG signals. Twenty-two ECoG recordings (22 datasets) of seven subjects who participated in an epilepsy surgery program were used in this study. As part of their presurgical evaluation, they had 63 to 126 subdural electrodes implanted on the surface of their cerebral cortices. The subjects performed various movement tasks in a self-paced manner with at least 50 repetitions of each task. The intervals between successive repetitions were not less than 3 sec. The time course for each repetition was recorded by EMG electrodes to form a trigger channel. The trigger channel and the ECoG signals were recorded at a sampling rate of 200 Hz. A detailed description of the subjects, the electrode locations, and the data collection can be found in Levine et al.[98]

In order to obtain some insights about the data that could be used for the development of a detection system for movement-related patterns, time-frequency analysis with ERD/ERS maps and short-time Fourier time courses were performed. Color Figure 14.15 shows an exemplary result of this time-frequency analysis for one channel. It illustrates that movement-related activity can occur in various frequency ranges in ECoG channels. Clearly, alpha and beta activity and evoked activity are present, but most interestingly, because it cannot be found in EEG, movement-related gamma synchronization can also be found. In four of the seven subjects, such synchronization was observed. The short-time Fourier time courses (Color Figure 14.15B) illustrate that the dominant frequencies are in the delta, beta, and gamma frequency range for this particular channel.

In general, all oscillatory patterns covered a very broad frequency range, which made it difficult to determine *a priori* which frequency components or features were most suitable for the detection of movement-related patterns. Therefore, a suitable feature extraction and selection method had to be applied to overcome this difficulty.

Wavelet packet analysis[99] was used to derive 18 features capturing the information contained in induced and evoked movement-related patterns.[16] A genetic algorithm (GA), a stochastic search and optimization technique based on evolutionary computation,[100] was employed to weight these features according to their importance for

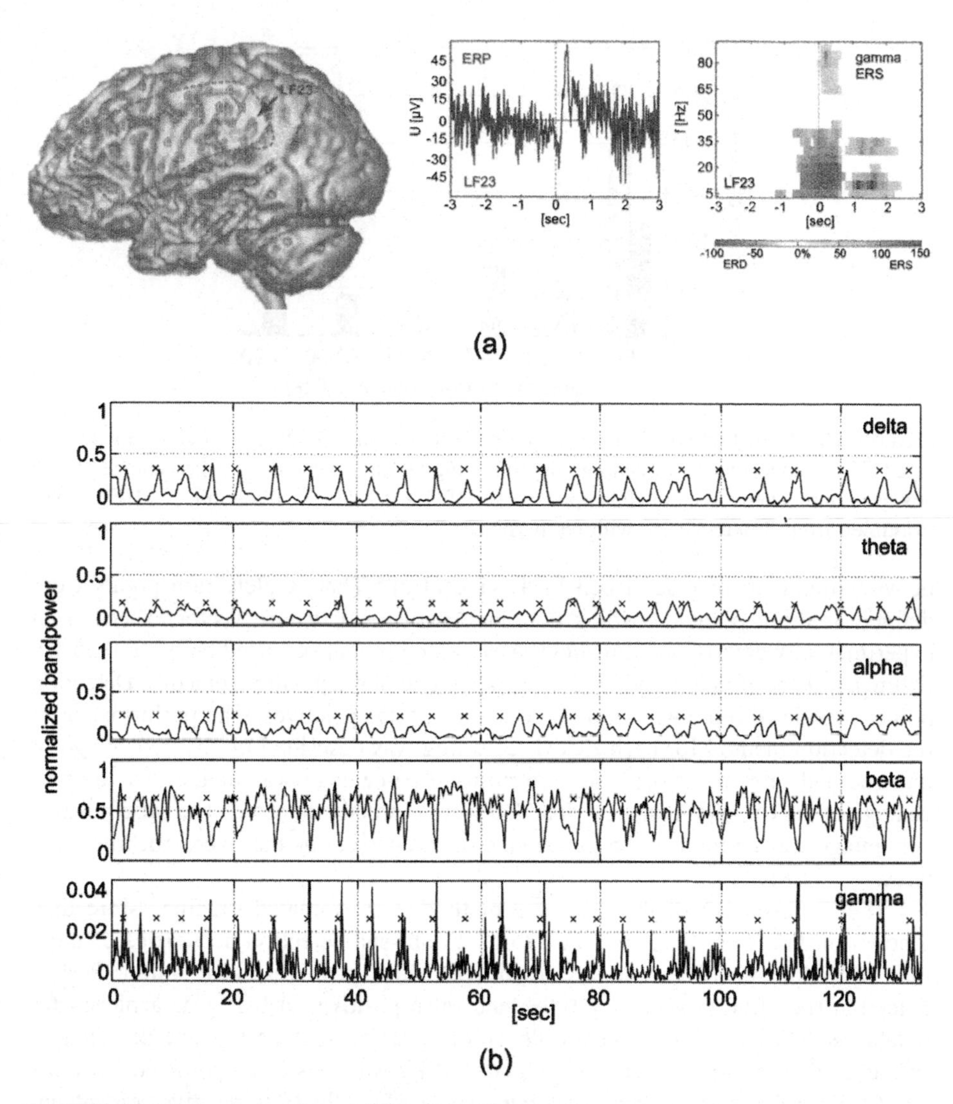

FIGURE 14.15 (see color figure) ERP, ERD/ERS map, and short-time Fourier time course of one exemplary ECoG channel. (a) ERP template calculated from 23 trials. This template was used for the cross correlation template matching (CCTM) method. The ERD/ERS maps represent averaged oscillatory activity in a frequency range from 5 to 100Hz. ERD is colored in red, and ERS is colored in blue. Movement onset is indicated by the vertical dash-dotted line. (b) The short-time Fourier time courses show ongoing normalized bandpower of movement-related patterns in the delta (<3.5 Hz), beta (12.5–30 Hz), and gamma (70–90 Hz) band. Theta (3.5–7.5 Hz) and alpha (7.5–12.5 Hz) bands do not show distinct peaks around movement onsets indicated by the crosses. (Modified from Reference 101, with permission.)

the detection performance and also to combine the six most important features in a linear fashion (associated with the largest weights) to obtain a one-dimensional feature signal.[16,101] The actual detection was performed by a simple threshold detector

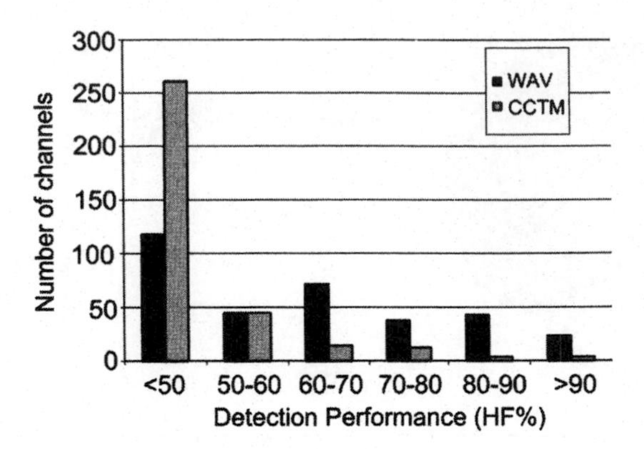

FIGURE 14.16 Histogram of the detection performance of all 339 ECoG channels investigated. The results are categorized into the following HF difference ranges (percentage of true positives minus percentage of false positives): HF% < 50, 50 ≤ HF% < 60, …, HF% > 90. (Modified from Reference 16, with permission.)

in way similar to that described by Levine et al.[8] That is, detection points were defined as those points at which the one-dimensional feature vector exceeded an experimentally determined threshold. Any detection that occurred between 0.25 sec before and 1 sec after a trigger point was defined as a true positive (or a hit). Detection points outside of this interval were counted as false positives. The performance of the detection method for each ECoG recording was described by the percentage of the true and false positives. The true positive percentage was defined by the percentage of the triggers in the test data that were correctly detected. The false positive percentage was defined as the percentage of the detections that were not true positives.

The performance of the proposed method was evaluated off-line. More than 2 hours of data were analyzed. For 9 of 22 datasets, detection accuracies of more than 90% true positives and less then 10% false positives were found. Perfect detection (i.e., true positives at 100% and false positives at 0%) was achieved for 6 datasets. The mean and standard deviation (SD) of the true positive percentages of the 22 datasets (ECoG channels) analyzed by wavelets and optimized by a GA was 95.5 ± 6.7%, and the corresponding mean ± SD of the false positive percentages was 9.2 ± 9%. These results show that the proposed method can classify movement-related patterns in ongoing EcoG very accurately. This is remarkable, since only single channels were used as input for the method and spatiotemporal features of the ECoG recordings were not employed.

Figure 14.16 depicts the histogram of the results of all ECoG channels investigated for the wavelet-based approach and a method based on cross-correlation template matching.[8] In the latter method, features are derived from ERP templates that are cross-correlated with the signal. Evidently, the wavelet method yielded improved results as compared with the cross-correlation template matching method. This can be seen as a consequence of the fact that the correlation template matching is based solely on the information contained in ERPs, while the wavelet-based approach employs the information contained in oscillatory activity as well. It is interesting to note that for almost all of the best performing channels, the features associated with gamma activity had a substantial impact. However, this result should be interpreted cautiously, since there are no studies available that report on the gamma ERS of imagery data that would be required for a practical BCI. On the other hand, it can be expected that gamma oscillations may also be present during motor imagery because of the great similarity in cortical activation patterns between real executed and imagined movements.

14.11 WHICH INFORMATION TRANSFER RATE IS POSSIBLE WITH MOTOR IMAGERY?

The practical usability of a BCI system to control, for example, a spelling device or a virtual keyboard, would require a high system performance which can be measured by the classification accuracy or the information transfer rate (ITR) in bits per minute. The latter includes the accuracy of classification, the number of possible targets (classes), and the speed of selection.[56] In general, the accuracy declines when the class number is increased.

In a recent study, a new experimental paradigm was investigated to determine the optimal decision speed (trial length) individually for a subject using the Graz BCI.[102] A simple game-like paradigm was implemented, in which the user had to move a falling ball into the correct goal ("basket") marked on the screen. The horizontal position of the ball was controlled via the BCI output signal and the falling speed could be adjusted by the investigator. Four male volunteers (paraplegic patients) participated in this study. None of them had any prior experience with BCI. Two bipolar EEG channels were recorded from electrode positions close to C3 and C4. Two different types of motor imagery (either right versus left hand motor imagery or hand versus foot motor imagery) were used, and band power within the alpha band (10–12 Hz) and the beta band (16–24 Hz) were classified. After several training runs without feedback, the best imagery strategy was chosen and a linear classifier was set up. Feedback was given to the participants in the form of a falling red ball. After a pause with a fixed length of 1 sec, the little red ball appeared at the top of the screen and began to fall downward with a constant speed. This speed (i.e., the time the ball took to cross the screen) was varied by the investigator across runs between 5 and 1.5 sec. The patient's task was to hit the highlighted basket (which changed sides randomly from trial to trial) as often as possible. Speed was increased run by run until the patient judged it as being too fast. The study attempted to find the optimal speed for a maximum information transfer rate. After each run patients were asked to rate their performance and to suggest whether the system operates too slowly or too fast. The highest information transfer rate of 17 bits per minute was reached with a trial length of 2.5 sec.

Theoretically, when the accuracy is 100% in a 2-class paradigm with motor imagery, an information transfer rate of 30 bits per minute is possible when the trial length is 2 sec (see also Section 14.5.1). A trial length shorter than 2 sec is problematical when oscillatory activity is used to control a BCI because desynchronization and synchronization processes of populations of neurons need time to develop. This time is in the order of seconds when alpha or mu rhythm is used for control.[21]

14.12 FUTURE DIRECTIONS

When the EEG is used as an input signal for a BCI system, multiple-channel recordings and special methods of preprocessing, such as, for example, independent component analysis (ICA), are recommended. Mu and central beta rhythms are especially suitable for ICA because both are spatially stable and can therefore be separated easily from other sources.[101,103] Also, measures such as phase coupling and instantaneous coherence[104] should be incorporated, when multiple-channel data are available. The near future will show whether ICA as a preprocessing method and phase-coupling measurements can increase the reliability and the speed of a BCI. In addition, new processing strategies, as described for instance in Section 14.10, could be of importance.

A clear-cut challenge for the future, furthermore, is to realize more effective BCI control paradigms, offering, for instance, a three-dimensional control over a neuroprostesis or the operation of a spelling device with a speed of at least 5 to 10 letters per minute. Principally, both applications mentioned should be realizable either by the detection of firing patterns in intracortical recordings or the analysis of cortical potential changes by ECoG electrode strips or grids.

The advantage of the ECoG over the EEG is the better signal-to-noise ratio and therefore the easier detection of gamma activity. Recently, this was reported on bursts of gamma activity between 60 and 90 Hz in ECoG recordings during self-paced limb and tongue movements.[28,29] These gamma bursts are short-lasting, display a high somatotopic specificity, and are embedded in the alpha and beta ERD lasting for some seconds. Examples of ERD/ERS time courses calculated in alpha and gamma bands during self-paced hand movement are displayed in Figure 14.2C. Based on these gamma bursts it was also possible to detect individual finger movements in the ongoing ECoG with satisfying accuracy (asynchronous BCI mode;[16] see also Section 14.10). Whether such gamma bursts also occur during motor imagery remains to be shown in ongoing research work.

Animal studies, focused on multiple-unit neuronal activity (the firing of groups of neurons) to perform two- and three-dimensional cursor control, are of special value for the realization of a multidimensional human BCI. The activity from a few primary motor cortex neurons in monkeys was used by Donoghue's group for two-dimensional cursor control.[19] In contrary, Andersen's group in Pasadena made use of recordings from monkey parietal cells to control cursor movements.[105] Just by thinking about a reaching movement, and without actual movement execution, the position of a cursor could be changed.[106] Nicolelis' group reported a tracking task in more dimensions performed by monkeys, and demonstrated thereby the possibility of mental control of a three-dimensional robotic prosthesis.[20] They showed that motor control signals, associated with an arm movement, appear simultaneously in large portions of the frontal and parietal cortices and that, theoretically, each of these distributed cortical signals may be used separately to generate hand trajectory signals in real-time applications. The feasibility of direct cursor control for the selection of icons or letters using an implanted neurotropic cortical electrode was already demonstrated by Kennedy et al.[107]

ACKNOWLEDGMENTS

We especially thank Professor Jon Wolpaw for providing the section on the Wadsworth BCI and Professor Simon Levine for support with ECoG data. We also thank B. Graimann, C. Keinrath, G. Krausz, G.R. Müller, A. Schlögl, D. Skliris, G. Supp, B. Wahl, and M. Wörtz for their help in preparing the manuscript. This research was partially supported by FWF and AUVA in Austria, DFG in Germany, and NIH in the United States.

REFERENCES

1. Vidal, J., Toward direct brain–computer communication, *Annu. Rev. Biophys. Bioeng.*, 157, 1973.
2. Wolpaw, J.R. et al., Brain–computer interfaces for communication and control, *Clin. Neurophysiol.*, 113, 767, 2002.
3. Petsche, H. et al., EEG coherence and musical thinking, *Music Percept.*, 11, 117, 1993.
4. Petsche, H. et al., Verbal and visual thinking in the EEG, in *Structural and Functional Organization of the Neocortex*, Albowitz, B., Albus, K., Kuhnt, U., Nothdurft, H.C., Eds., Springer-Verlag, Berlin, 1994, 445.
5. Farwell, L.A. and Donchin, E., Talking off the top of your head: toward a mental prosthesis utilizing event-related potentials, *Electroenceph. Clin. Neurophysiol.*, 70, 510, 1988.
6. Weiskopf, N. et al., Physiological self-regulation of regional brain activity using real-time functional magnetic resonance imaging (fMRI): methodology and exemplary data, *NeuroImage*, 19, 577, 2003
7. Birbaumer, N. and Kübler, A., The thought translation device (TTD) for completely paralyzed patients, *IEEE Trans. Rehab. Eng.*, 8, 190, 2000.
8. Levine, S.P. et al., A direct brain interface based on event-related potentials, *IEEE Trans. Rehab. Eng.*, 8, 180, 2000.
9. Pfurtscheller, G. and Neuper, C., Motor imagery and direct brain–computer communication, *Proc. IEEE Neural Eng.*, 89, 1123, 2001.
10. Nicolelis, M.A.L., Actions from thoughts, *Nature*, 409, 403, 2001.
11. Kübler, A. et al., Brain–computer communication: self-regulation of slow cortical potentials for verbal communication, *Arch. Phys. Med. Rehabil.*, 82, 1533, 2001.
12. Wolpaw, J.R. and McFarland, D.J., Multichannel EEG-based brain-computer communication, *Electroenceph. Clin. Neurophysiol.*, 90, 444, 1994.
13. Mason, S.G. and Birch, G.E., A brain-controlled switch for asynchronous control applications, *IEEE Trans. Biomed. Eng.*, 47, 1297, 2000.
14. Mason, S.G., and Birch, G.E., A general framework for brain–computer interface design, *IEEE Trans. Biomed. Eng.*, 11, 70, 2003.
15. Graimann, B. et al., Detection of movement-related desynchronization patterns in ongoing single channel electrocorticogram, *IEEE Trans. Neural Sys. Rehab. Eng.*, 11, 276, 2003.
16. Graimann, B. et al., Towards a direct brain interface based on human subdural recordings and wavelet packet analysis, *IEEE Trans. Biomed. Eng.*, 51, 954, 2004.
17. Blankertz, B., Curio, G., and Müller, K.R., Classifying single trial EEG: towards brain–computer interfacing, *Proc. Advances in Neural Processing Systems (NIPS 01)*, Vancouver, 2002.

18. Scherer, R. et al., An asynchronously controlled EEG-based virtual keyboard: improvement of the spelling rate, *IEEE Trans. Biomed. Eng.*, 51, 979, 2004
19. Serruya, M.D. et al., Instant neural control of a movement signal, *Nature*, 416, 141, 2002.
20. Wessberg, J. et al., Real-time prediction of hand trajectory by ensembles of cortical neurons in primates, *Nature,* 408, 361, 2000.
21. Pfurtscheller, G. and Lopes da Silva, F.H., Event-related EEG/MEG synchronization and desynchronization: basic principles, *Clin. Neurophys.*, 110, 1842, 1999.
22. Altenmüller, E.O. and Gerloff, C., Psychophysiology and the EEG, in *Electroencephalography*, 4th Ed., Niedermeyer, E. and Lopes da Silva, F., Eds., Williams & Wilkins, Baltimore, MD, 1998.
23. Rockstroh, B. et al., Cortical self-regulation in patients with epilepsies, *Epilepsy Res.,* 14, 63, 1993.
24. Birbaumer, N. et al., Slow cortical potentials of the cerebral cortex and behavior, *Physiol. Rev.,* 70, 1, 1990.
25. Speckmann, E. and Elger, C., Introduction to the neurophysiological basis of the EEG and DC potentials, in *Electroencephalography*, 4th Ed., Niedermeyer, E. and Lopes da Silva, F., Eds., Williams & Wilkins, Baltimore, MD, 1998, 1.
26. Braitenberg, V. and Schüz, A., *Anatomy of the Cortex*, Springer-Verlag, Berlin, 1991.
27. Pfurtscheller, G., and Neuper, C., Simultaneous EEG 10 Hz desynchronization and 40 Hz synchronization during finger movements, *NeuroReport*, 3, 1057, 1992.
28. Crone, N.E. et al., Functional mapping of human sensorimotor cortex with electrocorticographic spectral analysis. II. Event-related synchronization in the gamma band, *Brain*, 121, 2301, 1998.
29. Pfurtscheller, G. et al., Spatiotemporal patterns of beta desynchronization and gamma synchronization in corticographic data during self-paced movement, *Clin. Neurophysiol.*, 114, 1226, 2003.
30. Neuper, C. and Pfurtscheller G., Evidence for distinct beta resonance frequencies in human EEG related to specific sensorimotor cortical areas, *Clin. Neurophysiol.*, 112, 2084, 2001.
31. Neuper, C. and Pfurtscheller, G., Motor imagery and ERD, in *Event-Related Desynchronization, Handbook of Electroenceph. and Clin. Neurophysiol.*, Rev. Ed., Vol. 6, Pfurtscheller, G. and Lopes da Silva, F.H., Eds., Amsterdam, Elsevier, 1999, 303.
32. Pfurtscheller, G. et al., Brain oscillations control hand orthosis in a tetraplegic, *Neurosci. Lett.,* 292, 211, 2000.
33. Pfurtscheller, G. et al., "Thought" – control of functional electrical stimulation to restore hand grasp in a patient with tetraplegia, *Neurosci. Lett.*, 351, 1, 33, 2003.
34. Neuper, C. and Pfurtscheller, G., Event-related dynamics of cortical rhythms: frequency-specific features and functional correlates, *Int. J. Psychophysiol.,* 43, 41, 2001.
35. Suffcynski, P. et al., Computational model of thalamo-cortical networks: dynamical control of alpha rhythms in relation to focal attention, *Int. J. Psychophysiol.,* 43, 25, 2001.
36. Decety, J. et al., Mapping motor representations with positron emission tomography, *Nature,* 371, 600, 1994.
37. Rao, S.M. et al., Functional magnetic resonance imaging of complex human movements, *Neurology*, 43, 2311, 1993.
38. Roth, M. et al., Possible involvement of primary motor cortex in mentally simulated movement: a functional magnetic resonance imaging study, *NeuroReport,* 7, 1280, 1996.
39. Porro, C.A. et al., Primary motor and sensory cortex activation during motor performance and motor imagery: a functional magnetic resonance imaging study, *J. Neurosci.*, 16, 7688, 1996.

40. Annett, J., On knowing how to do things: a theory of motor imagery, *Brain Res. Cogn. Brain Res.*, 3, 65, 1996.
41. Grafton, S.T. et al., Premotor cortex activation during observation and naming of familiar tools, *NeuroImage*, 6, 231, 1997.
42. Buccino, G. et al., Action observation activates premotor and parietal areas in a somatotopic manner: an fMRI study, *Eur. J. Neurosci.*, 13, 400, 2001.
43. Klass, S.G. and Bickford, R.G., Observations on the rolandic arceau rhythm, *Electroenceph. Clin. Neurophysiol.*, 9, 570, 1957.
44. Chatrian, G.E., Petersen, M.C., and Lazarte, J.A., The blocking of the rolandic wicket rhythm and some central changes related to movement, *Electroenceph. Clin. Neurophysiol.*, 11, 497, 1959.
45. Beisteiner, R. et al., Mental representations of movements. Brain potentials associated with imagination of hand movements, *Electroenceph. Clin. Neurophysiol.*, 96, 183, 1995.
46. Cunningham, R., Iansek, R., and Stokes, M.J., Movement-related potentials associated with movement preparation and motor imagery, *Exp. Brain Res.*, 111, 429, 1996.
47. Schnitzler, A. et al., Involvement of primary motor cortex in motor imagery: a neuromagnetic study, *NeuroImage*, 6, 201, 1997.
48. Cassim, F. et al., Does post-movement beta synchronization reflect an idling motor cortex? *Neuroreport*, 12, 3859, 2001.
49. Hinterberger, T. et al., A brain–computer interface (BCI) for the locked-in: comparison of different EEG classifications for the thought translation device (TTD), *Clin. Neurophysiol.*, 114, 3, 416, 2003.
50. Stamm, J.S., Whipple, S., and Born, J., Effects of spontaneous cortical slow potentials on semantic information processing, *Int. J. Psychophysiol.*, 5, 11, 1987.
51. Stamm, J.S. and Rosen, S.C., The locus and crucial time of implication of prefrontal cortex in the delayed response task, in *Psychophysiology of the Frontal Lobes*, Pribram, K. and Luria, A., Eds., Academic Press, New York, 1973, 139.
52. Fetz, E., Operant conditioning of cortical unit activity, *Science*, 163, 955, 1969.
53. Hinterberger, T. et al., Brain areas activated in fMRI during self-regulation of slow cortical potentials (SCP), *Exp. Brain Res.*, 152, 113, 2003.
54. Birbaumer, N. et al., Slow brain potentials, imagery and hemispheric differences, *J. Neurosci.*, 39, 101, 1988.
55. Kotchoubey, B. et al., Is there a mind? Electrophysiology of unconscious patients, *News Physiol. Sci.*, 17, 38, 2002.
56. Wolpaw, J.R., McFarland, D.J., and Vaughan, T.M., Brain–computer interface research at the Wadsworth Center, *IEEE Trans. Rehab. Eng.*, 8, 222, 2002.
57. Sutter, E.E., The visual evoked response as a communicatioan channel, *Proc. IEEE/NSF Symp. Biosensors*, 95, 1984.
58. Donchin, E., Spencer, K.V. and Wijesinghe, R., The mental prosthesis: assessing the speed of a P300-based brain–computer interface, *IEEE Trans. Rehab. Eng.*, 8, 174, 2000.
59. Middendorf, M. et al., Brain–computer interfaces based on the steady-state visual-evoked response, *IEEE Trans. Rehab. Eng.*, 2, 211, 2000.
60. Wolpaw, J.R. et al., An EEG-based brain–computer interface for cursor control, *Electroenceph. Clin. Neurophysiol.*, 78, 252, 1991.
61. McFarland, D.J., Lefkowicz, A.T., and Wolpaw, J.R., Design and operation of an EEG-based brain–computer interface (BCI) with digital signal processing technology, *Behav. Res. Methods Instrum. Comput.*, 29, 337, 1997.
62. Miner, L.A., McFarland, D.J., and Wolpaw, J.R., Answering questions with an EEG-based brain–computer interface (BCI), *Arch. Phys. Med. Rehab.*, 79, 1029, 1998.

63. Wolpaw, J.R. et al., EEG-based communication: improved accuracy by response verification, *IEEE Trans. Rehab. Eng.*, 6, 326, 1998.
64. McFarland, D.J. et al., EEG-based brain–computer interface (BCI) communication: effects of target number and trial length on information transfer rate, *Soc. Neurosci. Abst.*, 26, 1228, 2000.
65. McFarland, D.J. et al., Spatial filter selection for EEG-based communication, *Electroenceph. Clin. Neurophysiol.*, 103, 386, 1997.
66. Ramoser, H., Wolpaw, J.R., and Pfurtscheller, G., EEG-based communication: evaluation of alternative signal prediction methods, *Biomed. Technik*, 42, 226, 1997.
67. Schalk, G. et al., EEG-based communication and control: presence of an error potential, *Clin. Neurophysiol.*, 111, 2138, 2000.
68. Vaughan, T.M. et al., EEG-based brain–computer interface: development of a speller, *Soc. Neurosci. Abstr.*, 27, 167, 2001.
69. Kalcher, J. et al., Graz Brain–Computer Interface II: towards communication between humans and computers based on online classification of three different EEG patterns, *Med. Biol. Eng. Comput.*, 34, 382, 1996.
70. Pregenzer, M. and Pfurtscheller, G., Frequency component selection for an EEG-based Brain–Computer Interface (BCI), *IEEE Trans. Rehab. Eng.*, 7, 413, 1999.
71. Pfurtscheller, G. et al., EEG-based discrimination between imagination of right and left hand movement, *Electroenceph. Clin. Neurophysiol.*, 103, 642, 1997.
72. Schlögl, A., Neuper C., and Pfurtscheller, G., Estimating the mutual information of an EEG-based Brain–Computer Interface, *Biomed. Technik*, 47, 3, 2002.
73. Pfurtscheller, G. et al., Separability of EEG signals recorded during right and left motor imagery using adaptive autoregressive parameters, *IEEE Trans. Rehab. Eng.*, 6, 316, 1998.
74. Neuper, C., Schlögl, A., and Pfurtscheller, G., Enhancement of left-right sensorimotor EEG differences during feedback-regulated motor imagery, *J. Clin. Neurophysiol.*, 16, 373, 1999.
75. Obermaier, B. et al., Hidden Markov Models for online classification of single trial EEG data, *Pattern Recog. Lett.*, 22, 1299, 2001.
76. Obermaier, B., Müller, G.R., and Pfurtscheller, G., 'Virtual Keyboard' controlled by spontaneous EEG activity, *IEEE Trans. Neural Sys. Rehab. Eng.*, 11, 422, 2003.
77. Guger, C., Ramoser, H. and Pfurtscheller, G., Real-time EEG analysis for a brain–computer interface (BCI) with subject-specific spatial patterns, *IEEE Trans. Rehab. Eng.*, 8, 447, 2000.
78. Müller-Gerking, J., Pfurtscheller, G., and Flyvbjerg, H., Designing optimal spatial filters for single-trial EEG classification in a movement task, *Clin. Neurophysiol.*, 110, 787, 1999.
79. Birbaumer, N. et al., Biofeedback of event-related slow potentials of the brain, *Int. J. Psychol.*, 16, 389, 1981.
80. Elbert, T. et al., Biofeedback of slow cortical potentials. Part I, *Electroenceph. Clin. Neurophysiol.*, 48, 293, 1980.
81. Kotchoubey, B. et al., Modification of slow cortical potentials in patients with refractory epilepsy: a controlled outcome study, *Epilepsia*, 42, 406, 2001.
82. Birbaumer, N. et al., A spelling device for the paralysed, *Nature*, 398, 297, 1999.
83. Müller, K.R. et al., An introduction to kernel-based learning algorithms, *IEEE Trans. Neural Networks*, 12, 181, 2001.
84. Perelmouter, J. et al., Language support program for thought-translation-devices, *Automedica*, 18, 67, 1999.

85. Cheng, M., Gao, X., and Gao, S., Design and implementation of a brain–computer interface with high transfer rates, *IEEE Trans. Biomed. Eng.*, 49, 1181, 2003.
86. Kennedy, P.R. and Bakay, R.A.E., Restoration of neural output from a paralyzed patient by a direct brain connection, *NeuroReport*, 9, 1707, 1998.
87. Salmoni, A.W., Schmidt, R.A., and Walter, C.B., Knowledge of results and motor learning: a review and critical reappraisal, *Psycholog. Bulletin*, 95, 355, 1984.
88. McFarland, D.J., McCane, L.M., and Wolpaw, J.R., EEG-based communication and control: short-term role of feedback, *IEEE Trans. Rehab. Eng.*, 6, 7, 1998.
89. Kuhlman, W.N., Functional topography of the human mu rhythm, *Electroenceph. Clin. Neurophysiol.* 44, 83, 1978.
90. Westphal, K.P. et al., EEG-blocking before and during voluntary movements: differences between the eyes closed and the eyes open condition, *Arch. It. Biol.*, 131, 25, 1993.
91. Koshino, Y. and Niedermeyer, E., Enhancement of rolandic mu-rhythm by pattern vision, *Electroenceph. Clin. Neurophysiol.*, 38, 535, 1975.
92. Cochin, S. et al., Perception of motion and EEG activity in human adults, *Electroenceph. Clin. Neurophysiol.*, 107, 287, 1998.
93. Rizzolatti, G., Fogassi, L., and Gallese, V., Neurophysiological mechanisms underlying the understanding and imitation of action, *Perspectives*, 2, 661, 2001.
94. Pfurtscheller, G. et al., Visually guided motor imagery activates sensorimotor areas in humans, *Neurosci. Lett.*, 269, 153, 1999.
95. Kübler, A. et al., Brain–computer communication: unlocking the locked-in, *Psychol. Bulletin*, 127, 358, 2001.
96. Neuper, C. et al., Clinical application of an EEG-based brain–computer interface: a case study in a patient with severe motor impairment, *Clin. Neurophysiol.*, 114, 399, 2003.
97. Müller, G.R., Neuper, C., and Pfurtscheller, G., Implementation of a telemonitoring system for the control of an EEG-based brain computer interface, *IEEE Trans. Neural. Systems Rehab. Eng.*, 11, 54, 2003.
98. Levine, S.P. et al., Identification of electrocorticogram patterns as the basis for a direct brain interface, *Clin. Neurophysiol.*, 16, 439, 1999.
99. Jensen, A. and LaCour-Harbo, A., *Ripples in Mathematics: The Discrete Wavelet Transform*, Springer, Berlin, 2001.
100. Goldberg, D.E., *Genetic Algorithms in Search, Optimization, and Machine Learning*, Addison-Wesley, Reading, MA, 1989.
101. Graimann, B., Movement-related patterns in ECoG and EEG: visualization and detection, T.U. Graz, Austria, Ph.D. thesis, 2002.
102. Krausz, G. et al., Critical decision speed and information transfer in the Graz brain–computer interface, *Appl. Psychophysiol. Biofeedback*, 28, 233, 2003.
103. Makeig, S. et al., Dynamic brain sources of visual evoked responses, *Science*, 295, 690, 2002.
104. Moller, E. et al., Instantaneous multivariate EEG coherence analysis by means of adaptive high-dimensional autoregressive models, *J. Neurosci. Meth.*, 105, 143, 2001.
105. Meeker, D. et al., Cognitive control signals for prosthetic systems, *Soc. Neurosci. Abstr.*, 27, 63, 2001.
106. Andersen, R.A. and Buneo, C.A., Intentional maps in posterior parietal cortex, *Annu. Rev. Neurosci.*, 25, 189, 2002.
107. Kennedy, P.R. et al., Direct control of a computer from the human central nervous system, *IEEE Trans. Rehab. Eng.*, 8, 198, 2000.

Index

A

B

C

D

E

F

G

H

I

J

K

N

O

P

R

S

T

U